Plants as Medicine and Aromatics

Since ancient times, plants serve as a valuable source of traditional herbal medicines. Unlike modern medicines, herbal medicines have consistently demonstrated health advantages, including a lack of serious adverse side effects, long-lasting curative impacts and overall cost-effectiveness. Even today, with various modern pharmaceutical medicines commonly available, plant-based medicines and aromatics are increasingly in demand throughout the health sector globally, where they are used not only for the treatment of disease, but also, preventatively for maintaining good health.

People are seeking alternatives to modern medical treatments turning to phytomedicine for primary health care. However, an inadvertent consequence of this increased demand for herbal medicines has resulted in medicinal plants being threatened due to their initial small population sizes, narrow distribution areas, habitat specificity, and increasingly destructive non-sustainable harvesting.

This book critically examines and reviews the status of medicinal plants and includes several important case studies of representative plant species. It contains information on aspects concerning phytochemistry, natural products, cultivation, conservation techniques, environmental interactions, and therapeutic features of medicinal aromatic plants.

FEATURES

- Evaluates plants as medicine and aromatics covering pharmacognosy and ecology of plants having therapeutic values.
- Discusses how plants can play a role in treatment of diseases and as potential therapeutics standards for maintaining good health.
- Presents conventional and contemporary approaches to conservation of such plants with commercial feasibility.

EXPLORING MEDICINAL PLANTS

Series Editor

Azamal Husen

Wolaita Sodo University, Ethiopia

Medicinal plants represent a rich source of bioactive compounds used in drug formulation and development, and they play a key role in traditional or indigenous health systems. As the demand for herbal medicines increases worldwide, supply is declining as most of the harvest is derived from wild-grown vegetation. Considering global interests and covering several important aspects associated with medicinal plants, the Exploring Medicinal Plants series comprises volumes valuable to academics, practitioners, and researchers interested in medicinal plants. Topics provide information on a range of subjects including diversity, conservation, propagation, cultivation, physiology, molecular biology, growth response under extreme environment, handling, storage, bioactive compounds, secondary metabolites, extraction, therapeutics, mode of action, and healthcare practices.

Led by Azamal Husen, PhD, this series is directed to a broad range of researchers and professionals, and consists of topical books exploring information related to medicinal plants. It includes edited volumes, references, and textbooks available for individual print and electronic purchases.

Sustainable Uses of Medicinal Plants, *Learnmore Kambizi and Callistus Bvenura*

Omics Studies of Medicinal Plants, *Ahmad Altaf*

Traditional Herbal Therapy for the Human Immune System, *Azamal Husen*

Environmental Pollution and Medicinal Plants, *Azamal Husen*

Herbs, Shrubs and Trees of Potential Me-dicinal Benefits, *Azamal Husen*

Phytopharmaceuticals and Biotechnology of Herbal Plants, *Sachidanand Singh, Rahul Datta, Parul Johri, and Mala Trivedi*

Exploring Poisonous Plants: Medicinal Values, Toxicity Responses, and Therapeutic Uses, *Azamal Husen*

Plants as Medicine and Aromatics: Pharmacognosy, Ecology and Conservation *Mohd Kafeel Ahmad Ansari, Bengu Turkyilmaz Unal, Munir Ozturk and Gary Owens*

Medicinal Plant Responses to Stressful Conditions *Arafat Abdel Hamed Abdel Latef*

Aromatic and Medicinal Plants of Drylands and Deserts: Ecology, Ethnobiology and Potential Uses *David Ramiro Aguill6n Gutiérrez, Cristian Torres Le6n, and Jorge Alejandro Aguirre Joya*

Plants as Medicine and Aromatics

Pharmacognosy, Ecology and Conservation

Edited by
Mohd Kafeel Ahmad Ansari, Bengu Turkyilmaz Unal,
Munir Ozturk, and Gary Owens

CRC Press
Taylor & Francis Group
Boca Raton London New York

CRC Press is an imprint of the
Taylor & Francis Group, an **informa** business

Designed cover image: © Shutterstock

First edition published 2023
by CRC Press
6000 Broken Sound Parkway NW, Suite 300, Boca Raton, FL 33487-2742

and by CRC Press
4 Park Square, Milton Park, Abingdon, Oxon, OX14 4RN

CRC Press is an imprint of Taylor & Francis Group, LLC

Library of Congress Cataloging-in-Publication Data

Names: Ansari, Mohd Kafeel Ahmad, editor.
Title: Plants as medicine and aromatics: pharmacognosy, ecology and conservation / edited by Mohd Kafeel Ahmad Ansari, Department of Biology, Faculty of Natural Sciences, University of Guyana, Turkeyen Campus, Georgetown, South America, Bengu Turkyilmaz Unal, Department of Biotechnology, Nigde Omer Halisdemir University, Turkiye, Munir Ozturk, Department of Botany, Ege University, Izmir, Turkiye, Gary Owens, Future Industries Institute, Mawson Lakes Campus, University of South Australia, Australia.
Description: First edition. | Boca Raton, FL : CRC Press, 2023. | Series: Exploring medicinal plants | Includes bibliographical references and index. |
Identifiers: LCCN 2022037963 (print) | LCCN 2022037964 (ebook) | ISBN 9781032117744 (hbk) | ISBN 9781032129396 (pbk) | ISBN 9781003226925 (ebk)
Subjects: LCSH: Medicinal plants. | Aromatic plants. | Herbs--Therapeutic use.
Classification: LCC QK99.A1 P633 2023 (print) | LCC QK99.A1 (ebook) | DDC 581.6/34--dc23/eng/20220916
LC record available at https://lccn.loc.gov/2022037963
LC ebook record available at https://lccn.loc.gov/2022037964

ISBN: 9781032117744 (hbk)
ISBN: 9781032129396 (pbk)
ISBN: 9781003226925 (ebk)

DOI: 10.1201/9781003226925

Typeset in Times New Roman
by Deanta Global Publishing Services, Chennai, India

This book is dedicated to.......................................

The late HAKEEM ABDUL HAMEED

The Founder-Chancellor of Jamia Hamdard University, Hakeem Abdul Hameed, a renowned physician of the Unani system of medicine in India, served the poorest of the poor and also many national and global celebrities and dignitaries. Hakeem Saheb was a philanthropist, who always kept public service as his driving force. He set up a multifaceted complex of educational institutions over a period of time, which finally got amalgamated in Jamia Hamdard University in 1989.

&

To my mentor

PROF. MUHAMMAD IQBAL

Muhammad Iqbal worked as Professor and Head of Botany Department, Dean of Science, Chairman of Admissions Committee, and Vice-Chancellor at Jamia Hamdard University, New Delhi, India. Earlier, he was also associated with Garhwal University, Srinagar, Aligarh Muslim University, Aligarh, and King Saud University, Riyadh. Specializing in developmental, environmental, and medicinal botany, he has produced over 425 publications, including 290 journal research papers, 72 book chapters, 9 books, and numerous technical reports, book reviews, and popular articles. His h-index and i10-index are around 50 and 185, respectively. He guided research work of 4 MPhil, 36 PhD, and 6 postdoctoral scholars. His research interests include radial growth in woody plants, plant-pollution interaction, stress crop physiology, phytoremediation, and traditional herbal medicine.

He acted as Invited Speaker or Section Chairman in numerous botanical conferences held in Austria, Azerbaijan, Bangladesh, Hong Kong, India, Japan, Jordan, Malaysia, Oman, Pakistan, Poland, Russia, Saudi Arabia, South Korea, Thailand, and Turkiye. He was on the panel of experts of various Institutions, Ministries, Commissions and Scientific Councils of India, Croatia, Hungary, Nepal, and Oman. The Prime Minister of India nominated him to the Prize Committee for the coveted 'Indira Gandhi Paryavaran Puraskar' (2006). He has been on the editorial boards of 20 reputed international research journals and acted as reviewer for over 70 scientific journals. He is a Series Editor for the Plant Biology, Sustainability, and Climate Change series of Elsevier.

Iqbal acted as President of the Academy of Environmental Biology (2006–2009), Vice-President of the Indian Botanical Society (2003–2004), and the International Society of Environmental Botanists (2013–18). He was a Visiting Scientist in South Korea (1998) and Poland (2003 & 2008) under the Scientists Exchange Program of Indian National Science Academy (INSA) and a Visiting Professor at King Saud University, Riyadh, Saudi Arabia. A Life Member of over two dozen scientific organizations and an elected Fellow of seven prestigious scientific organizations, including the Linnean Society of London (FLS) and the National Academy of Sciences, India (FNASc). He is a recipient of over a dozen academic awards, including the H.L. Chakravarty Award of the Indian Science Congress Association (1986), Best Scientist Award of the National Environmental Science Academy (2001), Prof. V. Puri Medal Award of the Indian Botanical Society (2002), Excellent Researcher Award of King Saud University, Riyadh (2010), Bharat Jyoti Samman of India International Friendship Society (2011), and the Lifetime Achievement Award of Scifax Information & Analytics, India (2020). Commentaries on his works and achievements have appeared in several Biographical Directories (Who's Who) published from India, the U.K., and the U.S.A.

Contents

Foreword

Medicinal and aromatic plants (MAPs) perform a vital role in serving human life as well as that of animals. Glocal demand (global + local = glocal) for plant raw materials (medicinal value) is increasing: Safety (undisclosed toxic contaminants), efficacy, affordability, availability of bioactive molecules according to consumer choice is gaining momentum. Several of the bioactives in MAPs act as adaptogens, anticancers, antidiabetics, antidiarrheals, bioavailability enhancers. brain tonics, antiasthmatics, antihypertensives, antiulcers, antiurolithiatics, hepatoprotectives, carminative stomachics, and purgatives. However, there is a vast gap between demand and supply due to plant material availability becoming rare in many cases. Since agrotechnology has not been developed for many of the important medicinal plants, most of these plants are collected from nature and there is a chance for the risk of undisclosed toxic contaminants.

The Dietary Supplement Health and Education Act of 1994 ("DSHEA"), is a 1994 statute of United States Federal legislation which defines and regulates dietary supplements of the following categories:

a) **Phytomedicines**: Plant-based pharmaceutical products with proven medical efficacy, and about 18% of the world's top 150 prescription drugs are from plant sources.
b) **Herbal remedies**: Covering a spectrum of basic formulae made by traditional healers that are sold over the counter as capsules, pills, or liquid tonics.
c) **Herbal teas and infusions**: These are hot water infusions made out of herbs and sold in the form of tea bags or in granulated form.
d) **Homeopathic drugs**: They are medicines derived from plant, mineral, or animal sources and used in extremely low concentrations.
e) **Ayurvedic drugs**: Ayurveda is a South Asian, holistic, therapeutic concept, based not only on medical therapy, but including all aspects of human life in its holistic philosophy.
f) **Dietary supplements**: A range of food supplements of plant and animal origin.
g) **Functional foods**: Any food or beverage which has some claim to enhance one's physical or mental health and well-being.
h) **Nutraceuticals**: Portmanteau word between "**nutr**ition" and "pharm**aceutical**". In USA, this term is becoming increasingly popular.
i) **Cosmeceuticals**: Herbal cosmetics enhance the health and beauty of the skin.

This book elucidates all these aspects in sufficient detail. The editors are well known for their research contribution in their areas of specialization related to MAPs and have succeeded in attracting a galaxy of learned contributors from different parts of the world, representing countries like Australia, Brazil, India, Japan, Morocco, Oman, Pakistan, Sudan, South America, Turkiye, and the UAE. I am sure this book will add significantly to our knowledge regarding MAPs and meet the interest of relevant scientists, researchers, pharmacists, and all those keen on the therapeutic potential of plants.

Prof. Majeti Narasimha Vara Prasad
Emeritus Professor
University of Hyderabad
Hyderabad-500 046, Telangana, India

Preface

Since the dawn of civilization, plants have been ubiquitously used as a valuable source of traditional, plant-based, "herbal" medicine. Unlike modern medicine, herbal medicines have several major advantages, including general absence of serious, adverse side effects, long-lasting curative impact, and overall cost effectiveness. Even today, plant-based medicines and aromatics are increasingly in demand throughout the health sector globally, which involves their use not only for the treatment of diseases but also as potential therapeutic standards for maintaining good health. Almost two-thirds of the world's population directly seek alternatives to Western medical treatments, which are free from side effects, and thus these consumers depend on phytomedicines for primary healthcare. However, an inadvertent consequence of this increased demand is that the occurrence of many medicinal and aromatic plants has become threatened due to their very small wild population size, narrow distribution area, habitat specificity, and destructive way of harvesting; only a few studies have been undertaken on their conservation, ecology, and pharmacognosy of individual MAPs.

Bearing this in mind, we have made an attempt, through this book, to bring together the diverse research trends currently active in the field of medicinal botany across the world. This book elucidates the varied facets of the current medicinal and aromatic plant research. Beginning with the core study of the current status of plants as medicine and aromatics, it covers the pharmacognosy, ecology, and conservation of plants having therapeutic values, unravels the effect of abiotic stress on secondary metabolites of medicinal plants, extends the chemical characterization of their therapeutic bioactives, the isolation of bioactive compounds, and risk assessment of harvesting of wild medicinal plants, highlighting the modern investigations on drug-active substances and quality assessment of medicinal plants. Thematically, conventional and contemporary approaches to conservation of such plants are covered, as well as their commercial feasibility; the book also explores the information available on MAPs and nutraceuticals and cosmaceuticals products.

The topics of interest of this book include biodiversiy, biotechnology, ecology, nanotechnology, and pharmacognosy, as well as *in-situ* and *ex-situ* conservation of aromatic and medicinal plants. Since plants are often very geographically isolated, this book also explores the aboriginal medicine traditions of countries and cultures. Hopefully, this book will prove to be a great source of information on recent research trends in medicinal plants and aromatics, catering to the needs of students and researchers keen on medicinal plants, as well as aromatics, and stimulating a general interest for the study of medicinal plants with a proper understanding of the associated cultural, country-wide, environmental, and scientific factors, as these plants cannot be studied, evaluated, and utilized fully and properly in isolation, and need to be looked at in a global perspective.

We sincerely thank our learned contributors from different parts of the world for their consistent cooperation during the compilation, editing, and printing of this work.

Mohd Kafeel Ahamd Ansari
Bengu Turkyilmaz Unal
Munir Ozturk
Gary Owens

Author/editor biographies

Prof. Dr Mohd Kafeel Ahamd Ansari (MSc, PhD, FCRF, ERF-Australia)
Mohd Kafeel Ahmad Ansari earned his PhD degree from the Department of Botany, School of Chemical and Life Sciences, Jamia Hamdard University, New Delhi, India. He is currently serving as an Associate Professor of Biology at the University of Guyana, Georgetown, South America. He received the prestigious Endeavour Postdoctoral Research Fellowship from the Government of Australia, a Foreign Citizen Research Fellowship from the Government of Turkiye, and the Dr D S Kothari Research Fellowship Award, from the Government of India. He has worked as a Visiting Scientist at the University of South Australia, Adelaide, Australia, in the Department of Biology, in Anadolu University, Turkiye and at the Department of Biosciences, Jamia Millia Islamia (Central University), New Delhi, India. He was associated with SP University, India as Associate Professor of Botany. Also, he has served as Australian Awards Ambassador for higher educational institutions of India.

He has more than 14 years of research and teaching experience in plant eco-physiology, plant stress physiology, industrial and environmental biotechnology, environmental microbiology, microbiology, molecular biology, medicinal and aromatical plant research, and plant–microbe–soil interactions. He has published over 35 research papers in peer-reviewed international journals, 11 conference proceedings, authored six book chapters, co-authored many book chapters, and currently editing three books with international publishers, including Springer Nature, and CRC Press (Taylor & Francis Group).

He has engaged with several professional societies of Germany, Hong Kong, Italy, India, and the USA. He has attended various national and international conferences. He also figured in various Scientific Committees globally. At present, he is serving as a member of many editorial boards and reviewer of a number of high-impact international scientific journals. In addition, he has been involved with a wide range of international research projects.

Associate Prof. Dr Bengu Turkyilmaz Unal (MSc, PhD)
Dr. Bengu Turkyilmaz Unal completed her undergraduate, graduate and doctoral degrees at the Ege University, Izmir, Turkiye. She worked for six years in the Botany Department of the same University and was appointed as Assistant Professor in 2011 at the Department of Organic Agriculture, Ulukisla Vocational School of Nigde Omer Halisdemir University. Since 2015 Dr. Unal has served as Associate Professor at the Biotechnology Department of the same University. Her scientific interests include plant biotechnology, plant stress physiology, plant eco-physiology, and organic agriculture. She has published 75 papers, including two books with Elsevier and Cambridge Scholars, eight book chapters and nearly 45 papers in high-impact-factor journals.

Prof. Dr Munir Ozturk (MSc, PhD, DSc, FIAS, FFPAS)
Professor Munir Ozturk earned his PhD in 1971 from the Ege University, Turkiye. He was awarded his DSc in 1975 from the same university. He became Assistant Professor in 1971 and then Associate Professor in 1975 and then he was appointed Full Professor in 1980. He has served in various prestigious academic positions include Chairman, Botany Department and Founder Director, Centre for Environmental Studies, Faculty of Science, Ege University, Izmir, Turkiye; Visiting Scientist, ICCBS, Karachi University, Pakistan; and Foreign Fellow at the Universiti Putra Malaysia, Malaysia.

His fields of scientific interest are medicinal and aromatic plants, eco-physiology, and management of plant diversity and conservation. He has published 46 books (authored, edited, co-edited) with Springer, Elsevier, Taylor & Francis, Cambridge Scholars. He has published 85 book chapters and 155 research papers in international journals. He has vast experience in teaching and research, covering more than 55 years.

Prof. Dr Gary Owens (MSc, PhD)
Dr Gary Owens has his completed undergraduate, graduate and doctorate work (1996) from the University of Adelaide, Australia. He worked with CSIRO-2001, Australia and CERAR-2003, UniSA, Australia. Gary also served as Program Director in 2007 at UniSA. Since 2013, Gary has been serving as Group leader of ECG, Future Industries Institute, UniSA, Australia. He has a broad, multidisciplinary background with more than 26 years' experience as an active teacher and researcher leading both international and multidisciplinary research projects, touching on all facets of plant science, environmental science, and engineering.

His research focus is on medicinal plants, engineered nanomaterials, plant stress physiology, phytoremediation and heavy metal bioavailability, speciation, and computer modeling. Gary has published 147 research papers in peer-reviewed international journals on plant stress physiology, phytoremediation, and cyanobacterial laccase. He is a member of the Royal Australian Chemical Institute, The Australian Nanotechnology Network. He is a reviewer for many international scientific journals worldwide.

List of Contributors

Abdelgadir A. Abdelgadir
Department of Pharmacognosy
Faculty of Pharmacy
University of Gezira
Wad Madani
Gezira State
Sudan

Abdelhamid Azeroual
Agri-Food and Health Laboratory (AFHL)
Faculty of Sciences and Techniques of Settat
Hassan First University
P.O. Box 382
26000 Settat
Morocco

Abdullah Adil Ansari
Department of Biology
Faculty of Natural Sciences
University of Guyana
Georgetown
Guyana
South America

Ana Helena Loos Moritz
Universidade Regional de Blumenau
Department of Pharmaceutical Sciences
Blumenau
SC
Brazil

Belgizar Karayiğit
Ankara University
Faculty of Science
Department of Biology
Ankara
Turkiye

Bengu Turkyilmaz Unal
Nigde Omer Halisdemir University
Art and Sciences Faculty
Biotechnology Department
Nigde
Turkiye

Bruna Gonçalves Lopes
Universidade Regional de Blumenau
Department of Pharmaceutical Sciences
Blumenau
SC
Brazil

Camila Confortin
Post-graduate Program in Sustainable Territorial Development
Universidade Federal do Paraná
Matinhos
Brazil

Cemil Islek
Biotechnology Department
Art and Sciences Faculty
Nigde Omer Halisdemir University
Nigde
Turkiye

Deepak Chandran
Department of Veterinary Sciences and Animal Husbandry
Amrita School of Agricultural Sciences
Amrita Vishwa Vidyapeetham University
Coimbatore
Tamil Nadu
India

Denis Maldonado
Department of Biology
Faculty of Natural Sciences
University of Guyana
Georgetown
Guyana
South America

Elhadi M. Ahmed
Department of Pharmacognosy
Faculty of Pharmacy
University of Gezira
Wad Madani
Gezira State
Sudan

Esra Koç
Ankara University
Faculty of Science
Department of Biology
Ankara
Turkiye

Fatih Uckaya
Department of Nursing
Faculty of Health Science
Alanya Alaaddin Keykubat University
Antalya
Turkiye

Feyza Candan
Manisa Celal Bayar University
Science and Letters Faculty
Biology Department
Manisa
Turkiye

Florencia Guerra
Department of Biology
Faculty of Natural Sciences
University of Guyana
Georgetown
Guyana
South America

Gary Owens
Environmental Contaminants Group
Future Industries Institute
University of South Australia
Mawson Lakes
Australia

Gomathinayagam Subramanian
Division of Agriculture
Faculty of Agriculture and Forestry
University of Guyana
Berbice Campus
Tain
Guyana
South America

Hafiz Ansar Rasul Suleria
School of Agriculture and Food
Faculty of Veterinary and Agricultural Sciences
The University of Melbourne
Parkville 3010
VIC
Australia

Huda Nafees
Department of Saidla
Faculty of Unani Medicine
Aligarh Muslim University
Aligarh
Uttar Pradesh
India

Huseyin Turker
Nigde Omer Halisdemir University
Art and Sciences Faculty
Biotechnology Department
Nigde
Turkiye

Kandikere Ramaiah Sridhar
Department of Biosciences
Mangalore University
Mangalagangotri
Mangalore
Karnataka
India

Khafsa Malik
Department of Botany
PMAS Pir Mehr Ali Shah Arid Agriculture University
Rawalpindi Pakistan

Lahcen Zidane
Plant, Animal Production and Agro-industry Laboratory, Department of Biology
Faculty of Sciences
Ibn Tofail University
Kenitra
Morocco

Linghong Shi
School of Agriculture and Food
Faculty of Veterinary and Agricultural Sciences
The University of Melbourne
Parkville 3010
VIC
Australia

Luiz Everson da Silva
Post-graduate Program in Sustainable Territorial Development
Universidade Federal do Paraná
Matinhos
Brazil

Maiko Inoue
Specified Nonprofit Corporation of Horticulture Therapy and Zen
Hamamatsu
Japan

Manthena Navabharath
Department of Biotechnology
Institute of Applied Sciences & Humanities
GLA University
Mathura
Uttar Pradesh
India

Meryem Uckaya
Department of Medical Laboratory Techniques
Vocational School of Health Services
Alanya Alaaddin Keykubat University
Antalya
Turkiye

Michele Debiasi Alberton
Universidade Regional de Blumenau
Department of Pharmaceutical Sciences
Blumenau
SC
Brazil

Mohammad Kamil
Lotus Holistic Health Institute
Abu Dhabi
United Arab Emirates

Mohammad Sohail Akhtar
College of Pharmacy and Nursing
University of Nizwa
Oman

Mohd Kafeel Ahmad Ansari
Department of Biology
Faculty of Natural Sciences
University of Guyana
Georgetown
Guyana
South America

Monalisha Samal
Bioactive Natural Product Laboratory
School of Pharmaceutical Education & Research
Jamia Hamdard
New Delhi, India

Mosa E O Ahmed
Department of Pharmacognosy
Faculty of Pharmacy
University of Alneelain
Khartoum
Khartoum State
Sudan

Muhammad Iqbal
Department of Botany
School of Chemical & Life Sciences
Jamia Hamdard University
New Delhi
India

Muhammad Zafar
Department of Plant Sciences
Quaid-i-Azam University Islamabad
Pakistan

Munir Ozturk
Botany Department and Centre for Environmental Studies
Ege University
Izmir
Turkiye

Mushtaq Ahmad
Department of Botany
PMAS Pir Mehr Ali Shah Arid Agriculture University
Rawalpindi Pakistan

Mustafa Keskin
Biology Department
Faculty of Science and Arts
Marmara University
Istanbul
Turkiye

Neelam Rashid
Department of Botany
PMAS Pir Mehr Ali Shah Arid Agriculture University
Rawalpindi Pakistan

Nouruddin Chaachouay
Interdisciplinary Research Laboratory in the Sciences, Education, and Training (IRLSET)
Hassan First University
50 Rue Ibnou Lhaytham B.P. 577
Settat
Morocco

Rabea Parveen
Department of Biosciences
Faculty of Natural Sciences
Jamia Millia Islamia (Central University)
New Delhi
India

Rajneesh Thakur
Department of Plant Pathology
Dr Yashwant Singh Parmar University of Horticulture and Forestry
Nauni
Solan
Himachal Pradesh
India

Rajini Kurup
Faculty of Health Sciences
University of Guyana
Georgetown
Guyana
South America

Rauoof Ahmad Rather
Division of Environmental Science
Sher-e-Kashmir University of Agricultural Sciences and Technology of Jammu
Jammu and Kashmir
India

Sageer Abass
Bioactive Natural Product Laboratory
School of Pharmaceutical Education & Research
Jamia Hamdard
New Delhi
India

Sana Nafees
All India Institute of Medical Sciences
New Delhi
India

Sayeed Ahmad
Bioactive Natural Product Laboratory
School of Pharmaceutical Education & Research
Jamia Hamdard
New Delhi
India

S. Nizamuddin
Department of Moalejat
Government Unani Medical College
Chennai
Tamil Nadu
India.

Shazia Sultana
Department of Plant Sciences
Quaid-i-Azam University Islamabad
Pakistan

Shinichiro Hayashi
Green Flask Inc.
Tokyo
Japan

Shoor Vir Singh
Department of Biotechnology
Institute of Applied Sciences & Humanities
GLA University
Mathura
Uttar Pradesh
India

Shyam Sundar Arputhanantham
College of Pharmacy and Nursing
University of Nizwa
Oman

Sirpaul Jaikishun
Department of Biology
Faculty of Natural Sciences
University of Guyana
Georgetown
Guyana

Süleyman Toprak
Biotechnology Department
Art and Sciences Faculty
Nigde Omer Halisdemir University
Nigde
Turkiye

Suvarna Jeppu Shreelalitha
Department of Biotechnology
St. Aloysius College
Mangalore
Karnataka
India

Varsha Srivastava
Bioactive Natural Product Laboratory
School of Pharmaceutical Education and Research
Jamia Hamdard
New Delhi
India

Vigasini Subbiah
School of Agriculture and Food
Faculty of Veterinary and Agricultural Sciences
The University of Melbourne
Parkville 3010
VIC
Australia

Volkan Altay
Department of Biology
Faculty of Science and Arts
Hatay Mustafa Kemal University
Hatay
Turkiye

Wanderlei do Amaral
Post-graduate Program in Sustainable Territorial Development
Universidade Federal do Paraná
Matinhos
Brazil

Wanrong Zhao
School of Agriculture and Food
Faculty of Veterinary and Agricultural Sciences
The University of Melbourne
Parkville 3010
VIC
Australia

Wegdan Aly Shehata
College of Pharmacy and Nursing
University of Nizwa
Oman

Zabta Khan Shinwari
Department of Plant Sciences
Quaid-i-Azam University Islamabad
Pakistan

Zihong Yang
School of Agriculture and Food
Faculty of Veterinary and Agricultural Sciences
The University of Melbourne
Parkville 3010
VIC
Australia

1 Medicinal Characterization and Phytochemical Constituents of Selected Native Plants of Guyana, South America

A Review

Sirpaul Jaikishun, Abdullah Adil Ansari, Denis Maldonado, Florencia Guerra, Gomathinayagam Subramanian, Rajini Kurup, and Mohd Kafeel Ahmad Ansari

CONTENTS

1.1 INTRODUCTION

Plants accumulate an array of secondary metabolite phytochemicals which are produced as by-products. These are generally naturally occurring, non-nutritive, biologically active compounds, often with inherent protective or disease-preventive properties. These metabolites possess healing and medicinal value which has been attributed to the plants' broad biochemical composition, including alkaloids, anthraquinones, essential oils, flavonoids, glycosides, resins, saponins, tannins, terpenes, and volatile oils (Robinson, 2006). Recently, interest in these compounds has increased with the goal of evaluating different plant extracts as potential sources of new antimicrobial drugs of global relevance (Bonjar and Farrokhi, 2004).

About 80% of the population on the African continent is still more inclined to use traditional medicine to treat their medical conditions than rely on Western medicines. Likewise, traditional medicine accounts for about 40% of health care in China. Furthermore, the cultural beliefs of people across Asia and Latin America continue to drive the use of traditional medicine as a treatment

DOI: 10.1201/9781003226925-1

for many different ailments. It has been estimated that 14–28% of all higher plant species have some medicinal relevance, with 74% of pharmacologically active plant-derived components being discovered after following up on the ethnomedicinal use of the plants (Ncube et al., 2008).

Methods for choosing which plants to investigate for their pharmacological effects include the random selection and plant collection for screening, phytochemical targeting those samples from members with proven biological active compounds, ethno-directed sampling based on previous traditional medicine uses of one plant, and chemotaxonomic approach that is based on a collection of specific plant components (Cotton, 1996). For example, antifungal phytochemicals could be prepared or formulated from the leaves, seeds, stem bark, or roots of plants of known fungicidal significance and could be applied in the form of extract, powders, cakes, or as plant exudates (Anjorin and Salako, 2009; Owino and Waudo, 1992). The phytochemicals found in plants are often dependent on the prevailing growing conditions, plant age at harvesting, extraction methods, storage conditions after extraction, and the age of the sample. The use of plant derivatives for fungal control is common in many developing countries before the advent of synthetic fungicides and is still attractive due to the lower cost implication relative to imported synthetic fungicides (Galvano et al., 2001).

According to El-Khateeb et al. (2013), the damage to crops caused by fungal plant pathogens requires the use of a range of fungal control agents. Among synthetic pesticides used to protect crops, fungicides were perceived until recently as being relatively safe. The use in crop protection of many synthetic fungicides that have various degrees of persistence has now been cautioned due to their carcinogenicity, teratogenicity, and other residual toxicities. Several of the synthetic fungicides are reported to cause adverse effects on treated soil ecosystems because of their non-biodegradable nature (Tegegne et al., 2008). Substantial use of synthetic pesticides causes problems of health and environmental hazards in the ecosystem. Therefore, holistically, for the benefit to human and plant health and that of the ecosystem, natural products with antimicrobial activity are the best bio-rational alternatives and the identification of new natural plant sources of such fungicides that can be grown and cropped in large amounts is an important endeavor to contribute to future food security (Tegegne et al., 2008).

1.2 IMPORTANCE OF MEDICINAL PLANTS

Medicinal plants are a source of readily available affordable health care and also an income source for indigenous and rural persons, who rely on them for these purposes. Studies of several plants have shown that phytochemicals found in medicinal plants are potentially more effective than synthetically produced drugs (Murtaza et al., 2015). Natural compounds or products derived from natural sources present a great variety of chemical structures, variation which is not usually found among synthetic compounds (Jantan et al., 2015). Plants have also been growing for millions of years and thus have adapted to withstand adverse conditions to produce unique secondary metabolites (Jantan et al., 2015). Some traditional medicinal plants are already known to contain important chemical compounds essential for modern medicine, or which provide a basis for their use in traditional medicine by local people who collect and use them. Pharmaceutically active ingredients such as taxol, vincristine, and morphine, isolated from yew, rosy periwinkle, and opium poppy, respectively, are some examples (Hassan, 2012). Some of the common plant species found in Guyana and their use in traditional medicine are discussed below.

1.2.1 *Cymbopogon citratus*

C. citratus (DC. ex Nees) Stapf., (Figure 1.1) and other *Cymbopogon* species are commonly known as lemongrass, and are coarse grasses with a strong, lemony taste. They are members of the grass family (Poaceae), which contains about 660 genera and 9,000 species; *C. citratus* is a tall, monocotyledonous, aromatic, perennial plant with slender, sharp-edged, green linear leaves that can grow up to 90 cm in height and 5 mm in width, with a pointed apex (Chisowa et al., 1998; Clayton, 1970;

FIGURE 1.1 *Cymbopogon citratus*

Ernst, 2008). Being a perennial plant, lemongrass is widely cultivated in the tropics and sub-tropics and has two different species, the East Indian *Cymbopogon flexuosus* and the West Indian *C. citratus*. *C. citratus* has been cultivated for many years for medicinal purposes in different countries (Lachman-White et al., 1992). Lemongrass has been used in folk remedies for treatment of coughs, elephantiasis, malaria, pneumonia, vascular disease, and "consumption" (tuberculosis). Research conducted on the essential oils of *C. citratus* and their antimicrobial properties identified bioactive compounds such as citronella and geranyl acetate (Hindumathy, 2011; Ojo and Anibijuwon, 2010).

1.2.2 Cajanus cajan

C. cajan (L.) Millsp. (pigeon pea) is a perennial member of the pea family (Fabaceae). It is an erect, branched, hairy shrub, 1–2 m tall. Leaves are lanceolate to elliptical in shape, with three leaflets. Flowers are yellow, about 1.5 cm long, in sparse peduncle racemes. The pea pod is hairy, 4–7 cm long, 1 cm wide, and contains two to seven seeds (Jorrin et al., 2021; Kumar et al., 2010). It is a multipurpose plant, rich in proteins. The leaves are used for rearing silkworms; the green pods are used as a vegetable; the pods and husks, green leaves, and tops are used as fodder for livestock and also as green manure (Kumar and Tyagi, 2013; Pal et al., 2011). Among its many medicinal uses, *C. cajan* is indicated in the relief of pain and as a sedative in Chinese traditional medicine with a history of use of over 3,000 years (Ahsan and Islam, 2009). According to Pal et al. (2011), *C. cajan* leaves are rich in flavonoids and stilbenes. They also contain saponins, conspicuous amounts of tannins, and moderate quantities of reducing sugars, resins, and terpenoids.

1.2.3 Plectranthus amboinicus

P. amboinicus (Lour.) Spreng (Figure 1.2), belonging to the Lamiaceae (mint) family, is a dicotyledonous perennial commonly used as food additives and fodder, and especially as medicine in treating a wide range of diseases (Lukhoba et al., 2006). It is a branched, fleshy, highly aromatic, pubescent herb with distinctive-smelling leaves, and is variously known as Indian borage, thick-leaved thyme, and Mexican mint. The plant consists of a fleshy stem about 30–90 cm tall, with simple, broad, ovate, and very thick leaves. The leaves are thickly studded with hairs and, on the lower surface, glandular hairs are particularly numerous, which give the plant a frosted appearance (Kaliappan and Viswanathan, 2008). *P. amboinicus* leaves have been widely used traditionally for various medicinal purposes such as treatment of coughs, cephalalgia, otalgia, anorexia, diarrhea, cholera, chronic asthma, malarial fever, and for cathartic purposes (Roshan et al., 2010).

FIGURE 1.2 *Plectranthus amboinicus*

Phytochemical studies have revealed that it contains various flavonoids in the forms of quercetin, apigenin, luteolin, salvigenin, and genkwanin, as well as volatile oils from the leaves. The principal component of the oil is thymol (Chiu et al., 2012; Rastogi and Mehrotra, 1990; Uma et al., 2011).

1.2.4 Cassia alata

C. alata (L.) Roxb. is a shrub and is a medicinal plant in the Fabaceae family (Figure 1.3). It has many common names such as candle busassih, ringworm bush, and acapulco. In Guyana, it is commonly known as the carrion crow bush or, in Guyanese creole, as "canny crow bush." The common height range is 1–3 m but in some cases, the plant can reach up to 5 m tall. The leaves are paripinnate, 30–60 cm long, with 8–20 leaflets (Hennebelle et al., 2009; Lachman-White et al., 1992). Each leaflet is oblong, elliptical, and rounded at both ends. The leaflets vary in size between 5 and 15 cm. The flowers are yellow and densely distributed in axillary racemes 20–50 cm long and 3–4 cm wide (Hennebelle et al., 2009; Irwin and Barneby, 1982; Langeland, 2003). This plant traditionally provided relief against typhoid, diabetes, malaria, asthma, ringworms, tinea infections, scabies, blotch, herpes, and eczema (Oladeji et al., 2020).

FIGURE 1.3 *Cassia alata*

FIGURE 1.4 *Senna bicapsularis*

1.2.5 Senna bicapsularis

S. bicapsularis (L.) Roxb. is a sprawling evergreen shrub of the Fabaceae family. It has many common names, such as butterfly bush, Christmas bush, and rambling cassia (Figure 1.4). In Guyana, it is commonly known as the money bush. It is commonly 4 m tall or wide. Leaves are alternate and even pinnately compound with 4–7 leaflets. The leaflets are up to 4 cm long, oblong with rounded tips. The leaf stalks have glands on the upper surface (Guerra et al., 2020). The flowers are either yellow or yellow-green, individually 3–4 cm, and arranged in 3- to 12-flowered racemes near the stem tips. The flowers have stamens with prominent, curved filaments. The fruit is a slender pod, brown in color when mature, cylindrical, and glabrous 7–12 cm long (Guerra et al., 2020; Langeland, 2003). Evidence has shown the presence of anthraquinone glycosides, including rhein, emodin, physion, obtusin, chryso-obtusin, chryso-obtusin-2-*O*-β-D-glucoside, obtusifolin, and chryso-obtusifolin-2-*O*-β-D-glucoside. Moreover, flavonoids and their glycosides, alkaloids, saponins, and steroids have been reported in the methanol and hexane extracts of tissues of this plant (Guerra et al., 2020; Mak et al., 2013). Extracts of this species are used as a remedy for various skin ailments, rheumatic disease, and as laxatives (Maity et al., 1997).

1.2.6 Pityrogramma calomelanos

P. calomelanos (L.) Link is a fern with common names, such as the silverback fern or silver fern, whereas, in Guyana, it is known mainly as wild maran. *Pityrogramma* is a genus containing some 17 species, occurring mainly in tropical America. The rhizomes of the plant are erect, bearing short tufts of fronds covered with scales. The scales are brown and narrow, growing 3–6 mm long. Rachises are black to chestnut brown, shiny and glabrous (Prado, 2005). Sori are approximately 3 mm long, situated along the veins on the lower surface of the lamina between the costule and the margin. The white powdery covering of the lower surface of fronds is a distinct feature of this species. It is cultivated widely as an ornamental and the plants growing in Asia are usually considered to have been naturalized from cultivation (Cao et al., 2017). Its traditional medicinal uses include as an astringent, analgesic, antihemorrhagic, antihypertensive, and antipyretic. A decoction of the rhizomes is used as an anthelmintic (Sajeev et al., 2015; Vasuda, 1999).

1.3 ANTIMICROBIAL ACTIVITY OF SELECTED MEDICINAL PLANTS

In 2012, Jafari and colleagues reported the antibacterial activity of methanolic lemongrass extract and essential oil against the four reference bacterial species (*Staphylococcus aureus*, *Bacillus cereus*, *Escherichia coli*, and *Pseudomonas aeruginosa*) (Jafari et al., 2012). Their results indicated that increasing concentrations of the methanolic extract of the lemongrass plant prevented bacterial growth of *S. aureus*, *B. cereus*, and *E. coli*. It was discovered that a 1000 mg/mL concentration of the essential oil presented inhibitory effects toward *S. aureus*, *B. cereus*, and *E. coli*. No inhibitory effect of either lemongrass preparation toward the growth of *P. aeruginosa* was observed. The inhibitory effects of the methanolic extract on bacterial growth suggested that inhibitory effects against fungi could also be a possibility (Jafari et al., 2012). Bhattacharjee et al. (2006) showed that methanolic extracts of the leaves and seeds of *Argemone mexicana* expressed greater antibacterial activity than that of water extracts. Although an aqueous decoction of the plant parts or a preparation achieved by boiling the plant in water was employed to treat an infection, the preparation of an extract with an organic solvent was far more effective at inhibiting bacterial growth. These observations may be attributed to two reasons: firstly, the activity of biologically active components could be enhanced in the presence of methanol; secondly, the greater extractability of the active constituents in methanol could have resulted in qualitative and/or quantitative differences in composition of the extract responsible for the antibacterial activity exhibited. The findings of this study were the basis for testing the effects of several solvents for extraction (hexane, methanol, and water) on the biochemicals extracted and hence the medical efficacy of the extracts. Both disk-diffusion and agar-well-diffusion methods were implemented when assessing the sensitivity of two Gram-positive (*S. aureus* and *B. subtilis*) and two Gram-negative (*E. coli* and *P. aeruginosa*) pathogenic, multi-drug resistant bacteria (Bhattacharjee et al., 2006; Jafari et al., 2012; Nair et al., 2005).

Manjamalai (2010) analyzed the phytochemical constituents of some selected medicinal plants, namely *Tridax procumbens*, *Wedelia chinensis*, and *P. amboinicus*. The solvents methanol and ethanol were used for phytochemical extraction with subsequent antibacterial activity analysis to determine the minimum inhibitory concentration (MIC) and the size of the zones of inhibition. *P. amboinicus* extract exhibited 11 major peaks in the quantitative phytochemical analysis performed using gas chromatography-mass spectrometry (GC-MS) (Manjamalai, 2010), compared with *T. procumbens* (15 major peaks) and *W. chinensis* (ten major peaks). The research of Onawunmi et al. (1984) highlighted the importance of synergistic antibacterial effects of biological agents when tested together compared with when tested alone. Studies on *C. citratus* antibacterial properties revealed bioactivity in two of the three main components of the oils. Whereas the α-citral (geranial) and β-citral (neral) components individually elicited antibacterial action toward the Gram-negative and Gram-positive bacteria, the third component, myrcene, did not show any observable antibacterial activity on its own. In combination with the other two main components, however, myrcene's antibacterial activities were enhanced.

El-Khateeb et al. (2013) evaluated the effects of plant extracts (*Vitis vinifera* cv. Sultana, *Vitis vinifera* cv. Roumy Ahmer, *Ziziphus spina-christi* cv. Willd, *Punica granatum* cv. Baladi, and *Ficus carica* cv. Sultani) on five plant pathogenic fungi, namely *Alternaria solani*, *Botrytis cinerea*, *Botrytis fabae*, *Fusarium oxysporum*, and *Fusarium solani*, isolated from diseased potato plants, diseased fruits of tomato, diseased broad bean plants, and naturally infected plants of tomato and globe artichoke, respectively. Phytochemical screening of crude methanolic extracts revealed the presence of alkaloids, anthraquinones, carbohydrates, saponins, steroids, tannins, and terpenes in the extracts, although saponins were detected in only *Ziziphus* (jujube) extracts. All methanolic leaf extracts showed extensive antifungal activity against the different fungal pathogens tested. However, fungal sensitivity varied according to the species. The analyses identified ten polyphenol compounds in the methanolic extract of *Ziziphus* leaves, with the main antifungal component found in this extract being phenol. Arunkumar and Muthuselvam (2009) researched the phytochemical composition and activities of *Aloe vera* against clinical pathogens. The antifungal phytochemical activity against *Aspergillus flavus* and *Aspergillus niger* was analyzed using the agar-diffusion method. In the analysis, 26 bioactive phytochemicals compounds were identified in the ethanolic extract of *A. vera*. Qualitative analysis gave positive results for flavonoids, tannins, terpenoids, and saponins. When comparing extracts produced using different solvents, the highest antifungal activities were observed in the acetone extract. Of the two fungal organisms tested, greater growth suppression was observed in *A. flavus*.

Ojo and Anibijuwon (2010), studied the antimicrobial activities of extracts of three medicinal plants from Southwestern Nigeria, namely *Hoslundia opposita*, *Lantana camara*, and *C. citratus*. The poor antibacterial activity observed with ethyl acetate extracts indicated that this solvent was not a good extractant of antibacterial plant components. However, the methanolic extract proved to be far more effective and probably the best extractant for antibacterial activity from these plants, although the aqueous and ethanolic extracts also produced zones of inhibition, suggesting significant inhibitory effects toward the test organisms. At 20 μg/mL, the antibacterial effect of methanol and aqueous extracts of *C. citratus* showed growth inhibition of *Klebsiella pneumoniae*, *Streptococcus mutans*, and *E. coli*, which had zones of inhibition ranging from 4.00 to 10.00, while *P. aeruginosa* and *Staphylococcus aureus* were insensitive to the extracts. Qualitative phytochemical tests of *C. citratus* showed that only tannins and steroids were present, with alkaloids, flavonoids, and saponins not being detected. However, in stark contrast to this study, another study by Gills (1992) revealed that *C. citratus* contained many alkaloids. This discrepancy indicated that plants and their natural constituents were also subject to variations, due, for example, to differences in their geographic locations and the climatic conditions under which they live. When alcoholic or water extracts of *C. citratus* were examined for antibacterial properties, the phytochemical constituents were screened against four Gram-negative bacteria, namely *E. coli*, *K. pneumoniae*, *P. aeruginosa*, and *Proteus vulgaris*, and two Gram-positive bacteria, *B. subtilis* and *S. aureus* (Hindumathy, 2011). The MIC of the extracts tested against the bacteria ranged between 150 mg/mL and 50 mg/mL. The alcoholic extracts were found to be generally more effective than the water extracts in terms of inhibiting microbial growth. Phytochemical studies of the plant extracts showed the presence of tannins, saponins, alkaloids, phenolics, and steroids in both the water and alcoholic extracts, and both extracts showed the presence of glycosides (Hindumathy, 2011; Ojo and Anibijuwon, 2010). The methanol and aqueous extracts of *C. citratus* inhibited all the tested organisms, although the percentage inhibition varied with the type of leaf extracts and extract concentration, as well as the species of bacterium. The results showed that inhibition of microbial growth was greater at high concentrations of the methanol extracts and that less inhibition was observed as the concentration decreased. A zone of inhibition was visible in the water extract tests but it was smaller than with the methanol extract. This implies that the antibacterial compound(s) in the plant extracts were either more efficacious or were extracted to higher concentrations in ethanol than in water, with the inhibitory compound(s) apparently being more soluble in alcohol than in water.

Trivedi and Singh (2014) tested ethyl acetate, aqueous, and methanolic extracts, as well as the essential oils, of *C. citratus* for antifungal activity in vitro using the agar-well-diffusion method and the Poisoned Food Technique, where plant extracts were incorporated into fungal culture media. The antifungal activity of plant extracts was evaluated against *A. flavus* and *Mucor* sp. The ethyl acetate extract showed the highest activity against both fungi whereas the aqueous and methanol extracts showed the least activity. It was also observed that active extracts inhibited spore germination and that the agar-well-diffusion method offered a fast, cost-effective, and generally accurate result in comparison to the Poisoned Food Technique. It was thus recommended that *C. citratus* could be used as an antifungal agent in syrups and tonics, using the plant extracts as a preservative for drug safety.

Mohana and Raveesha (2007) performed a study on antifungal activity of several plant extracts against a number of fungi which attacked plants or stored food. Aqueous extracts of eight plant species at a concentration of 10% (w/v) were screened for antifungal activity against *Fusarium solani* and *A. flavus* measured as mycelial dry weight, spore germination, and colony growth, using the Poisoned Food Technique. Results for extracts of *Decalepis hamiltonii* Wight &Arn. showed significant antifungal activity at different concentrations using the Poisoned Food Technique against ten species of *Aspergillus*, among others. In addition, *D. hamiltonii* was extracted using different solvents, namely petroleum ether, benzene, chloroform, methanol, and ethanol, to identify the solvent extract with the highest activity. Ezeifeka et al. (2004) evaluated the antimicrobial activities of *C. cajan*, *Centella asiatica* (gotu kola), and *Xylopia aethiopica* against the bacteria *S. aureus*, *P. aeruginosa*, *E. coli*, and the fungus *Candida albicans*. The antimicrobial tests of the plant extracts were tested using the agar-well-diffusion inhibition test and the paper-disk-diffusion inhibition test. The ethanolic extract of each of the plants showed various levels of antimicrobial activity when tested by each method, whereas the aqueous extracts showed antimicrobial activity only when tested by the paper-disk-diffusion method. The ethanolic extract of both leaves and seeds of *C. cajan* showed the highest antimicrobial activity. The extracts of *C. cajan* showed greater inhibition of the fungus *C. albicans*; it is likely that such extracts could be used to treat infections and diseases caused by this organism (Ezeifeka et al., 2004).

Microbial inhibition by different natural products, such as plant extracts, have been investigated over the years. The methanolic extracts of *Thymus vulgaris* (thyme) and *Zingiber officinale* (ginger) were highly active and showed both fungistatic and fungicidal activities against tomato phytopathogenic fungi (Al-Rahmah et al., 2013). The fungicidal activities of methanolic extracts of five plant species against three tomato phytopathogenic fungi were evaluated. The extracts were from the plants *L. camara*, *Salvadora persica*, *T. vulgaris*, *Z. officinale*, and *Ziziphus spina-christi*, whereas the fungi were the damping-off pathogens, *Fusarium oxysporum*, *Pythium aphanidermatum, and Rhizoctonia solani*. Of the five plant methanolic extracts, three were effective against the phytopathogenic fungi, of which both *T. vulgaris* and *Z. officinale* extracts were highly active and showed both fungistatic and fungicidal properties against the fungi, with MICs of 4 mg/mL and minimal fungicidal concentrations (MFCs) of 8 mg/mL for each fungus, except for *F. oxysporum* which was less sensitive and for which the MFC of the *Z. officinale* extract was 16 mg/mL. *S. persica* methanolic extracts showed moderate fungicidal activity, whereas those of *L. camara* and Z. *spina-christi* were effective against only *P. aphanidermatum*, which was inhibited completely at 10 mg/mL of *L. camara* extract. The *T. vulgaris* methanolic extract was composed mostly of thymol (38.73%), carvacrol (19.31%), β-cimene (10.13%), and α-terpinolene (5.94%), whereas that of *Z. officinale* was composed mostly of gingerol (46.85%), cedrene (8.39%), zingiberene (7.41%) and α-curcumene (7.32%). Hence, such plant extracts might contribute to the development of safer, alternative fungicides for the control of tomato damping-off fungi (Al-Rahmah et al., 2013). Investigations have also been carried out on the effectiveness of extracts of *Azadirachta indica* (neem) and *Persea americana* (avocado pear) against rice sheath blight disease caused by *R. solani* in Guyana; both extracts showed antifungal activities, although *P. americana* was the more active (Jagessar et al., 2015). In further research, the fungicidal activity of aqueous leaf extract of a number of plant species were tested against the phytopathogenic fungus *Alternaria brassicae*, and it was found that, whereas the majority of the plant extracts were active, plant extracts from *Solanum nigrum* were the most effective, with activity being comparable to synthetic chemical fungicides (Sheema and Durai, 2014).

Kumar and Tyagi (2013) conducted research on the antifungal activity of different extracts of *Bergenia stracheyi*. The Poisoned Food Technique was employed using different concentrations of extracts against six test fungi. The six fungal species chosen for the research were *Alternaria alternata*, *A. niger*, *Colletotrichum gloeosporoides*, *F. oxysporum*, *Ganoderma lucidum*, and *R. solani*. The solvents used for extraction were chloroform, ethanol, ethyl acetate, and petroleum ether. The antifungal activity was determined by calculating the minimum inhibitory concentration (MIC) and zone of inhibition. The ethanol extract of *Bergenia stracheyi* was the most effective one against all the test fungi. Of the other extracts, the petroleum ether, chloroform, and ethyl acetate extracts exhibited antifungal activity but insufficient to inhibit total fungal growth at the lower concentrations.

According to Ogunjobi and Abiala (2013), the antimicrobial activities of extracts of *Cassia (Senna) alata* and *Phyllanthus amarus* against *E. coli*, *S. aureus*, *Salmonella typhimurium*, *A. niger*, *A. flavus*, and *C. albicans* were variable. The results showed that the ethanol and water extracts of *S. alata* and *P. amarus* were effective against the test organisms. The largest zone of inhibition was recorded in the water extract of *C. alata* against *A. niger* at 27.2 mm, whereas the smallest was achieved for this extract against *S. typhimurium* at 10.1 mm with ethanol for *C. alata*. The greatest antifungal activity in the ethanol extract of *P. amarus* was recorded against *A. niger* at 18.2 mm. Therefore, both *C. alata* and *P. amarus* were shown to be good sources of antimicrobial agents against fungal pathogens common to both humans and plants (Ogunjobi and Abiala, 2013).

There has been research on the antifungal activity of plants as of *Cassia* genus. In particular, one study highlighted how the methanolic extracts of three *Cassia* genera, *Cassia alata*, *Cassia fistula*, and *Cassia tora*, acted against three pathogenic fungi (Phongpaichit et al., 2004). The research determined that the *C. alata* extract was the most effective against *Trichophyton rubrum* and *Microsporum gypseum* with a 50% inhibition concentration with respect to antifungal properties (Phongpaichit et al., 2004). Owoyale et al. (2006) investigated the antifungal activity of alcoholic extracts of *C. alata*. The methanolic extract showed greater inhibitory activity than the ethanolic and petroleum ether extracts. The unidentified active antifungal compounds showed the lowest activity against the fungi *Mucor* sp., *Rhizopus* sp., and *A. niger* at a concentration of 70 μg/mL and the highest activity against these fungi at a concentration of 860 μg/mL. Moreover, phytochemical screening indicated that the most active antifungal component of the extract was a flavonoid glycoside. Sule et al. (2010) investigated the phytochemical properties and in vitro antifungal activity of *C. alata* crude stem bark extract, obtained by Soxhlet distillation, on the test dermatophytes, *Microsporium canslaslomyces*, *Trichophyton verrucosum*, *Trichophyton mentagrophytes*, and *Epidermophyton floccosum*. The crude bark ethanol extract exhibited clear antifungal effects on each of the fungi. The extract showed the highest inhibition toward *T. verrucosum* and *E. floccosum*, with zones of inhibition of 21.00 and 20.05 mm, respectively. The MIC for the extract on each of the fungi was 5 mg/mL. The result showed that the stem bark extract was fungicidal against each of the fungi at concentrations of 5 mg/mL and 10 mg/ml, except for the fungus *E. floccosum*, for which the extract was fungicidal only at a concentration of 5 mg/mL. Phytochemical analysis revealed the presence of various secondary metabolites, namely alkaloids, anthraquinones, carbohydrates, saponins, steroids, tannins, and terpenes, indicating that the crude bark extract of *C. alata* was a rich, potential source of antimicrobial agents (Sule et al., 2010).

Souza et al. (2012) investigated the possible enhancement of aminoglycoside antibiotic activity by *Plectranthus calomelanos*. Methanol and ethanolic extracts were evaluated for antifungal, antibacterial, and modulatory activity against the bacteria *E. coli*, *P. aeruginosa*, and *S. aureus*, and the fungi *C. albicans*, *Candida krusei* and *Candida tropicalis*. The ethanolic and methanolic extracts of *P. calomelanos* showed high activity against *S. aureus* when combined with aminoglycoside antibiotics and against species of the genus *Candida* in conjunction with benzoylmetronidazole. Luciano-Montalvo et al. (2013) screened extracts of Caribbean herbal remedies for antimicrobial activities against five bacterial species, using an agar-disk-diffusion assay. Aqueous extracts of *P. calomelanos*, *Tapeinochilos ananassae*, and *Syzygium jambos*, as well as the juice of *Gossypium barbadense*, showed > 20% of the growth inhibition against several bacteria which was achieved with the positive control, which was the antibiotic streptomycin. For *P. calomelanos* which is one of the plants of interest in the current chapter, the percentage inhibition was 2–20%, compared with 100% inhibition by the positive control; none of the concentrations of this plant extract tested showed bactericidal activity.

1.4 CONCLUSION

The local biodiversity in tropical countries like Guyana is very rich and includes many plants with medicinal benefits that have been used by indigenous communities for centuries. However, many of these plants require identification and characterization as to their medicinal attributes and phytochemical composition. These plants are abundantly distributed across the tropical areas and can be harvested with potentially minimal effect on the ecosystem. Phytochemicals relevant to specific diseases and infections need to be isolated for further analyses and clinical trials, which may lead to the development of novel drugs or crop protection chemicals.

REFERENCES

Ahsan, R., & Islam, M. (2009). In vitro antibacterial screening and toxicological study of some useful plants (*Cajanus cajan*). *European Journal of Scientific Research*, *41*, 227–232.

Al-Rahmah, A., Mostafa, A., Abdel-Megeed, A., Yakout, S., & Hussein, S. (2013). Fungicidal activities of certain methanolic plant extracts against tomato phytopathogenic fungi. *African Journal of Microbiology Research*, *7*(6), 517–524.

Anjorin, S., & Salako, E. (2009). The status of pesticide plants and materials identification in Nigeria. *Nigerian Journal of Plant Protection*, *23*, 25–32.

Arunkumar, S., & Muthuselvam, M. (2009). Analysis of phytochemical constituents and antimicrobial activities of *Aloe vera* L. against clinical pathogens. *World Journal of Agricultural Sciences*, *5*(5), 572–576.

Bhattacharjee, I., Chatterjee, S. K., Chatterjee, S., & Chandra, G. (2006). Antibacterial potentiality of *Argemone mexicana* solvent extracts against some pathogenic bacteria. *Memórias do Instituto Oswaldo Cruz*, *101*(6), 645–648.

Bonjar, G. S., & Farrokhi, P. R. (2004). The anti-bacillus activity of some plants is used in traditional medicine of Iran. *Nigerian Journal of Natural Products and Medicine*, *8*, 34–39.

Cao, H., Chai, T.-T., Wang, X., Morais-Braga, M. F. B., Yang, J.-H., Wong, F.-C., Wang, R., Yao, H., Cao, J., & Cornara, L. (2017). Phytochemicals from fern species: Potential for medical applications. *Phytochemistry Reviews*, *16*(3), 379–440.

Chisowa, E. H., Hall, D. R., & Farman, D. I. (1998). Volatile constituents of the essential oil of *Cymbopogon citratus* Stapf grown in Zambia. *Flavour and Fragrance Journal*, *13*(1), 29–30.

Chiu, Y.-J., Huang, T.-H., Chiu, C.-S., Lu, T.-C., Chen, Y.-W., Peng, W.-H., & Chen, C.-Y. (2012). Analgesic and anti-inflammatory activities of the aqueous extract from *Plectranthus amboinicus* (Lour.) Spreng. both in vitro and in vivo. *Evidence-Based Complementary and Alternative Medicine*, *2012*, 508137.

Clayton, W. (1970). Flora of tropical East Africa. Gramineae (Part 1). *Flora of tropical East Africa. Gramineae (Part 1)*. London: Royal Botanic Gardens

Cotton, C. M. (1996). *Ethnobotany: Principles and applications* London: John Wiley and Sons.

El-Khateeb, A. Y., Elsherbiny, E. A., Tadros, L. K., Ali, S. M., & Hamed, H. B. (2013). Phytochemical analysis and antifungal activity of fruit leave extract on the mycelial growth of fungal plant pathogens. *Journal of Plant Pathology and Microbiology*, *4*(9), 1–6.

Ernst, E. (2008). Chiropractic: A critical evaluation. *Journal of Pain and Symptom Management*, *35*(5), 544–562.

Ezeifeka, G., Orji, M., Mbata, T., & Patrick, A. (2004). Antimicrobial activities of *Cajanus cajan*, *Garcinia kola* and *Xylopia aethiopica* on pathogenic microorganisms. *Biotechnology*, *3*(1), 41–43.

Galvano, F., Piva, A., Ritieni, A., & Galvano, G. (2001). Dietary strategies to counteract the effects of mycotoxins: A review. *Journal of Food Protection*, *64*(1), 120–131.

Gills, L. (1992). *Ethnomedical uses of plants in Nigeria*. Benin City: Uniben Press.

Guerra, F., Ansari, A. A., Kurup, R., & Subramanian, G. (2020). Antifungal activity of *Senna alata*, *Senna bicapsularis* and *Pityrogramma calomelanos*. *Journal of Complementary and Alternative Medical Research*, *10*(3), 11–21.

Hassan, B. A. R. (2012). Medicinal plants (importance and uses). *Pharmaceutica analytica acta*, *3*(10), 1000–1139.

Hennebelle, T., Weniger, B., Joseph, H., Sahpaz, S., & Bailleul, F. (2009). Senna alata. *Fitoterapia*, *80*(7), 385–393.

Hindumathy, C. (2011). In vitro study of antibacterial activity of Cymbopogon citratus. *International Journal of Biotechnology and Bioengineering*, *5*(2), 48–52.

Irwin, H. S., & Barneby, R. C. (1982). *The American Cassiinae: A synoptical revision of Leguminosae tribe Cassieae Subtribe Cassiinae in the new world*. New York: New York Botanical Gardens Bronx.

Jafari, B., Ebadi, A., Aghdam, B. M., & Hassanzade, Z. (2012). Antibacterial activities of lemongrass methanol extract and essence on pathogenic bacteria. *American–Eurasian Journal of Agricultural and Environment Science*, *2*, 1042–1046.

Jagessar, R., Nankishore, A., & Hafeez, A. (2015). An investigation of the effectiveness of plant extracts from *Azadirachta indica* (neem) and *Persea americana* (pear) against rice sheath blight diseases induced by *Rhozocotonia Solani*. *Americal Journal of Research Communication*, *3*, 123–139.

Jantan, I., Bukhari, S. N. A., Seyed Mohamed, M. A., Wai, L. K., & Mesaik, M. A. (2015). The Evolving Role of Natural Products from the Tropical Rainforests as a Replenishable Source of New Drug Leads. In O. Vallisuta, & S. Olimat (Eds.), *Drug Discovery and Development*. London: IntechOpen.

Jorrin, B., Maluk, M., Atoliya, N., Kumar, S. C., Chalasani, D., Tkacz, A., Singh, P., Basu, A., Pullabhotla, S. V., & Kumar, M. (2021). Genomic diversity of pigeon pea (Cajanus cajan L. Millsp.) endosymbionts in India and selection of potential strains for use as agricultural inoculants. *Frontiers in Plant Science*, *12*, 2021.680981.

Kaliappan, N. D., & Viswanathan, P. K. (2008). Pharmacognostical studies on the leaves of *Plectranthus amboinicus* (Lour) Spreng. *International Journal of Green Pharmacy*, *2*(3), 182.

Kumar, H., Bajpai, V. K., Dubey, R., Maheshwari, D., & Kang, S. C. (2010). Wilt disease management and enhancement of growth and yield of *Cajanus cajan* (L) var. Manak by bacterial combinations amended with chemical fertilizer. *Crop Protection*, *29*(6), 591–598.

Kumar, V., & Tyagi, D. (2013). Antifungal activity evaluation of different extracts of Bergenia stracheyi. *International Journal of Current Microbiology and Applied Sciences*, *2*(7), 69–78.

Lachman-White, D. A., Adams, C. D., & Trotz, U. O. D. (1992). *A guide to the medicinal plants of Coastal Guyana* (Vol. 8). London: Commonwealth Science Council.

Langeland, K. (2003). *Merry Christmas Senna*. Florida: UF/IFAS Agronomy Department and Center for Aquatic and Invasive Plants.

Luciano-Montalvo, C., Boulogne, I., & Gavillán-Suárez, J. (2013). A screening for antimicrobial activities of Caribbean herbal remedies. *BMC Complementary and Alternative Medicine*, *13*(1), 1–9.

Lukhoba, C. W., Simmonds, M. S., & Paton, A. J. (2006). Plectranthus: A review of ethnobotanical uses. *Journal of Ethnopharmacology*, *103*(1), 1–24.

Maity, T. K., Mandal, S. C., Mukherjee, P. K., Saha, K., Das, J., Saha, B., & Pal, M. (1997). Evaluation of the hepatoprotective potential of *Cassia Tora* leaf extract. *Natural Product Sciences*, *3*(2), 122–126.

Mak, Y. W., Chuah, L. O., Ahmad, R., & Bhat, R. (2013). Antioxidant and antibacterial activities of hibiscus (*Hibiscus rosa-sinensis* L.) and Cassia (*Senna bicapsularis* L.) flower extracts. *Journal of King Saud University – Science*, *25*(4), 275–282.

Manjamalai, A. (2010). Sardar Sathyajith Singh, R., Guruvayoorappan, C., & Berlin Grace, V. M. Analysis of phytochemical constituents and antimicrobial activity of some medicinal plants in Tamil Nadu, India. *Global Journal of Biotechnology and Biochemistry*, *5*(2), 120–128.

Mohana, D., & Raveesha, K. (2007). Anti-fungal evaluation of some plant extracts against some plant pathogenic field and storage fungi. *Journal of Agricultural Technology*, *4*(1), 119–137.

Murtaza, G., Mukhtar, M., & Sarfraz, A. (2015). A review: Antifungal potentials of medicinal plants. *Journal of Bioresource Management*, *2*(2), 4.

Nair, R., Kalariya, T., & Chanda, S. (2005). Antibacterial activity of some selected Indian medicinal flora. *Turkish Journal of Biology*, *29*(1), 41–47.

Ncube, N., Afolayan, A., & Okoh, A. (2008). Assessment techniques of antimicrobial properties of natural compounds of plant origin: Current methods and future trends. *African Journal of Biotechnology*, *7*(12), 1797–1806.

Ogunjobi, A., & Abiala, M. (2013). Antimicrobial activity of *Senna alata* and Phyllanthus amarus. *Global Journal of Pharmacology*, *7*(2), 198–202.

Ojo, O., & Anibijuwon, I. (2010). Studies on extracts of three medicinal plants of south-western Nigeria: Hoslundia opposita, Lantana camara and Cymbopogon citratus. *Advances in Natural and Applied Sciences*, *4*(1), 93–99.

Oladeji, O. S., Adelowo, F. E., Oluyori, A. P., & Bankole, D. T. (2020). Ethnobotanical description and biological activities of Senna alata. *Evidence-Based Complementary and Alternative Medicine*, *2020*, 2580259.

Onawunmi, G. O., Yisak, W.-A., & Ogunlana, E. (1984). Antibacterial constituents in the essential oil of *Cymbopogon citratus* (DC.) Stapf. *Journal of Ethnopharmacology*, *12*(3), 279–286.

Owino, P., & Waudo, S. (1992). Medicinal plants of Kenya: Effects of *Meloidogyne incognita* and the growth of okra. *Afro-Asian Journal of Nematology*, *2*, 64–66.

Owoyale, J. A., Olatunji, G. A., & Oguntoye, S. (2006). Antifungal and antibacterial activities of an alcoholic extract of *Senna alata* leaves. *Journal of Applied Sciences and Environmental Management*, *9*(3). https://doi.org/10.4314/jasem.v9i3.17362.

Pal, D., Mishra, P., Sachan, N., & Ghosh, A. K. (2011). Biological activities and medicinal properties of *Cajanus cajan* (L) Millsp. *Journal of Advanced Pharmaceutical Technology and Research*, *2*(4), 207.

Phongpaichit, S., Pujenjob, N., Rukachaisirikul, V., & Ongsakul, M. (2004). Antifungal activity from leaf extracts of *Cassia alata* L., *Cassia fistula* L. and *Cassia tora* L. *Songklanakarin Journal of Science and Technology*, *26*(5), 741–748.

Prado, J. (2005). Flora da Reserva Ducke, Amazonas, Brasil: Pteridophyta Chave para as famílias. *Rodriguésia*, *56*(86), 27–28.

Rastogi, R. P., & Mehrotra, B. (1990). *Compendium of Indian medicinal plants*. New Delhi: Central Drug Research Institute and Public and Information Directorate.

Robinson, N. (2006). Integrated traditional Chinese medicine. *Complementary Therapies in Clinical Practice, 12*(2), 132–140. https://doi.org/10.1016/j.ctcp.2006.01.006.

Roshan, P., Naveen, M., Manjul, P., Gulzar, A., Anita, S., & Sudarshan, S. (2010). *Plectranthus amboinicus* (Lour) Spreng: An overview. *Pharmaceutical Research, 4*, 1–15.

Sajeev, S., Raj, P., Adarsh, D., & Hegde, S. (2015). Phytoconstituents of *Nephrolepis hirsutula* and *Pityrogramma calomelanos*, two medicinal ferns of the Western Ghats. *Indian Fern Journal, 32*, 244–256.

Sheema, K., & Durai, M. (2014). Fungicidal activity of aqueous leaf extract on Alternaria brassicae. *International Journal of Pharmacognosy and Phytochemical Research, 6*(4), 801–805.

Souza, T. M., Morais-Braga, M. F., Costa, J. G., Saraiva, A. A., & Coutinho, H. D. (2012). Enhancement of antimicrobial activity of antibiotics and antifungals by the use of natural products from *Pityrogramma Calomelanos* (L.) Link. *Archives of Biological Sciences, 64*(1), 43–48.

Sule, W., Okonko, I., Joseph, T., Ojezele, M., Nwanze, J., Alli, J., & Adewale, O. (2010). In vitro antifungal activity of *Senna alata* Linn. crude leaf extract. *Research Journal of Biological Sciences, 5*(3), 275–284.

Tegegne, G., Pretorius, J., & Swart, W. (2008). Antifungal properties of *Agapanthus africanus* L. extracts against plant pathogens. *Crop Protection, 27*(7), 1052–1060.

Trivedi, B., & Singh, P. (2014). Antifungal activity of *Cymbopogon citratus* leaves extracts against herbal drug contaminants. *World Journal of Pharmacy and Pharmaceutical Sciences, 3*(6), 768–775.

Uma, M., Jothinayaki, S., Kumaravel, S., & Kalaiselvi, P. (2011). Determination of bioactive components of *Plectranthus amboinicus* (Lour) Spreng by GC–MS analysis. *New York Science Journal, 4*(8), 66–69.

Vasuda, S. (1999). Economic importance of pteridophytes. *Indian Fern Journal, 16*, 130–152.

2 *Ex-situ* Conservation of Medicinal and Aromatic Plants Using *In Vitro* Techniques

Bengu Turkyilmaz Unal, Huseyin Turker, and Munir Ozturk

CONTENTS

2.1 INTRODUCTION

The use of medicinal and aromatic plants (MAPs) dates back to ancient times, where humans routinely used MAPs to relieve their psychological stress, treat diseases, and heal wounds with natural bioactive substances obtained from nature (Máthé, 2015). Even today, MAPs are still widely used by people for self-treatment and national health services (Ozturk and Hakeem, 2018, 2019a, 2019b). MAPs have a wide variety of active ingredients, which have been included in many commercial products, ranging from drugs used in the treatment of diseases in traditional and modern medicine, to significant components in the food and cosmetics industry. Nearly 40,000 plant species are used ethnomedicinally. Several of these have been commonly used traditionally in large amounts following direct collection from nature (Máthé, 2015). However, overuse of MAPs by collecting directly from nature is posing a problem, as these MAPs are now facing the threat of extinction because of indiscriminate collection worldwide. As a consequence, the number of threatened species is increasing every year (FAO, 2020), where the pressure on existing populations has increased with the increasing global demand for herbs as well as the local demand for traditional herbal medicines (Baričevič and Kušar, 2006; Ozturk et al., 2020). Uncontrolled overuse, habitat loss, and climate change directly affect ecosystems (Baričevič et al., 2015).

Natural ecosystems have a significant role in the protection of plants and habitats in a particular region, and it is known that each species has a specific niche in its related habitat (Ozturk and Altay, 2017). The ecological niche is defined as a combination of environmental factors which allows an organism to protect its vital activity in a population, following the effects of an organism on the environment (Chase and Leibold, 2003). Therefore, the ecodiversity of MAPs should be carefully evaluated and followed ecogeographically, and their status within the population determined. Factors leading towards the extinction of a particular species can be natural (environmental) or anthropogenic (e.g., demographic increase, intensive industrial, and agricultural activities) (Zlobin, 2012). It is well known that extinction of species(s) is usually caused by direct (over-collecting, grazing) and/or indirect (e.g., tillage, deforestation, fires, introduction of alien and even invasive species to the relevant area, global warming, climate change) effects (Zlobin, 2012). It is not known exactly how many species and/or which species are threatened, as the Red List(s) or Red Data Books have not been published by all countries (Heywood, 2011). It is estimated that around 15,000 MAP

DOI: 10.1201/9781003226925-2

species are threatened (Baričevič et al., 2015). The pharmaceutical industry and other industries are evaluating several medicinal and aromatic plants such as *Harpagophytum* spp., found in southern Africa; *Prunus africana* distributed in Madagascar and Central and southern Africa; *Warburgia salutaris* found in the Brazilian rainforest, and *Saussurea costus* distributed in the Himalayas, all of which are in need of urgent attention due to their current scarcity (Abensperg-Traun, 2009; Baričevič et al., 2015; Schippmann et al., 2006). MAPs like *Chamaelirium luteum*, *Cimicifuga racemosa*, *Cypripedium* spp., *Dioscorea villosa*, *Echinacea tennesseensisve*, *Hydrastis canadensis*, *Ligusticum porteri*, *Mitchella repens*, *Panax ginseng*, *Panax quinquefolius*, *Piper methysticum*, *Sanguinaria canadensis*, *Taxus brevifolia*, and *Trillium* spp. are also excessively collected in the United States (Baričevič et al., 2015, Burkhart et al., 2012, McKenzie et al., 2009, Simberloff, 2012), contributing to their scarcity as well.

2.2 PLANT GROWTH REGULATORS USED IN CONSERVATION AND *IN VITRO* STUDIES USING BIOTECHNOLOGICAL METHODS

Globally, the use of MAPs is increasing rapidly due to the increasing demand for natural health products and herbal medicines (Cole et al., 2007; Kundu et al., 2018). The latest estimates indicate that the current extinction rate of MAPs is 1,000 times higher than the previously predicted trend. Thus, our planet is potentially facing a significant loss of medicines on an annual basis due to this situation (Kundu et al., 2018). It is estimated that nearly 20% of the plant resources will become extinct due to the constantly increasing world population (Ross, 2005). In view of this, several recommendations have been presented on the sustainable use and conservation of the medicinal and aromatic plants, but only specific suggestions have been evaluated for a few conservation methods and sustainable use of such plants. These include *ex-situ* conservation in botanical gardens and *in situ* conservation in nature reserves (Kundu et al., 2018; Uprety et al., 2012).

To address these shortcomings, biotechnological developments are now enabling us to develop processes that result in the biosynthesis of natural products, their elicitation in plant tissue culture, and ultimately the desired level and uniformity of the targeted bioactive compound. According to Engelmann (2011), developments in the field of biotechnology are contributing significantly towards the protection of MAPs. These biotechnological methods are preferred by scientists as they do not replace traditional conservation methods, but supplement and modify existing methods (Coelho et al., 2020; Engelmann, 2011).

Plant protection biotechnology includes the management, characterization, and sustainable use of plant genetic resources as well as the conservation of plant genetic resources. The protection of cultured plant germplasm using *in vitro* tissue culture technology provides opportunities for an easy evaluation of plant material (Coelho et al., 2020; Engelmann, 2011; Reed et al., 2011). *In vitro* propagation techniques are now accepted as an effective alternative to traditional plant propagation methods, because they allow for the production of a large number of plants in a short time frame and within a limited area (Engelmann, 2011). This technique is also very useful for the conservation and sustainable use of endemic and endangered MAPs which cannot be produced by conventional methods or have excessive dormancy (Coelho et al., 2020). Another advantage of conservation using *in vitro* techniques is that with a very small amount of starting plant material (explants) taken from the mother plants, thousands of plants can be propagated without harming the MAP population level in their natural habitat (Figure 2.1).

In *in vitro* propagation of MAPs, the steps include callus induction, shoot development, root formation, and acclimatization stages (Figure 2.2). Plant growth regulators (PGRs) are generally used in the callus induction, shoot development, and root formation stages (Baksha et al., 2005; Turker and Turkyilmaz Unal, 2022). PGRs regulate various biochemical, morphological, and physiological processes when added to the nutrient medium in *in vitro* studies. PGRs are also known as phytohormones (Srivastava, 2002). In fact, *in vitro* growth and shoot formation cannot be obtained without

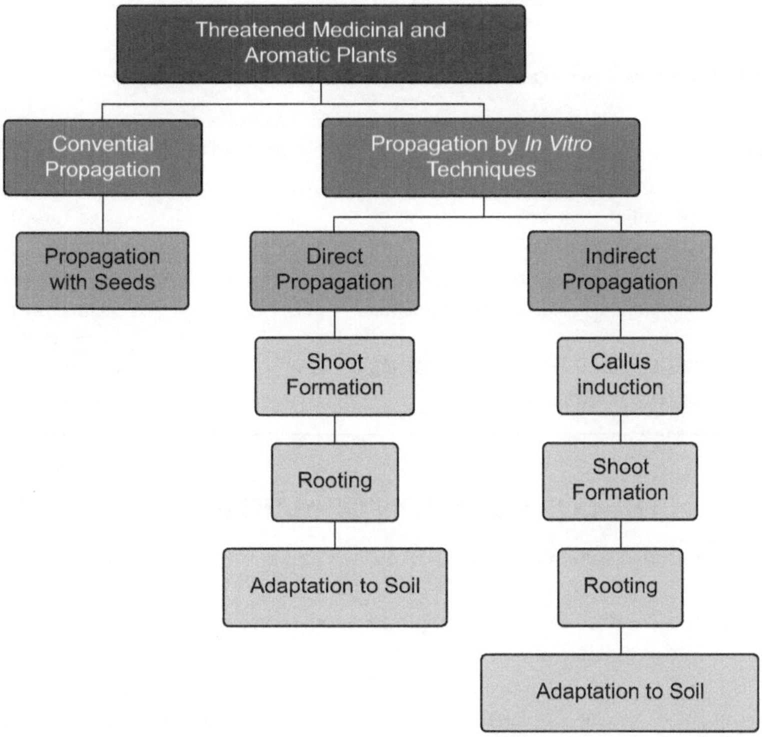

FIGURE 2.1 Schematic representation of the conservation of MAPs.

FIGURE 2.2 *Ex-situ* conservation of *Hypericum bilgehan-bilgilii in vitro* (Turker and Turkyilmaz Unal, 2022)

PGRs being added to the nutrient medium at appropriate concentrations, as reported for *Hypericum bilgehan-bilgilii* and many other species (Turker and Turkyilmaz Unal, 2022). Insufficient or excessive amounts of PGRs added to *in vitro* cultures can cause morphological and physiological abnormalities (Bouza et al., 1994). Table 2.1 shows various hormones and their role in plants. Here, we will present evolving *in vitro ex-situ* preservation techniques for MAPs in the light of new biotechnological methods.

TABLE 2.1
Plant Growth Hormones and Their Function in Plant Tissue Culture

Types of Hormones	Duties	Examples of Hormones	References
Auxins	Cell division, cell elongation, root formation, root development, and root differentiation	IAA, IBA, NAA, 2,4-D	Anis and Ahmad, 2016; Turker and Turkyilmaz Unal, 2022
Cytokinins	Cell division, shoot formation, shoot development, and shoot proliferation	BAP, Kin, BA, IPA, TDZ, Z, mT	Anis and Ahmad, 2016; Turker and Turkyilmaz Unal, 2022; Uzun et al., 2020
Gibberellins	Breaking dormancy, seed germination, growth, and flowering	GA_3	Turker and Turkyilmaz Unal, 2021

2.3 STUDIES ON THE CONSERVATION OF MAPS BY *IN VITRO* TECHNIQUES

Some studies of *ex-situ* conservation of MAPs using *in vitro* techniques are summarized in Table 2.2.

In vitro ex-situ studies of critically endangered species are particularly important. Some of the studies carried out in this direction are outlined below.

Ajuga xylorrhiza is a rare, endemic, critically endangered perennial plant distributed in the southeastern Anatolian region of Turkiye. For an efficient micropropagation protocol, explants obtained from the shoots were cultured on Murashige & Skoog (MS) basal medium supplemented with different concentrations of kinetin (Kin), benzyl aminopurine (BAP), and thidiazuron (TDZ). Optimal shoot formation occurred in 0.125 mg/L BAP medium. The best shoot length was found in the presence of kinetin. MS medium containing indole acetic acid (IAA) and naphthylene acetic acid (NAA) was used for rooting. The best results were obtained in the medium supplemented with 0.5 mg/L IAA. The acclimatization success rate was 100% (Işıkalan et al., 2020).

The cotyledon, leaf, and cotyledon node explants of the critically endangered, endemic species *Centaurea amaena* collected from Turkiye were cultured on MS media, using various concentrations of BAP, TDZ, meta-topolin (mT), and NAA. A simple and effective *in vitro* micropropagation procedure was developed for this species (Uzun et al., 2020).

Reji et al. (2016) used nodal explants of the critically endangered *Hypericum hookerianum* (Hypericaceae), a medicinally important plant in India. They reported the highest number of shoots in MS medium containing Kin, which rooted best in MS medium supplemented with indole butyric acid (IBA) supplemented. The authors developed a successful micropropagation protocol with a 90% acclimation rate.

Another critically endangered Turkish endemic plant, *Hypericum bilgehan-bilgili*, was investigated by Turker and Turkyilmaz Unal (2022). The nodal parts of this plant were preferred as explant sources and cultured in MS medium containing various concentrations of 2,4 -dichlorophenoxy acetic acid (2,4-D) and Kin. The shoots which developed *in vitro* were transferred to MS medium containing 0.5 or 1.0 mg/L IBA for rooting. The best rooting in this study has been recorded in MS containing 1.0 mg/L IBA. A successful in vitro ex-situ conservation protocol has been developed with an acclimatization rate of 80%.

2.4 CONCLUSIONS

The latest developments in biotechnology have provided very important opportunities for plant growth and development, which have become significantly more important globally (Alatar, 2015). It is well known that many valuable MAPs are now threatened with extinction due to excessive and

TABLE 2.2
Previous Studies Concerned with MAP Conservation Via *In Vitro* Techniques

Species	Threat Status of the Species	Explant Type	Phytohormone(s) Used	Reference
Adhatoda vasica	Rare	Petiole	TDZ, NAA, 2iP	Mandal and Laxminarayana, 2014
Allium wallichii	Threatened	Seedling	BA, IAA, IBA, Kin, Z, NAA	Wawrosch et al., 2001b
Atropa acuminata	Endangered	Nodal segment, petiole and root	IAA, IBA, NAA, 2,4-D, Kin	Maqbool et al., 2016; Rajput and Agrawal, 2020
Baliospermum montanum	Endangered	Nodal segment	Kin, BAP, IBA, IAA	Sasikumar et al., 2009
Buchanania lanzan	Vulnerable	Seedling	BAP, IBA, NAA	Shende and Rai, 2005
Celastrus paniculatus	Rare	Nodal segment	BAP, IAA, IBA	De Silva and Senarath, 2009
Curcuma caesia	Endangered	Rhizome bud	BA, IAA, IBA	Shahinozzaman et al., 2013
Elaeocarpus sphaericus (Rudraksha)	Threatened	Nodal segment	BA, Kin, NAA	Saklani et al., 2015
Embelia ribes	Threatened	Nodal segment	IAA, NAA, TDZ, BA	Preetha et al., 2012
Garcinia indica	Threatened	Immature seed	BAP, NAA, IBA, GA_3	Joshi et al., 2015
Gentiana kurroo	Critically endangered	Apical meristem	Kin, BAP, IAA, NAA	Kaushal et al., 2014
Gymnema sylvestre	Threatened	Nodal segment	TDZ, BA, IBA, IAA, NAA	Shah et al., 2013
Hypericum empetrifolium	Not evaluated	Shoot	BA, NAA, IBA, IAA	Sarropoulou et al., 2018
Hypericum gaitii	Endemic	Apical and nodal segment	Kin, BAP, NAA, IBA, IAA	Swain et al., 2016
Hypericum heterophyllum	Endemic	Seed	BA, 2,4-D, IAA	Ayan and Cirak, 2006
Hypericum richeri ssp. *transsilvanicum*	Threatened	Leaf, nodal segment	BA, IBA, IAA, NAA	Coste et al., 2012
Hypericum umbellatum	Threatened	Leaf, nodal segment	BA, IBA, IAA, NAA	Coste et al., 2012
Lilium nepalense	Threatened	Mature bulb	BAP, IAA, IBA, Kin, Z, NAA	Wawrosch et al., 2001a
Picrorhiza kurroa	Endangered	Nodal segment	BAP, 2,4-D, NAA	Jan et al., 2010
Psoralea corylifolia	Rare	Seed	BA, BAP, Kin, IBA, IAA, NAA, 2iP	Verma et al., 2012
Thymus broussonetii	Vulnerable	Seed	Kin, BAP, NAA, IBA, IAA	Aicha et al., 2014
Thymus hyemalis	Endangered	Seed	Kin, BAP, NAA, IBA	Nordine et al., 2013
Thymus pallidus	Endangered	Seed	Kin, Adenin, NAA, IBA, IAA, GA_3	Ansari et al., 2020
Thymus satureioides	Endangered	Seed	BAP, Kin, NAA, IAA, IBA	Aicha et al., 2013
Tylophora indica	Endangered	Shoot	BAP, Kin, 2,4-D	Sellathurai and Rathinavel, 2012
Vanda coerulea	Endangered	Seed	BAP, Kin, TDZ	Nag and Kumaria, 2018

unsustainable use (Ozturk et al., 2011). The application of biotechnological conservation approaches to MAPs allows the development of cell, tissue, and organ culture systems for their *in vitro* growth and regeneration, and the creation of *in vitro ex-situ* conservation protocols (Ozturk et al., 2020). Plant biotechnology plays a significant role in the *in vitro ex-situ* conservation and micropropagation of rare, endangered, or critically endangered medicinally and aromatically important plant genetic resources.

Most of the medicinally and aromatically important species populations are declining in their natural habitats due to the lack of conservation policies or sustainable conservation strategies (Ramakrishnan et al., 2017). Micropropagation of such plants by *in vitro* techniques has enabled its use as an innovative and reliable biotechnological tool to protect such plants from possible extinction. The secondary metabolite profile of many of these medicinally and aromatically important plant species is not known scientifically, and people will continue to use such plants in light of existing traditional knowledge (Ramakrishnan et al., 2017). However, at the current rate of increased usage, it seems likely that the valuable secondary metabolites and plant genetic resources could be lost forever from nature due to increased demand. Thus, more sustainable *in vitro* conservation protocols are needed for the protection of such plant resources which are facing extinction. In addition, to meet the demands of the pharmaceutical industry, we need plant material of uniform development, which is not found in nature. This stresses the fact that the conservation of MAPs through *in vitro* techniques is almost compulsory for sustainable propagation. Here, we have only included basic information on the use of MAPs which are currently most endangered. We stress that, for the sustainable use of such raw materials for herbal medicines, *ex-situ* conservation is urgently required, and large-scale (global) *in vitro* culture protocols need to be developed and adopted.

REFERENCES

Abensperg-Traun, M. 2009. CITES, sustainable use of wild species and incentive-driven conservation in developing countries, with an emphasis on southern Africa. *Biological Conservation*, 142(5), 948–963.

Aicha, N., Mohamed, H. and Abdelmalek, E. 2014. *In vitro* clonal propagation through direct shoot organogenesis of *Thymus broussonetii*–a vulnerable aromatic and medicinal plant species. *International Journal of Pharmaceutical Research and Bio-Sciences*, 3, 425–439.

Aicha, N., Rachida, T. C. and El Abdelmalek, M. 2013. Micropropagation of *Thymus satureioides* Coss. an endangered medicinal plant of Morocco. *Journal of Agricultural Technology*, 9, 487–501.

Alatar, A. A. 2015. Thidiazuron induced efficient in vitro multiplication and ex vitro conservation of *Rauvolfia serpentina*–a potent antihypertensive drug producing plant. *Biotechnology and Biotechnological Equipment*, 29(3), 489–497.

Anis, M. and Ahmad, N. 2016. *Plant tissue culture: Propagation, conservation and crop improvement.* Springer, Singapore, p.621.

Ansari, Z. N. E., Boussaoudi, I., Benkaddour, R., Hamdoun, O., Lemrini, M., Martin, P., Badoc, A. and Lamarti, A. 2020. Conservation of *Thymus pallidus* Cosson ex Batt. by shoot tip and axillary bud in vitro culture. *Journal of Plant Biotechnology*, 47(1), 53–65.

Ayan, A. K. and Cirak, C. 2006. In vitro multiplication of *Hypericum heterophyllum*, an endemic Turkish species. *American Journal of Plant Pathology*, 1, 76–81.

Baksha, R., Jahan, M. A. A., Khatun, R. and Munshi, J. L. 2005. Micropropagation of *Aloe barbadensis* Mill. through in vitro culture of shoot tip explants. *Plant Tissue Culture and Biotechnology*, 15, 121–126.

Baričevič, D. and Kušar, A. 2006. Conservation of medicinal and aromatic plants in Europe-a review of current progress. 4th Conference on Medicinal and Aromatic Plants of South-East European Countries. 9th National Symposium 'Medicinal Plants-Present and Perspectives'. 3rd National Conference of Phytotherapy, Proceedings. Iaşi, Romania, 28–31 May 2006, Association for Medicinal and Aromatic Plants of Southeast European …, 24–30.

Baričevič, D., Máthé, Á. and Bartol, T. 2015. Conservation of wild crafted medicinal and aromatic plants and their habitats. In: Máthé, Á. (Ed.), *Medicinal and aromatic plants of the world.* Springer, Dordrecht, pp. 131–444.

Bouza, L., Jacques, M. and Miginiac, E. 1994. *In vitro* propagation of *Paeonia suffruticosa* Andr. cv.'Mme de Vatry': Developmental effects of exogenous hormones during the multiplication phase. *Scientia Horticulturae*, 57, 241–251.

Burkhart, E. P., Jacobson, M. G. and Finley, J. 2012. A case study of stakeholder perspective and experience with wild American ginseng (*Panax quinquefolius*) conservation efforts in Pennsylvania, USA: Limitations to a CITES driven, top-down regulatory approach. *Biodiversity and Conservation*, 21(14), 3657–3679.

Chase, J. and Leibold, M. 2003. *Ecological niches: Linking classical and contemporary approaches*. Chicago, IL: University of Chicago Press.

Coelho, N., Gonçalves, S. and Romano, A. 2020. Endemic plant species conservation: Biotechnological approaches. *Plants*, 9(3), 345.

Cole, I. B., Saxena, P. K. and Murch, S. J. 2007. Medicinal biotechnology in the genus *Scutellaria*. *In Vitro Cellular and Developmental Biology – Plant*, 43(4), 318–327.

Coste, A., Halmagyi, A., Butiuc-Keul, A. L., Deliu, C., Coldea, G. and Hurdu, B. 2012. *In vitro* propagation and cryopreservation of Romanian endemic and rare *Hypericum* species. *Plant Cell, Tissue and Organ Culture (PCTOC)*, 110(2), 213–226.

De Silva, M. and Senarath, W. 2009. Development of a successful protocol for in vitro mass propagation of *Celastrus paniculatus* willd. – A valuable medicinal plant. *Tropical Agricultural Research*, 21(1), 21–29.

Engelmann, F. 2011. Use of biotechnologies for the conservation of plant biodiversity. *In Vitro Cellular and Developmental Biology – Plant*, 47(1), 5–16.

FAO. 2020. The state of the World's biodiversity for food and agriculture. In *FAO commission on genetic resources for food and agriculture assessments*; Bélanger, J., Pilling, D., Eds. Rome, Italy: FAO. http://www.fao.org/3/CA3129EN/CA3129EN.pdf (accessed on 30 November 2021).

Heywood, V. H. 2011. Ethnopharmacology, food production, nutrition and biodiversity conservation: Towards a sustainable future for indigenous peoples. *Journal of Ethnopharmacology*, 137(1), 1–15.

Işikalan, Ç., Orcan, P., Akbaş, F., Namli, S., Kuru, İ. S. and Buluş, Ş. 2020. Effect of cytokinins on *in vitro* propagation of *Ajuga xylorrhiza* Kit Tan (critically endangered), endemic to Turkey. *In Vitro Cellular and Developmental Biology – Plant*, 56(6), 911–914.

Jan, A., Thomas, G., Shawl, A. S., Jabeen, N. and Kozgar, M. I. 2010. Improved micropropagation protocol of an endangered medicinal plant-*Picrorhiza kurroa* Royle ex Benth. promptly through auxin treatments. *Chiang Mai Journal of Science*, 37, 304–313.

Joshi, P., Hegde, A. and Hegde, V. 2015. *In vitro* propagation of *Garcinia indica* Choisy from seedling explants. *International Journal of Current Research*, 7, 24676–24678.

Kaushal, S., Sidana, A. and Dev, K. 2014. *In vitro* plant production through apical meristem culture of *Gentiana kurroo* Royle. *Journal of Medicinal Plants Studies*, 3, 4–9.

Kundu, S., Salma, U. and Gantait, S. 2018. Cryopreservation of medicinal herbs: Major breakthroughs, hurdles and future. In: Kumar, N. (Ed.), *Biotechnological approaches for medicinal and aromatic plants*. Springer, Singapore, pp. 353–381.

Mandal, J. and Laxminarayana, U. 2014. Indirect shoot organogenesis from leaf explants of *Adhatoda vasica* Nees. *SpringerPlus*, 3, 1–8.

Maqbool, F., Singh, S., Kaloo, Z. A. and Jan, M. 2016. A rapid micropropagation protocol of Atropa acuminata Royle ex Lindl.——A threatened medicinal plant species of Kashmir Himalaya. *IndianJournal of Biotechnology*, 15, 576–580.

Máthé, Á. 2015. Introduction: Utilization/significance of medicinal and aromatic plants. In: Máthé, Á. (Ed.). *Medicinal and aromatic plants of the world*. Springer, Dordrecht, pp. 1–12.

Mckenzie, M., Kirakosyan, A. and Kaufman, P. B. 2009. Risks associated with overcollection of medicinal plants in natural habitats. In: Kirakosyan, A., Kaufman, P.B. (Eds.) *Recent Advances in Plant Biotechnology*. Springer, New York. pp. 363–387.

Nag, S. and Kumaria, S. 2018. *In vitro* propagation of medicinally threatened orchid *Vanda coerulea*: An improved method for the production of phytochemicals, antioxidants and phenylalanine ammonia lyase activity. *Journal of Pharmacognosy and Phytochemistry*, 7, 2973–2982.

Nordine, A., Bousta, D., El Khanchoufi, A. and El Meskaoui, A. 2013. An efficient and rapid *in vitro* propagation system of Thymus *hyemalis* lange, a wild medicinal and aromatic plant of Mediterranean region. *International Journal of Pharma Bioscience and Technology*, 1, 118–129.

Ozturk, M. and Altay, V. 2017. An overview of the endemic medicinal and aromatic plants of Turkiye-conservation and sustainable use. *Full Text Proceedings Book*, 10, 11.

Ozturk, M., Egamberdieva, D. and Pešić, M. 2020. *Biodiversity and biomedicine: Our future*. Academic Press, Cambridge, MA, USA. pp. 608.

Ozturk, M. and Hakeem, K. R. 2018. *Plant and human health, volume 1: Ethnobotany and physiology*. Springer, Cham., Switzerland. pp. 805.

Ozturk, M. and Hakeem, K. R. 2019a. *Plant and human health, volume 2: Phytochemistry and molecular aspects*. Springer, Cham., Switzerland. pp. 697.

Ozturk, M. and Hakeem, K. R. 2019b. *Plant and human health, volume 3: Phytochemistry and therapeutic uses*. Springer, Cham., Switzerland. pp. 385.

Ozturk, M., Okmen, M., Guvensen, A., Celik, A. and Gucel, S. 2011. Land degradation, urbanisation and biodiversity in the Gediz Basin-Turkiye. In: Ozturk et al (Eds.). *Urbanisation, land use, land degradation and environment, nam proceedings*. New Delhi: Daya Publishing House, 74–93.

Preetha, T., Hemanthakumar, A. and Krishnan, P. 2012. Effect of plant growth regulators on high frequency *in vitro* multiplication of a vulnerable woody medicinal climber *Embelia ribes* Burm. f. *Journal of Medicinal Plants Research*, 6, 4011–4018.

Rajput, S. and Agrawal, V. 2020. Micropropagation of *Atropa acuminata* Royle ex Lindl.(a critically endangered medicinal herb) through root callus and evaluation of genetic fidelity, enzymatic and non-enzymatic antioxidant activity of regenerants. *Acta Physiologiae Plantarum*, 42, 1–17.

Ramakrishnan, R., Periyasamy, R., Lakkakula, S., Subramani, P., Arockiam, S. and Manikandan, R. 2017. *In vitro* propagation and conservation of useful endangered medicinal plants with anticancer activity. *Journal of Molecular Biology and Biotechnology*, 2, 8.

Reed, B. M., Sarasan, V., Kane, M., Bunn, E. and Pence, V. C. 2011. Biodiversity conservation and conservation biotechnology tools. *In Vitro Cellular and Developmental Biology – Plant*, 47(1), 1–4.

Reji, I. V., Bayyapureddy, A., Pushparaj, S. P. and Seeni, S. 2016. A high efficiency *in vitro* regeneration protocol and clonal uniformity analysis in *Hypericum hookerianum* Wight & Arn., a lesser known plant of ethnomedicinal and economic importance. *Brazilian Journal of Botany*, 39(1), 377–386.

Ross, I. A. 2005. *Medicinal plants of the world, volume 3: Chemical constituents, traditional and modern medicinal uses*. Springer, HumanaTotowa, NJ. pp. 648.

Saklani, K., Singh, S., Purohit, V. K., Prasad, P. and Nautiyal, A. 2015. *In vitro* propagation of Rudraksha (*Elaeocarpus sphaericus* (Gaertn.) K. Schum): A biotechnological approach for conservation. *Physiology and Molecular Biology of Plants*, 21(4), 611–615.

Sarropoulou, V., Krigas, N., Grigoriadou, K. and Maloupa, E. 2018. Asexual propagation and ex situ conservation of *Hypericum empetrifolium* Willd. Subsp. *empetrifolium* (Hypericaceae), an East Mediterranean medicinal plant with ornamental value. *Journal of Medicinal Plants*, 6, 235–241.

Sasikumar, S., Raveendar, S., Premkumar, A., Ignacimuthu, S. and Agastian, P. 2009. Micropropagation of *Baliospermum montanum* (Willd.) Muell. Arg.—A threatened medicinal plant. *Indian Journal of Biotechnology*, 8, 223–226.

Schippmann, U., Leaman, D. and Cunningham, A. 2006. A comparison of cultivation and wild collection of medicinal and aromatic plants under sustainability aspects. *Frontis*, 75–95.

Sellathurai, T. and Rathinavel, S. 2012. *In vitro* micropropagation of *Tylophora indica* (Burm. f) Merril through shoot tip explants. *Plant Cell Biotechnology and Molecular Biology*, 13, 65–68.

Shah, S. N., Amjad, M. and Ansari, S. 2013. Micropropagation of *Gymnema sylvestre* R.Br. *Journal of Medicinal Plants Research*, 2, 18–28.

Shahinozzaman, M., Ferdous, M., Faruq, M., Azad, M. and Amin, M. 2013. Micropropagation of black turmeric (*Curcuma caesia* Roxb.) through in vitro culture of rhizome bud explants. *Journal of Central European Agriculture*, 14(3), 110–114.

Shende, S. and Rai, M. 2005. Multiple shoot formation and plant regeneration of a commercially-useful tropical plant, *Buchanania lanzan* (Spreng). *Plant Biotechnology*, 22(1), 59–61.

Simberloff, D. 2012. Sustainability of biodiversity under global changes, with particular reference to biological invasions. In: Weinstein, M.P., Turner, R.E. (Eds.). *Sustainability Science*. Springer, New York. pp. 139–157.

Srivastava, L. M. 2002. *Plant growth and development: Hormones and environment*. AcademicPress, USA. p. 846.

Swain, D., Lenka, S., Hota, T. and Rout, G. R. 2016. Micro-propagation of *Hypericum gaitii* Haines, an endangered medicinal plants: Assessment of genetic fidelity. *The Nucleus*, 59(1), 7–13.

Turker, H. and Turkyilmaz Unal, B. 2021. The effect of different gibberellic acid concentrations on seed germination in endemic species *Hypericum bilgehan-bilgılii* under *in vitro* conditions. *8th International KOP Regional Development Symposium*. Nevsehir, Turkey.

Turker, H. and Turkyilmaz Unal, B. 2022. Development of a micropropagation protocol for endangered *Hypericum bilgehan-bilgilii* Başkose & Savran (Hypericaceae) species, local endemic to Turkey. *Pakistan Journal of Botany*, 54(3), 1089–1095.

Uprety, Y., Asselin, H., Dhakal, A. and Julien, N. 2012. Traditional use of medicinal plants in the boreal forest of Canada: Review and perspectives. *Journal of Ethnobiology and Ethnomedicine*, 8, 1–14.

Uzun, S., Ekinci, E., Özkatan, H. and Atasagun, B. 2020. *In vitro* shoot regeneration of a *Centaurea amaena* Boiss & Balansa–A critically endangered and endemic plant. *Applied Ecology and Environmental Research*, 18(2), 2285–2293.

Verma, P., Mathur, A. K., Jain, S. P. and Mathur, A. 2012. In vitro conservation of twenty-three overexploited medicinal plants belonging to the Indian sub continent. *The Scientific World Journal*, 2012, Article ID 929650.

Wawrosch, C., Malla, P. R. and Kopp, B. 2001a. Clonal propagation of *Lilium nepalense* D. Don, a threatened medicinal plant of Nepal. *Plant Cell Reports*, 20(4), 285–288.

Wawrosch, C., Malla, P. R. and Kopp, B. 2001b. Micropropagation of *Allium wallichii* Kunth, a threatened medicinal plant of Nepal. *In Vitro Cellular and Developmental Biology – Plant*, 37(5), 555–557.

Zlobin, Y. A. 2012. Rare plant species: Floristic, phytocenotic, and population approaches. *Biology Bulletin Reviews*, 2(3), 226–237.

3 Conservation and Sustainable Management of Wild Medicinal and Aromatic Plant Species

Rajneesh Thakur and Rauoof Ahmad Rather

CONTENTS

3.1 INTRODUCTION

Aromatic and medicinal plants are considered the key sources of new pharmaceuticals and therapies all over the world, and there has been a considerable increase in demand due to their effective therapeutic properties (Parveen et al., 2020a, 2020b). Over 1,300 medicinal plants are found in Europe, with 90% of them coming from natural wild sources. About 118 of the top 150 pharmaceutical drugs in the United States originate from natural sources (Hussain et al., 2021). In impoverished countries, 80 percent of people rely completely on herbal therapies, whereas in developed countries, 25 percent of medicines are made from wild plant species. The usage of aromatic and medicinal plants is quickly rising around the world due to growing interest in organic medicinal products, and the effective bioactive components of herbal medicines (Table 3.1) (Bano et al., 2021; Ole et al., 2007; Parveen et al., 2022).

DOI: 10.1201/9781003226925-3

TABLE 3.1
Some Famous Medicinal Plants With Their Active Ingredients and Therapeutic Uses

S.No.	Plant	Bioactive Compounds	Therapeutic Uses
1	*Artemisia annua* L.	Artemotil	Antimalarial activity
2	*Callistemon citrinus* (Curtis) Skeels	Leptospermone	Antityrosinemia activity
3	*Camptotheca acuminats* Decne.	Camptothecin	Treatment of liver problem, digestive problem; anticancer effect
4	*Catharanthus roseus* (L.) G.Don	Vincristine, vinblastine	Anticancer effect
5	*Adhatoda vasica* Nees.	Vasicine, vasicinone	Bronchodilator activity
6	*Bacopa monnieri* (L.) Pennell.	Bacosides	Antidepressant, anticonvulsive, and antioxidant activities
7	*Tylophora indica* R.Br.	Tylophorine	Treatment of asthma, high blood pressure, and as a bronchodilator
8	*Holarrhena antidysenterica* (L.) Wall.	Conessine	Antimalarial activity
9	*Ocimum sanctum* L.	Monoterpenes, sesquiterpenes	Treatment of immunomodulatory and respiratory diseases
10	*Asparagus racemosus* Willd.	Shatavarin	Treatment of immunomodulatory and respiratory diseases
11	*Aloe vera* (L.) Burm.f.	Aloin	Anticancer activity
12	*Withania somnifera* (L.) Dunal.	Withanolides	Cytotoxicity against human breast cancer cell lines
13	*Tinospora cordifolia* (Thunb.) Miers	Tinosporic acid, cordifoliosides	Anticancer activity, treatment of immunomodulatory and respiratory diseases
14	*Emblica officinalis* Gaertn.	Polyphenols	Anticancer activity
15	*Abutilon indicum* (fruit)	Myricetin, pentadecanoic acid, 14-methyl-methyl ester	Anti-inflammatory effect
16	*Thespesia populnea* (seed)	Palmitic acid, stearic acid	Anti-inflammatory, analgesic, and antipyretic effects
17	*Datura metel* (stem)	(6R,9R)-3-oxo-α-ionol-9-*O*-β-D-glucopyranoside, solavetivone-14-*O*-β-D-glucopyranoside	Antiproliferative activities against human cancer HepG2 and HeLa cell lines
18	*Abelmoschus manihot* (flower)	Quercetin, isoquercetin, hyperin, hibifolin, myricetin	Anti-inflammatory, antibacterial, antioxidant, anticonvulsant, cardioprotective, and neuroprotective activities
19	*Solanum surattense*	Hexadecanoic acid, ethyl ester	Antioxidant, hypocholesterolemic nematicidal, pesticidal, anti-androgenic flavor, hemolytic, 5-alpha reductase inhibitor activities
20	*Mucuna pruriens*	Phytol, vitamin E linoleic acid arachidonic acid	Anticancer activity, antibronchitic, anticoronary, anti-ischemic, antidiabetic activities; treatment of cardiovascular diseases, cancer and coronary heart diseases

(Continued)

TABLE 3.1 (CONTINUED)
Some Famous Medicinal Plants With Their Active Ingredients and Therapeutic Uses

S.No.	Plant	Bioactive Compounds	Therapeutic Uses
21	*Senna auriculata*	Levodopa, quercetin cyanidin spermidine, *N*8-acetylspermidine, hesperidin, vitexin-2-*o*-rhamnoside	Anti-aging, treatment of rheumatoid arthritis, diabetes, male infertility and nervous disorders antioxidant activity antidiabetic activities, antihyperglycemic activity, effective inhibition of glucosidase enzyme
22	*Eclipta alba*	Wedelolactone, luteolin, apigenin,	Anti-HCV replication activity
23	*Saussurea lappa*	Costunolide	Anticancer activity
24	*Picrorhiza kurroa*	Kutkoside, picroside I, picroside II	Treatment of hepatic, respiratory, leishmanial, inflammatory conditions
25	*Scoparia dulcis*	Hydroxamic acid, acteoside	Antibacterial, antifungal, antimicrobial, hepatoprotective, sedative effects
26	*Gymnema sylvestre*	Gymnemic acid	Antidiabetic, anti-sweetener, anti-inflammatory effects
27	*Averrhoa bilimbi*	Quercetin	Antihyperglycemic and antioxidant activities
28	*Aegle marmelos*	Eugenol, cineol, cuminaldehyde, citronellal, lupeol, skimmianine, citral, aegeline, marmesinin	Antioxidant, anti-inflammatory activities
29	*Salacia oblonga*	Mangiferin	Antidiabetic effects
30	*Valeriana wallichii*	Acacetin	Antioxidant activity, neuroprotective effect in Parkinson's disease
31	*Berberis asiatica*	Rutin	Antidiabetic and anticancer properties; for treatment of hypercholesterolemia and hypertension
32	*Acorus calamus*	Alpha (α)-asarone, beta (β)-asarone	Antitumor and chemopreventive activities
33	*Cyperus rotundus*	Vitexin, orientin, (+)-nootkatone	Anti-inflammatory activity; prevention of platelet-associated, treatment of cardiovascular diseases
34	*Vitex negundo*	Vitedoamine	Anti-inflammatory activity

Chinapolaiah et al., 2019; Das et al., 2019; Dua et al., 2013; Naikawadi et al., 2016; Ngamkitidechakul et al., 2010; Prasathkumar et al., 2021.

3.2 MEDICINAL PLANTS AND THEIR USES

Between 50,000 and 80,000 species of flowering plants are utilized for therapeutic purposes around the world, according to the International Union for Conservation of Nature (IUCN) and the World Wildlife Fund (WWF). Of these, 15,000 species are threatened with extinction due to overharvesting and habitat loss (Bentley, 2010), but 20% of their wild populations have already been exhausted due to human population expansion and plant consumption. Medicinal plants and aromatic species are becoming increasingly rare around the world, especially in China (Heywood and Iriondo,

2003), India (Hamilton, 2008), Nepal (Hamilton, 2008; Rather et al., 2022a), Tanzania (Zerabruk and Yirga, 2012), and Uganda (Zerabruk and Yirga, 2012).

3.3 WORLDWIDE DIVERSITY OF MEDICINAL PLANTS

The distribution of medicinal plants around the world is not uniform (Rafieian-Kopaei, 2013). For example, China and India have the highest numbers of medicinal plants used, with 11,146 and 7,500 species, respectively, followed by Colombia, South Africa, and the US, with 16 nations having the percentage of their native plants being medicinal varying from 7% in Malaysia to 44% in India (Srujana et al., 2012).

Overexploitation, uncontrolled collection, unrestrained deforestation, and niche loss are all factors that contribute to species rarity. Natural characteristics of individual plant species such as niche specificity, dispersion range, size of population, taxonomic variety, rate of growth, and rate of reproduction, all contribute to the threat of extinction (Rather et al., 2022c, 2022d).

Medicinal plants that have long been overexploited by man and are likely to lose their natural habitats, putting not only the individual species but also entire plant groups and ecosystems at risk (Okigbo et al., 2008). This situation necessitates a quick response to protect wild plant populations. In this context, it is important to have reliable data on the distribution and intensity of usage of such plants, particularly in understudied areas, to achieve this goal.

3.4 THREATS TO MEDICINAL PLANTS

Habitat destruction, bioprospecting for alternate sources, and overexploitation of recognized medicinal species are all threats to herbal medicines. The Forest Service, the National Center for the Preservation of Medicinal Herbs (NCPMH), many herb organizations, and scientists all work to protect endangered medicinal species.

The word "biopiracy" was coined for private companies' practice of patenting natural therapies and selling them for a big profit, with little or no profit going back to the place of origin or the indigenous and local communities (Padder et al., 2021).

3.5 ENDANGERED MEDICINAL PLANTS

Due to overexploitation and habitat loss, a large number of rare and endangered species are on the verge of extinction; frequent multiplication and organized cultivation will help ensure the sustained availability of these species. Collection of the following Indian species from wild sources should be banned: bankakri (*Podophyllum hexandrum*), chirayata (*Swertia chirayata*), gaozaban (*Anchusa strigosa)*, glory lily, agnisikha (*Gloriosa superba*), halad-vachnag, mamira (*Coptis teeta)*, henbane *(Hyoscyamus niger*), Indian belladonna (*Atropa acuminata*), Indian barberry, daru haridra (*Berberis aristata*), Indian gentian, pashanbheda (*Gentiana kurroa*), Indian ginseng *(Panax pseudo-ginseng* var. *himalaicus*), Indian hing (*Ferula jaeskaena*), Indian rhubarb, ghandini (*Rheum emodi*), Indian yew (*Taxus wallichiana*). jata mansi (*Nardostachys grandiflora*), jeevaka (*Microstylis nucifera*), jeevanti (*Leptadenia reticulata*), kababchini (*Piper cubeba*), kadu bonsha (*Gynocardia odorata*), kala zeera (*Brunium persicum*), kali musli (*Curculigo orchioides*), kiramar (*Aristolochia bracteata*), kshira kankoli (*Lilium polyphyllum*), kutki (*Picrorhiza kurroa*), monkshood, bachang *(Aconitum* sp.), nirpani (*Iphigenia indica*), peet chandan, kasturi mazil (*Holostemma annualare*), pshkar mool (*Inula racemosa*), ratnajyot (*Onosma bracteatum*), rewadchini (*Rheum astrata*), rishvake (*Microstylis wallichii*), safed musli (*Chlorophytum* spp.), salam panja (*Orchis latifolia*), saleb misri (*Eulophia campestris*), sarpagandha (*Rauwolfia serpentina*), scholar, lalthang (*Physochlaena praelta*), shila-pushpa (*Didymocarpus pedicellata*), somlata (*Ephedra gerardiana*), sundew (*Drosera* sp.), suranjan-1- talah, sweet flag, vekhand (*Acorus* spp.), talispatra (*Taxus baccata*), tara-tutiya, guggul, golden collyrium (*Colchicum luteum*), wormwood (*Artemisia annua*), yogispada (*Saussurea sacra*).

3.6 EXAMPLES OF NORTH AMERICAN NATIVE SPECIES AT RISK

3.6.1 Panax quinquefolius (American ginseng)

Annual sales of both wild and cultivated ginseng in North America often exceed $25 million, with wild ginseng being the more prized kind. The decrease in population size and near-extinction of wild ginseng is a serious issue since it is so much more profitable than the cultivated ginseng (U.S. Forest Service, 2001). This species is protected by the Convention on International Trade in Endangered Species (CITES).

3.6.2 *Taxus species (yew)*

The popular, productive, and effective cancer medicine taxol is made from yew trees, including *Taxus chinensis* (Chinese yew) and *Taxus brevifolia* (Pacific yew). CITES has designated a number of yew species that need to be protected against overharvesting for international trade.

3.6.3 Cimicifuga racemosa (black cohosh)

Colds, discomfort, rheumatism, and menopause have all been traditionally treated with this buttercup relative. Wild harvesting accounts for nearly all of the black cohosh supply. Overharvesting and niche decline have been cited as threats to the species by the US Forest Service and other institutions (U.S. Forest Service, 2001).

3.6.4 Hydrastis canadensis (goldenseal)

This plant has a long tradition of usage as a stimulant and cure for a range of ailments, especially hemorrhoids. More than 60 million goldenseal plants are gathered without ever being replanted every year, according to information from National Center for the Preservation of Medicinal Herbs (NCPMH), and goldenseal has already been categorized as a rare, challenged, or critically endangered species by a range of international bodies, including CITES (http://www.ncpmh.org) (U.S. Forest Service, 2001).

3.7 CONSERVATION AND SUSTAINABLE UTILIZATION

To prevent the extinction of these plant resources, wild species and ecosystems are being safeguarded. We must accept and put in place sustainable usage and cultivation practices for commercially important species. The Convention on Biological Diversity (CBD), the Convention on International Trade in Endangered Species of Wild Fauna and Flora (CITES), the National Center for the Preservation of Medicinal Herbs (NCPMH), the International Standard for Sustainable Wild Collection of Medicinal and Aromatic Plants (ISSCMAP), the Medicinal Plant Working Group, and others have all worked on developing such techniques (see their websites).

For example, the International Standard for Sustainable Wild Collection of Medicinal and Aromatic Plants (ISSC-MAP) provides the following guidelines:

- All relevant laws or regulations must be observed, including federal and state threatened species regulations, public property usage guidelines, collection bans, licensing requirements, and quantity limitations.
- Understanding and respecting Indigenous and local communities' rights, traditions, and practices.
- Developing and implementing responsible and sustainable management approaches.
- Establishing and implementing responsible corporate practices.

OVEREXPLOITATION AND HARMFUL BIOPROSPECTING ARE PROPOSED TO BE CURTAILED BY CREATING STRATEGIES SUCH AS:

- Ensuring that Indigenous and local communities are treated fairly and equally in the distribution of income produced by incorporating their primitive traditions' techniques; and working to ensure that people and institutions interested in using these resources obtain prior consent from Indigenous and local neighborhood representatives.
- Altering the way risk assessments are performed in relation to any planned development on sacred places or land, as well as waterways inhabited by Indigenous and local populations.
- Assisting governments in implementing laws or other methods to keep the recognition, conservation, and preservation of Indigenous knowledge and related activities.
- Remembering the value of medicinal and aromatic plants and our enduring natural biodiversity in our legislation, and also our land and resource management strategies, by implementing the essential norms, management methods, and conceptual frameworks.

3.8 CONSERVATION STRATEGIES FOR MEDICINAL PLANTS

There are basically three scientific strategies for conserving medicinal plants' genetic diversity.

- Legislation
- Conservation *in situ*
- Conservation *ex situ*

3.8.1 Legislation

There are no specific legislation or regulations in place in India to protect medicinal plants that thrive in forests. The following regulations concerning forestry also cover conservation of plants.

a. 1927 The Forest Act
b. 1972 The Wildlife (Protection) Act, amended in 1991 as the Wildlife (Protection) Amendment Act
c. 1980 Forest Conservation Act
d. 1986 Environmental Protection Act
e. 1988 National Forest Policy
f. 2002 National Biodiversity Act

To preserve Indigenous tribes and other traditional forest dwellers, the Scheduled Tribes and Other Traditional Forest Dwellers Act was passed in 2006, which included:

a. The protection of wildlife in its native surroundings, or the site where that develops, known as *in situ* conservation.
b. Establishment of gene libraries, genome reserves, and ecological reserves, nature reserves, sacred sites, and sacred groves.
c. Vegetation diversification will be kept in the environment throughout time, whether it is genetic variation, life forms, or environments.
d. In discrete, relevant, biologically diverse zones, it is vital to preserve inter- and intra-specific heritable variation.

3.8.2 In Situ or On-Site Conservation

This arises when a particular species or community of an ecological system is protected and conserved in its native habitat, and it is considered to be one of the most cost-effective strategies for maintaining biological and genetic variation.

National parks, wildlife reserves, holy groves, and other protected zones are examples of *in situ* conservation initiatives. The concept of establishing conservation areas and architecture has played a significant part in all legislative decision-making processes aimed at conserving the natural ecosystem on a global and regional scale.

3.8.2.1 Biosphere Reserves

Data from a survey by India's Ministry of Environment and Forests designated 14 biosphere reserves.

3.8.2.2 National Parks

Pin Valley National Park and the Great Himalayan Park, both in Himachal Pradesh, have lately been formed in the Lahaul-Spiti and Kullu areas, respectively.

The Parks Department should develop governmental legislation on indigenous herbal conservation and deployment in natural reserves. It must include:

- Development of approaches and techniques for procuring native herbs within the safeguarded areas, and determination of goals and plans for monitoring and reporting on medicinal plants in conservation areas.
- Means to preserve the long-term viability of all of the country's commercial medicinal herbs by constructing additional reserve areas and improving the existing ones. The International Union for the Conservation of Nature should come up with cost-effective and politically feasible solutions to promote the protection of natural areas and their native organisms.
- Medicinal plant collection methods and techniques must be developed in protected areas. Locations which are particularly important for medicinal herbs must be identified, then goals and techniques be set for recording and evaluating the native herbs in those zones.
- The Parks Department should investigate how many medicinal plants are protected under the country's Protected Areas Programme. It should construct new conservation lands and extend the existing ones to ensure conservation of all of the nation's medicinal herbs.
- The National Park Service must devise social and financial ways to encourage the protection of the environmental regions and the wild species therein, and the park management must ensure that medicinal plants' cultivation and harvesting are successfully integrated into the organization's work plan. Species that are threatened by population collapse should be reintroduced to their natural habitats.

3.8.3 Ex Situ or Off-Site Conservation

Ex-situ conservation, taking place outside the natural niche of the plant, involves conservation of indigenous herbs carried out in botanical gardens, parks, and other appropriate locations, as well as through lengthy collection of vegetation propagules in gene banks (grain banks, spore banks, DNA libraries, etc.) and by means of tissue culture as well as cryogenics (Parveen et al., 2020a).

3.8.3.1 Field Gene Bank (Field Repository/Clonal Repository)

Gene banks store seeds (active collections are stored at +4°C to 0°C; base collections at −20°C). The gene banks in India have continuously been utilized to preserve medicinal and aromatic plants.

The National Bureau of Plant Genetic Resources (NBPGR) of the Indian Council of Agricultural Research (ICAR) is headquartered in New Delhi. It is primarily responsible for the protection of genetic variability in naturally grown and farmed plants. The Council of Scientific and Industrial Research (CSIR), which is part of the Ministry of Science and Technology of India, runs the Central Institute of Medicinal and Aromatic Plants (CIMAP) of Lucknow, Uttar Pradesh, Palode, Thiruvananthapuram, TBGRI (Tropical Botanical Gardens Research Institute) (Kerala).

3.8.3.1.1 Seed Gene Bank

Gene bank maintenance is secured at the Seed Genebank Database. Seed retention for agricultural and horticultural crops and their wild counterparts is conducted in the NGB (National Gene Bank) in New Delhi. The Base Collection is kept at −20°C in modules. The seeds are processed at a low relative humidity of 4–6% before being carefully wrapped in humidity-resistant aluminum-cushioned pouches. Seeds that have been kept in storage for 50 to 100 years are still viable. Only seed samples with a viability of more than eighty five percent are stored in most crops. According to FAO/IPGRI gene bank criteria, the seeds in the gene bank database are kept under the best possible conditions.

3.8.3.1.2 National Active Germplasm Sites

National active germplasm sites (NAGS) are the most significant part of the network. The NAGS are currently housed in ICAR institutions and SAUs (crop-based organizations for a certain crop or range of commodities). This is an essential part of the national network for the preservation of agro-biodiversity. The NAGS provide care of the replication, assessment, management, and preservation of active collections, and also their dissemination to legal users, on a national and worldwide level.

The active collections are stored in compartments at a temperature of +4°C and a relative humidity of 35–40%. At some of these temperatures, seeds might persist for 15 to 50 years. For such short-term storage, the water content of the seeds is reduced to 8–10%. To promote the development and possible genotype preservation operations, the NBPGR maintains a set of numerous regional stations across the country in diverse agro-climatic regions.

3.8.3.1.3 Cryopreservation (at −165°C to −196°C in Liquid Nitrogen)

Cryopreservation is a way of preserving living organisms by liquid-nitrogen freezing.

a) **Seed Preservation**. The plant species are separated into two groups based on how quickly their seeds can be dehumidified (Roberts, 1972). Species with "orthodox" maturities are tolerant of desiccation and can be stored for a long time at very low temperatures. The second type of plant is referred to as "recalcitrant," which refers to seeds which are damaged during drying and hence cannot be stored at subfreezing degrees.
b) **Pollen Preservation**. This strategy was created to help with the regulation of group flowering in plants, especially fruit trees. Pollen conservation is also recommended as a method for conserving genetic information (Withers, 1991). Pollen can be collected in large numbers and frozen for preservation in a small space.

When germplasm is exchanged via pollen, plant quarantine analysis is decreased. Pollen cryopreservation has recently evolved in a range of species (Towill, 1985), and pollen cryobanks for fruit-tree species have been developed in numerous countries (Alexander, 1993).

3.8.3.1.4 Botanical Gardens/Arboreta

A botanical garden or arboretum is a well-managed collection of wild plants used for scientific research, conservation, display, and educational purposes. They act as germplasm collections for rare and endangered native and foreign species in particular (Sharma and Goel, 1994). Botanic Garden Conservation Worldwide (BGCI) is a non-profit organization devoted to botanical garden management and collaborative partnerships. It was founded in London, England in 1987. Botanical Garden Conservation International (BGCI) is a global organization with 500 botanical gardens from different countries as members.

There are approximately 1,846 botanical gardens worldwide, according to the BGCI database. Around 140 botanical gardens exist in India, with 33 botanical gardens each linked to a specific university botany department. Another 30 botanical gardens are actively involved in protecting and restoring the environment. The Tropical Botanical Gardens & Research Institute (TGBRI) in

Kerala is a wonderful example of *ex situ* plant biodiversity protection in India, since it is located in a degraded forest area of the Western Ghat Mountains.

3.9 CONCLUSION

Natural resources, such as medicinal and fragrant plants, are quite highly valued. Due to increased demand for herbal drugs, natural remedies, and bioactive components from traditional medicines, the use of aromatic and medicinal plants is rapidly increasing around the world. The extinction of some species and a shortage of a number of herbs have resulted from unplanned development and overexploitation of traditional treatments obtained from non-managed environmental assets. Plant production and maintenance in botanical gardens, parks, and other appropriate areas, as well as long-term storage of plant propagules, are examples of *ex situ* (beyond the natural niche) and *in situ* (within the natural niche) herbal medicinal conservation options (pollen bank, seed bank, DNA libraries, etc.).

REFERENCES

Alexander, M.P., Ganeshan, S. (1993). Pollen storage. In: *Advances in Horticulture. Fruit Crops–Part I.* Chadha, K.L., et al. editors. New Delhi: Malhotra Publishing House, 481–496.

Bano, H., Rather, R.A., Bhat, J.I.A., Bhat, T.T., Azad, H., Bhat, S.A., Hamid, F., Bhat, M.A. (2021). Effect of pre-sowing treatments using phytohormones and other dormancy breaking chemicals on seed germination of *Dioscorea deltoidea* Wall. Ex Griseb.: An endangered medicinal plant species of North Western Himalaya. *Ecology, Environment and Conservation*, 27:253–260.

Bentley, R. (2010). *Medicinal Plants.* London: Domville-Fife Press, 23–46.

Chinapolaiah, A., Manjesh, G.N., Thondaiman, V., Rao, V.K., Rao, H., Kumar, S. (2019). Variability in L-dopa and other biochemical composition of Mucuna pruriens (L.) an underutilized tropical legume. *Industrial Crops and Products*, 138:111447. doi:10.1016/j.indcrop.2019.06.010.

Das, B.K., Swamy, A.V., Koti, B.C., Gadad, P.C. (2019). Experimental evidence for use of Acorus calamus (asarone) for cancer chemoprevention. *Heliyon*, 5(5):e01585. doi:10.1016/j.heliyon.2019.e01585.

Dua, V.K., Verma, G., Singh, B., Rajan, A., Bagai, U., Agarwal, D.D., Gupta, N.C., Kumar, S., Rastogi, A. (2013). Anti-malarial property of steroidal alkaloid conessine isolated from the bark of Holarrhena antidysenterica. *Malaria Journal*, 12(1):194.

Hamilton, A.C. (2008). Medicinal plants in conservation and development: Case studies and lessons learned. In: *Medicinal Plants in Conservation and Development.* Kala, C.P., editor. Salisbury: Plantlife International Publisher, 1–43.

Heywood, V.H., Iriondo, J.M. (2003). Plant conservation: Old problems, new perspectives. *Biological Conservation*, 113(3):321–335.

Hussain, S.M., Hussain, K., Malik, A.J., Hussaini, A.M., Farwah, S., Rashid, M., Rather, R.A. (2021). Development of a novel in-vitro protocol for micro propagation of tomato male sterile line (Shalimar FMS-1) of Kashmir Valley India. *Acta Scientific Agriculture*, 5(4):61–69.

Naikawadi, V.B., Ahire, M.L., Lahiri, A., Nikam, T.D. (2016). In vitro propagation and cell cultures of memory tonic herb Evolvulus alsinoides: A best source for elicited production of scopoletin. *Applied Microbiology and Biotechnology*, 100(8):3463–3476.

Ngamkitidechakul, C., Jaijoy, K., Hansakul, P., Soonthornchareonnon, N., Sireeratawong, S. (2010). Antitumour effects of Phyllanthus emblica L.: Induction of cancer cell apoptosis and inhibition of in vivo tumour promotion and *in vitro* invasion of human cancer cells. *Phytotherapy Research*, 24(9):1405–1413.

Okigbo, R.N., Eme, U.E., Ogbogu, S. (2008). Biodiversity and conservation of medicinal and aromatic plants in Africa. *Biotechnology and Molecular Biology Reviews*, 3(6):127–134.

Ole, I.B., Saxena, P.K., Murch, S.J. (2007). Medicinal biotechnology in the genus *Scutellaria. In Vitro Cellular & Developmental Biology*, 43(4):318–327.

Padder, S.A., Mansoor, S., Bhat, S.A., Baba, T.R., Rather, R.A., Wani, S.M., Popescu, S.M., Sofi, S., Aziz, M.A., Hefft, D.I., Alzahrani, O.M. (2021). Bacterial endophyte community dynamics in apple (Malus domestica Borkh.) germplasm and their evaluation for scab management strategies. *Journal of Fungi*, 7(11):923.

Parveen, A., Ahmad, M., Parveen, B., Parveen, R., Iqbal, M. (2022). Unani system of medicine, its origin, evolution and Indianization: A critical appraisal. *Indian Journal of Traditional Knowledge*, 21(4):in press.

Parveen, A., Parveen, R., Akhatar, A., Parveen, B., Siddiqui, K.M., Iqbal, M. (2020a). Concepts and quality considerations in Unani system of medicine. *Journal of AOAC International*, 103(3):609–633.

Parveen, B., Parveen, A., Parveen, R., Ahmad, S., Ahmad, M., Iqbal, M. (2020b). Challenges and opportunities for traditional herbal medicine today, with special reference to its status in India. *Annals of Phytomedicine*, 9(2):97–112.

Prasathkumar, M., Anishaa, S., Dhrisyaa, C.M., Beckya, R., Sadhasivama, S. (2021). Therapeutic and pharmacological efficacy of selective Indian medicinal plants – A review. *Phytomedicine Plus*, 2: 1–29.

Rafieian-Kopaei, M. (2013). Medicinal plants and the human needs. *Journal of Herbmed Pharmacology*, 1:1–2.

Rather, R.A., Bano, H., Firoz, A., Mohammed, A.H., Bhat, M.A., Padder, S.A., Nafees, H., Hakeem, K.R. (2022a). The assessment of morphological diversity of *Colchicum luteum* L., an economically important threatened medicinal plant of Kashmir Himalaya. *Sustainability*, 14(3):1327.

Rather, R.A., Bano, H., Padder, S.A., Baba, T.R., Ara, S., Lone, F.A., Nazir, S. (2022d). Impact of anthropogenic pressure on physico-chemical characteristics of forest soils of Kashmir Himalaya. *Bulletin of Environmental Contamination and Toxicology*, 108(6): 1088–1097.

Rather, R.A., Bano, H., Perveen, K., Bukhari, N.A., Padder, S.A., Baba, T.R., Qureshi, A., Khan, N.A., Khan, A.H., Samaraweera, H. (2022c). Antifungal potential of Colchicum luteum and determination of colchicine content using HPLC for application as a fungicide. *Journal of King Saud University – Science*, 34(3):101–876.

Roberts, E.H. (1972). Storage of environment and the control of viability. In: Roberts, E.H. (eds) *Viability of Seed*. Springer, Dordrecht, 14–58. https://doi.org/10.1007/978-94-009-5685-8_2

Sharma, S.C., Goel, A.K. (1994). Role of botanic gardens in modern times. *Indian Journal of Forestry*, 17:230–238.

Srujana, S.T., Babu, K.R., Rao, B.S.S. (2012). Phytochemical investigation and biological activity of leaves extract of plant Boswellia serrata. *Journal of Pharmaceutical Innovation*, 1:22–46.

Towill, L.E. (1985). Low temperature and freeze/vacuum–drying preservation of pollen. In: *Cryopreservation of Plant Cells and Organs*. Kartha, K.K., et al. editors. Boca Raton, FL: CRC Press, 171–198.

Withers, L.A. (1991). Biotechnology and plant genetic resources conservation. In: *Plant Genetic Resources Conservation and Management*. Paroda, R.S., et al. editors. New Delhi: IBPGR Regional Office, 273–297.

Zerabruk, S., Yirga, G. (2012). Traditional knowledge of medicinal plants in Gindeberet district, Western Ethiopia. *South African Journal of Botany*, 78:165–169.

4 Use of Plants as Medicines and Aromatics by Indigenous Communities of Morocco

Pharmacognosy, Ecology and Conservation

Noureddine Chaachouay, Abdelhamid Azeroual, Mohd Kafeel Ahmad Ansari, and Lahcen Zidane

CONTENTS

LIST OF ABBREVIATIONS

ADAMPM: Association for the Development of Aromatic and Medicinal Plants of Morocco
AH: Anno Hegirae
MAD: Moroccan dirham
NAMAP: National Agency of Medicinal and Aromatic Plants
NFCAMPM: National Federation of Cooperatives of Aromatic and Medicinal Plants of Morocco.
UNESCO: United Nations Educational, Scientific and Cultural Organization

4.1 INTRODUCTION

Ever since ancient times, man has been looking for drugs in nature (Srivastava, 2018). The use of medicinal plants to cure human ailments was based, in ancient times, merely on experience, observation, hit-and-miss experimentation, and then a little documentation (Ahmed, 2017; Beyene et al., 2016; Petrovska, 2012).

DOI: 10.1201/9781003226925-4

The ancient cultures are known for systematic collection of information on herbs and their rich herbal pharmacopeias. Although some of the therapeutic properties attributed to plants have proven to be erroneous, medicinal plant therapy is based on the empirical findings of hundreds and thousands of years (Gurib-Fakim, 2009; Khuroo, 2011).

Plants have been the main basis of various sophisticated traditional medicine practices that have been in existence for thousands of years and continue to provide mankind with new remedies (Gurib-Fakim, 2009; Khuroo, 2011). Traditional medicine is defined by the World Health Organization (Wachtel-Galor and Benzie, 2011) as "knowledge, skill, and practices used to maintain health, as well as to prevent, diagnose, improve, or treat physically and psychologically disease, different theoretical beliefs, and practices indigenous to various cultures."

Traditional uses of plants by Moroccans to treat various diseases are widespread due to shortage of pharmaceutical products, prices for conventional drugs which are unaffordable for small farmers and cattle herders, prohibitive distances between health service stations, the frequent emergence of certain diseases, and the occurrence of drug-resistant microbes and helminths (Chaachouay et al., 2020). People in this country, especially those from tribal communities and living in rural areas, rely for their health care needs on traditional medicines rather than on modern systems that are out of their reach. The sensitivity of socio-cultural status, historic belief, and the availability of resources generated from medicinal plants have all contributed to the change to empirical medical methods. For ages, Indigenous people have used the distinctive approach of their traditional system of medicine, with the Islamic, Arabic, and African systems of medicine being among the best known.

In recent decades, however, the public healthcare system has abandoned the use of medicinal and aromatic herbs, and numerous plant species remain unacknowledged with respect to their therapeutic potential. Furthermore, lack of proper coordination among agencies dealing with medicinal plants has led to overexploitation of these plants, which are becoming increasingly endangered. The main objective of the present chapter is to provide an overview of how the Indigenous Moroccan population use herbs as medicines and aromatics as a fundamental health care modality, as well as to understand human strategies and attitudes toward pharmacognosy, plant ecology, and conservation.

4.2 GLOBAL IMPORTANCE OF MEDICINAL AND AROMATIC PLANTS

Plant-based medication was common in ancient cultures and continues to be practiced by many tribal and rural peoples even today (Anis et al., 1986, 2000; Atique et al., 1985a, 1985b, 1992). Medicinal plants are a valuable source of ingredients for the production of pharmacopoeial, non-pharmacopoeial, and synthetic formulations (Iqbal et al., 2018). Extracts of so many plants are useful in a crude form, e.g., *Allium sativum*, *Annona squamosa*, *Ficus acemose*, *Gymnema sylvestre*, and *Piper nigrum* are used as antidiabetic phytomedicines (Amin et al., 2015; Ashraf et al., 2011; Atique et al. 1985a, 1985b; Eidi et al., 2006; Singh et al., 2008), *Ocimum basilicum,* and *Lepidium sativum* for hypertension and as a tranquilizer (Aftab et al., 2006; Saba et al., 1997, 1999a, 1999b, 1999c), *Bacopa monniera* as a memory stimulant (Ali et al. 1997a, 2000), *Podophyllum hexandrum* as an anti-cancer drug (Iqbal et al., 2004), and *Papaver rhoeas* tincture as an analgesic (Mills, 2021), among others. With the discovery of numerous new molecules from herbs for treating dreaded illnesses like diabetes, tuberculosis, and cancer, plus the relative safety of these products, the global interest in medicinal plant products has increased in recent years (Iqbal et al., 2018; Parveen et al., 2020b). Similarly, consumers favor cosmetics that contain aromatic plant products, resulting in increased demand for the raw material. Furthermore, many plants, such as *Zingiber officinale* Roscoe, *Aloe vera* (L.) Burm.f., *Camellia sinensis* (L.) Kuntze, *Rosmarinus officinalis* L., *Thymus saturejoides* Coss., and *Lavandula dentata* L., are regarded as an important source of nutrients and hence used to treat deficiency disorders. Some plants and their derivatives are also used in natural dyes, pest control and in the preparation of food items, tea, toothpastes, perfumes and other similar items. In many areas of Morocco, different kinds of medicinal herbs are used to keep ants, flies, mice and fleas away from homes and offices. Nowadays, medicinal herbs also provide important raw materials for pharmaceutical manufacturing (Parveen et al., 2020b; Ulrich-Merzenich, 2014).

4.3 CONSERVATION ASPECTS OF HERBS

Traditional herbal medicine uses remedies derived from plants, animals, metals, and minerals. However, numerous species are threatened by issues ranging from human activities to global climatic change. Efforts are made to conserve the threatened species by different means such as *in-situ* conservation of plant biodiversity, *ex-situ* conservation through gene banks, mass production of selected species by *in vitro* cultural techniques, cryopreservation of plant cells and meristems via slow cooling or vitrification, etc. (Anis et al., 2014; Beigh et al., 2011; Jackson and Sutherland, 2000; Olorode, 2004; Tahtamouni et al., 2015). Efforts are also in progress to upgrade cultivation of medicinal plants (Ahmad et al., 2005, 2006a, 2006b; Ahmed et al., 2017) and to increase production of their ingredients of therapeutic significance by regulating plant physiology via biochemical and biotechnological means (Ali et al., 1997b; Beigh et al., 2002b; Saba et al., 2000a, 2000b; Srivastava et al., 2002). Application of abiotic stress has also been used frequently to increase secondary metabolite production (Ali et al., 2008; Iqbal et al., 2011a, 2011b; Saifullah et al., 2016; Sirousmehr et al., 2014). Of late, nanobiotechnology has also stepped into the arena of phytomedicine (Ansari et al., 2012; Nasrollahzadeh et al., 2019; Pandey and Pandey, 2013). Morocco enjoys a great diversity of climates and habitats hosting the second-highest terrestrial biodiversity in the Mediterranean basin, after Turkiye. The general trend for natural resources in Morocco is toward biodiversity degradation and loss. Over 600 endangered species have been identified across the country, and, for some of them, notably those found near cities and in the Central Rif region, the rate of degradation is irreversible. *Thymus saturejoides* Coss., *Thymus vulgaris* L., *Artemisia absinthium* L., *Artemisia herba-alba* Asso, *Rosmarinus officinalis* L. *Laurus nobilis* L., *Origanum elongatum* (Bonnet) Emb. & Maire, *Olea europaea* L., *Mentha pulegium* L., and *Matricaria recutita* L. are among the most frequently used medicinal and aromatic plants in Morocco.

Morocco is a party to several international treaties such as Biodiversity, Climate Change, Desertification, Endangered Species, Hazardous Wastes, Marine Dumping, Ozone Layer Protection, Ship Pollution, and Wetlands. Over 154 ecologically or biologically significant sites have been recognized in this country, comprising approximately 2.5 million hectares of land and coastal habitats, as well as humid zones. Eleven national parks (Toubkal National Park, Tazekka National Park, Souss-Massa National Park, Iriqui National Park, Al Hoceima National Park, Talassemtane National Park, Ifrane National Park, Eastern High Atlas National Park, Khenifiss National Park, Khenifra National Park, and Dakhla National Park) have consequently been created. Also, 24 humid zones are classified under the RAMSAR list, and three natural areas in Morocco are included in UNESCO's international network of biosphere reserves. These protected zones aim to ensure with conservation of species through a participative, ecosystem-based approach.

Along with protected areas, the Government of Morocco has implemented various programs on rehabilitation and recovery of endangered species, as well as a strategy on education and public awareness-raising with regard to environment and sustainable development. Management and development plans for some vulnerable areas and species have been implemented and gene banks established (Convention on Biological Diversity, 2020). The Department of Forestry has been working with cooperatives and developing trade fairs in natural forests, including the forestry co-management initiative in the Kenitra region. Another good example is the "Rosemary Co-management Project" carried out in the eastern region, close to Oujda. A "contract project" was undertaken jointly by the Department of Forestry and a women's cooperative to exploit 22,000 ha of wild rosemary to extract and sell essential oils, and manage the area sustainably in return. Positive impacts of these efforts in terms of employment and well-being of the local people and ecological recovery are quite visible.

4.4 TRANSMISSION OF MEDICINAL PLANT KNOWLEDGE IN MOROCCO

The methods of acquisition and transmission of data, which are different in each culture, shape the local information defined as "a collective body of knowledge, practices, and beliefs, evolving by adaptive processes and delivered down through generations by cultural transmission" (Mathez-Stiefel and Vandebroek, 2012). Conditioning and imprinting, imitative learning, engaging with knowledgeable relatives, elders, neighbors, and the natural environment, active teaching and learning, or a combination of these methods are all used to transmit cultural knowledge (Teixidor Toneu, 2017). The relative relevance of these channels in the transfer of medicinal plant knowledge varies depending on demographic and socio-economic factors, which impact the evolution of medicinal floras since they allow for more-or-less conservative imparting of knowledge. (Bellakhdar, 1997; Chaachouay et al., 2019; Teixidor Toneu, 2017). Elders have a greater understanding of medicinal plants than their younger counterparts (Chaachouay et al., 2019; Montanari, 2013). Medicinal plant knowledge is frequently passed down from parents to their children (Ramet et al. 2018). Additionally, formal training, community location, and infrastructure correlate with the ethnobotanical knowledge of Moroccan people. Plant information was equally transmitted vertically and horizontally, implying that it is first gained within the family and then transformed as a result of future interactions with other knowledge sources, such as peers of the same age.

4.5 ECOLOGICAL AND FLORISTIC DIVERSITY IN MOROCCO

Morocco formed a land bridge between Europe and Africa. It sits at a floristic crossroads since it has two long (about 2800-km) sea coasts, one on the Atlantic Ocean and the other on the Mediterranean Sea. Morocco has the widest plains and highest peaks of North Africa; four mountain ranges, the Rif, the Middle Atlas, the High Atlas, and the Anti-Atlas, form a semicircle around the coastal and middle plains separating them from the Saharan desert. Morocco's plant diversity is favored by topographic and climatic heterogeneity, as well as a long history of human activity (Médail and Diadema, 2009; Molina-Venegas et al., 2017), and mountainous areas are considered to be biodiversity hotspots within the Mediterranean (Medail and Quezel, 1997). The flora of Morocco is one of the most diverse in the Mediterranean region and the richest in North Africa. Forests, matorrals, steppes, and grasslands are the various types of vegetation that make up the Moroccan vegetable carpet. The steppes encompass almost three million hectares, whereas the woodlands cover approximately five million hectares. On the ecological level, these ecosystems are distributed across several types of bioclimate, such as humid, sub-humid, semi-arid, Saharan arid climate, and high-mountain climate. Moroccan vascular flora includes about 4,500 species and subspecies distributed among 930 genera and 130 families. Rare, threatened, or/and endemic flora represents a significant proportion: 951 are endemics, 463 rare, 1,284 threatened, and 36 vulnerable (Fennane, 1997; Fennane et al., 1998; Taleb and Fennane, 2020) (Figure 4.1).

4.6 PRODUCTION AND TRADE IN MEDICINAL AND AROMATIC PLANTS IN MOROCCO

The aromatic and medicinal plant sector in Morocco is full of potential thanks to the diversity of species. According to figures provided by the National Agency of Medicinal and Aromatic Plants (NAMAP), annual production has reached 140,000 tons, providing alternative income to local communities, and generating an average of some 500,000 working days/year. The production of medicinal plants includes both wild-grown plants and cultivated fresh or dried plants. Currently, Morocco is ranked the 12th largest exporter of medicinal plants globally, with 52,000 tons of plants and 5,000 tons of oils. The main destinations of Moroccan plant exports are the European Union and United States markets, but opening to other markets (Japan, Canada, Switzerland, Spain, and Germany) has allowed export of increasing volumes (NAMAP 2021) (Table 4.1).

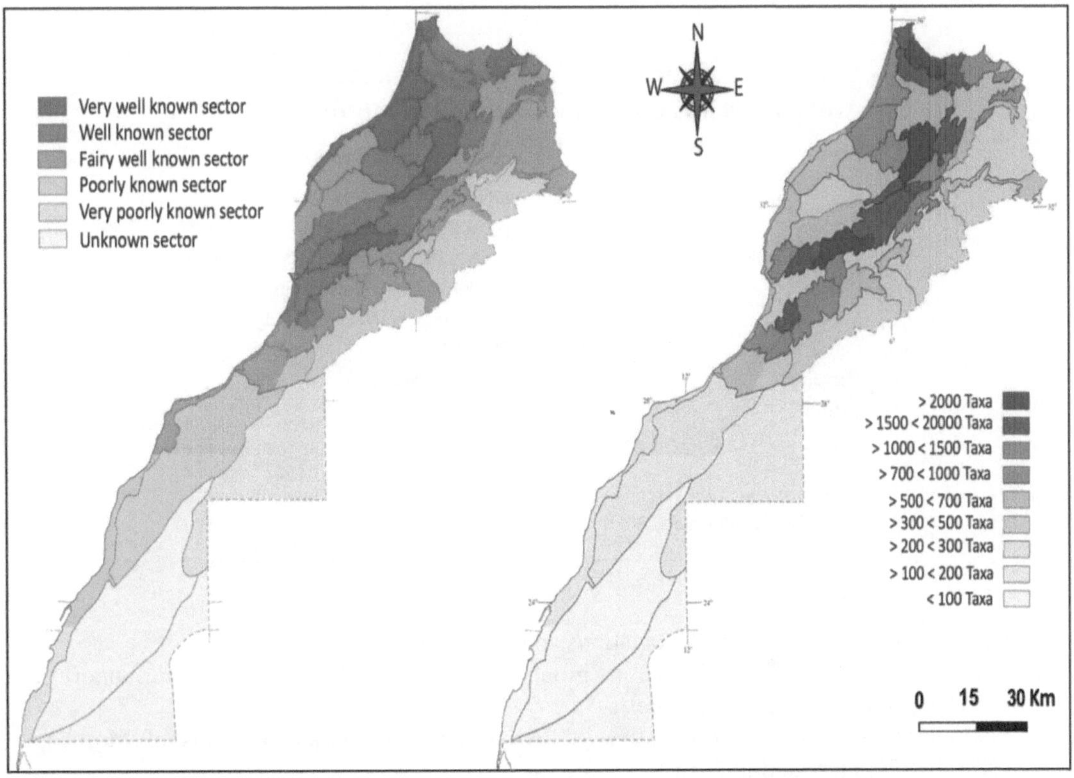

FIGURE 4.1 Current state of the floristic knowledge of Morocco.[a]. [a]FLORE DU MAROC, https://www.floramaroccana.fr/connaissance-floristique.html, 2016

More than 50% of these exports are to the food sector (carob, spices, flavors, etc.), whereas 35% are destined for perfumery and cosmetics, and around 5% are exploited for their medicinal properties. Export revenues are estimated to be in the range of 615 million MAD (Moroccan dirham). *Rosmarinus officinalis* L. is the most requested product, with a 63% share of the exported medicinal plants, followed by *Thymus vulgaris* L. with 24%, *Origanum compactum* Benth., and *Origanum elongatum* (Bonnet) Emb. and Maire (DWF, 2018). *Artemisia herba-alba* Asso. is also of great commercial interest in Morocco. Its essential oil content varies from 1% to 1.5% of dry material. Its essence is used for the industry of cosmetology and perfumery industries. Morocco controls 90% of the global market for this plant (Taleb, 2017). Plants commonly used in the production of essential oils and aromatic extracts in Morocco include *Rosmarinus officinalis*, *Artemisia herba-alba*, *Cedrus atlantica* (Endl.) Manetti ex. Carrière, *Myrtus communis* L., *Thymus vulgaris*, *Thymus satureoides* Coss, *Laurus nobilis*, *Origanum compactum*, *Ormenis mixta* (L.), Dumort., and *Mentha pulegium* L.

4.7 ORGANIZATION OF THE MEDICINAL AND AROMATIC PLANT SECTOR IN MOROCCO

Since 1995, the organization of the plant sector in Morocco has undergone a major evolution, with the creation of the following agencies:

- Association for the Development of Aromatic and Medicinal Plants of Morocco (ADAMPM) (created in 1995).
- Moroccan Society of Aromatic and Medicinal Plants (created in 2006).
- Moroccan Association of Herbalists (created in 2006).

TABLE 4.1
Geographical Distribution of Wild Medicinal and Aromatic Plants Currently Exploited in Morocco.[a]

Plants	Production Areas
Rosemary	Al Hoceima, Oujda, Taza, Boulman
Thyme	Agadir, Azilal, Marrakech
Carob	Azilal, Beni Mallal, Taza, Khenifra, Sefrou
Lichen	Beni Mallal, Khenifra, Ifran, Taounat
Mugwort	Agadir, Errachidia, Ouarzazate, Oujda
Laurel sauce	Tétouan, Beni Mellal
Verveine	Marrakech, Sale, Khemisset
Saffron	Taroudant, Taliouine
Caper	Fez, Bouaarfa, Figuig

[a] Department of Water and Forests; http://www.eauxetforets.gov.ma/SitePages/Index.aspx, 2018

- Creation of several women's associations and cooperatives.
- Creation of several economic interest groups and local cooperatives, with the support of the Department of Water and Forests.
- National Federation of Cooperatives of Aromatic and Medicinal Plants of Morocco (NFCAMPM) (created in 2015).

4.8 STRATEGY FOR THE DEVELOPMENT OF MEDICINAL AND AROMATIC PLANTS IN MOROCCO

Since 2009, the High Commission for Water, Forests and the Fight Against Desertification has put in place a strategy for the development of the plant sector based on the following pillars:

- Consolidation of the current knowledge and its development to approach the market professionally.
- Optimization of production and marketing with a view to a better valorization of the Moroccan medicinal and aromatic plants.
- Regulation, organization and encouragement of the sector to prepare a framework that is both adequate and stimulating for professionals and protectors of the resource.
- Promotion and animation of the plant sector while creating positive synergies with other sectors.
- Promotion of local populations and the preservation and sustainable management of the natural resource products.

Above all, it is a question of preparing the plant sector to make transition from a sector supplying unprocessed raw materials to a genuine industrial sector, offering high-quality product lines with a high added value, for both local and international markets.

4.9 THE PHARMACOPOEIA OF MOROCCO

A pharmacopeia is a lawfully binding collection of standards and quality specifications for medicines used in a country or region (Uddin et al., 2016). As a public tool, it maintains the quality of medicines by collecting the recommended procedures for analysis and specifications for the determination of pharmaceutical substances, excipients, and dosage forms and, in most cases, consists of a general part (methods, tests, and general requirements) and a specific part in the form of monographs for pharmaceutical substances (Organization 2006). Morocco is one of the Mediterranean countries with a long history of medicinal plant use. Indeed, the Moroccan pharmacopeia arose from knowledge passed down by several ethnic communities, including Arabs from the Middle East, Andalusians, and European Jews, who settled in the nation after migrating from other parts of the world (Bellakhdar, 1997).

Many of the reported species are aromatic or medicinal herbs used in Morocco's broad and widely practiced traditional medicine system. Approximately 231 indigenous plant species in Morocco are known to have bioactive characteristics and each is used to treat several ailments by the local inhabitants (Bellakhdar et al., 1991; Ouarghidi et al., 2013). Traditional herbalists employ a variety of methods to extract the beneficial phytochemical components from the chosen plant. The majority of botanically based medicines are taken orally as water, tea, milk, or other beverages containing diluted or concentrated chemical components.

Heating a raw plant in a liquid medium not only helps to extract and concentrate the curative ingredients, but also helps to remove poisons and impurities before consumption (Azaizeh et al., 2006; Halberstein, 2005). It is not unexpected that various edible plant components have been employed as sources of both sustenance and medicine among various ethnic groups since there is no clear division between food and medication (Azaizeh et al., 2006). The seeds of *Trigonella foenum-graecum* L., for example, could be used as a meal or as a way to prevent excessive blood pressure and heart disease (Chaachouay et al., 2021; Yadav and Baquer, 2014; Zia et al., 2001). The oil as well as the fruits of olive, *Olea europaea* L., is an important food for nutrition and as an antioxidant or diabetes preventive in many societies (El and Karakaya, 2009). Numerous plant species have been utilized to cure a variety of pathological disorders, e.g., *Origanum vulgare* L. is a perennial plant that has been used for its antidiabetic, anti-inflammatory, anti-ulcer, and antihypotensive activities, and to treat liver illnesses (Begnini et al., 2014; Oniga et al., 2018).

4.10 LEGAL FRAMEWORK OF MOROCCAN HERBALISM AND PHYTOTHERAPY

The majority of Moroccans still employ the ancient medical approach to address their primary healthcare needs today. Particularly in rural areas, Moroccans continue to rely on this ancient medical strategy to address their primary healthcare and subsistence needs. Phytotherapy is an important aspect of primary health care in Morocco, especially for the poorest people. Indeed, traditional medicine in this society has always occupied an important place (Elachouri et al. 2021). Herbalists have been practicing since time immemorial and constitute a branch of pharmacy. In Morocco, at the beginning of the last century, the practice of the herbalist profession was not subject to any regulatory restrictions. At that time, the country did not have any legislation that could make such an exercise subject to rules of law. The sale of medicinal herbs as well as medicines was not subject to any regulation.

The concern to align the legislation of the French protectorate in Morocco with the French legislation led to the establishment on 19 September 1913 of steps to regulate the practice of pharmacy. Subsequently, clauses concerning the exercise of the medical professions in general, and of pharmacy in particular, were included, providing for and regulating the existence and practice of the profession of herbalist in Morocco, although the sale of medicinal plants remains shared between the two circuits. The first is an archaic and informal circuit, grouping sellers of simple drugs and often toxic, generally walking from city to city and from one weekly souk to another (Dahir, 1923), whereas the second herbalist sector, which consists of modern herbalists, has the advantage of being regulated by Dahir 1-59-367 of February 19, 1960 (Dahir, 1960). This law has governed the practice of doctors, pharmacists, dental surgeons, herbalists, and midwives for many years.

4.11 CONCLUSION

In the review presented in this chapter, we have shown that Morocco contains a diverse range of medicinal and aromatic plants, with the majority of them still needing to be fully investigated for their therapeutic properties and hidden potential. Regrettably, the overall trend for these natural resources in Morocco is toward biodiversity degradation and loss. In a country like Morocco, where biodiversity constitutes a major source of local people's livelihood, biodiversity loss will continue unless conservation measures are applied rapidly with a participative approach involving the local communities Combining the efforts of ethnobiologists, ethnobotanists, pharmacists, anthropologists, and chemists to analyze and validate the use of traditional medicinal plants, using current scientific methodologies and creative approaches, could be a reasonable strategy. Moreover, attention must be focused on the depletion of plant resources on Earth due to the consistently increasing demand for medicinal and aromatic herbs.

4.12 CONFLICT OF INTERESTS

The authors declared that they have no conflict of interests with respect to the authorship and/or publication of this paper.

REFERENCES

Aftab, K. 2006. Blood pressure lowering action of active principle from Ocimum basilicum. *Planta Medica*, 72(11): 72P_001, DOI: 10.1055/s-2006-949801.

Ahmad, A., Khan, I., Abrol, Y.P. and Iqbal, M. 2005. Role of sulphate transporter system in sulphur efficiency of mustard genotypes. *Plant Science*, 169(5): 842–846.

Ahmad, S., Fazli, I.S., Jamal, A., Kamaluddin, Maaz, M., Iqbal, M. and Abdin, M.Z. 2006a. Cultivation of Babchi (*Psoralea corylifolia* L., Fabaceae) for the production of furocoumarin (psoralen) of therapeutic value. In: Abdin, M.Z. and Abrol, Y.P. (eds.), *Traditional Systems of Medicine*, 465–470. Narosa Publishing House, New Delhi.

Ahmad, S., Jamal, A., Fazli, I.S., Iqbal, M., Kamaluddin and Abdin, M.Z. 2006b. Nutritional approach for improved production of Atrilal (*Ammi majus* L.). In: Abdin, M.Z. and Abrol, Y.P. (eds.), *Traditional Systems of Medicine*, 534–545. Narosa Publishing House, New Delhi.

Ahmed, A.I., Aref, I.M., El-Atta, H.E. and Iqbal, M. 2017. Enhancing propagation and growth of *Balanites aegyptiaca* through seed treatment. *Journal of Environmental Biology*, 38(4): 617–622.

Ahmed, M.F. 2017. Evaluation of antioxidant activity of Stephania Japonica and Mikania cordata. PhD Thesis. East West University.

Ali, G., Purohit, M., Saba, Iqbal, M. and Srivastava, P.S. 1997a. Morphogenic response and isozymes of *Bacopa monniera* (L.) Wettst cultures grown under salt stress. *Phytomorphology*, 47: 97–106.

Ali, G., Mughal, M.H., Srivastava, P.S. and Iqbal, M. 1997b. Micropropagation of *Tribulus terrestris* L., an important medicinal plant. *Journal of Plant Biology*, 40(3): 202–205.

Ali, G., Iqbal, M. and Srivastava, P.S. 2000. Experimental studies on Brahmi, the innocuous little herb. *Hamdard Medicus*, 43: 41–45.

Ali, S.T., Mahmooduzzafar, Abdin, M.Z. and Iqbal, M. 2008. Ontogenetic changes in foliar features and psoralen content of *Psoralea corylifolia* Linn. exposed to SO_2 stress. *Journal of Environmental Biology*, 29(5): 661–668.

Amin, M.M., Bhakta, S. and Das, S.K. 2015. Anti-diabetic potential of *Ficus racemosa*: Current state and prospect especially in the developing countries. *Journal of Bioscience and Agriculture Research*, 5(2): 65–72.

Anis, M. and Iqbal, M. 1986. Antipyretic utility of some Indian plants in traditional medicine. *Fitoterapia*, 57: 52–55.

Anis, M., Sharma, M.P. and Iqbal, M. 2000. Herbal ethnomedicine of the Gwalior-forest division in Madhya Pradesh, *India. Pharmaceutical Biology*, 38(4): 241–253.

Anis, M., Varshney, A., Fatima, N., Khan, M.I. and Ahmad, M.R. 2014. Tissue culture technology in conservation and propagation of medicinal plants. In: Iqbal, M. and Ahmad, A. (eds.), *Current Trends in Medicinal Botany*, 271–287. IK International Publishing House, New Delhi.

Ansari, S.H., Islam, F. and Sameem, M. 2012. Influence of nanotechnology on herbal drugs: A review. *Journal of Advanced Pharmaceutical Technology and Research*, 3(3): 142–146.

Ashraf, R., Rafeeq Alam, K. and Ashraf, I. 2011. Garlic (Allium sativum) supplementation with standard antidiabetic agent provides better diabetic control in type 2 diabetes patients. *Pakistan Journal of Pharmaceutical Sciences*, 24(4): 565–570.

Atique, A., Iqbal, M. and Ghouse, A.K.M. 1985a. Use of *Annona squamosa* and *Piper nigrum* against diabetes. *Fitoterapia*, 56: 190–192.

Atique, A., Iqbal, M. and Ghouse, A.K.M. 1985b. Ethnobotanical study of cluster fig (*Ficus racemosa*). *Fitoterapia*, 56: 236–240.

Atique, A. and Iqbal, M. 1992. Traditional single drug administration for the psychosomatic dermal diseases: Psoriasis. *Hamdard Medicus*, 35: 76–81.

Azaizeh, H., Bashar, S., Khalid, K. and Omar, S. 2006. The state of the art of traditional Arab herbal medicine in the Eastern region of the Mediterranean: A review. *Evidence-Based Complementary and Alternative Medicine* 3(2): 229235.

Begnini, K.R., Nedel, F., Lund, R.G., Carvalho, P.H.de A., Rodrigues, M.R.A., Beira, F.T.A. and Del-Pino, F.A.B. 2014. Composition and antiproliferative effect of essential oil of Origanum vulgare against tumor cell lines. *Journal of Medicinal Food*, 17(10): 1129–1133.

Beigh, S.Y., Nawchoo, I.A. and Iqbal, M. 2002a. Herbal drugs in India: Past and present uses. *Journal of Tropical Medicinal Plants*, 3: 197–204.

Beigh, S.Y., Iqbal, M. and Nawchoo, I.A. 2002b. Seed germination and seedling survival of *Aconitum heterophyllum*, an endangered medicinal plant of the northwest Himalaya. *Indian Journal of Plant Physiology*, 7(2): 109–113.

Beigh, S.Y., Nawchoo, I.A. and Iqbal, M. 2003. Plants in traditional medicine of Kashmir Himalayas. *Journal of Economic and Taxonomic Botany*, 27(1): 99–104.

Beigh, S.Y., Arshi, A., Nawchoo, I.A., Aref, I.M. and Iqbal, M. 2011. Conservation and reproductive biology of *Aconitum heterophyllum* and *Podophyllum hexandrum*. In: Ahmad, A., Siddiqi, T.O. and Iqbal, M. (eds.), *Medicinal Plants in Changing Environment*, 140–164. Capital Publishing Company, New Delhi.

Bellakhdar, J. 1997. Contribution à l'étude de la pharmacopée traditionnelle au Maroc: La situation actuelle, les produits, les sources du savoir (enquête ethnopharmacologique de terrain réalisée de 1969 à 1992). PhD Thesis. Université Paul Verlaine-Metz.

Bellakhdar, J., Claisse, R., Fleurentin, J. and Chafique, Y. 1991. Repertory of standard herbal drugs in the Moroccan pharmacopoea. *Journal of Ethnopharmacology*, 35(2): 123–143.

Benkhnigue, O., Chaachouay, N., Khamar, H., El Azzouzi, F., Douira, A. and Zidane, L. 2022. Ethnobotanical and ethnopharmacological study of medicinal plants used in the treatment of anemia in the region of Haouz-Rehamna (Morocco). *Journal of Pharmacy and Pharmacognosy Research*, 10(2): 279–302.

Beyene, B., Belachew, B. and Habitamu, D. 2016. Review on application and management of medicinal plants for the livelihood of the local community. *Journal of Resources Development and Management*, 22(1): 33–39.

Chaachouay, N., Benkhnigue, O., Fadli, M., El Ibaoui, H. and Zidane, L. 2019a. Ethnobotanical and ethnopharmacological studies of medicinal and aromatic plants used in the treatment of metabolic diseases in the Moroccan Rif. *Heliyon*, 5(10): e02191.

Chaachouay, N., Benkhnigue, O., Fadli, M., El Ibaoui, H., El Ayadi, R. and Zidane, L. 2019b. Ethnobotanical and ethnopharmacological study of medicinal and aromatic plants used in the treatment of respiratory system disorders in the Moroccan Rif. *Ethnobotany Research and Applications*, 18: 1–16.

Chaachouay, N., Benkhnigue, O., El Ibaoui, H., El Ayadi, R. and Zidane, L. 2019c. Medicinal plants used for diabetic problems in the Rif, Morocco. *Ethnobotany Research and Applications*, 18: 1–19.

Chaachouay, N., Douira, A., Hassikou, R., Brhadda, N., Dahmani, J., Belahbib, N., Ziri, R. and Zidane, L. 2020a. Etude floristique et ethnomédicinale des plantes aromatiques et médicinales dans le Rif (Nord du Maroc). PhD Thesis. Département de Biologie - Université Ibn Tofail, Kénitra. https://tel.archives-ouvertes.fr/tel-03376377 (18 octobre 2021).

Chaachouay, N., Benkhnigue, O. and Zidane, L. 2020b. Ethnobotanical study aimed at investigating the use of medicinal plants to treat nervous system diseases in the Rif of Morocco. *Journal of Chiropractic Medicine*, 19(1): 70–81.

Chaachouay, N., Benkhnigue, O., Douira, A. and Zidane, L. 2021a. Poisonous medicinal plants used in the popular pharmacopoeia of the Rif, northern Morocco. *Toxicon*, 189: 2432.

Chaachouay, N., Douira, A. and Zidane, L. 2021b. Herbal medicine used in the treatment of human diseases in the Rif, Northern Morocco. *Arabian Journal for Science and Engineering*, 47: 131–153. https://doi.org/10.1007/s13369-021-05501-1.

Chaachouay, N., Orch, H. and Zidane, L. 2021c. Cystitis treatment with phytotherapy within the Rif, Northern Morocco. *Future Journal of Pharmaceutical Sciences*, 7(1): 1–9.

Convention on Biological Diversity. 2020. Morocco, biodiversity facts; Secretariat of the Convention on Biological Diversity, 413, Saint Jacques Street, suite 800, Montreal QC H2Y 1N9 Canada. https://www.cbd.int/countries/profile/?country=ma.

Dahir. Dahir du 27 février 1923, portant réglementation de l'exercice des profes- sions de pharmacien, chirurgien-dentiste, herboriste et sage-femme, N°543.

Dahir. Dahir n° 1-59-367 du 19 février 1960 portant réglementation de l'exercice des professions de pharmacien, chirurgien-dentiste, herboriste et sage-femme, n° 2470.

Eidi, A., Eidi, M. and Esmaeili, E. 2006. Antidiabetic effect of garlic (Allium sativum L.) in normal and streptozotocin-induced diabetic rats. *Phytomedicine*, 13(9–10): 624–629.

El, S.N. and Sibel, K. 2009. Olive tree (Olea europaea) leaves: Potential beneficial effects on human health. *Nutrition Reviews*, 67(11): 632–638.

Elachouri, M., Kharchoufa, L., Fakchich, J., Lorigooini, Z., Subhasis, P. and Subhash, M. 2021. Ancestral phytotherapeutic practices in Morocco: Regards on history, current state, regulatory and safety of commonly used herbal medicine. *Arabian Journal of Chemical and Environmental Research*, 8(1): 133–149.

Fennane, M. 1997. Étude nationale sur la biodiversité. *Ministère de l'Environnement, Programme des Nations Unies pour l'Environnement PNUE* (GEF 6105/92. Vol. 1, 88 p).

Fennane, M., Ibn Tattou, M. and Valdés, B. 1998. Catalogue des plantes vasculaires rares, menacées ou endémiques du Maroc. Herbarium Mediterraneum Panormitanum.*Bocconea,* 8: 5–243. ISSN 1120-4060,

Gurib-Fakim, A. 2009. Natural product networks in Africa and their contribution to cooperation and economic development. *Regional and Interregional Cooperation to Strengthen Basic Sciences in Developing Countries*, 103–112.

Halberstein, R.A. 2005. Medicinal plants: Historical and cross-cultural usage patterns. *Annals of Epidemiology*, 15(9): 686–699.

Iqbal, M., Beigh, S.Y. and Nawchoo, I.A. 2004. Variability in morphology and active constituents of *Podophyllum hexandrum*: A Himalayan herb known for its anti-cancer activity. *Journal of Tropical Medicinal Plants*, 5: 33–36.

Iqbal, M., Ahmad, A. and Siddiqi, T.O. 2011a. Characterization of controversial plant drugs and effect of changing environment on active ingredients. In: Ahmad, A., Siddiqi, T.O. and Iqbal, M. (eds.), *Medicinal Plants in Changing Environment*, 1–10. Capital Publishing Company, New Delhi.

Iqbal, M., Mahmooduzzafar, Nighat, F. and Aref, I.M. 2011b. Composition of seed oils of *Peristrophe bicalyculata* and *Ruellia tuberosa* as affected by coal-smoke stress. *Journal of Food, Agriculture and Environment*, 9(3 & 4): 1101–1104.

Iqbal, M., Parveen, R., Parveen, A., Parveen, B. and Aref, I.M. 2018. Establishing the botanical identity of plant drugs based on their active ingredients under diverse growth conditions. *Journal of Environmental Biology*, 39(1): 123–136.

Jackson, W. and Sutherland, L.A. 2000. International Agenda for Botanic Gardens in Conservation. Botanic Gardens Conservation International, U.K. The North Carolina, ISBN: 0 9520275 93.

Khuroo, M.A. 2011. Phytochemical investigation of Rhododendron lepidotum dissertation. PhD Thesis. Indian Institute of Integrative Medicine.

Mathez-Stiefel, S.-L. and Vandebroek, I. 2012. Distribution and transmission of medicinal plant knowledge in the Andean highlands: A case study from Peru and Bolivia. *Evidence-Based Complementary and Alternative Medicine*, 2012, Article ID 959285, doi:10.1155/2012/959285

Médail, F. and Diadema, K. 2009. Glacial refugia influence plant diversity patterns in the Mediterranean Basin. *Journal of Biogeography*, 36(7): 1333–1345.

Medail, F. and Quezel, P. 1997. Hot-spots analysis for conservation of plant biodiversity in the Mediterranean Basin. *Annals of the Missouri Botanical Garden*, 84(1): 112–127.

Mills, J.H. 2021. From "papaber errat" to "tincture of opium": Poppies, opiates, and pain in early modern Scotland (ca. 1664–1785). *The Social History of Alcohol and Drugs*, 35(1): 91–114.

Molina-Venegas, R., Abelardo, A., Sébastien, L. and Arroyo, J. 2017. Climatic and topographical correlates of plant palaeo-and neoendemism in a Mediterranean biodiversity hotspot. *Annals of Botany*, 119(2): 229–238.

Montanari, B. 2013. The future of agriculture in the high Atlas Mountains of Morocco: The need to integrate traditional ecological knowledge. In: Mann, S. (eds) *The Future of Mountain Agriculture Geography*, . Springer, Berlin, Heidelberg. https://doi.org/10.1007/978-3-642-33584-6_5..

NAMAP. 2021. *National Agency of Medicinal and Aromatic Plants*, 159. Taounate Principale, Taounate Morocco. https://www.anpma.ma.

Nasrollahzadeh, M., Sajadi, S.M. and Iqbal, M. 2019. Basic chemistry and biomedical significance of nanomaterials. In: Husen, A. and Iqbal, M. (eds.), *Nanomaterials and Plant Potential*, Springer, Cham. https://doi.org/10.1007/978-3-030-05569-1_2.

Olorode, O. 2004. Conservation of plant genetic resources. *African Journal of Traditional, Complementary and Alternative Medicines*, 1(1): 4–14.

Oniga, I., Puşcaş, C., Silaghi-Dumitrescu, R., Olah, N.-K., Sevastre, B., Marica, R., Marcus, I., Sevastre-Berghian, A.C., Benedec, D. and Pop, C.E. 2018. Origanum vulgare ssp. vulgare: Chemical composition and biological studies. *Molecules*, 23(8): 2077.

Orch, H., Chaachouay, N., Douiri, E.M., Faiz, N., Zidane, L. and Douira, A. 2021. Use of medicinal plants in dermato-cosmetology: An ethnobotanical study among the population of Izarène. *Jordan Journal of Pharmaceutical Sciences*, 14(3): 323–340.

Ouarghidi, A., Martin, G.J., Powell, B., Esser, G. and Abbad, A. 2013. Botanical identification of medicinal roots collected and traded in Morocco and comparison to the existing literature. *Journal of Ethnobiology and Ethnomedicine*, 9(1): 1–13.

Pandey, A. and Pandey, G. 2013. Usefulness of nanotechnology for herbal medicines. *Plant Archives*, 13(2): 617–621.

Parveen, A., Parveen, R., Akhatar, A., Parveen, B., Siddiqui, K.M. and Iqbal, M. 2020a. Concepts and quality considerations in Unani system of medicine. *Journal of AOAC International*, 103(3): 609–633. https://doi.org/10.5740/jaoacint.19-0284.

Parveen, B., Parveen, A., Parveen, R., Ahmad, S., Ahmad, M. and Iqbal, M. 2020b. Challenges and opportunities for traditional herbal medicine today, with special reference to its status in India. *Annals of Phytomedicine*, 9(2): 97–112.

Parveen, A., Ahmad, M., Parveen, B., Parveen, R. and Iqbal, M. 2022. The traditional system of Unani medicine, its origin, evolution and Indianization: A critical appraisal. *Indian Journal of Traditional Knowledge*, in press.

Petrovska, B.-B. 2012. Historical review of medicinal plants' usage. *Pharmacognosy Reviews*, 6(11): 1.

Ramet, A., Benyei, P., Parada, M., Aceituno-Mata, L., García-del-Amo, D. and Reyes-García, V. 2018. Grandparents' proximity and children's traditional medicinal plant knowledge: Insights from two schools in intermediate-rural Spain. *Journal of Ethnobiology*, 38(2): 187–204.

Saba, Ali, G., Iqbal, M. and Srivastava, P.S. 1997. Detection of alkaloid in a medicinally potential herb, *Lepidium sativum* (Linn). *Indian Drugs*, 34: 286–288.

Saba, Iqbal, M. and Srivastava, P.S. 1999a. Detection of hyoscyamine in *Lepidium sativum. Hamdard Medicus*, 42: 72–75.

Saba, Iqbal, M. and Srivastava, P.S. 1999b. Three new phytoconstituents from *Lepidium sativum. Die Pharmazie*, 51: 851–853.

Saba, Mughal, M.H., Ali, M., Iqbal, M. and Srivastava, P.S. 1999c. A steryl ester from *Lepidium sativum. Phytochemistry*, 50: 1375–1377.

Saba, Iqbal, M. and Srivastava, P.S. 2000a. In vitro micropropagation of *Silybum marianum* L. from various explants, and sylibin content. *Journal of Plant Biochemistry and Biotechnology*, 9(2): 81–87.

Saba, Pande, D., Iqbal, M. and Srivastava, P.S. 2000b. Effect of $ZnSO_4$ and $CuSO_4$ on regeneration and lepidine content in *Lepidium sativum* L. *Biologia Plantarum*, 43(2): 253–256.

Saifullah, Khan, M.N., Iqbal, M., Naeem, A., Bibi, S., Waraich, E.A. and Dahlawi, S. 2016. Elemental sulfur improves growth and phytoremediative ability of wheat grown in lead-contaminated calcareous soil. *International Journal of Phytoremediation*, 18(10): 1022–1028.

Singh, V.K., Umar, S. and Iqbal, M. 2008. *Gymnema sylvestre* for diabetics. *Journal of Herbs, Spices and Medicinal Plants*, 14(1–2): 88–106.

Sirousmehr, A., Arbabi, J. and Asgharipour, M. 2014. Effect of drought stress levels and organic manures on yield, essential oil content and some morphological characteristics of sweet basil (Ocimum basilicum). *Advances in Environmental Biology*, 8(4): 880–885.

Srivastava, A.K. 2018. Significance of medicinal plants in human life. In: A. Tewari, S. Tiwari (eds.) *Synthesis of Medicinal Agents from Plants*, 1–24. Elsevier, ISBN 9780081020715. https://doi.org/10.1016/B978-0-08-102071-5.00001-5.

Srivastava, P.S., Ali, G. and Iqbal, M. 2002. Micropropagation of *Bacopa* and effects of heavy metals on growth performance. In: Nandi, S.K., Palmi, L.M.S. and Kumar, A. (eds.), *Role of Plant Tissue Culture in Biodiversity, Conservation and Economic Development*, 325–344. Gyanodaya Prakashan, Nainital.

Tahtamouni, R.W., Shibli, R.A., Al-Abdallat, A.M., Al-Qudah, T.S., Laila Younis, L., AlBaba, H. and Al-Ruwaiei, H. 2015. In vitro conservation and cryopreservation of medicinal and aromatic plants: A review. *Jordan Journal of Agricultural Science*, 11: 147–167.

Taleb, M.-S. 2017. Aromatic and medicinal plants in Morocco: Diversity and socio-economic role. *World Academy of Science, Engineering and Technology International Journal of Agricultural and Biosystems Engineering*, 11(12): 742–746.

Taleb, M.-S. and Fennane, M. 2020. Forests, Steppes and Grasslands in Morocco: Diversity, ecological importance and socio-economic role.

Teixidor Toneu, I. 2017. The evolution of Medicinal Floras: Insights from Moroccan medicinal plant knowledge transmission. PhD Thesis. University of Reading.

Uddin, Md.-S., Al Mamun, A., Mamunur, R. and Asaduzzaman, Md. 2016. In-process and finished products quality control tests for pharmaceutical capsules according to pharmacopoeias. *Journal of Pharmaceutical Research International*, 9(2): 1–9.

Ulrich-Merzenich, G.S. 2014. Combination screening of synthetic drugs and plant-derived natural products — Potential and challenges for drug development. *Synergy*, 1(1): 59–69.

Wachtel-Galor, S. and Iris, F.F.-B. 2011. Herbal medicine. *Lester Packer, D*(Doctoral dissertation, Ph D 2011).

Yadav, U.C.S. and Baquer, N.Z. 2014. Pharmacological effects of Trigonella foenum-Graecum L. in health and disease. *Pharmaceutical Biology*, 52(2): 243–254.

Zia, T., Nazrul Hasnain, S. and Hasan, S.K. 2001. Evaluation of the oral hypoglycaemic effect of Trigonella foenum-Graecum L.(methi) in normal mice. *Journal of Ethnopharmacology*, 75(2–3): 191–195.

5 Medicinal and Aromatic Plants from Sudan

Traditional Uses, Pharmacology, and Phytoconstituents

Mosa E. O. Ahmed, Abdelgadir A. Abdelgadir, and Elhadi M. Ahmed

CONTENTS

5.1 INTRODUCTION

The presence of vegetation on the Earth is indispensable for the survival of mankind, because oxygen, an essential requirement to enable breathing, is released by green plants. They also provide materials to be used as food, medicine, and building blocks for shelter. Ancient man identified a variety of naturally growing plants that could be used to cure different diseases and disorders of themselves and their livestock (Atique et al., 1985a, 1985b, 1986; Atique and Iqbal, 1992; Beigh et al., 2003a, 2003b). As a result, a huge body of folklore about the medicinal utility of various plant species was gradually built up in different parts of the world (Anis and Iqbal, 1986, 1994; Anis et al., 2000; Beigh et al., 2002b, 2008). Later, the active ingredients present in those plants and also the various environmental factors affecting the biosynthesis of those ingredients of therapeutic value could be identified through extensive scientific research (Ahmad et al., 2004, 2007; Ali et al. 2008; Iqbal et al., 2011, 2018; Saba et al. 1999a, 1999b, 1999c, 2000; Verma et al. 2006). Of late, cultivation of medicinal plants has also attracted attention of the pharmaceutical sector, giving rise to numerous laboratory experiments and field studies (Ahmad et al. 2005; Beigh et al., 2002a, 2005; Husen et al., 2017Yousuf et al., 2016).

Due to their popular use in traditional herbal medicines, three-quarters of the plants used in traditional medicine have contributed one or more active ingredients to the modern prescription drugs. Of the 120 active constituents isolated from higher plants and extensively used in conventional medicine today, 80% exhibited a positive correlation between their conventional therapeutic use and traditional uses. At least 7,000 active medicinal products in the modern pharmacopoeia are obtained from plants (Abdelgadir et al., 2021).

Traditional Ecological Knowledge (TEK) or Indigenous Knowledge (IK) represent

> The body of knowledge, practice, and beliefs evolved from adaptive processes that are unique to a given culture or society, existing within, or acquired by local people over time through accumulation

DOI: 10.1201/9781003226925-5

of experiences, society–nature relationships, community practices and institutions, and by passing it down through generations.

This information is typically local, holistic, integrative and rooted in cultural traditions (Hegazy et al., 2020). Scientists have recently recognized the value of exploring IK uses of plants and animals to discover new medicines to fight the emergence of new diseases (Aziz et al., 2018; Parveen et al., 2020a, 2020b). In fact, more than 80% of "modern" medicines were originally extracted from wild plants and microorganisms, and most of these were first verified and used in traditional medicine by natives in different regions of the world (Hegazy et al., 2020).

The World Health Organization (WHO) has defined "Traditional medicine" as health practices, approaches, knowledge and beliefs incorporating plant-, animal- and mineral-based medicines, spiritual therapies, manual techniques and exercises, applied singularly or in combination to treat, diagnose, and prevent illnesses or maintain well-being (Chaitanya et al., 2021). More than 80% people in Africa rely on traditional medicine for their primary health care demands (WHO, 2002) and depending on various plant and animal sources of drug in treatment (Chaitanya et al., 2021). In Tanzania, for instance, the number of traditional healers was between 30,000 and 40,000, compared with only 600 medical doctors (Lako et al., 2020). Likewise, in Malawi, there were approximately 17,000 traditional medical practitioners, compared with just 35 medical doctors in the whole country (Lako et al., 2020). Therefore, traditional herbal medicine still plays a significant and active role in health care throughout Africa.

This chapter specifically considers the plants used traditionally by the local communities in central Sudan. It provides authenticated, local plant names with their corresponding scientific names, and summarizes the plant's habitat and active constituents responsible for their pharmacological activities, as well as their approved medicinal uses globally.

5.2 GLOBAL REPUTATION OF TRADITIONAL MEDICINE

Traditional medicine is still widely used in almost every country by different communities, and due to its potential value, several countries, including India, China, Korea, and many African and Latin American nations, have made dedicated efforts to preserve and utilize their traditional system of medicines (Parveen et al., 2020a, 2020b, 2022).

In 1976, the World Health Assembly (WHA) recognized a number of measures to create awareness of the potential resources built up by traditional practitioners. In 1977, it advised countries to use their traditional medicine approaches and later involved a comprehensive approach to support the use of herbal medicines (Khalid et al., 2012).

In March 1988, an international consultation group for the conservation of medicinal plants assembled in Chiang Mai, Thailand, under the aegis of WHO in association with the International Union for Conservation of Nature and Natural Resources (IUCN) and the World Wildlife Fund (WWF). One outcome of this conference was the approval of the Chiang Mai Declaration, entitled "Save Plants, Save Nature". Thereafter, in May 1988, the 41st WHA validated the Chiang Mai declaration and its call for international co-operation and coordination in establishing conservation programs to secure availability of various medicinal plants for future generations (Khalid et al., 2012).

5.3 TRADITIONAL SUDANESE MEDICINE

Sudanese traditional medicine constitutes a distinctive combination of the Indigenous cultures of Islamic, Arabic, and African traditions, and provides treatments for many epidemic and endemic diseases (Khalid et al., 2012).

In Sudan, only 11% of the population has access to conventional health care and thus traditional remedies come to the rescue of low-income groups of people living in rural areas or towns. Poor people, who cannot afford to pay for too-expensive, conventional medicines, also depend on traditional remedies prescribed by unregistered traditional practitioners (Khalid et al., 2012).

Most ethnic people have effective traditional remedies of plant origin. Traditional health practices are usually transmitted orally from knowledgeable members of the community to the younger generations. To preserve this valuable knowledge, the Medicinal and Aromatic Plants Research Institute (MAPRI) has taken the initiative to document valuable ethnobotanical data and traditional knowledge. MAPRI has published several studies on the ethnomedicine of different places in Sudan, covering the Erkowit, Nuba Mountains, White Nile Province, Kordofan, Khartoum State, and Ingassana areas (Ahmed et al., 2020).

5.4 DISTRIBUTION OF MEDICINAL AND AROMATIC PLANTS IN SUDAN

Sudan is a relatively large African country with a total area of 1,861,484 km^2, sharing boundaries with seven other countries, *viz*., Central African Republic, Chad, Egypt, Eritrea, Ethiopia, Libya, and South Sudan. After the independence of South Sudan, Sudan became the third largest African country after Algeria and the Democratic Republic of the Congo (CIA, 2021). The weather of Sudan is hot and dry; arid desert with the rainy season (April to November) varying with a wide range of bioclimatic regions from desert and semi-desert in the north to the tropical with a short, rainy season (semi-arid and semi-humid) in central Sudan (CIA, 2021) (Khalid et al., 2012). The diversity of the Sudanese climate results in a very rich flora.

In recognition of the growing importance of traditional medicines, the Sudanese government has placed significant importance on the cultivation of medicinal plants, which have demand in world markets. On the other hand, it has also imported some propagated species, which are suitable for growth in the Sudanese environment, e.g., *Datura inermis* and *Hyocyamus niger*. When these plants are grown on a large scale, they could be an important source of foreign currency, in addition to meeting the local market needs (Musa, 2009).

Studies have shown that of the 3,137 recorded flowering species in Sudan, 15% are endemic to the country and are distributed across 170 families and 1,280 genera. However, the country is still not in a position to develop products from its enormous resources and currently lacks the ability to contribute sufficiently to the rapidly growing national and global markets (Adam et al., 2020).

Table 5.1 presents data on 86 plants belonging to 43 families; the most commonly used species included *Trigonella foenum-graecum*, *Solenostemma argel*, *Nigella sativa*, *Hibiscus sabdariffa*, and *Acacia* spp.

All these plants are known to contain rich mixtures of biochemicals including phenolics, flavonoids, terpenes, and sterols. The most represented plant families are the Lamiaceae (seven species), Fabaceae (seven species), Apiaceae (six plant species), and Asteraceae (five plant species). In Sudanese traditional medicine, plants are used extensively for treating conditions like constipation, malaria, gout, and jaundice. Most of the reported plants exhibit anti-hemorrhoid, anti-hyperglycemic, antihypertensive, aphrodisiac, carminative, and anti-GIT (gastro-intestinal tract) spasm properties. In addition, some of the reported plants have documented nutritional value, such as anise and Abuliali (*Detarium microcarpum*). Several clinical and pharmacological uses of the studied plants, such as antioxidant, antimalarial, antidiarrhea, antimicrobial, antibacterial, antifungal, and anti-inflammatory activities are also shown in Table 5.1. Some of the reported plants, e.g., Hab-ALaziz (*Cyperus rotundus*) and *Acacia* spp., have also shown promising uses as pharmaceutical aids.

Aromatic plants rich in essential oils, such as *Mentha* spp. (Adam et al., 1998), *Cymbopogon* spp. (Warrag et al., 2014), fennel (Parejo et al., 2004), and ginger (Awad et al., 2012; Eisa et al., 2020; Osman et al., 2020), find the greatest use as spices and flavors with medicinal antispasmodic, anti-inflammatory, expectorant, and antimicrobial activities. Medicinal plants (wild or cultivated), such as gum arabic, senna, hibiscus, *Lawsonia*, *Capsicum*, coriander, nigella, *Tamarindus*, and *Adansonia*, are collected normally in small quantities for the local market and/or in large quantities

TABLE 5.1
Overview of Some Medicinal and Aromatic Plants Used in central Sudan

No.	Plant Scientific Name, (Family), Plant Part Used*	Vernacular Name(s)	English Name(s)	Habitat and/or Distribution	Pharmacological Uses	Main Folkloric, Pharmaceutical Preparation and Dosage Form	Active Constituent(s)/class and References
1	*Cymbopogon schoenanthus* (Poaceae) H	Maharib, Halfa bar المحريب, الحلفابر	Camel's hay	Northern & central Sudan	Antioxidant and antiacetylcholinesterase	Decoction of the whole plant used as diuretic and antigout.	Essential oils (limonene (10.5–27.3%), β-phellandrene (8.2–16.3%), δ-terpinene (4.3–21.2%), and α-terpineol (6.8–11.0%) (Khadri et al., 2008)
2	*Cyperus rotundus* (Cyperaceae) S, R	Alseidaa السعدة	Nut grass	Sudan	Antimalarial	Herbal tea and powder of rhizome used as diuretic, antispasmodic, to treat gum bleeding, antiemetic, and hypotensive agent.	Sesquiterpene (patchoulenone, caryophyllene α-oxide, 10,12-peroxycalamenene and 4,7-dimethyl-1-tetralone) (Thebtaranonth et al., 1995)
3	*Haplophyllum tuberculatum* (Rutaceae) H	Alhazza الحزا	Plant of the mosquito (Plant Use, 2020)	Northern Sudan	Antibacterial and antifungal	Maceration of whole plant used for menstrual irregularities	Essential oils (β-phellandrene, limonene, (*Z*)-β-ocimene, β-caryophyllene, myrcene, and α-phellandrene) (Al-Burtamani et al., 2005)
4	*Ziziphus spina-christi* (Rhamnaceae) L, B, S, R	Sidr, Nabq السدر,النبق	Christ's thorn	Northern & central Sudan	1. Aqueous extract of root bark has an anti-nociceptive activity in mice and rats. CNS depressant effect in mice. 2. Stem bark showed antidiarrheal effects in rats (Ads et al., 2018).	1. Leave maceration used as hairdressing, antiseptic, and antidandruff. 2. Seed oil used for rheumatoid arthritis and as astringent.	1. Leaf contains saponin glycoside. 2. Bark contain cyclopeptide alkaloid (Wallis, 2002) 1. From butanol extract of the leaves four triterpenoidal saponin glycosides were isolated and named christinin-A-D. 2. 36 different chemical compounds were identified in the stem bark (Ads et al., 2018)
5	*Mentha* sp. (Lamiaceae) H	Na'naa baladi النعناع البلدى	Spearmint	Manyregionsin Sudan	Anti-fungal	Decocted extract of the whole plant used as carminative. Boiling vapor used in sinusitis (aromatic stimulant)	Essential oils (carvacrol, and thymol (Adam et al., 1998)
6	*Carum carvi* (Apiaceae) F	الكراوية Karawia	Caraway	Sudan, Egypt	Anti-bacterial	Maceration used as antispasmodic, carminative, and lactogoge	Essential oils (carvone, limonene, germacrene D, and *trans*-dihydrocarvone) (Iacobellis et al., 2005)

(Continued)

TABLE 5.1 (CONTINUED)
Overview of Some Medicinal and Aromatic Plants Used in central Sudan

No.	Plant Scientific Name, (Family), Plant Part Used*	Vernacular Name(s)	English Name(s)	Habitat and/or Distribution	Pharmacological Uses	Main Folkloric, Pharmaceutical Preparation and Dosage Form	Active Constituent(s)/class and References
7	*Pimpinella anisum* (Apiaceae) F	Yanson اليانسون	Anise	Several parts of the Sudan	Anti-fungal	Maceration of fruit used as carminative, antispasmodic, and nutritive	Essential oil (anethole) (Kubo and Himejima, 1991)
8	*Lavandula officinalis* (Lamiaceae) Fl, H	Khuzama, Daram الخزامة, الضرم	Lavender	Sudan, alsham	Bactericidal	Induce relaxation and calmness (aromatherapy). Rheumatic pain	V. Oil (broneol, linalol) (Evans and Trease, 2002a) Euginol, 1,8-cineol and citronellol (Mohaddeseh et al., 2006)
9	*Thymus vulgaris* (Lamiaceae) H.	Zatar الزعتر	Thyme	Egypt, alsham	Anti-oxidant	Herbal tea used as carminative	Essential oils (Eugenol, thymol, carvacrol, linalool (0.471 mg/g), α-terpineol (0.291 mg/g), and 1,8-cineole and 4-allylphenol) (Lee et al., 2005)
10	*Lawsonia inermis* (Lythraceae) L	Henna الحناء	Henna	Several parts of the Sudan	Fungicidal	Poultice used for skin and hair staining, febrifuge	2-hydroxy-1,4-naphthoquinone (Lawsone) (Tripathi et al., 1978)
11	*Eucalyptus globules* (Myrtaceae) L	Alban البان	Eucalyptus	Widespread	Pediculicide	Oil used as massage for joint pain, antiseptic, and insect repellant	Monoterpenoids [1,8-cineole, *l*-phellandrene, (−)-α-pinene, 2-β-pinene, *trans*-pinocarveol, γ-terpinene, and 1-α-terpineol] and terpenoids (β-eudesmol and geranyl acetate) (Yang et al., 2004)
12	*Eruca sativa* (Brassicaceae) S.	Gergir, Kuthaa الجرجير, الكثا	Arugula	Widely cultivated	Antioxidant	The seeds used as general tonic, to treat hair fall and as aphrodisiac.	Flavonoids, carotenoids, ascorbic acid, fibers, glucosinolates (glucoerucin) (Barillari et al., 2005a)
13	*Cyperus esculentus*	Hab-ALaziz, Hab Alzalom, Loz Alard	Cyperus, Earth almond, Tiger nut	Kordofan	1. Protection against liver damage in rats.	General tonic, aphrodisiac, stimulant, and removal of black spot from skin.	β-Sitosterol (Abu-Mustafa et al., 1960)
14	(Cyperaceae) tuber	حب العزيز,حب الزلم,لوز الارض		Darfur	2. Antioxidant activity. 3. Active against *Salmonella typhimurium.*	Tuber starch has lower swelling power than potato and maize, indicating tiger nut starch offer outstanding binding properties without conceding drug release characteristics and would be vital in pharmaceutical formulation (Adenowo and Kazeem, 2020)	1. Tuber contains vitamins A, C, and E as well as various amino acids. 2. Tuber has low amount of antinutrients such as saponins, tannins, oxalates, phytate, and cyanogenic glycosides (Adenowo and Kazeem, 2020)

(*Continued*)

TABLE 5.1 (CONTINUED)
Overview of Some Medicinal and Aromatic Plants Used in central Sudan

No.	Plant Scientific Name, (Family), Plant Part Used*	Vernacular Name(s)	English Name(s)	Habitat and/or Distribution	Pharmacological Uses	Main Folkloric, Pharmaceutical Preparation and Dosage Form	Active Constituent(s)/class and References
					4. Reduce in growth of atherosclerotic lesion. 5. Alleviates bone and joint pains (Adenowo and Kazeem, 2020)		
15	*Raphanus sativus* (Brassicaceae) S	Alfegel الفجل	Radish	Wide and cultivated in Sudan	Antioxidant	General tonic, stomachic, asthma, and bile stone	Glucosinolates (glucoraphasatin and glucoraphenin) (Barillari et al., 2005b)
16	*Petroselinum crispum* (Apiaceae) H	Albagdonis البقدونس	Parsley	Many regions in Sudan	Antioxidant	Macerate of the whole plant used as antispasmodic and diuretic.	Phenolic acids (caffeic acid) (Hyun-Jin, 2008)
	Zea mays (Poaceae) corn silk	Shawashi alzoraa شواشي الدرة	Corn silk	Several parts of the Sudan	Antioxidant	Decoction used as antispasmodic and diuretic	Volatile oil (*cis*-α-terpineol (24.22%), 6,11-oxidoacor-4-ene (18.06%), citronellol (16.18%), *trans*-pinocamphone (5.86%), eugenol (4.37%), neo-iso-3-thujanol (2.59%), and *cis*-sabinene hydrate (El-Ghorab et al., 2007)
17	*Oscimium basilicum* (Lamiaceae) Fl, S, L	Al reehan الريحان	Sweet basil	Wild plant in central & northern Sudan	Anti-oxidant	Aphrodisiac, insect repellant, and carminative	Essential oils (linalool; 3.94 mg/g), (estragole; 2.03 mg/g), methyl cinnamate (1.28 mg/g), (eugenol; 0.896 mg/g), and 1,8-cineole (0.288 mg/g)) (Lee et al., 2005)
18	*Elettaria cardamomum* (Zingiberaceae) F	Alhabahan, Hab Alhal, Heel الهبهان, حب الهال, هيل	Cardamon	India and Australia	Antifungal	Fruit macerate improve memory, aphrodisiac, and vitiligo with olive oil and henna	Essential oil (1,8-cineole and α-terpinyl acetate) (Rahman et al., 1999)
19	*Cinnamomum verum* (Lauraceae) B	Girfaa, Darseni القرفة, الدارسين	Cinnamon	China	Antifungal	General tonic	Cyclic monoterpene (β –phellandrene) and phenol (carvacrol) (Gourine et al., 2010)
20	*Pistacia lentiscus* (Anacardiaceae) Voil (resin)	Mistica, Mastaki المستكا, المصطكى	Lentisque	Greece	Antioxidant	Macerate used as expectorant and to improve memory, diuretic, antidiarrheic in children (side effect: vocal muscle relaxation)	Monoterpenes, and sesquiterpenes (α-pinene/α-thujene, spathulenol, bicyclogermacrene, and δ-3-carene) (Gourine et al., 2010)

(Continued)

TABLE 5.1 (CONTINUED)
Overview of Some Medicinal and Aromatic Plants Used in central Sudan

No.	Plant Scientific Name, (Family), Plant Part Used*	Vernacular Name(s)	English Name(s)	Habitat and/or Distribution	Pharmacological Uses	Main Folkloric, Pharmaceutical Preparation and Dosage Form	Active Constituent(s)/class and References
21	*Coriandrum sativum* (Apiaceae) F	Kasbara الكسبرة	Coriander	Northern Sudan	Bactericide	Fruit maceration. Spicy and carminative	Aliphatic (2*E*)-alkenals and alkanals ((2*E*)-Dodecenal (C_{12}), and (2*E*)-undecenal (C_{11})) (Kubo et al., 2004)
22	*Foeniculum vulgare* (Apiaceae) F	Shamar الشمار	Sweet fennel	Several parts of the Sudan	Antioxidant	Fruit macerate used as carminative	Phenolic compounds (3-caffeoylquinic acid, 4-caffeoylquinic acid, 1,5-*O*-dicaffeoylquinic acid, rosmarinic acid, eriodictyol-7-*O*-rutinoside, quercetin-3-*O*-galactoside, kaempferol-3-*O*-rutinoside, and kaempferol-3-*O*-glucoside) (Parejo et al., 2004)
23	*Capsicum frutescens* (Solanaceae) F	Shataa الشطة	Hot pepper, chilli	Several parts of the Sudan	Antioxidant	Febrifuge	Carotenoids (provitamin A), flavonoids, phenols (capxanthin, lutein and zeaxanthin), vitamin C (Howard et al., 2000)
24	*Datura stramonium* (Solanaceae) R, Fl, S	Saikaran السيكران	Stramonium, Jimsonweed	Central Sudan	1. Complete protection time (mosquito repellency) 2. Analgesic and anti-asthamatic 3. The boiled extract of seed has rapid onset of effects and may be useful for treatment of organophosphorus poisoning (Soni et al., 2012)	Sedative and antispasmodic	Alkaloids (hyoscine, hyoscyamine, atropine) (Evans and Trease, 2002b) 2. Two new tropane alkaloids, 3-phenylacetoxy-6, 7-epoxynortropane and 7-hydroxyapoatropine were tentatively identified. 3. Minor alkaloids are reported for the first time in *D. stramonium* are tigloidin, aposcopolamine, apoatropin, hyoscyamine *N*-oxide and scopolamine *N*-oxide17-20. 6 α-ditigloyloxytropane and 7-hydroxyhyoscyamine (Soni et al., 2012)
25	*Lepidium sativum* (Brassicaceae) S	Al Rashad الرشاد	Garden cress	Wild plant in central Sudan	Chemoprotective	Oral powder used for Peptic ulcer. Poultice used for abscess and Tinea capitis	Glucotropaeolin (GT) and benzylisothiocyanate(BITC) (Kassie et al., 2002)

(Continued)

TABLE 5.1 (CONTINUED)
Overview of Some Medicinal and Aromatic Plants Used in central Sudan

No.	Plant Scientific Name, (Family), Plant Part Used*	Vernacular Name(s)	English Name(s)	Habitat and/or Distribution	Pharmacological Uses	Main Folkloric, Pharmaceutical Preparation and Dosage Form	Active Constituent(s)/class and References
26	*Carthamus tinctorius*	Osfur, Gurtom	Safflower, false saffron	Saudi Arabia, Egypt, South Asia, India, Chinaand Iran	1. Clinically used to manage of menstrual cramps, post-partum hemorrhage, whooping cough and chronic bronchitis, rheumatism, and sciatica.	Skin staining, diuretic, headache, and edema.	Flavanoids (carthamone)
	(Asteraceae) Fl	العصفر,القرطم			2. Purgative, analgesic and antipyretic.		Safrole (el Ghazali et al., 1998)
27					3. Water extract of the flower is an anticoagulant, vasodilator, antihypertensive, antioxidative, neuroprotective, immunosuppressive, anticancer agent with inhibitory impacts on the synthesis of melanin (Delshad et al., 2018)		1. Oilseed consists of 70% linoleic acid, 10% oleic acid, and minor amounts of stearic acid. 2. Flavonoids, phenylethanoid glycosides, coumarins, fatty acids, and steroids identified from various parts of the plant (Delshad et al., 2018)
28	*Tamarix aphylla*	Tarfaa, Aathil	Tamarisk	Wild plant, on Nile bank in Sudan	All plant parts have antifungal, antimicrobial. Antioxidant, cytotoxicity, antipyretic, analgesic, anti-inflammatory and antihyperglycemic agent (Jasiem et al., 2019)	After burning, ash is used to treat burn wounds and hemorrhoids	1. Polyphenolic compounds
	(Tamaricaceae) H	الطرفة, الأثل					2. Tamarix ellagic acid Ellagatannin (El Ghazali et al., 1998). Flowers contains rhamnocitrin, cyanidin 3-O-glycoside, cyanidin, delphinidin glycosides, kaempferol, quercetin, methylated flavonols, rhamnazin, rhamnetin, rhamnocitrin, kaempferide, tamarixetin, kaempferol 7,4'-dimethyl ether, and dillenetin, flavonols (Jasiem et al., 2019)
29	*Origanum vulgare*	Bardagosh, Mardagosh	Sweet marjoram	Egypt	Antibacterial	The herbal tea is used for weight loss. Vapor is used for migraine.	Essential oils (thymol (33%), gamma-terpinene (26%), and p-cymene (11%) (Faleiro et al., 2005)
	(Lamiaceae) H	البردقوش, المردقوش					

(*Continued*)

TABLE 5.1 (CONTINUED)
Overview of Some Medicinal and Aromatic Plants Used in central Sudan

No.	Plant Scientific Name, (Family), Plant Part Used*	Vernacular Name(s)	English Name(s)	Habitat and/or Distribution	Pharmacological Uses	Main Folkloric, Pharmaceutical Preparation and Dosage Form	Active Constituent(s)/class and References
30	*Piper nigrum* (Piperaceae) F	Filfil Abiet الفلفل الابيض	White piper	India	Antioxidant	Aphrodisiac, carminative	Essential oil and oleoresin (β-caryophylline, limonene, β-pinene, sabinene) and piperine (Kapoor et al., 2009)
31	*Quercus* spp. (Fagaceae) S	Baloat, Sendian البلوط,السنديان	Oak	Syria	Gastroprotective	Hemorrhoids, varicose vein, fistula, ulcer, enurises	Tannins (pedunculagin, castalagin, phillyraeoidin A, and acutissimin) (Khennouf et al., 2003)
32	*Tamarindus indica* (Caesalpinaceae) F, L	Aradeeb, Tamr hindi العرديب,تمر هندى	Tamarind	Northern and central Sudan	Antioxidant	Fruit macerate used as smooth laxative, antimalarial, and to treat jaundice.	Phenolic derivatives {2-hydroxy-3',4'-dihydroxyacet ophenone(TA0), methyl-3,4-dihydroxybenzoate (TA1,) 3,4-dihydroxyphenylacetate (TA2) and (-)-epicatechin (TA3)} (Tsuda et al., 1994)
33	*Glycyrrhiza glabra* (Fabaceae) R	Erg Alsos عرق السوس	Licorice	Egypt	Antioxidant	Root macerate used to treat respiratory tract infections and ulcers.	Phenolic derivatives (hispaglabridin B, isoliquiritigenin, and paratocarpin) (Chin et al., 2007)
34	*Chamomilla recutita* (*Matricaria recutita)* (Asteraceae) Fl	Babong, Ein elget البابونج, عين القط	Roma chamomile	Mediterranean sea basin	Antiproliferative and apoptotic	Flower macerate used as carminative, antigout agent, and to induce calmness	Glycoside (apigenin 7-*O*-glucoside) (Srivastava and Gupta, 2007)
35	*Punica granatum* (Lythraceae) F, S, B	Roman الرمان	Pomegranate	Egypt	Antioxidant	Seed used as anthelmentic. Fruit juice used for ulcer healing. Fruit peel used to stop nasal bleeding	Juices contain (hydrolyzable tannins (punicalagin), anthocyanins, ellagic acid derivatives) (Gil et al., 2000)
36	*Ruta graveolens* (Rutaceae) H	Al sazab السذاب	Rue	Mediterranean sea	Antifungal	Macerate of whole plant used as antiepileptic, urinary tract Infection (UTI), aphrodisiac, and to treat back pain.	Leaf contain: furanocoumarins (7-hydroxycoumarin, 4-hydroxycoumarin, and 7-methoxycoumarin) Novel quinolone alkaloid (1-methyl-2-[6'-(3' ',4' '-met hylenedioxyphenyl)hexyl]-4-quinolone) (Oliva et al., 2003)

(Continued)

TABLE 5.1 (CONTINUED)
Overview of Some Medicinal and Aromatic Plants Used in central Sudan

No.	Plant Scientific Name, (Family), Plant Part Used*	Vernacular Name(s)	English Name(s)	Habitat and/or Distribution	Pharmacological Uses	Main Folkloric, Pharmaceutical Preparation and Dosage Form	Active Constituent(s)/class and References
37	*Salvia officinalis* (Lamiaceae) H	Al Meremeia المرامية	Sage	Syria	Antioxidant	Macerate of whole plant used as skin lightening, carminative	Glycosides (ethyl β-d-glucopyranosyl tuberonate,1-*O*-(2,3,4-trihydroxy-3-methyl)butyl-6-*O*-feruloyl-β-d-glucopyranoside, *p*-hydroxybenzoic acid, (−)-hydroxyjasmonic acid, caffeic acid, and 4-hydroxyacetophenone 4-*O*-[5-*O*-(3,5-dimethoxy-4-hydroxybenzoyl)-β-d-apiofrunosyl]-(1→2)-β-d-glucopyranoside) (Wang et al., 2000) abietane diterpenoid(12-*O*-methyl carnosol) (Miura et al., 2002)
38	*Rosmarinus officinalis* (Lamiaceae) H	Ecleel Algabal, Hasaleban اكليل الجبل,حصالبان	Rosemary	Mediterranean Sea	Anti-inflammatory	Memory stimulant, antibacterial, and antifungal agent, and to treat headache.	Tritepenoids, Prunol, oleanic acid, and micromeric acid) (Gianmario et al., 2007)
39	*Detarium microcarpum* (Caesalpiniaceae) F	Abu-laili ابوليلة	Sweet dattock (Plant use, 2020)	Nubba Mountain Bahr el ghazal	Antifungal and inhibition of acetylcholinesterase	Fruit macerate used as antihypertensive (Sweet pulp used as sugar substitute)	Diterpenes of clerodane type (3,4-dihydroxyclerodan-13*E*-en-15-oic acid and 5α,8α (2-oxokolavenic acid)). New clerodane diterpenes (3,4-epoxyclerodan-13*E*-en-15-oic acid, 5α,8α(2-oxokolavenic acid) and 3,4-dihydroxyclerodan-13*Z*-en-15-oic acid) (Cavin et al., 2006)
40	*Hyphaene thebaica* (Arecaceae) F	Al doam الدوم	Palm tree, Gingerbread tree	Wild plant in northern and central Sudan	Antioxidant	Fruit decoction used as antihypertensive	Flavonoids (quercetin glucoside, kaempferol rhamnoglycoside and dimethyloxy-quercetin) (Eldahshan et al., 2009).
41	*Hagenia abyssinica* (Rosaceae) S, F	Shaw makkada شاو مكادة كوحبشي	African redwood, brayera, cussohagenia, or kousso	Sudan (equatorial)	Anti-tumour	Powdered fruit used as purgative, anthelmintic, and for treatment of gardiasis	Kosins (alpha-kosin, kosotoxin and protokosin) (Woldemariam et al., 1992)

(Continued)

TABLE 5.1 (CONTINUED)
Overview of Some Medicinal and Aromatic Plants Used in central Sudan

No.	Plant Scientific Name, (Family), Plant Part Used*	Vernacular Name(s)	English Name(s)	Habitat and/or Distribution	Pharmacological Uses	Main Folkloric, Pharmaceutical Preparation and Dosage Form	Active Constituent(s)/class and References
42	*Catharanthus roseus* (Apocynaceae) H	Winka وينكا	Madagascar periwinkle	Madagascar	Antioxidant	Whole plant macerate used to treat blood cancer	Alkaloids (vindoline, catharanthine, vinblastine and vincristine) (Ferreres et al., 2008) Flavonol glycosides (di- and trisaccharides of kaempferol, quercetin and isorhamnetin) (Guo et al., 2007)
43	*Boswellia* sp. (Burseraceae) V. oil (resin), stem bark	Murr hegazee, Luban gawee المر الحجازى, لبان البخور	Olibanum tree, Frankincense	Sudan (Kassala state), Somalia and Yaman	1. Inhibition of phosphodiesterase I and xanthine oxidase. 2-Prolyl endopeptidase inhibition	Antiseptic, anthelmentic, antitumor, to treat abscesses and hemorrhoids	Triterpenoids (11-oxo-β-boswellic acid, 3-oxotirucallenoic Acid, acetylketo-β-boswellic Acid and β-sitosterol) Glycosides of stilbene type Triterpenoids (3-Acetoxy-27-hydroxy-20(29)-lupen (i) and β-boswellic acid(ii)) (Atta-ur-Rahman et al., 2005)
44	*Tinospora bakis* (Menispermaceae) R	Irg al-hagar عرق الحجر	-	Ingassana hill in East and South of Sudan.	1. Antidiabetic. 2. Immunomodulatory effect (Farah et al., 2019)	Macerate of root used for wound healing. The macerated roots are used for headache (Farah et al., 2019)	The root contains alkaloids, including protoberberine type alkaloid palmatine, and 2–4% columbin, a diterpenoid furanolactone (PROTA4U, 2020)
45	*Capparis decidua* (Capparaceae) F	Tundub الطندب	-	central Sudan	Antidiabetic, anthelmintic, antibacterial, antifungal, analgesic, anti-nociceptiv e, antirheumatic, hypolip idemic, antiatherosclerot ic, anti-tumor, antigiardial, a ntioxidant, anti-inflamma tory, hepatoprotective, and anticonvulsant activities (Sonaina et al., 2020)	Asthma, poultice used for treatment of headache	Alkaloids (14-N-acetylisocodonocaprine and 15-N-acetlycapparisine) (Ahmad et al., 1992) 1. Hygroscopic alkaloids isolated from roots (capparine, cappariline and capparinine) 2. Oxygenated heterocyclic capparises terpenolides 1,4-cyclohexadien-6-one) (Neelkamal, 2009)

(Continued)

TABLE 5.1 (CONTINUED)
Overview of Some Medicinal and Aromatic Plants Used in central Sudan

No.	Plant Scientific Name, (Family), Plant Part Used*	Vernacular Name(s)	English Name(s)	Habitat and/or Distribution	Pharmacological Uses	Main Folkloric, Pharmaceutical Preparation and Dosage Form	Active Constituent(s)/class and References
46	*Hibiscus sabdariffa* (Malvaceae) Fl, S, St, L	Kurkadai	Red sorrel, hibiscus	Several parts of the Sudan	Antioxidant and antimutagenic	Fruit macerate used to treat hypertension and scorpion sting. Seed powder used for ulcer healing, general tonic, as antimalarial	V. Oil (camphor and linalool) (Rosa et al., 2006)
		الكركدى	Roselle			Cold cough.	Vitamin E and some stigmasterol derivatives (Jirovetz et al., 1992)
47	*Linum usitatissimum*	Al ketan	Linseed,	Mediterranean Sea	Inhibits tumor metastasis	Poultice used as febrifuge and for rheumatoid arthritis.	Phytoestrogen of lignin (secoisolariciresinol glycosides, matairesinol, isolariciresinol, and pinoresinol) (Sicilia et al., 2003a)
	(Linaceae) S	الكتان	flax				Secoisolariciresinol diglycoside (SDG) and flaxseed oil (FO) (Chen et al., 2006)
48	*Abrus precatorius*	Hub el-arous	Precatory bean, rosary pea, 'John Crow'	Bahar el ghazal	Immunomodulatory	Female contraceptive	1. Flavonoids (Pebrellin; abrectorin and demethoxycentaureidin 7-O-rutinoside) (Bhardwaj et al., 1980)
	(Fabaceae) S.	حب العروس	bead, saga tree, Indian licorice.				Agglutinin (Tripathi and Maiti, 2005)
49	*Citrullus colocynthis*	Hunzal	Bitter melon,	Wild plant in several parts of the Sudan	Antiallergic	Seed oil used as massage for rheumatoid arthritis, vitaligo, external pathogen, hemorrhoid and diabetes	Glycoside (alpha-elaterin (cucurbitacin E), 1,2-dehydroelatericin A (cucurbitacin I) and cucurbitacin L (Lavie et al., 1964)
	(Cucurbitaceae) F, S	الحنظل	bitter apple, bitter cucumber				
							(Cucurbitane-type triterpene glycosides, cucurbitacin E 2-*O*-beta-D-glucopyranoside, its aglycone, cucurbitacin E and colocynthosides A and B) (Yoshikawa et al., 2007)
50	*Cucurbita maxima* (Cucurbitaceae) S	Garaa	Pumpkin	Several parts of the Sudan	Antioxidant	Seed used orally as anthelmentic. Seed oil used for skin infection, hair nutrition, prostatitis. and benign prostatic hypertrophy (BPH)	Phytoestrogen of lignan (secoisolariciresinol glycosides) (Stevenson et al., 2007)
		القرع					Tocopherol (α-tocopherol, γ-tocopherol, and δ-tocophe).
							Unsaturated fatty acid (stearic, oleic, linoleic, gadoleic acid) (Sicilia et al., 2003b)

(Continued)

TABLE 5.1 (CONTINUED)
Overview of Some Medicinal and Aromatic Plants Used in central Sudan

No.	Plant Scientific Name, (Family), Plant Part Used*	Vernacular Name(s)	English Name(s)	Habitat and/or Distribution	Pharmacological Uses	Main Folkloric, Pharmaceutical Preparation and Dosage Form	Active Constituent(s)/class and References
51	*Trigonella foenum-graecum* (Fabaceae) S	Helbaa الحلبة	Fenugreek	Several parts of the Sudan	Acetylcholinesterase enzyme inhibitor	Seed powder used orally as antispasmodic, antidiarrhea agent, treatment of dysentery, lactogoge. poultice used for treatment of tumors and hemorrhoids	Alkaloid (trigonelline) (SatheeshKumar et al., 2010)
52	*Calotropis procera* (*Apocynaceae*) Fl. Latex	Oshar العشر	Sodom apple, milk weed	Wild plant in several parts of the Sudan	Antibacterial	Latex used externally to treat scorpion sting and rheumatoid arthritis.	Cardenolide {proceragenin (7β,14β-dihydroxy-5α-card -20(22)-enolide)} (Akhtar et al., 1992)
53	*Azadirachta indica* (Meliaceae) L, B, S	Neem النيم	Neem, Margosa, Indian lilac	Several parts of the Sudan	Antimutagenic	Bark macerate used as antimalaria and vitiligo. Leaf macerate used as anthelmentic, for measles, and insecticide.	Prenylated flavanones (5,7,4'-trihydroxy-8-prenylflavanone, 5,4'-dihydroxy-7-methoxy-8-prenylfl avanone, 5,7,4'-trihydroxy-3',8-diprenylflavanone, and 5,7,4'-trihydroxy-3',5'-diprenylflavanone) (Nakahara et al., 2003)
54	*Balanites aegyptiaca* (Zygophyllaceae) B, F	Hegleeg, La'loob, Balah alsahraa, Tamr el-abeed الهجليج,اللالوب,بلح الصحراء و تمر العبيد	Desert date	Wild plant in several parts Sudan	Anti-tumor	Bark used to treat vitiligo, rheumatoid arthritis and gonorrhea. Fruits used for diabetes, constipation, and gardiasis. Seed oil used for diabetes wounds.	Fruit contain: (spirostanol glycosides (balanitin-3, and a new sapogenol(6-methyldiosgenin acetate), a new furostanol saponin (balanitoside) (Hosny et al., 1992) Steroidal saponins: (balanitin-6 (28%) and balanitin-7 (72%) (Gnoula et al., 2008)
55	*Acacia nilotica* (Fabaceae) Aerial parts	Garad, Sunut القرض, السنط	Acacia	Wild plant in several parts of central and northern Sudan	Local anti-inflammatory	Fruit macerate used to treat malaria, hemorrhoids, cough, dysentery and as a wound antiseptic.	Aerial part contain:sex hormone (3-beta-acetoxy-17 beta-hydroxy-androst-5-ene) (Chaubal et al., 2011)

(Continued)

TABLE 5.1 (CONTINUED)
Overview of Some Medicinal and Aromatic Plants Used in central Sudan

No.	Plant Scientific Name, (Family), Plant Part Used*	Vernacular Name(s)	English Name(s)	Habitat and/or Distribution	Pharmacological Uses	Main Folkloric, Pharmaceutical Preparation and Dosage Form	Active Constituent(s)/class and References
56	*Cassia angustifolia* *Cassia italica* (Caesalpiniaceae) L, S	Sanamakaa السنمكة	Senna, *Alexandrian Senna*	Central and northern Sudan	Laxative	Leaf and seed powder used orally as laxative	Several compounds were isolated from the leaves (tinnevellin glycoside(I), sorhamnetin 3-O-β-gentiobioside (II), Vicenin-2 (III), emodin-8-O-beta-D-glucopyranoside(IV), robigenin (V), aloe-emodin(VI), D-pinitol (VII) (Qiu-ping et al., 2007) Anthraquinone (sennosides A + B) (Hietala et al., 1987)
57	*Solenostemma argel* (Asclepiadaceae) L	Hargel الحرجل	Argel	Northern Sudan	Topical anti-inflammatory	Leaf powder and macerate used as carminative, antispasmodic, antidiabetic	New glycosides of pregnene type (solenoside A) kaempferol-3-*O*-D-glucoside and nicotiflorin) 14 Beta,15 alpha-dihydroxy-delta(4)pregnene-3,20 dione and kaempferol-3-*O*-rutinoside (Innocenti et al., 2005)
58	*Ambrosia maritima* (Asteraceae) H	Dimsissa, Afsenteen الدمسيسة, الافسنتين	Blood weeds bitter weeds rage weeds	Wild plant in central and northern Sudan	Molluscicide	Decocted extract of whole plant used for diabetes, hypertension. Powder used to treat renal pain and stones.	Sesquiterpene (damsin, ambrosin and hymenin) (Picman et al., 1986)
59	*Artemisia annua* (Asteraceae) H	Sheeh الشيح	Sweet wormwood	Egypt & North Africa	1. Antimalarial 2. Anticancer 3.Antiviral 4. Immunosuppressive	Whole plant used for treatment of gardiasis, diabetes, athelmentic, headache, and hypertension.	1. A sesquiterpene lactone (artemisinin and oil-soluble derivatives of artemisinin (sodium artesunate and sodium artelinate) (Klayman, 1993) 2. Sesquiterpene lactone (artemisinin) (Firestone and Sundar, 2009) 3. sterols (β-sitosterol and stigmasterol (Abid Ali Khan et al., 1991) 4. Artemisinin (Noori et al., 2004) Essential oil (camphor, germacrene D, *trans*-pinocarveol, β-selinene, β-caryophyllene and ketone) (Juteau et al., 2002)

(Continued)

TABLE 5.1 (CONTINUED)
Overview of Some Medicinal and Aromatic Plants Used in central Sudan

No.	Plant Scientific Name, (Family), Plant Part Used*	Vernacular Name(s)	English Name(s)	Habitat and/or Distribution	Pharmacological Uses	Main Folkloric, Pharmaceutical Preparation and Dosage Form	Active Constituent(s)/class and References
					5. Antibacterial and antioxidant		5. Artemisinin derivatives (artemisinin, artemisinic acid, arteannuin, anhydrodihydro-artemisinin, β-arteether and α-arteether (Galal et al., 2005)
					6. Antifungal		
60	*Ammi visnaga*	Khella baladia	Visnaga	Egypt	1. Vasodilator	Decoction used as diuretic, anti-ulcer, and for renal stoneS	Visnagin (4-methoxy-7-methyl-5*H*-furo [3,2-*g*] [1]-benzopyran-5-one) (Duarte et al., 1995)
	(Apiaceae) S.	الخلة البلدية			2. Inhibition of calcium oxalate nucleation		Khellin and visnagin (Abdel-Aal et al., 2009)
61	*Nigella sativa*	Kammon aswad	Black cumin	Northern Sudan and Ethiopia	1. Anti-inflammatory	Seeds used for weight gain and to treat diabetes, hypertension, gardiasis, hair loss, and ulcers	1. Thymoquinone (Chehl et al., 2009)
	(Ranunculaceae) S	الكمون الاسود			2. Antihyperglycemic		2. Thymoquinone (Pari and Sankaranarayanan, 2009)
					3. Lipid peroxidation		3. Seed oil (thymoquinone) (Hosseinzadeh et al., 2007)
					4. Antihyperlipidemic		4. Thymoquinone (Al-Naqeep et al., 2009)
					5. Anticonvulsant		5. Thymoquinone (Hosseinzadeh and Parvardeh, 2004)
					6. Anti-oxidant and anti-inflammatory		6. Thymoquinone (2-Isopropyl-5-methyl-1, 4-benzoquinone) (Ragheb et al., 2009)
					7. Antibacterial		7. Essential oil (*p*-cymene and thymoquinone) (Kokoska et al., 2008)
					8. Antimutagenic		8. Phenolic compound (vanillic acid) (Bourgou et al., 2008)
62	*Zingiber officinale*	Genzabeel	Ginger	China, Nigeria	1. Antioxidant, anti-inflammatory, and antitumor	Macerat uesed as general tonic, expectorant, aphrodisiac, anthelmintic, and carminative.	1. Gingerols and shogaols (Sang et al., 2009)
	(Zingiberaceae) Rh	الجنزبيل			2. Gastroprotective effect		2. Gallic acid and cinnamic (Nanjundaiah et al., 2011)
					3. Antifungal		3. Diarylheptenones (gingerenone A, gingerenone B, isogingerenone B and gingerenone C) (Endo et al., 1990)
					4. Anti-inflammatory and antitumorigenic		4. [6]-Gingerol (Jeong et al., 2009)
					5. Anti skin cancer		5. [6]-Gingerol (Nigam et al., 2009)
					6. Anticarcinogenic		6. Non-volatile pungent principles (gingerols, shogaols, paradols and zingerone) (Kundu et al., 2009)
					7. Antibacterial activity		7. Highly alkylated gingerols ([10]-gingerol and [12]-gingerol) (Park et al., 2008)

(*Continued*)

TABLE 5.1 (CONTINUED)
Overview of Some Medicinal and Aromatic Plants Used in central Sudan

No.	Plant Scientific Name, (Family), Plant Part Used*	Vernacular Name(s)	English Name(s)	Habitat and/or Distribution	Pharmacological Uses	Main Folkloric, Pharmaceutical Preparation and Dosage Form	Active Constituent(s)/class and References
63	*Croton zambesicus* (Euphorbiaceae) H.	Umm geliagla, Umm guleela امقليقلة, ام غليلة	………	Nubba Mountain	Vasorelaxant	Herbal tea used to treat hypertension, diabetes, dandruff, pulmonary infection and malaria	Trachylobane diterpenes (Baccelli et al., 2007)
64	*Guiera senegalensis* (Combretaceae) L	Ghibaish الغبيش	Moshi medicine (Plant for a future, 2020)	Kordofan	1. Antifungal 2. antimalarial	Herbal tea used to treat hypertension, diabetes, spasm, as a febrifuge, as a shower for muscle relaxation.	1. Methoxylated naphthyl butenone, {guieranone A (1)} (Silva and Gomes, 2003) 2. Beta-carboline alkaloids {harmalan (dihydroharman), harman and tetrahydroharman} (Fiot et al., 2006) 3. 5-methyldihdroflavasperone (Ahmed et al., 2022)
65	*Nauclea latifolia* (Rubiaceae) H	Karmadodaa كرمدودة	Smith	Nuba Mountain	Anti-GST (glutathione S-transferases) and antifungal	MaceratE used for hypertension and cold cough	Strictosamide, naucleamides A, naucleamide F, quinovic acid-3-*O*-beta-rhamnosylpyranoside, and quinovic acid 3-*O*-beta-fucosylpyranoside (Ata et al., 2009)
66	*Kigelia africana* (Bignoniaceae) F	Umm shotor ام شطور	Kigelia	Sudan	Anti-inflammatory	After burning, ash poultice used to treat breast cancer	Verminoside, iridoid and polyphenols (verbascoside) (Picerno et al., 2005)
67	*Aristolochia bracteolate* (Aristolochiaceae) H	Um galagel ام جلاجل	Scorpion root	Widespread in Sudan	1. Antibacterial, antifungal, anti-arthritis, hypotensive, hypothermia, antioxidant, anti-inflammatory, antihyperglycemic and antihyperlipidemic activities. 2. Hexane extract of *A. bracteolata* showed *in vitro* antiplasmodial activity on chloroquine sensitive *Plasmodium falciparum* MRC-2 strain (Mathew et al., 2020)	Poultice for tumor and scorpion sting	16 major groups classified by their chemical structures, including aristolochic acids and esters, aristolactams, aporphines, protoberberines, isoquinolines, benzylisoquinolines, amides, lignans, biphenyl ethers, coumarins, tetralones, terpenoids, benzenoids (Nandhini et al., 2017)

(Continued)

TABLE 5.1 (CONTINUED)
Overview of Some Medicinal and Aromatic Plants Used in central Sudan

No.	Plant Scientific Name, (Family), Plant Part Used*	Vernacular Name(s)	English Name(s)	Habitat and/or Distribution	Pharmacological Uses	Main Folkloric, Pharmaceutical Preparation and Dosage Form	Active Constituent(s)/class and References
68	*Striga hermonthica* (Scrophulariaceae) H	Al-buda البودا	Witchweed	Sudan (parasitic on sorghum)	Antimalarial	Macerate of whole plant used to treat cancer	Saponins, tannins, flavonoids, volatile oils and cardiac glycosides (Okpako and Ajaiyeoba, 2004)
	Sterculia setigera (Sterculiaceae) B	Tirtir and Baroat.– (Al Safi, 2006) ترتر	Karaya gum tree (Jstor, 2022)	Sudan	Dichlormethane: Ethanol extract revealed remarkable tyrosinase inhibitory activity and high antiproliferative activity against human colon adenocarcinoma HT29 (Khansa et al., 2021)	(i) Antihypertensive. (ii) Bark used in treating jaundice, and bilharzia (Al Safi, 2006)	Five known compounds namely procyanidins as dimer B, trimer C1 and tetramer, 3,4-dimethoxyphenol β-D-apiofuranosyl (1″→6″)- β -D-glucopyranoside and (+)-catechin were identified, in addition to the isolation and characterization of lupeol (Khansa et al., 2021)
69	*Salvadora persica* (Salvadoraceae) St	Arak الاراك	Tooth brush tree, chewing stick (miswak)	Central and Eastern Sudan	1. Antibacterial 2. Antimycotic	Cleaning and mouth antiseptic	1. Volatile oil (benzyl nitrile, eugenol, thymol, isothymol, eucalyptol, isoterpinolene, and beta-caryophyllene) (Alali and Al-Lafi, 2003) 2. Chlorine, trimethylamine, and alkaloid resin, and sulfur compounds (al-Bagieh et al., 1994)
70	*Lupinus termis* (Fabaceae) S	Turmos الترمس	Lupin	Several parts of the Sudan	Food protein supplement	Diabetes, skin infection	Lysine, and methionine (Gabrial and Morcos, 1976)
71	*Adansonia digitata* (Bombacaceae) F, R, S, B	Tabaldi, Gunglaiz التبلدى, القنقليز	Baobab	Kordofan	Food supplement	Fruit juice used to treat diarrhea. Poultice of root used topically to increase penile size	Protein: valine (phenylalanine, tyrosine, isoleucine, lysine, arginine, threonine, cysteinem, methionine, and tryptophan) and mineral (calcium, iron, potassium, magnesium, manganese, molybdenum, phosphorus, and zinc)} (Yazzie et al., 1994)
72	*Brassica nigra* (Brassicaceae) S	Khardal الخردل	Black Mustard	Mediterranean Sea and several parts of the Sudan	Vasodilation	Cold cough, vasodilator.	Mustard oil contains (allyl-isothiocyanate) (Bánvölgyi et al., 2004)
73	*Ricinus communis* (Euphorbiaceae) S	Khairwi الخروع	Castor	Several parts of the Sudan	Purgative 1. Inhibits or prevents bacterial and fungal infections. 2. Cytotoxic, antioxidant, insecticidal, anti-asthmatic, anti-inflammatory (Ribeiro et al., 2016).	Seed oil used as laxative, hair food. With same volume of fenugreek oil, castor oil used topically to increase buttock size.	83 compounds were isolated from different tissues, which includes alkaloids, terpenoids, flavonoids, benzoic acid derivatives, coumarins, tocopherols, terpenoids, and fatty acids (Ribeiro et al., 2016).

(Continued)

TABLE 5.1 (CONTINUED)
Overview of Some Medicinal and Aromatic Plants Used in central Sudan

No.	Plant Scientific Name, (Family), Plant Part Used*	Vernacular Name(s)	English Name(s)	Habitat and/or Distribution	Pharmacological Uses	Main Folkloric, Pharmaceutical Preparation and Dosage Form	Active Constituent(s)/class and References
74	*Aloe* sp. (Aloaceae) H	Sabar الصبار	Asploe	India, South America, Central America, the Caribbean, Australia, and Africa (Kar and Bera, 2018)	1. Antioxidant 2. Analgesics, anti-bacterials, and anti-virals. 3. Antiseptic 4. Anti-inflammatory (Kar and Bera, 2018)	Juice used topically for wound healing, to treat hemorrhoids, and orally to treat diabetes.	β-coumaric acid, ascorbic acid, pyrocatechol and cinnamic acid (Lawrence et al., 2009) 1. Anthraquinone, aloin and emodin. 2. Four plant steroids; cholesterol, campesterol, β-sisosterol, and lupeol. 3. Eight enzymes: aliiase, alkaline phosphatase, amylase, bradykinase, carboxy-peptidase, catalase, cellulase, lipase, and peroxidase. 4. Minerals: calcium, chromium, copper, selenium, magnesium, manganese, potassium, sodium, and zinc (Kar and Bera, 2018)
75	*Eugenia caryophyllus* (Myrtaceae) F.	Guronful القرنفل	Clove	India	1. Antioxidant 2. Anti-herpes simplex virus 3. Antimicrobial, antioxidant, antifungal, antiviral activity, anti-inflammatory, cytotoxic, insect repellent. and anesthetic	Topical analgesic and carminative	1. Leaf contains essential oils (eugenol, β-caryophyllene, α-humulene and eugenyl acetate) (Jirovetz et al., 2006) 2. Essential oil (eugenol) (Tragoolpua and Jatisatienr, 2007) 3. Essential oil phenylpropanoids (carvacrol, thymol, eugenol, and cinnamaldehyde) (Chaieb et al., 2007)
76	*Santalum album* (Santalaceae)St.	Sandal الصندل	Sandal wood	India	Anti-*Helicobacter pylori*	Scent	Six new sesquiterpenes {(Z)-2beta-hydroxy-14-hydro-beta-santalol, (Z)-2-alpha-hydroxy-albumol (2), 2R-(Z)-campherene-2,13-diol, (Z)-campherene-2-beta,13-diol, (Z)-7-hydroxynuciferol, and (Z)-1beta-hydroxy-2-hydrolanceol, together with five known compounds, (Z)-alpha-santalol, (Z)-beta-santalol, (Z)-lanceol, alpha-santaldiol, and beta-santaldiol} (Ochi et al., 2005)
77	*Acacia seyal* (Fabaceae) St, F	Talih الطلح	Gum Arabic	Sudan	1. Antioxidant. 2. Attenuate the development of nephropathy in type I diabetes in rat 3. Inhibits absorption of glucose in the intestine. 4. Reduces body weight and fat deposition (Musa et al., 2018).	Fume used for rheumatoid arthritis Fruit macerate used as antiseptic. Used as a carrier of drugs (Musa et al., 2018).	Polysaccharides (rhamnose and glucuronic acid, arabinose and 4-O-methyl glucuronic acid), The amino acids (hydroxyproline and serine) (Williams and Phillips, 2009)

(Continued)

TABLE 5.1 (CONTINUED)
Overview of Some Medicinal and Aromatic Plants Used in central Sudan

No.	Plant Scientific Name, (Family), Plant Part Used*	Vernacular Name(s)	English Name(s)	Habitat and/or Distribution	Pharmacological Uses	Main Folkloric, Pharmaceutical Preparation and Dosage Form	Active Constituent(s)/class and References
78	*Hordeum vulgare* (Poaceae) F	Shaeir baladi شعير بلدى	Barley	Several parts of the Sudan	Diuretic, fatty acid synthase inhibition, glucose tolerance effect, laxative, hypocholesterolemic, lowering of the fasting plasma glucose and glycosylated hemoglobin levels (Kumari et al., 2016)	Diabetes and renal stones	New cyanogenic glycoside(2-β-glucopyranosyl-oxy-3-methyl-(2*R*)-butyronitrile) (Erb et al., 1979)
	Tribulus terrestris (Zygophyllaceae) H	Dressa الضريسة	Puncture vine	Wild plant widespread in Sudan	Cytotoxic	Macerate used as diuretic and astringent.	Furostanol saponins from fruits (terrestroside A, and terrestroside B, together with three known compounds, chloromaloside E, terrestrinin B and terrestroneoside A) (Wang et al., 2009)
79	*Grewia tenax* (Tiliaceae) F, R	Guddaim and Basham. (Al Safi, 2006) (قضيم)	White cross berry	Northern and central Sudan (Elmuez et al., 2014).	Fruit juice has antioxidant activity (Suliman et al., 2018)	Fruit is used to treat malaria and anemia. Root is used to treat tonsillitis –(Al Safi, 2006)	Proximate composition of fruit is carbohydrate was 66.59%, while moisture, crude fiber, ash, crude protein and crude fat were 11.72%, 9.41%, 4.12%, 7.68% and 0.48%, respectively (Elmuez et al., 2014).
80	*Prunus mahaleb* (Rosaceae) S	Mahlab (محلب)	Lucie cherry	Epirus (cultivated, not wild species) (University of Ioannina, 2020)	Sedative and vasodilator(University of Ioannina, 2020) Cytoprotective and hypolipidemic effect (Ferramosca et al., 2019)	Scenting, anti-diabetes, renal stone, gout, rheumatoid arthritis, cough treatment.	Herniarin and herniarin glycoside Dihydrocoumarin, coumarin (Al-Said and Hifnawy, 1986) Coumarin, amygdalin, prunasin, flavonoids, isoflavones, bioflavonoids, anthocyanins, proanthocyanins, vitamins, trace elements (University of Ioannina, 2020)

(Continued)

TABLE 5.1 (CONTINUED)
Overview of Some Medicinal and Aromatic Plants Used in central Sudan

No.	Plant Scientific Name, (Family), Plant Part Used*	Vernacular Name(s)	English Name(s)	Habitat and/or Distribution	Pharmacological Uses	Main Folkloric, Pharmaceutical Preparation and Dosage Form	Active Constituent(s)/class and References
81	*Terminalia brownii* (Combretaceae) St.	Sobag, Darout, Shaf, Subaraya (صباغ، دروت، صبارية، شاف)	Red pod terminalia (Pl@nt use, 2022)	Wild plant widespread in Sudan	Crude extract and its solvent fractions showed a significant antihyperglycemic activity in streptozocin-induced diabetic mice (Alema et al., 2020)	Vapor used for rheumatoid arthritis	80% methanolic leaf and stem, bark extract revealed the presence of tannins, saponins, flavonoids, polyphenols, terpenoids, steroids, phytosterols, and coumarins (Alema et al., 2020)
82	*Cymbogon citratus* (Poaceae) H.	Hashishut al-lemon حشيشة اليمون	Lemon grass	India and Kongo	Free radical scavengers and antioxidants	Macerate of the whole plant used to treat renal stones, as lactogoge and antitumor agent.	V. oil (Iso-orientin, isoscoparin, swertiajaponin, isoorientin 2′ ′-*O*-rhamnoside, orientin, chlorogenic acid, and caffeic acid) (Cheel et al., 2005)
83	*Anogeissus leiocarpus* (Combretaceae) S,B	Sahab صهب	African birch; Bambara: Ngálăma (ACEPRD 2020)	Sudan (southern Kordofan (Nuba Mountain), southern Darfur, and Blue Nile State).	Antifungal	Seed powder used orally for giardiasis. Oral macerate of bark used for hemorrhoids	Five hydroethanolic extracts (Batawila et al., 2005) The root contains: Al, K, Fe, V, Na, Sc, Ca, Br, Zn, Mg, La, Cs, and C.
84							The bark contains L, Sn, Co, Mn, Au and 3,3,4-trimethoxyflavellagic acid while the stem is reported to contain 4-*O*-β-D-glucoside (0.2 ppm) and fluoride (16.4) (ACEPRD, 2020)
85	*Cinnamomum camphora* (Lauraceae) V. Oil	Kafur الكافور	Camphor	Egypt	Anti-inflammatory agent, antibacterial (Chen et al., 2020)	Oil used in massage for joint pain	Sesquiterpene (cadinenol and epicubenol) (Takaoka, 1976)
86	*Acacia senegal* (Fabaceae) Gum.	Hashab الهشاب	Gum Arabic	Wild plant, widespread in Sudan	1. Antioxidant. 2. Attenuates the development of nephropathy in type I diabetes rat 3. Inhibits absorption of glucose in the intestine. 4. Reduces body weight and fat deposition (Musa et al., 2018).	Gum used for ulcer, diarrhea, carminative. Used as a carrier of drugs (Musa et al., 2018).	Polysaccharides (rhamnose and glucuronic uronic acid, arabinose and 4-*O*-methyl glucuronic acid), The amino acids (hydroxyproline and serine) (Williams and Phillips, 2009)

Plant part used: B: bark, F: fruit, H: herb, L: leaf, Rh: Rhizome, R: root, S: seed, St: stem.

for international markets (Eman, 2007). However, there are concerns with the interrelated issues of quality, safety, and efficacy, when such plants are incorporated in modern medical practices.

5.5 CONCLUSION

In this chapter, 86 medicinal plant species, belonging to 51 families have been described. The most common species are from the family Lamiaceae and Apiaceae. The data presented in this chapter demonstrate the traditional uses of the most commonly used Sudanese traditional medicinal plants to meet primary health care needs. Information has been provided on each plant's scientific name, vernacular name and/or names, habitat and/or distribution, main folkloric uses, method of preparation, dosage form, and recent research related to the pharmacology and phytochemistry of these plants. These plants have shown different biological activities, mainly as antimicrobial, anticancer, antidiabetic, anti-inflammatory, antihypertensive, antimalarial, antioxidant, and antispasmodic agents. However, there are many Sudanese medicinal plants which have yet to be screened at the molecular level and are likely to provide novel molecules in future research.

Clinical trials that evaluate the quality, safety, and efficacy of herbal products and their specified use in combination with conventional therapy will certainly produce add-on therapeutic value to prevent or control diseases. Moreover, there is a need for standardization and rationalization of herbal medicines by promoting authentication, registration, and licensing of medicinal plants.

REFERENCES

Abdel-Aal, E.A., Daosukho, S., El-Shall, H., 2009. Effect of supersaturation ratio and Khella extract on nucleation and morphology of kidney stones. *J. Cryst. Growth* 311(9), 2673–2681. https://doi.org/10.1016/j.jcrysgro.2009.02.027.

Abdelgadir, A.A., Ahmed, M.E.O., Ahmed, E.M., 2021. Traditional uses of herbal medicines in Khartoum and Gezira state (Central Sudan). *AJMAP* 7(1), 29–73. https://doi.org/10.48347/IMIST.PRSM/ajmap-v7i1.22273.

Abid Ali Khan, M.M., Jain, D.C., Bhakuni, R.S., Zaim, M., Thakur, R.S., 1991. Occurrence of some antiviral sterols in *Artemisia annua*. *Plant Sci.* 75(2), 161–165. https://doi.org/10.1016/0168-9452(91)90230-6.

Abu-Mustafa, E., Fayez, M., Gad, A., Osman, F., 1960. Notes-isolation of β-sitosterol from Chufa (*Cyperus esculentus* L.) tubers. *J. Org. Chem.* 25(7), 1269–1269. https://doi.org/10.1021/jo01077a629.

ACEPRD Africa Centre for Excellence in Phytomedicine Research and Development (ACEPRD) University of Jos, 2020. *Anogeissus leiocarpus* (DC) Guill and Perr. *Retrieved from* http://aceprd.unijos.edu.ng/plant-info/9e534126-f04b-4014-8ca1-2c36d75cded5. *Accessed online in 12/12/2020.*

Adam, K., Sivropoulou, A., Kokkini, S., Lanaras, T., Arsenakis, M., 1998. Antifungal activities of *Origanumvulgare* subsp. *hirtum, Mentha spicata, Lavandula angustifolia*, and *Salvia fruticosa* essential oils against human pathogenic fungi. *J. Agric. Food Chem.* 46(5), 1739–1745. https://doi.org/10.1021/jf9708296.

Adam Manar, A., Ahmed, A.A., Yagi, A., Yagi, S., 2020. Ethnobotanical investigation on medicinal plants used against human ailments in Erkowit and Sinkat areas, Eastern Sudan. *Biodivers.* 21(7). https://doi.org/10.13057/biodiv/d210748.

Adenowo Abiola, F., Kazeem Mutiu, I., 2020. Tiger nut as A functional food, pharmacological and industrial agent: A mini review. *Ann.Sci. Technol.* 5(1), 31–38.

Ads, E.N., Rajendrasozhan, S., Hassan, S.I., Sharawy, S.M.S. Humaidi, J.R., 2018. Phytochemical screening of different organic crude extracts from the stem bark of *Ziziphus spina-christi* (L.). *Biomed. Res.*, 29(8),1645–1652. ISSN 0970-938X. www.biomedres.info

Ahmad, A., Khan, I., Anjum, N.A., Diva, I., Abdin, M.Z., Iqbal, M., 2005. Effect of timing of sulphur fertilizer application on growth and yield of rapeseed. *J. Plant Nutr.* 28(6), 1049–1059.

Ahmad, S., Abdin, M.Z., Fazli, I.S., Jamal, A., Maaz, M., Iqbal, M., 2004. Variability in xanthotoxin concentration in organs of *Ammi majus* and yield at various phenological stages. *J. Med. Arom. Plant Sci.* 26, 8–11.

Ahmad, S., Jamal, A., Fazli, I.S., Alam, T., Khan, M.A., Kamaluddin, I.M., Abdin, M.Z., 2007. Impact of sulphur and nitrogen application on seed xanthotoxin yield in *Ammi majus* L. Korean. *J. Crop Sci.* 52, 153–161.

Ahmad, V.U., Ismail, N., Arif, S., Amber, A.-R., 1992. Two new N-acetylated spermidine alkaloids from *Capparis decidua*. *J. Nat. Prod.* 55(10), 1509–1512. https://doi.org/10.1021/np50088a018.

Ahmed, I.M., Tahir, Y.F., Nour, S.M., Suliman, M.A., 2020. Traditional use of medicinal plants among the Barti tribe community in Fangoga area, Sennar State, Sudan. *Trop. Plant Res.*, 7(2), *517, 521.* ISSN (Online): 2349-1183; ISSN (Print): 2349–9265 https://doi.org/10.22271/tpr.2020.v7.i2.063.

Ahmed, M.E.O., Senabe, J., Yahaya, E.S., Fouche, G., Steenkamp, P., Steenkamp, V., 2022. Isolation and antioxidant activity of 5-methyldihydroflavasperone from ethanol leaf extract of *Guiera senegalensis* JF Gmel. *Journal of Medicinal Plants for Economic Development.* 6(1), a137. https://doi.org/10.4102/jomped.v6i1.137

Akhtar, N., Malik, A., Ali, S.N., Kazmit, S.U., 1992. Proceragenin, an antibacterial cardenolide from *Calotropis procera. Phytochemistry* 31(8), 2821–2824. https://doi.org/10.1016/0031-9422(92)83639-G.

Al Safi, A., 2006. *Traditional Sudanese Medicine: A Primer for Health Care Providers, Researchers, and Students.* Chapter, academia.edu.

Alali, F., Al-Lafi, T., 2003. GC-MS analysis and bioactivity testing of the volatile oil from the leaves of the toothbrush tree *Salvadora persica* L. *Nat. Prod. Res.* 17(3), 189–194. https://doi.org/10.1080/1057563021000040790.

Al-Bagieh, N., Idowu, A., Salako, N.O., 1994. Effect of aqueous extract of miswak on the in vitro growth of *Candida albicans. Microbios* 80(323), 107–113.

Al-Burtamani, S.K.S., Fatope, M.O., Marwah, R.G., Onifade, A.K., Al-Saidi, S.H., 2005. Chemical composition, antibacterial and antifungal activities of the essential oil of *Haplophyllum tuberculatum* from Oman. *J. Ethnopharmacol.* 96(1–2), 107–112. https://doi.org/10.1016/j.jep.2004.08.039.

Alema, N.M., Periasamy, G., Sibhat, G.G., Tekulu, G.H., Hiben, M.G., 2020. Antidiabetic activity of extracts of *Terminalia brownii* Fresen. Stem Bark in mice. *J. Exp. Pharmacol.* 12, 61–71.

Ali, S.T., Mahmooduzzafar, Abdin, M.Z., Iqbal, M., 2008. Ontogenetic changes in foliar features and psoralen content of *Psoralea corylifolia* Linn. exposed to SO_2 stress. *J. Environ. Biol.* 29(5), 661–668.

Al-Naqeep, G., Ismail, M., Allaudin, Z., 2009. Regulation of low-density lipoprotein receptor and 3-hydroxy-3-methylglutaryl coenzyme A reductase gene expression by thymoquinone-rich fraction and thymoquinone in HepG2 Cells. *J. Nutrigenet. Nutrigenomics* 2(4–5), 163–172. https://doi.org/10.1159/000227264.

Al-Said, M.S., Hifnawy, M.S., 1986. Dihydrocoumarin and certain other coumarins from *Prunus mahaleb* seeds. *J. Nat. Prod.* 49(4), 721–721. https://doi.org/10.1021/np50046a040.

Anis, M., Iqbal, M., 1986. Antipyretic utility of some Indian plants in traditional medicine. *Fitoterapia* 57, 52–55.

Anis, M., Iqbal, M., 1994. Medicinal plantlore of Aligarh, India. *Int. J. Pharmacogn.* 32(1), 59–64.

Anis, M., Sharma, M.P., Iqbal, M., 2000. Herbal ethnomedicine of the Gwalior-forest division in Madhya Pradesh, *India. Pharm. Biol.* 38(4), 241–253.

Ata, A., Udenigwe, C.C., Matochko, W., Holloway, P., Eze, M.O., Uzoegwu, P.N., 2009. Chemical constituents of *Nauclea latifolia* and their anti-GST and anti-fungal activities. *Nat. Prod. Commun.* 4(9), 1185–1188. https://doi.org/10.1177/1934578X0900400905.

Atique, A., Iqbal, M., 1992. Traditional single drug administration for the psychosomatic dermal diseases: Psoriasis. *Hamdard Med.* 35, 76–81.

Atique, A., Iqbal, M., Ghouse, A.K.M., 1985a. Use of *Annona squamosa* and *Piper nigrum* against diabetes. *Fitoterapia* 56, 190–192.

Atique, A., Iqbal, M., Ghouse, A.K.M., 1985b. Ethnobotanical study of cluster fig (*Ficus racemosa*). *Fitoterapia* 56, 236–240.

Atique, A., Iqbal, M., Ghouse, A.K.M., 1986. Folk-medicinal uses of *Ficus bengalensis* Linn. and *Punica granatum* Linn. in northern Uttar Pradesh. *Bull. Med. Ethnobot. Res.* 6, 42–46.

Atta-ur-Rahman, Naz, H., Fadimatou, Makhmoor, T., Yasin, A., Fatima, N., Ngounou, F.N., Kimbu, S.F., Sondengam, B.L., Choudhary, M.I., 2005. Bioactive constituents from *Boswellia papyrifera. J. Nat. Prod.* 68(2), 189–193. https://doi.org/10.1021/np040142x.

Awad, E.M., Ahmed, E.M., Elhadiya, T.M., 2012. Investigation of anti-nociceptive activity of Zingabeel (*Zingiber offficinale*) on acetic acid induced writhing in rats. *Gezira J. Health Sci.* 8(1), 11–16.

Aziz, M.A., Khan, A.H., Adnan, M., Ullah, H., 2018. Traditional uses of medicinal plants used by Indigenous communities for veterinary practices at Bajaur Agency, Pakistan. *J. Ethnobiol. Ethnomed.* 14(1), 11. https://doi.org/10.1186/s13002-018-0212-0.

Barillari, J., Canistro, D., Paolini, M., Ferroni, F., Pedulli, G.F., Iori, R., Valgimigli, L., 2005a. Direct antioxidant activity of purified glucoerucin, the dietary secondary metabolite contained in rocket (*Eruca sativa* Mill.) seeds and sprouts. *J. Agric. Food Chem.* 53(7), 2475–2482. https://doi.org/10.1021/jf047945a.

Barillari, J., Cervellati, R., Paolini, M., Tatibouët, A., Rollin, P., Iori, R., 2005b. Isolation of 4-methylthio-3-butenyl glucosinolate from *Raphanus sativus* sprouts (Kaiware Daikon) and its redox properties. *J. Agric. Food Chem.* 53(26), 9890–9896. https://doi.org/10.1021/jf051465h.

Baccelli, C., Navarro, I., Block, S., Abad, A., Morel, N., Quetin-Leclercq, J., 2007. Vasorelaxant activity of diterpenes from *Crotonzambesicus* and synthetic trachylobanes and their structure–activity relationships. *J. Nat. Prod.* 70(6), 910–917. https://doi.org/10.1021/np0603976.

Bánvölgyi, Á., Pozsgai, G., Brain, S.D., Helyes, Z.S., Szolcsányi, J., Ghosh, M., Melegh, B., Pintér, E., 2004. Mustard oil induces a transient receptor potential vanilloid 1 receptor-independent neurogenic inflammation and a non-neurogenic cellular inflammatory component in mice. *Neuroscience* 125(2), 449–459. https://doi.org/10.1016/j.neuroscience.2004.01.009.

Batawila, K., Kokou, K., Koumaglo, K., Gbéassor, M., de Foucault, B., Bouchet, P., Akpagana, K., 2005. Antifungal activities of five Combretaceae used in Togolese traditional medicine. *Fitoterapia* 76(2), 264–268. https://doi.org/10.1016/j.fitote.2004.12.007.

Beigh, S.Y., Iqbal, M., Nawchoo, I.A., 2002a. Seed germination and seedling survival of *Aconitum heterophyllum*, an endangered medicinal plant of the northwest Himalaya. *Indian J. Plant Physiol.* 7(2), 109–113.

Beigh, S.Y., Nawchoo, I.A., Iqbal, M., 2002b. Herbal drugs in India: Past and present uses. *J. Trop. Med. Plants* 3, 197–204.

Beigh, S.Y., Nawchoo, I.A., Iqbal, M., 2003a. Plants in traditional medicine of Kashmir Himalayas. *J. Econ. Taxon. Bot.* 27(1), 99–104.

Beigh, S.Y., Nawchoo, I.A., Iqbal, M., 2003b. Traditional veterinary medicine among the tribes of Kashmir Himalaya. *J. Herbs Spices Med. Plants* 10(4), 121–127.

Beigh, S.Y., Nawchoo, I.A., Iqbal, M., 2005. Cultivation and conservation of *Aconitum heterophyllum*, a critically endangered medicinal herb of the north-west Himalayas. *J. Herbs Spices Med. Plants* 11, 47–56.

Beigh, S.Y., Nawchoo, I.A., Iqbal, M., 2008. Alkaloid contents of *Aconitum heterophyllum* growing at different altitudes of the Kashmir Himalaya. *J. Indian Bot. Soc.* 87, 252–255.

Bhardwaj, D.K., Bisht, M.S., Mehta, C.K., 1980. Flavonoids from Abrus precatorius. *Phytochemistry* 19(9), 2040–2041.

Bourgou, S., Ksouri, R., Bellila, A., Skandrani, I., Falleh, H., Marzouk, B., 2008. Phenolic composition and biological activities of Tunisian *Nigella sativa* L. shoots and roots. *C. R. Biol.* 331(1), 48–55. https://doi.org/10.1016/j.crvi.2007.11.001.

Cavin, A.-L., Hay, A.E., Marston, A., Stoeckli-Evans, H., Scopelliti, R., Diallo, D., Hostettmann, K., 2006. Bioactive diterpenes from the fruits of *Detariummicrocarpum*. *J. Nat. Prod.* 69(5), 768–773. https://doi.org/10.1021/np058123q

Chaubal, R., Mujumdar, A., Misar, A., Deshpande, V., Deshpande, N., 2011. Structure-activity relationship study of androstene steroids with respect to local anti-inflammatory activity. *Arzneim. Forsch.* 56, 394–398. https://doi.org/10.1055/s-0031-1296740.

Chaieb, K., Hajlaoui, H., Zmantar, T., Kahla-Nakbi, A.B., Rouabhia, M., Mahdouani, K., Bakhrouf, A., 2007. The chemical composition and biological activity of clove essential oil, *Eugenia caryophyllata* (*Syzigium aromaticum* L. Myrtaceae): A short review. *Phytother. Res.* 21(6), 501–506. https://doi.org/10.1002/ptr.2124.

Chaitanya, M.V., Baye, H.G., Ali, H.S., Usamo, F.B., 2021, March 4. Traditional African medicine, in H. A. El-Shemy (ed.), *Natural Medicinal Plants*. IntechOpen, London. https://doi.org/10.5772/intechopen.96576. Available from: https://www.intechopen.com/online-first/75561.

Cheel, J., Theoduloz, C., Rodríguez, J., Schmeda-Hirschmann, G., 2005. Free radical scavengers and antioxidants from lemongrass (*Cymbopogon citratus* (DC.) Stapf.). *J. Agric. Food Chem.* 53(7), 2511–2517. https://doi.org/10.1021/jf0479766.

Chen, J., Tang, C., Zhou, Y., Zhang, R., Ye, S., Zhao, Z., Lin, L., Yang, D., 2020. Anti-inflammatory property of the essential oil from *Cinnamomum camphora* (Linn.) Presl Leaves and the evaluation of its underlying mechanism by using metabolomics analysis. *Molecules* 25(20), 4796. https://doi.org/10.3390/molecules25204796.

Chen, J., Wang, L., Thompson, L.U., 2006. Flaxseed and its components reduce metastasis after surgical excision of solid human breast tumor in nude mice. *Cancer Lett.* 234(2), 168–175. https://doi.org/10.1016/j.canlet.2005.03.056.

Chehl, N., Chipitsyna, G., Gong, Q., Yeo, C.J., Arafat, H.A., 2009. Anti-inflammatory effects of the *Nigella sativa* seed extract, thymoquinone, in pancreatic cancer cells. *HPB* 11(5), 373–381. https://doi.org/10.1111/j.1477-2574.2009.00059.x.

Chin, Y.-W., Jung, H.-A., Liu, Y., Su, B.-N., Castoro, J.A., Keller, W.J., Pereira, M.A., Kinghorn, A.D., 2007. Anti-oxidant constituents of the roots and stolons of licorice (*Glycyrrhiza glabra*). *J. Agric. Food. Chem.* 55, 4691–4697. https://doi.org/10.1021/jf0703553.

CIA (Central Intelligence Agency). World factbook. Sudan. 2021. Available online: https://www.cia.gov/the-world-factbook/countries/sudan. Retrieved 2021-8.

Duarte, J., Pérez-Vizcaíno, F., Torres, A.I., Zarzuelo, A., Jiménez, J., Tamargo, J., 1995. Vasodilator effects of visnagin in isolated rat vascular smooth muscle. *Eur. J. Pharmacol.* 286(2), 115–122. https://doi.org/10.1016/0014-2999(95)00418-K.

Delshad, E., Yousefi, M., Sasannezhad, P., Rakhshandeh, H., Ayati, Z., 2018. Medical uses of *Carthamus tinctorius* L. (Safflower): A comprehensive review from traditional medicine to modern medicine. *Electron. Phys.* 10(4), 6672–6681. http://doi.org/10.19082/6672.

Eisa, A.A., Babiker, H.E., Taj Eldin, I.M., El Tahir, E.M., Ahmed, E.M., 2020. Early stage clinical trial of ginger (*Zingiber officinale*) as and add-on antiepileptic therapy in children with generalized epilepsies. *Khartoum J. Pharm. Sci.* 1(1), 30–34.

Eldahshan, O.A., 2009. Potential antioxidant phenolic metabolites from doum palm leaves. *Afr. J. Pharm. Pharmacol.* 3, 158–164.

El Ghazali, G., Khalid, H.E.S., el Tohami, M.S., Yagi, S.M.A., Abdalla, W.E., 1998. Common medicinal plant in Khartoum state, National Centre for Research, Khartoum, 83–84.

El-Ghorab, A., El-Massry, K.F., Shibamoto, T., 2007. Chemical composition of the volatile extract and antioxidant activities of the volatile and nonvolatile extracts of Egyptian corn silk (*Zea mays* L.). *J. Agric. Food Chem.* 55(22), 9124–912. https://doi.org/10.1021/jf071646e

Elmuez Alsir, A. Aboagarib, Yang, R., Hua, X., Siddeeg, A., 2014. Chemical compositions, nutritional properties and volatile compounds of Guddaim (*Grewia Tenax*. Forssk) Fiori fruits. *J. Food Nutr. Res.*, 2(4), 187–192. https://doi.org/10.12691/jfnr-2-4-9.

Eman, A., 2007. *Quality Assurance of 13 Selected Sudanese Medicinal Plants*. PhD thesis, Faculty of Pharmacy, University of Gezira.

Endo, K., Kanno, E., Oshima, Y., 1990. Structures of antifungal diarylheptenones, gingerenones A, B, C and isogingerenone B, isolated from the rhizomes of *Zingiber officinale*. *Phytochemistry* 29(3), 797–799. https://doi.org/10.1016/0031-9422(90)80021-8.

Erb, N., Dietmar Zinsmeister, H., Lehmann, G., Nahrstedt, A., 1979. A new cyanogenic glycoside from Hordeum vulgare. *Phytochemistry* 18(9), 1515–1517. https://doi.org/10.1016/S0031-9422(00)98486-3.

Evans, W.C., 2002a. *Trease and Evans. Pharmacognosy*. 16th edition, Chapter 22. Volatile oils and resins, Wiley, New York, 263 page.

Evans, W.C., 2002b. *Trease and Evans. Pharmacognosy*. 16th edition, Chapter 26. Volatile oils and resins, Wiley, New York, 353 page.

Faleiro, L., Miguel, G., Gomes, S., Costa, L., Venâncio, F., Teixeira, A., Figueiredo, A.C., Barroso, J.G., Pedro, L.G., 2005. Antibacterial and antioxidant activities of essential oils isolated from *Thymbra capitata* L. (Cav.) and *Origanum vulgare* L.). *J. Agric. Food Chem.* 53(21), 8162–8168. https://doi.org/10.1021/jf0510079.

Farah Hayat, M., Khalid, H.E., El Hussein, A.M., Osman, H.M., 2019. Toxic activity of *Tinospora bakis* (Irg al-hagar) Roots in Wistar albino Rats. *Eur. J. Med. Plants* 26(3), 1–9.

Ferramosca, A., Treppiccione, L., Di Giacomo, M., Aufiero, V.R., Mazzarella, G., Maurano, F., Gerardi, C., Rossi, M., Zara, V., Mita, G., Bergamo, P., 2019. *Prunus mahaleb* fruit extract prevents chemically induced colitis and enhances mitochondrial oxidative metabolism via the activation of the Nrf2 pathway. *Mol Nutr Food Res.*, 63(22),1900350. https://doi.org/10.1002/mnfr.201900350

Ferreres, F., Pereira, D.M., Valentão, P., Andrade, P.B., Seabra, R.M., Sottomayor, M., 2008. New phenolic compounds and antioxidant potential of *Catharanthus roseus*. *J. Agric. Food. Chem.* 56(21), 9967–9974. https://doi.org/10.1021/jf8022723.

Fiot, J., Sanon, S., Azas, N., Mahiou, V., Jansen, O., Angenot, L., Balansard, G., Ollivier, E., 2006. Phytochemical and pharmacological study of roots and leaves of *Guiera senegalensis* J.F. Gmel (Combretaceae). *J. Ethnopharmacol.* 106(2), 173–178. https://doi.org/10.1016/j.jep.2005.12.030.

Firestone, G.L., Sundar, S.N., 2009. Anticancer activities of artemisinin and its bioactive derivatives. *Expert Rev. Mol. Med.* 11, e32. https://doi.org/10.1017/S1462399409001239.

Gabrial, G.N., Morcos, S.R., 1976. The use of *Lupinus termis* L. cultivated in Egypt, as a food protein supplement. *Z. Ernährungswiss* 15(4), 333–339. https://doi.org/10.1007/BF02020501.

Galal, A.M., Ross, S.A., Jacob, M., ElSohly, M.A., 2005. Antifungal activity of artemisinin derivatives. *J. Nat. Prod.* 68(8), 1274–1276. https://doi.org/10.1021/np050074u.

Gianmario, A., 2007. Characterization of topical antiinflammatory compounds in *Rosmarinus officinalis*. *J. Agric. Food Chem.* 55(5), 1718–1723.

Gil, M.I., Tomás-Barberán, F.A., Hess-Pierce, B., Holcroft, D.M., Kader, A.A., 2000. Antioxidant activity of Pomegranate juice and its relationship with phenolic composition and processing. *J. Agric. Food. Chem.* 48(10), 4581–4589. https://doi.org/10.1021/jf000404a.

Gourine, N., Yousfi, M., Bombarda, I., Nadjemi, B., Gaydou, E., 2010. Seasonal variation of chemical composition and antioxidant activity of essential oil from *Pistacia atlantica* Desf. Leaves. *J. Am. Oil Chem. Soc.* 87(2), 157–166. https://doi.org/10.1007/s11746-009-1481-5.

Gnoula, C., Mégalizzi, V., De Nève, N., Sauvage, S., Ribaucour, F., Guissou, P., Duez, P., Dubois, J., Ingrassia, L., Lefranc, F., Kiss, R., Mijatovic, T., 2008. Balanitin-6 and -7: Diosgenyl saponins isolated from Balanites aegyptiaca Del. display significant anti-tumor activity *in vitro* and *in vivo*. *Int. J. Oncol.* 32(1), 5–15. https://doi.org/10.3892/ijo.32.1.5.

Guo, X., Yang, L., Yu, J., Tang, Z., Zu, Y., 2007. Alkaloid variations in *Catharanthus roseus* seedlings treated by different temperatures in short term and long term. *J. For. Res.* 18(4), 313–315. https://doi.org/10.1007/s11676-007-0063-3.

Hala, M.A.A., 2009. *Quality of Medicinal Plants Traditionally Used in Sudan as Affected by Ionizing Radiation Treatments.* PhD thesis. Retrieved online on: 5/8/2021

Hegazy, A.K., Hosni, H.A., Lovett-Doust, L., Kabiel, H.F., Badawy, E.M., Mwavu, E.N., 2020. Indigenous knowledge of wild plants collected in Darfur, Sudan. *Ethnobot. Res. Appl.* 19, 1–19. http://doi.org/10.32859/era.19.47.1-19.

Hietala, P., Marvola, M., Parviainen, T., Lainonen, H., 1987. Laxative potency and acute toxicity of some anthraquinone derivatives, senna extracts and fractions of senna extracts. *Pharmacol. Toxicol.* 61(2), 153–156. https://doi.org/10.1111/j.1600-0773.1987.tb01794.x.

Hosny, M., Khalifa, T., Çaliş, I., Wright, A.D., Sticher, O., 1992. Balanitoside, a furostanol glycoside, and 6-methyl-diosgenin from *Balanites aegyptiaca. Phytochemistry* 31(10), 3565–3569. https://doi.org/10.1016/0031-9422(92)83728-H.

Hosseinzadeh, H., Parvardeh, S., 2004. Anticonvulsant effects of thymoquinone, the major constituent of *Nigella sativa* seeds, in mice. *Phytomedicine* 11(1), 56–64. https://doi.org/10.1078/0944-7113-00376.

Hosseinzadeh, H., Parvardeh, S., Asl, M.N., Sadeghnia, H.R., Ziaee, T., 2007. Effect of thymoquinone and *Nigella sativa* seeds oil on lipid peroxidation level during global cerebral ischemia-reperfusion injury in rat hippocampus. *Phytomedicine* 14(9), 621–627. https://doi.org/10.1016/j.phymed.2006.12.005.

Howard, L.R., Talcott, S.T., Brenes, C.H., Villalon, B., 2000. Changes in phytochemical and antioxidant activity of selected pepper cultivars (*Capsicum* species) as influenced by maturity. *J. Agric. Food Chem.* 48(5), 1713–1720. https://doi.org/10.1021/jf990916t.

Husen, A., Iqbal, M., Siddiqui, S.N., Sohrab, S.S., Masresha, G., 2017. Effect of indole-3-butyric acid on clonal propagation of mulberry (*Morus alba* L.) stem cuttings: Rooting and associated biochemical changes. *Proc. Natl Acad. Sci. India (Biol. Sci.)* 87, 161–166.

Hyun-Jin, K., Feng Chen, J.C., Xi, W., Yueming, J., Jiang, Y., 2008. Chapter 15. Identification and evalutation of antioxidant phenolic compounds in parsley (*Petroselinum crispum* var. *neapolitanum*) and radish (*Raphanus sativus* L.) sprout". *Funct. Food and Health.* https://doi.org/10.1021/bk-2008-0993.ch015.

Iacobellis, N.S., Lo Cantore, P., Capasso, F., Senatore, F., 2005. Antibacterial activity of *Cuminum cyminum* L. and *Carum carvi* L. essential oils. *J. Agric. Food Chem.* 53(1), 57–61. https://doi.org/10.1021/jf0487351.

Ikram, A.M,, Tahir, Y.F., Nour, S.M., Suliman, M.A., 2020. Traditional use of medicinal plants among the Barti tribe community in Fangoga area, Sennar State, Sudan. *Trop. Plant Res.* 7(2), 517–521. https://doi.org/10.22271/tpr.2020.v7.i2.063.

Innocenti, G., Dall'Acqua, S., Sosa, S., Altinier, G., Della Loggia, R., 2005. Topical anti-inflammatory activity of *Solenostemma argel* leaves. *J. Ethnopharmacol.* 102(2), 307–310. https://doi.org/10.1016/j.jep.2005.06.007.

Iqbal, M., Ahmad, A., Siddiqi, T.O., 2011. Characterization of controversial plant drugs and effect of changing environment on active ingredients. In: Ahmad, A., Siddiqi, T.O. & Iqbal, M. (eds.), *Medicinal Plants in Changing Environment*, 1–10. Capital Publishing Company, New Delhi.

Iqbal, M., Parveen, R., Parveen, A., Parveen, B., Aref, I.M., 2018. Establishing the botanical identity of plant drugs based on their active ingredients under diverse growth conditions. *J. Environ. Biol.* 39(1), 123–136.

Jasiem, T.M., Nasser, N.M., Albazaz, H., 2019. *Tamarix aphylla* L.: A review. *Res. J. Pharm. Technol.* 12(7), 3219–3222. https://doi.org/10.5958/0974-360x.2019.00541.9.

Jeong, C.-H., Bode, A.M., Pugliese, A., Cho, Y.-Y., Kim, H.-G., Shim, J.-H., Jeon, Y.-J., Li, H., Jiang, H., Dong, Z., 2009. [6]-gingerol suppresses colon cancer growth by targeting leukotriene A_4 hydrolase. *Cancer Res.* 69(13), 5584–5591. https://doi.org/10.1158/0008-5472.CAN-09-0491.

Jirovetz, L., Buchbauer, G., Stoilova, I., Stoyanova, A., Krastanov, A., Schmidt, E., 2006. Chemical composition and antioxidant properties of clove leaf essential oil. *J. Agric. Food. Chem.* 54, 6303–6307. https://doi.org/10.1021/jf060608c.

Jirovetz, L., Jaeger, W., Remberg, G., Espinosa-Gonzalez, J., Morales, R., Woidich, A., Nikiforov, A., 1992. Analysis of the volatiles in the seed oil of *Hibiscus sabdariffa* (Malvaceae) by means of GC-MS and GC-FTIR. *J. Agric. Food Chem.* 40(7), 1186–1187. https://doi.org/10.1021/jf00019a021.

Jstor (Journal storage). 2022. Global plants. *Sterculia setigera* Del. [family STERCULIACEAE]. Avialable online https://plants.jstor.org/stable/10.5555/al.ap.upwta.5_293. Accessed on 5/11/2022.

Juteau, F., Masotti, V., Bessière, J.M., Dherbomez, M., Viano, J., 2002. Antibacterial and antioxidant activities of *Artemisia annua* essential oil. *Fitoterapia* 73(6), 532–535. https://doi.org/10.1016/S0367-326X(02)00175-2.

Kapoor, I.P.S., Singh, B., Singh, G., De Heluani, C.S., De Lampasona, M.P., Catalan, C.A.N., 2009. Chemistry and in vitro antioxidant activity of volatile oil and oleoresins of black pepper (Piper nigrum). *J. Agric. Food Chem.* 57(12), 5358–5364. https://doi.org/10.1021/jf900642x.

Kar, S.K., Bera, T.K., 2018. Phytochemical constituents of *Aloe vera* and their multifunctional properties: A comprehensive review. *Int. J. Pharm. Sci. Res.* 9(4), 1416–1423. https://doi.org/10.13040/ijpsr.0975-8232.9(4).1416-23.

Kassie, F., Rabot, S., Uhl, M., Huber, W., Qin, H., Helma, C., Schulte-Hermann, R., Knasmüller, S., 2002. Chemoprotective effects of garden cress (*Lepidium sativum*) and its constituents towards 2-amino-3-methyl-imidazo[4,5-f]quinoline (IQ)-induced genotoxic effects and colonic preneoplastic lesions. *Carcinogenesis* 23(7), 1155–1161. https://doi.org/10.1093/carcin/23.7.1155.

Khadri, A., Serralheiro, M.L.M., Nogueira, J.M.F., Neffati, M., Smiti, S., Araújo, M.E.M., 2008. Antioxidant and antiacetylcholinesterase activities of essential oils from *Cymbopogon schoenanthus* L. Spreng. determination of chemical compositionby GC–mass spectrometry and 13C NMR. *Food Chem.* 109(3), 630–637. https://doi.org/10.1016/j.foodchem.2007.12.070.

Khalid, H., Abdalla, W.E., Abdelgadir, H., Opatz, T., Efferth, T., 2012. Gems from traditional north-African medicine: Medicinal and aromatic plants from Sudan. *Nat. Prod. Bioprospect.* 2(3), 92–103. https://doi.org/10.1007/s13659-012-0015-2.

Khansa A., Yagi, S., Elbashir, A.A., Schohn, H., Tzanova, T., Mohammed, I., Gunes, Ak, Mahomoodally, M.F., Stefanucci, A., Mollica, A., Zengin, G., 2021. Chemical constituents and biological activities of African medicinal tree *Sterculia setigera Delile* stem bark. *S. Afr. J. Bot.* 143, 274–281 https://doi.org/10.1016/j.sajb.2020.10.008. ISSN 0254-6299, Retrieved from: https://www.sciencedirect.com/science/article/pii/S0254629920311194

Khennouf, S., Benabdallah, H., Gharzouli, K., Amira, S., Ito, H., Kim, T.-H., Yoshida, T., Gharzouli, A., 2003. Effect of tannins from *Quercus suber* and *Quercus coccifera* Leaves on ethanol-induced gastric lesions in mice. *J. Agric. Food Chem.* 51(5), 1469–1473. https://doi.org/10.1021/jf020808y.

Klayman, D.L., 1993. *Artemisia annua*, 242–255. https://doi.org/10.1021/bk-1993-0534.ch017.

Kokoska, L., Havlik, J., Valterova, I., Sovova, H., Sajfrtova, M., Jankovska, I., 2008. Comparison of chemical composition and antibacterial activity of *Nigella sativa* seed essential oils obtained by different extraction methods. *J. Food Prot.* 71(12), 2475–2480. https://doi.org/10.4315/0362-028X-71.12.2475.

Kubo, I., Himejima, M., 1991. Anethole, a synergist of polygodial against filamentous microorganisms. *J. Agric. Food Chem.* 39(12), 2290–2292. https://doi.org/10.1021/jf00012a040.

Kubo, I., Fujita, K., Kubo, A., Nihei, K., Ogura, T., 2004. Antibacterial activity of coriander volatile compounds against *Salmonella choleraesuis*. *J. Agric. Food Chem.* 52(11), 3329–3332. PMID: 15161192. doi: 10.1021/jf0354186.

Kumari, R., Manju, S., Mita, K. 2016. Yava (*Hordeum vulgare* Linn.): A review. *Int. Res. J. Pharm.* 7(3), 5–9. http://dx.doi.org/10.7897/2230-8407.07319

Kumari, R., Manju, S., Mita, K., 2016. Yava (*Hordeum vulgare* linn.): A review. *Int. Res. J. Pharm.* 7(3). https://doi.org/10.7897/2230-8407.07319

Kundu, J.K., Na, H.-K., Surh, Y.-J., 2009. Ginger-derived phenolic substances with cancer preventive and therapeutic potential. *Forum Nutr.* 61, 182–192. https://doi.org/10.1159/000212750.

Lako, J.D.W., Sube, K.L.L., Lumori, C.S.G., Yengkopiong, J.P., Utong, J.A.M., Binyason, S.A., Ngerja, Y.S.L., Moilinga, M.K., Lado, T.F., Kheiralla, A.H., 2020. Diversity and distribution of medicinal plants in the republic of South Sudan. *WJARR* 7(1), 018–031. https://doi.org/10.30574/wjarr.2020.7.1.0165.

Lavie, D., Willner, D., Merenlender, Z., 1964. Constituents of *Citrullus colocynthis* (L.) Schrad. *Phytochemistry* 3(1), 51–56. https://doi.org/10.1016/S0031-9422(00)83994-1.

Lawrence, R., Tripathi, P., Jeyakumar, E., 2009. Isolation, purification and evaluation of antibacterial agents from *Aloe vera*. *Braz. J. Microbiol.* 40(4), 15–906.

Lee, S.-J., Umano, K., Shibamoto, T., Lee, K.-G., 2005. Identification of volatile components in basil (*Ocimum basilicum* L.) and thyme leaves (Thymus vulgaris L.) and their antioxidant properties. *Food Chem.* 91(1), 131–137. https://doi.org/10.1016/j.foodchem.2004.05.056.

Mathew, L.S., Mtewab, A.G., Ajayi, C.O., Deyno, S., Weisheit, A., Tolo, C.U., Deng, A.L., Ogwang, P.E., 2020. Ethnopharmacology, pharmacology and phytochemistry of *Aristolochia bracteolata* Lam: A review of an antimalarial plant. *East Afr. Sci.* 2020(2).

Miura, K., Kikuzaki, H., Nakatani, N., 2002. Antioxidant activity of chemical components from sage (*Salvia officinalis* L.) and thyme (*Thymus vulgaris* L.) measured by the oil stability index method. *J. Agric. Food Chem.* 50(7), 1845–1851. https://doi.org/10.1021/jf011314o.

Mohaddeseh, M., Fereshteh, S., Mohammad, M.F., 2006. Bactericidal effects of essential oils from clove, lavender and geranium on multi-drug resistant isolates of Pseudomonas aeruginosa. *Iran. J. Biotechnol.* 4(2), 137–140.

Mosa, A.E.O., Abdelgadir, A.A., Ahmed, E.M., 2021. Traditional use of medicinal plants in Central Sudan. *AJMAP* 7(1). https://doi.org/10.48347/imist.prsm/ajmap-v7i1.22273.

Musa, H.H., Ahmed, A.A., Musa, T.H., 2018. Chemistry, biological, and pharmacological properties of gum arabic. In: Mérillon, J.M., Ramawat, K. (eds.), *Bioactive Molecules in Food. Reference Series in Phytochemistry.* Springer, Cham. https://doi.org/10.1007/978-3-319-54528-8_11-1.

Musa, H. A. A. 2009. Quality of medicinal plants traditionally used in Sudan as affected by ionizing radiation treatments. PhD thesis. Available from INIS: http://inis.iaea.org/search/search.aspx?orig_q=RN:42016217; Available from INIS in electronic form. Also available from Sudan Atomic Energy Commission, Khartoum (SD).

Nakahara, K., Roy, M.K., Ono, H., Maeda, I., Ohnishi-Kameyama, M., Yoshida, M., Trakoontivakorn, G., 2003. Prenylated flavanones isolated from flowers of *Azadirachta indica* (the Neem Tree) as antimutagenic constituents against heterocyclic amines. *J. Agric. Food Chem.* 51(22), 6456–6460. https://doi.org/10.1021/jf034666z.

Nandhini, D.U., Rajasekar, M., Venmathi, T., 2017. A review on worm killer: *Aristolochia bracteolate. J. Pharmacogn. Phytochem.* 6(2), 6–9E-ISSN: 2278-4136, P-ISSN: 2349-8234. Available online: phytojournal. Com. Accessed on 5/11/2022..

Nanjundaiah, S.M., Annaiah, H.N.M., Dharmesh, S.M., 2011. Gastroprotective effect of ginger rhizome (*Zingiber officinale*) extract: Role of gallic acid and cinnamic acid in H^+, K^+ -ATPase/*H. pylori* inhibition and anti-oxidative mechanism. *Evid. Based Complement. Alternat. Med.* 2011, 1–13. https://doi.org/10.1093/ecam/nep060.

Neelkamal, C., 2009. Evaluation of hypolipidaemic activity of *Capparis decidua. Int. J. Biomed. Sci.* 5(1), 70–73.

Nigam, N., Bhui, K., Prasad, S., George, J., Shukla, Y., 2009. [6]-gingerol induces reactive oxygen species regulated mitochondrial cell death pathway in human epidermoid carcinoma A431 cells. *Chem. Biol.* 181(1), 77–84. https://doi.org/10.1016/j.cbi.2009.05.012.

Noori, S., Naderi, G.-A., Hassan, Z.M., Habibi, Z., Bathaie, S.Z., Hashemi, S.M.M., 2004. Immunosuppressive activity of a molecule isolated from *Artemisia annua* on DTH responses compared with cyclosporin A. *Int. Immunopharmacol.* 4(10–11), 1301–1306. https://doi.org/10.1016/j.intimp.2004.05.003.

Ochi, T., Shibata, H., Higuti, T., Kodama, K., Kusumi, T., Takaishi, Y., 2005. Anti- *Helicobacterpylori* compounds from *Santalumalbum. J. Nat. Prod.* 68(6), 819–824. https://doi.org/10.1021/np040188q.

Okpako, L.C., Ajaiyeoba, E.O., 2004. *In vitro* and *in vivo* antimalarial studies of *Striga hermonthica* and *Tapinanthus sessilifolius* extracts. *Afr. J. Med. Med. Sci.* 33(1), 5–73.

Oliva, A., Meepagala, K.M., Wedge, D.E., Harries, D., Hale, A.L., Aliotta, G., Duke, S.O., 2003. Natural fungicides from *Ruta graveolens* L. Leaves, including a new quinolone alkaloid. *J. Agric. Food Chem.* 51(4), 890–896. https://doi.org/10.1021/jf0259361.

Osman, A.M.E., Taj Eldin, I.M., Elhag, A.M., Elhassan, M.M.A., Ahmed, E.M.M., 2020. *In-vitro* anticancer and cytotoxic activity of ginger extract on human breast cell lines. *Khartoum J. Pharm. Sci.* 1(1), 26–29.

Parejo, I., Viladomat, F., Bastida, J., Schmeda-Hirschmann, G., Burillo, J., Codina, C., 2004. Bioguided isolation and identification of the nonvolatile antioxidant compounds from fennel (*Foeniculum vulgare* Mill.) waste. *J. Agric. Food Chem.* 52(7), 1890–1897. https://doi.org/10.1021/jf030717g.

Pari, L., Sankaranarayanan, C., 2009. Beneficial effects of thymoquinone on hepatic key enzymes in streptozotocin–nicotinamide induced diabetic rats. *Life Sci.* 85(23–26), 830–834. https://doi.org/10.1016/j.lfs.2009.10.021.

Park, M., Bae, J., Lee, D.-S., 2008. Antibacterial activity of [10]-gingerol and [12]-gingerol isolated from ginger rhizome against periodontal bacteria. *Phytother. Res.* 22(11), 1446–1449. https://doi.org/10.1002/ptr.2473.

Parveen, A., Ahmad, M., Parveen, B., Parveen, R., Iqbal, M., 2022. Unani system of medicine, its origin, evolution and Indianization: A critical appraisal. *Indian J. Trad. Know.* In Press.

Parveen, A., Parveen, R., Akhatar, A., Parveen, B., Siddiqui, K.M., Iqbal, M., 2020. Concepts and quality considerations in Unani system of medicine. *J. AOAC Int.* 103(3), 609–633.

Parveen, B., Parveen, A., Parveen, R., Ahmad, S., Ahmad, M., Iqbal, M., 2020. Challenges and opportunities for traditional herbal medicine today, with special reference to its status in India. *Ann. Phytomed.* 9(2), 97–112.

Picerno, P., Autore, G., Marzocco, S., Meloni, M., Sanogo, R., Aquino, R.P., 2005. Anti-inflammatory activity of Verminoside from *Kigeliaafricana* and evaluation of cutaneous irritation in cell cultures and reconstituted human epidermis. *J. Nat. Prod.* 68(11), 1610–1614. https://doi.org/10.1021/np058046z.

Picman, A.K., Arnason, J.T., Lambert, J.D.H., 1986. Hymenin, another sesquiterpene lactone in *Ambrosia maritima. J. Nat. Prod.* 49(3), 556–556. https://doi.org/10.1021/np50045a044.

Pl@nt use, 2020. *Detarium microcarpum (PROTA).* Retrieved from https://uses.plantnet-project.org/en/Haplophyllum_tuberculatum_(PROTA). Accessed online in 13/12/2020.

Pl@nt use, 2022. *Terminalia brownii* (PROTA). Retrived from https://uses.plantnet-project.org/en/Terminalia_brownii_(PROTA). Accessed oline in 4/11/2022.

PROTA4U, 2020. *Tinospora bakis* (A. Rich.) Miers. Retrieved from https://prota4u.org/database/protav8.asp?g=pe&p=Tinospora+bakis+(A.Rich.)+Miers. Accessed online in 13/12/2020.

Qiu-Ping, W., Zhu-Ju, W., Mei-Hong, F., Qiang-Feng, G., 2007. Chemical constituents from the leaves of *Cassia angustifolia. Zhong Yao Cai* 30(10), 1250–1252.

Ragheb, A., Attia, A., Eldin, W., Elbarbry, F., Gazarin, S., Shoker, A., 2009. The protective effect of thymoquinone, an anti-oxidant and anti-inflammatory agent, against renal injury: A review. *Saudi J. Kidney Dis. Transpl.* 20(5), 741–752.

Rahman, A.-U., Choudhary, M., Farooq, A., Ahmed, A., Iqbal, M., Demirci, B., Demirci, F., Can Baser, K., 1999. Antifungal activities and essential oil constituents of some spices from Pakistan. In: *Proceedings of the 3rd International Electronic Conference on Synthetic Organic Chemistry*, 1773. https://doi.org/10.3390/ecsoc-3-01773.

Ribeiro Paulo, R., de Castro, R.D., Fernandez, L.G., 2016. Chemical constituents of the oilseed crop *Ricinus communis* and their pharmacological activities: A review. *Ind. Crops Prod.* 91, 358–376. https://doi.org/10.1016/j.indcrop.2016.07.010.

Rosa, R.M., Melecchi, M.I.S., Abad, F.C., Simoni, C.R., Caramão, E.B., Henriques, J.A.P., Saffi, J., de Paula Ramos, A.L.L., 2006. Antioxidant and antimutagenic properties of *Hibiscus* tiliaceus L. methanolic extract. *J. Agric. Food Chem.* 54(19), 7324–7330. https://doi.org/10.1021/jf061407b.

Saba, Iqbal, M. & Srivastava, P.S., 1999a. Three new phytoconstituents from *Lepidium sativum. Die Pharmazie* 51, 851–853.

Saba, Iqbal, M. & Srivastava, P.S., 1999b. Detection of hyoscyamine in *Lepidium sativum. Hamdard Med.* 42, 72–75.

Saba, Mughal, M.H., Ali, M., Iqbal, M., Srivastava, P.S., 1999c. A steryl ester from *Lepidium sativum. Phytochemistry* 50, 1375–1377.

Saba, Pande, D., Iqbal, M., Srivastava, P.S. 2000. Effect of $ZnSO_4$ and $CuSO_4$ on regeneration and lepidine content in *Lepidium sativum* L. *Biol. Plant.* 43(2), 253–256.

SatheeshKumar, N., Mukherjee, P.K., Bhadra, S., Saha, B.P., 2010. Acetylcholinesterase enzyme inhibitory potential of standardized extract of *Trigonella foenum graecum* L and its constituents. *Phytomedicine* 17(3–4), 292–295. https://doi.org/10.1016/j.phymed.2009.06.006.

Sang, S., Hong, J., Wu, H., Liu, J., Yang, C.S., Pan, M.-H., Badmaev, V., Ho, C.-T., 2009. Increased growth inhibitory effects on human cancer cells and anti-inflammatory potency of Shogaols from *Zingiber officinale* relative to gingerols. *J. Agric. Food. Chem.* 57, 10645–10650. https://doi.org/10.1021/jf9027443.

Sicilia, T., Niemeyer, H.B., Honig, D.M., Metzler, M., 2003a. Identification and stereochemical characterization of lignans in flaxseed and pumpkin seeds. *J. Agric. Food Chem.* 51(5), 1181–1188. https://doi.org/10.1021/jf0207979.

Sicilia, T., Niemeyer, H.B., Honig, D.M., Metzler, M., 2003b. Identification and stereochemical characterization of lignans in flaxseed and pumpkin seeds. *J. Agric. Food Chem.* 51(5), 1181–1188. https://doi.org/10.1021/jf0207979.

Silva, O., Gomes, E.T., 2003. Guieranone A, a naphthyl butenone from the leaves of *Guiera senegalensis* with antifungal activity. *J. Nat. Prod.* 66(3), 447–449. https://doi.org/10.1021/np0204904.

Sonaina, N., Hussain, M.A., Khan, A., Muhammad, G., Tahir, M.N., 2020. *Capparis decidua* Edgew (Forssk.): A comprehensive review of its traditional uses, phytochemistry, pharmacology and nutrapharmaceutical potential. *Arab. J. Chem.* 13(1), 1901–1916.ISSN 1878-5352, https://doi.org/10.1016/j.arabjc.2018.02.007 (https://www.sciencedirect.com/science/article/pii/S1878535218300388)

Soni, P., Siddiqui, A.A., Dwivedi, J., Soni, V., 2012. Pharmacological properties of *Datura stramonium* L. as a potential medicinal tree: An overview. *Asian Pac. J. Trop. Biomed.* 2(12), 1002–1008. https://doi.org/10.1016/s2221-1691(13)60014-3.

Srivastava, J.K., Gupta, S., 2007. Antiproliferative and apoptotic effects of chamomile extract in various human cancer cells. *J. Agric. Food Chem.* 55(23), 9470–9478. https://doi.org/10.1021/jf071953k.

Stevenson, D.G., Eller, F.J., Wang, L., Jane, J.-L., Wang, T., Inglett, G.E., 2007. Oil and tocopherol content and composition of pumpkin seed oil in 12 cultivars. *J. Agric. Food. Chem.* 55(10), 4005–4013. https://doi.org/10.1021/jf0706979.

Suliman, E., Zahra, A., Zidan, N.S., Foudah, S.H.I., 2018. Chemical compositions, antioxidant, and nutritional properties of the food products of Guddaim (*Grewia tenax*). *Int. J. Pharm. Res. Allied Sci.* 7(3), 172–182. Available online www.ijpras.com.

Takaoka, D., 1976. Sesquiterpene alcohols in camphor oil. *Phytochemistry* 15(3), 425–426.

Thebtaranonth, C., Thebtaranonth, Y., Wanauppathamkul, S., Yuthavong, Y., 1995. Antimalarial sesquiterpenes from tubers of Cyperus rotundus: Structure of 10,12-Peroxycalamenene, a sesquiterpene endoperoxide. *Phytochemistry* 40(1), 125–128. https://doi.org/10.1016/0031-9422(95)00260-E.

Tragoolpua, Y., Jatisatienr, A., 2007. Anti-herpes simplex virus activities of *Eugenia caryophyllus* (Spreng.) Bullock & S.G. Harrison and essential oil, eugenol. *Phytother. Res.* 21(12), 1153–1158. https://doi.org/10.1002/ptr.2226.

Tripathi, R.D., Srivastava, H.S., Dixit, S.N., 1978. A fungitoxic principle from the leaves of *Lawsonia inermis* Lam. *Experientia* 34(1), 51–52. https://doi.org/10.1007/BF01921895.

Tripathi, S., Maiti, T.K., 2005. Immunomodulatory role of native and heat denatured agglutinin from *Abrus precatorius. Int. J. Biochem. Cell Biol.* 37(2), 451–462. https://doi.org/10.1016/j.biocel.2004.07.015.

Tsuda, T., Watanabe, M., Ohshima, K., Yamamoto, A., Kawakishi, S., Osawa, T., 1994. Antioxidative components isolated from the seed of tamarind (*Tamarindus indica* L.). *J. Agric. Food Chem.* 42(12), 2671–2674. https://doi.org/10.1021/jf00048a004.

University of Ioannina, School of Health Science, Faculty of Medicine, Department of Pharmacology, 2020. *Prunus mahaleb.* Retrieved from: http://mediplantepirus.med.uoi.gr/pharmacology_en/plant_details.php?id=342.

Verma, R.B., Mahmooduzzafar, Siddiqi, T.O., Iqbal, M., 2006. Foliar response of *Ipomea pes-tigridis* L. to coal-smoke pollution. *Turk. J. Bot.* 30, 413–417.

Wallis, T.E., 2002. *Textbook of Pharmacognosy*, Churchul Ltd, London.

Wang, J., Zu, X., Jiang, Y., 2009. Five furostanol saponins from fruits of *Tribulus terrestris* and their cytotoxic activities. *Nat. Prod. Res.* 23(15), 1436–1444. https://doi.org/10.1080/14786410902940990.

Wang, M., Kikuzaki, H., Zhu, N., Sang, S., Nakatani, N., Ho, C.-T., 2000. Isolation and structural elucidation of two new glycosides from sage (*Salvia officinalis* L.). *J. Agric. Food Chem.* 48(2), 235–238. https://doi.org/10.1021/jf99076.

Warrag, N.M., Tag Eldin, I.M., Ahmed, E.M., 2014. Effect of *Cymbopgon proximus* (Mahareb) on ethylene glycol-induced nephrolithiasis in rats. *Afr. J. Pharm. Pharmacol.* 8(17), 443–450.

Williams, P.A., Phillips, G.O., 2009. *Handbook of Hydrocolloids.* 2nd edition, Woodhead Publishing, Cambridge, UK.

Woldemariam, T.Z., Fell, A.F., Linley, P.A., Bibby, M.C., Phillips, R.M., 1992. Evaluation of the anti-tumour action and acute toxicity of kosins from *Hagenia abyssinica. J. Pharmaceut. Biomed.* 10(8), 555–560. https://doi.org/10.1016/0731-7085(92)80080-7.

World Health Organization. Programme on Traditional Medicine, 2002. *WHO Traditional Medicine Strategy 2002–2005*, World Health Organization, Geneva, Switzerland.

Yang, Y.-C., Choi, H.-Y., Choi, W.-S., Clark, J.M., Ahn, Y.-J., 2004. Ovicidal and adulticidal activity of *Eucalyptus globulus* Leaf oil terpenoids against *Pediculus humanus capitis* (Anoplura: Pediculidae). *J. Agric. Food Chem.* 52(9), 2507–2511. https://doi.org/10.1021/jf0354803.

Yazzie, D., VanderJagt, D.J., Pastuszyn, A., Okolo, A., Glew, R.H., 1994. The amino acid and mineral content of baobab (*Adansonia digitata* L.) leaves. *J. Food Compos. Anal.* 7(3), 189–193. https://doi.org/10.1006/jfca.1994.1018.

Yoshikawa, M., Morikawa, T., Kobayashi, H., Nakamura, A., Matsuhira, K., Nakamura, S., Matsuda, H., 2007. Bioactive saponins and glycosides. XXVII. Structures of new cucurbitane-type triterpene glycosides and antiallergic constituents from *Citrullus colocynthis. Chem. Pharm. Bull. (Tokyo)* 55(3), 428–434. https://doi.org/10.1248/cpb.55.428.

Yousuf, P.Y., Ganie, A.H., Khan, I., Qureshi, M.I., Ibrahim, M.M., Sarwat, M., Iqbal, M., Ahmad, A., 2016. Nitrogen-efficient and nitrogen-inefficient Indian mustard showed differential expression pattern in response to elevated CO_2 and low nitrogen. *Front. Plant Sci.* 7, 1074. https://doi.org/3389/fpls.2016.01074.

6 Rhus coriaria
A Review of Phytochemical and Biopharmaceutical Properties

Süleyman Toprak and Cemil İşlek

CONTENTS

6.1 INTRODUCTION

Like macronutrients, which are necessary for the survival of all living things, micronutrients are also essential for many important biochemical processes, such as processing and storing energy, and signal transmission. Furthermore, just as the deficiency or excess of macronutrients affects the body's normal functioning, imbalances in micronutrients can also lead to system disruptions (Savarino et al., 2021). In addition, secondary metabolites, the importance of which is just beginning to be understood, also show regulatory effects on the functioning of body systems. After industrialization, many diseases such as deficient, inappropriate or unbalanced nutrition, hypertension, obesity, neurodegenerative, and systemic illnesses became more common. Although the foods that we currently eat may make us sick, it is also food that will help us alleviate the same diseases. Plants are important here because they have significant potential regulatory and beneficial effects and are an important source of minerals, vitamins, and secondary metabolites that are required in our daily diet. Several studies have revealed the regulatory, balancing, and healing properties of plants used as seasonings and spices in world cuisines (Gottardi et al., 2016; Dini, 2018), where sumac is an important component.

Sumac (*Rhus coriaria* L., of the family Anacardiaceae) grows to ~3 m in height, with imparipinnate leaves and densely branched bunches of 15–20-cm-long flowers, producing its fruit in groups, hairy on top, with the color turning from green to red or red-brown when mature. It is a shrubby tree, mentioned by Dioscorides and İbn-iSina, and used medicinally to treat some chronic diseases, such as strokes, and has diuretic and gas-reducing effects (Davis, 1997; Mamıkoğlu, 2015; Tanker, 2007). The clustered fruits and leaves of the sumac plant are shown in Figure 6.1 and the dried, ground spice

DOI: 10.1201/9781003226925-6

FIGURE 6.1 The clustered fruits and leaves of the sumac plant (flickr.com).

FIGURE 6.2 Dried and ground fruits of the sumac plant (flickr.com).

form is shown in Figure 6.2. The plant, used in meals as sumac syrup which is obtained by straining the fruits soaked in water, can also be consumed by adding the dried fruits to table salt (Abu-Reidah, 2014; Tiryaki, 2010). The natural environment of the sumac shrub or small tree, which is also called somalık, tit, tetri, tutku, tatari, and somak, is shrubland, rocky soils, and coastal locations. The extract obtained from the roots and leaves by the decoction method is used to treat stomach conditions, and the extract obtained from the ripe fruits is used to reduce kidney stones. Fresh leaves are put on the soles of the shoes to remove cracks in the skin of the feet. Chewing fruits in gum is used to treat stomach cramps and mouth sores (Güner, 2012; Tuzlacı, 2006; Tuzlacı, 2011). Regular use also has a protective effect on atherosclerosis, oxidative stress, and on liver enzymes associated with a high-fat diet (Setorki et al., 2012). Sumac is also used to treat conjunctivitis, dysentery, liver diseases, and anorexia, and hair and skin problems such as burns and dermatitis (Ali-Shtayeh et al., 2013). In Azerbaijan, the sumac plant has been used to treat diabetes and hypertension and is effective in treating constipation due to the laxative effect of its fruits (Hasanova, 2000).

6.2 GENERAL COMPOSITION, MINERAL AND VITAMIN CONTENT OF *R. CORIARIA*

Dried sumac fruit contains 6–11.8% water, 1.0% essential oil, 2.3–2.6% protein, 14.6–22.5% fiber, and 1.50–2.66% waste material. The water-soluble extract accounts for 63.8% and the fixed oil 17.4% (Kizil and Turk, 2010, Alsamri et al., 2021). Analysis by inductively coupled plasma–atomic emission spectroscopy shows that the mineral composition of sumac fruits includes K, Ca, Mg, P, Fe, Na, Zn, Mn, Cu, and Al as dominant elements (Özcan, 2003; Raut and Karuppayil, 2014). The fruit of the sumac plant contains 49.7 ppm zinc (Özcan, 2004). The fruit pericarp of sumac from the Mersin region contained 3,577.46 ppm calcium and 781.16 ppm phosphorus. Sumac pericarp from the Hatay region contained significant amounts of minerals, such as magnesium (700.07 ppm). Pericarp from the Kastamonu region also contained higher concentrations of iron than sumac from other regions (610.99 ppm). When the cores of sumac fruits from the İskenderun region were investigated, copper, aluminum, and iron were found in trace amounts; however, there was a high concentration of calcium (1,062.14 ppm) (Özcan, 2007). The data collected from these studies have shown that the mineral composition of sumac fruits may vary according to geographical regions and environmental factors (Kossah et al., 2009; Alsamri et al., 2021).

6.2.1 Fruits: Main Source of Organic Chemical Components of R. coriaria

The organic chemical content of *R. coriaria* fruits has previously been determined in several studies. The metabolites detected included polyphenols (flavonols, phenolic acids, and anthocyanins) and organic acids (malic, citric, fumaric, and tartaric acids) (Mavlyanov et al., 1997). Anthocyanins, pentagaloyl glucose and phenolics (such as gallic acid) were found by researchers (Koşar et al., 2007). In another study, malic, citric, fumaric, and tartaric acids, and fatty acids (mainly oleic, linoleic, and palmitic acids) were identified (Kossah et al., 2009). Likewise, other investigators reported oleic, linoleic, palmitic, and stearic acids (Kizil and Turk, 2010). For the first time, Shabana et al. (2011) reported protocatechuic acid and isoquercitrin in addition to previously found compounds (phenol acids, flavonoids [including kaempferol, and quercetin], gallic acid, and methyl gallate) -L-rhamnoside. Dalar and co-workers reported the presence of gallic acid, volatiles (β-pinene, octanal, limonene, nonanal, β-caryophyllene, and humulene), and fatty acids (oleic, palmitic, and linoleic acids) (Dalar et al., 2018). In another study, alkaloids, tannins, saponins, terpenoids, flavonoids, and polyphenols were found in sumac fruits by researchers (Batiha et al., 2020).

6.2.2 Leaves: Another Significant Chemical Resource From R. coriaria

Many researchers have also examined the metabolite composition of the leaves of *R. coriaria*, which contain gallic acid (Perkin and Allen, 1896); the flavonols quercetin, myricetin, and kaempferol (El-Sissi et al., 1966); polyphenolic components of gallotannins (El-Sissi et al., 1971); methyl gallate (El-Sissi et al., 1972); polyisoprenoids consisting of 10 to 13 C5 isoprene units, polyprenol homologs and undecaprenols as the dominant prenols (Mamatkulova et al., 2012);linoleic acid and tridecaprenol ester (Mamatkulova et al., 2014); and 7-methyl-cyanidin-3-galactoside and gallic acid (Romeo et al., 2015).

6.3 PHARMACOLOGICAL EFFECTS OF *R. CORIARIA*

Many health problems, especially oxidative stress, cardiovascular diseases, diabetes, obesity, and cancers, arise due to environmental factors, stress, malnutrition, and a sedentary lifestyle. Despite the progresses made by modern medicine and the availability of effective treatments, the number of people suffering from these issues is rapidly increasing. The vitamins, minerals, and phytochemicals contained in plants on ingestion play a regulatory role in our body. *R. coriaria*, a pharmacologically valuable herb, has many proven antimicrobial, antidiabetic, antioxidant, anticancer, cardioprotective, and anti-inflammatory properties.

6.3.1 Antimicrobial Effects

Like with other therapeutic plants, there have been many studies into sumac's antimicrobial effects toward bacteria, viruses, and fungi. Although, these studies have yielded striking results, it is clear that there is still a long way to go in future research. Some results on antibacterial, antiviral, and antifungal effects have been gathered under these headings.

6.3.1.1 Antibacterial Effects

Many studies have shown that exposure to *R. coriaria* results in positive health effects due to detrimental effects on pathogenic bacteria. Ripe sumac fruits have greater antibacterial effects than those of unripe fruits (Ali-Shtayeh et al., 2013). Sumac extracts exhibited antibacterial activity against *Enterococcus faecalis*, *Pseudomonas aeruginosa*, *Acinetobacter baumannii*, and *Staphylococcus aureus*. In addition, the development of promastigotes and amastigotes was inhibited, with IC_{50} values (half-maximal inhibitory concentration) of 147 µg/mL and 233 µg/mL, respectively (Ashoori et al., 2020). The antibacterial properties of ethanolic sumac extracts are generally more effective against Gram-positive bacteria, such as *Bacillus cereus*, *Bacillus megaterium*, *Bacillus subtilis* (ATCC 6633), *Bacillus thuringiensis,* and *S. aureus* (ATCC25923), than against Gram-negative bacteria, such as *Escherichia coli* (ATCC2592) (Ertürk, 2010; Fazeli et al., 2007; Nasar-Abbas and Halkman, 2004). Another study on three Gram-negative and three Gram-positive bacterial species showed that *B. subtilis* was the most sensitive of the Gram-positive bacteria tested with a MIC (minimum inhibitory concentration) of 0.5 µg/mL. The MIC values of sumac toward different Gram-negative bacterial species were similar, with the effective range being between 10 and 20 µg/mL (Raodah, 2014). Ethanolic sumac extracts significantly reduced the bacterial population in minced meat compared with the control group (Radmehr and Abdolrahimzade, 2009). In a study in which the size of bacterial growth inhibition zones around the test disks were measured by the disk diffusion method, the antibacterial activity of the various fractions of the sumac extract was tested. It was revealed that *S. aureus* and *E. coli* were strongly inhibited in the ethyl acetate fraction (Ahmadian-Attari et al., 2016). In a study where sumac fruit extract was used as a stabilizing agent for the synthesis of copper nanoparticles, it was reported that the nanoparticles obtained showed good antibacterial activity against *P. aeruginosa*, *E. coli*, *S. aureus*, and *Bacillus cereus* (Ismail, 2020). Concentrations of 5 mg/mL and 10 mg/mL of aqueous extracts of sumac fruit were tested on various bacterial species. Compared with *E. coli*, *Proteus vulgaris*, and *Shigella* spp., the highest antibacterial effects were shown on *S. aureus, S. aureus* (methicillin-resistant, MRSA), *Streptococcus aureus*, and *P. aeruginosa* (Gabr and Alghadir, 2019). It has been determined that methyl gallate obtained from the methanolic extract of sumac fruit is the most effective bioactive component against *Streptococcus mutans*, which is a bacterium that forms a biofilm. *R. coriaria* extracts applied to *S. mutans* inhibited biofilm formation by 77% at 6 mg/mL. It has been reported that application of a sumac extract reduced the *S. mutans* biofilm biomass on a polystyrene surface by 68–93% and inhibited the decrease in pH level by 97% (Kacergius et al., 2017).

6.3.1.2 Antiviral Effects

In a study evaluating their antiviral activities, it was observed that bioflavonoids isolated from sumac could inhibit respiratory tract viruses (influenza A, influenza B, and measles) and herpes viruses (HSV-1, HSV-2 and varicella zoster virus [VZV]) (Lin et al., 1999). Significant inhibition of the fish-pathogenic infectious hematopoietic necrosis virus and viral hemorrhagic septicemia virus was observed following exposure to the sumac extract (Kang et al., 2012). It has been found that urushiol in the sumac plant exhibits reverse transcriptase inhibitory activity on human immunodeficiency virus type 1 (HIV-1) (Kadokura et al., 2015). It has been reported that the sumac extract, which has a strong antiviral effect against HSV-1 and HSV-2, affects the viral envelope and impairs absorption (Reichling et al., 2009). *R. coriaria* fruit extract has antiviral activity against herpes simplex virus 1, the most common cause of herpes infections (Parsania et al., 2017).

6.3.1.3 Antifungal Effects

Three new aromatic phytochemicals (coriarianaphthyl ether, coriariaoic acid, and coriariantracenyl ester) were extracted with ethanol from sumac seeds. When all three were applied separately to *Aspergillus flavus* (MTCC-277), *Candida albicans* (MTCC3958), and *Penicillium citrinum* (MTCC-3395), the coriariantracenyl ester had an antifungal effect against three fungi at a concentration of 25 μg/mL (Onkar et al., 2011). *R. coriaria* could also be used to control tomato anthracnose on tomato plants and fruits caused by *Colletotrichum acutatum*. An aqueous extract of sumac fruits also revealed significant antifungal activity against tomato anthracnose caused by *C. acutatum*. This suggested that *R. coriaria* could be a cost-effective and environmentally friendly alternative to chemical fungicides in the treatment of tomato anthracnose disease (Najjar et al., 2017; Alsamri et al., 2021). Significant antifungal activity of a sumac methanolic extract (19 μg/mL) was observed against four fungal species (*Microsporum canis*, *Microsporum gypseum*, *Trichophyton equinum* and *Trichophyton mentagrophyte*) that cause dermatophytosis in humans and animals. In addition, Arabian horses with dermatophytosis treated with a cream containing 5% sumac every day for 10 days were completely cured of this infection by 28 days after the treatment (Sayed-Ahmed et al., 2019).

6.3.2 Antidiabetic Effects

Poor diets and lifestyles have increased the incidence of type 2 diabetes globally. Although our bodies can tolerate diabetes to a certain degree, dysfunction of the hormone-producing organs becomes inevitable as the conditions worsen. The sumac plant can have a marked pharmacological effect here. For example, sumac fruit prepared at different concentrations and using different extraction solvents (methanol, ethyl acetate or n-hexane) has shown an inhibitory effect on the glycoside hydrolase enzyme (α-amylase), whose function is to achieve hypoglycemic activity. Extracts prepared with ethyl acetate at 250 mg/mL showed the highest rate of inhibition at $93.8 \pm 0.01\%$. Thus, ethyl acetate extract can increase glucose tolerance (Giancarlo et al., 2006). Daily intake of 6 g sumac powder by patients with type 2 diabetes reduced fasting serum insulin levels and insulin resistance (Ardakani et al., 2016). In an *in-vivo* study on streptozotocin-induced diabetic rats, a hypoglycemic effect of lyophilized sumac fruit was observed. In this study, extracts at one of three concentrations (250, 500 and 1000 mg/kg body weight) were administered orally to diabetic rats for 21 days and some parameters were monitored. Examination of the collated data indicated that serum glucose and glycosylated HbA1c levels decreased, insulin levels increased, C-peptide levels did not change, and α-glycosidase enzyme activity was significantly inhibited (Doğan and Çelik, 2016). In addition, the PTP1B enzyme, which is diabetic by nature, was inhibited by sumac, preventing it from reducing insulin sensitivity (Baumgartner et al., 2010). *R. coriaria* induces an antidiabetic effect through various mechanisms. *R. coriaria* extracts significantly reduced blood glucose and cholesterol levels (Ahangarpour et al., 2017). Diet supplemented with *R. coriaria* given to women with type 2 diabetes increased total antioxidant capacity. A significant decrease in levels of serum glucose, HbA1c, and Aapo-B and a substantial increase in total antioxidant capacity were observed in patients with type 2 diabetes who consumed 3.0 g of sumac powder daily for three months (Shidfar et al., 2014). The result of all these studies revealed that sumac fruit can be used as an herbal drug for the treatment of type 2 diabetes and its complications.

6.3.3 Antioxidant Properties and Anticancer Effects

Thanks to its free radical inhibition property, an ethyl acetate extract of sumac fruits gave effective results even in response to butylatedhydroxyanisole (BHA), butylatedhydroxytoluene (BHT), and rosemary extract. Free radical scavenging activities of ethyl acetate fractions of sumac were observed to increase from 32.93% to 93.44% (Bozan et al., 2002). The methanolic extract of sumac fruits prevented hydroperoxide formation in peanut oil at 1.0, 3.0, and 5.0% (w/v) concentrations and increased oxidative stress tolerance (Ozcan, 2003). Aqueous sumac extracts showed an effective radical scavenging effect for DPPH (1,1-diphenyl-2-picryl-hydrazyl) free radicals with an EC_{50} (half-maximal effective concentration) of 36.4 µg/mL and DMPD (*N,N*-dimethyl-*p*-phenylenediamine) cation radicals with an EC_{50} of 44.7 µg/mL (Bursal and Köksal, 2011). Daily oral administration of 2 mL/kg bodyweight of sumac to rat pups with induced necrotizing enterocolitis prevented intestinal tissue damage which was caused by increased free radical scavenging activity, TNF-α (tumor necrosis factor-alpha) and IL-6 (Interleukin 6) (Isik et al., 2019). *R. coriaria* and nanoparticles synthesized with *R. coriaria* showed a significant antioxidant capacity as determined by the ABTS and DPPH tests (Ibrahim et al., 2019). Angiogenesis involves vessel formation by vascular endothelial growth factor (VEGF) and other inducer biochemicals, so that preventing angiogenesis is among the most critical cancer treatment strategies (Rajabi and Mausa, 2017). In a study examining the cytotoxicity and antiangiogenic properties of sumac extract on umbilical vein endothelial cells (HUVEC) and retinoblastoma cells, it was determined that the cytotoxic and antiangiogenic effects of the extract increased, depending on the effective concentration. Sumac extract's inhibitory effect on cancer cells and its negligible cytotoxic impact on healthy tissues have been found to aid in the treatment of cancer and metastasis as a result of these two variables (Mirian et al., 2015). Sumac contains gallotannins, and gallotannins strongly inhibit the JAK/STAT pathway in colon cancer, showing that it has anticancer properties (Houssein, 2020). *R. coriaria* exhibited anti-breast cancer activity by suppressing metastasis, angiogenesis, and tumor growth by inhibiting the STAT3, NFκB, and nitric oxide pathways (El Hasasna et al., 2016). *R. coriaria* inhibited cell viability of breast cancer cells, namely lines MDA-MB-231 (305 µg/mL IC_{50} at 48 hours), T47D (261 µg/mL IC_{50} at 48 hours), and MCF-7 (510 µg/mL IC_{50} at 48 hours). In addition, sumac extract induced senescence and autophagy by activating p38 and ERK1/2 pathways in triple-negative breast cancer cells (El Hasasna et al., 2015). A study was conducted to determine whether sumac extract (RCE) has an effect on cell viability, using the MTS (a type of cell viability assay) test. After 48 hours of treatment with RCE, a 30–65% reduction in the viability of cervical cancer cells (HeLa) was observed, in direct proportion to increasing concentration. A significant reduction of approximately 15–35% on the migration of HeLa cells by RCE was also reported after 24 hours of incubation. Based on both results, it can be said that RCE has an inhibitory and anti-migration effect on uterine cervical cancer cells (Abdallah et al., 2019).

6.3.4 Cardioprotective Effects

Herbal medicines and alternative medicines have been used to manage cardiovascular risk factors. It is well known that *R. coriaria* is rich in bioactive compounds which can improve cardiovascular health. It was recently reported that the ethanolic extract of *R. coriaria* fruits relaxes isolated rat aorta in a dose-dependent manner (Anwar et al., 2018). It has also been reported that gallotannin derivatives in aqueous and methanolic extracts of *R. coriaria* fruits reduce blood urea nitrogen and blood pressure (Regazzoni et al., 2013).The ability of sumac, which contains many polyphenols with high resin-binding capacity, to reduce lipid absorption in the gastrointestinal tract is the source of its potential in lowering blood lipids (Beretta et al., 2009). In previous studies, which highlighted its serum-cholesterol-lowering effect, sumac showed a high inhibitory effect on xanthine oxidase activity (Candan, 2003). Capsules prepared with powdered sumac fruits were given to 72 dyslipidemic patients for 1 month as a hypolipidemic agent and the changes in serum total-C (total cholesterol), LDL-C (low-density lipoprotein–cholesterol), HDL-C (high-density lipoprotein–cholesterol), and TG (triglyceride) levels were monitored. Whereas a significant decrease was observed in total-C, LDL-C, and TG levels, no significant difference was observed in HDL-C as a result of short-term use of sumac (Sabzghabaee et al., 2014). Ischemia is a cardiovascular disease state that manifests itself when the blood that carries oxygen to the tissues is reduced or completely cut off by a blockage and the tissue is left without oxygen (Süleyman et al., 2018). If migration of vascular smooth muscle cells to the site of the obstruction can be prevented, then ischemia can also be prevented. The effects of sumac on the migration of vascular smooth muscle cells were thus investigated in an *in-vitro* study, where extracts of dried sumac with high tannin content were prepared, Sprague–Dawley rats' neck arteries' vascular smooth muscle cells were extracted and grown. For ten days, the cells were treated with various doses of sumac extract (5, 1, 0.5, or 0.1% per mL in DMEM broth) for ten days. The findings showed that whereas cell loss occurred in the 1% and 5% extracts, migration of vascular smooth muscle cells was 62% reduced in the 0.1% and 0.5% extracts without resulting in cell loss. This suggested that extracts of sumac at specific concentrations can be used as an adjunct in the treatment of ischemia (Zargham and Zargham, 2008). Depending on the dose, hydrolyzable gallotannins, obtained from the aqueous-alcoholic extract of sumac leaves, normalizes coronary perfusion pressure and reduces the pressure in the left ventricle after ischemia. Thus, sumac might have cardioprotective properties and be preventive against post-ischemic myocardial damage (Beretta et al., 2009).

6.3.5 Anti-Inflammatory Activities

Although the effectiveness of sumac against diseases caused by inflammation has been known anecdotally for a long time, many recent scientific studies have proven this phenomenon. Anti-inflammatory responses have been observed in rats with necrotizing enterocolitis (NEC), where sumac may provide an alternative treatment method due to its protection of intestinal tissue (Işık et al., 2019). In a study on male Wistar rats, sumac application reduced inflammatory cytokines (Roshankhah et al., 2020). Sumac, which plays a role in inhibiting the production of skin pro-inflammatory mediators, exhibited a protective effect against keratinocyte inflammation (Khalilpour et al., 2019). IL-5, IL-8, and TNF-α in breast cancer cells decreased significantly after sumac application (El Hasasna et al., 2016). The injured rats treated with 5 or 10 mg/mL concentrations of sumac extract showed decreased MMP-8 (matrix metalloproteinases) and MPO (myeloperoxidase) enzyme activities, while increased collagen and hydroxyproline accumulation was observed 6 and 10 days after the injury. It is thought that the healing time of wounds also changes in response to these treatments (Gabr and Alghadir, 2019).

6.3.6 Toxicity

Janbaz et al. (2014) showed the antisecretory and antidiarrheal effects of *R. coriaria* extract in mice, reporting that this extract was safe even at 5 g/kg. A study investigating the protective and therapeutic effects of *R. coriaria* extracts on streptozotocin-induced diabetic rats, observed that optimum tolerance was achieved even at very high doses (1000 mg/kg) (Janbaz et al., 2014), suggesting low toxicity. After daily application of the extract for three days, no signs of toxicity or death were observed (Doğan and Çelik, 2016). Oral application of *R. coriaria* seed extract (300 mg/kg) was also shown to improve some blood parameters in type 2 diabetic mice without adverse side effects (Ahangarpour et al., 2017). These findings suggest that, overall, sumac and its extracts are very safe, making them attractive for medical use or further drug discovery studies (Alsamri et al., 2021).

6.4 CONCLUSION

R. coriaria is a herb rich in beneficial phytochemicals and, to date, its importance has not been fully understood compared with alternative medicines. Its advantages can be listed as low toxicity, easy availability over a wide area, and natural (non-synthetic) origin. It is a plant that can be used, especially in the pharmaceutical industry, because toxicity is rarely seen and it exhibits beneficial action on human health. Sumac is a powerful antioxidant that has many more potential applications beyond those of a simple spice and a flavoring agent. However, it is currently not included in any commercial pharmaceutical preparations for therapeutic purposes. Thus, more studies should be undertaken on *R. coriaria*, and its bioactive ingredients should be isolated and evaluated for use in many emerging industrial areas, such as alternative medicines.

REFERENCES

Abdallah, S., Abu-Reidah, I., Mousa, A. and Abdel-Latif, T. 2019. *Rhus coriaria* (Sumac) Extract Reduces Migration Capacity of Uterus Cervix Cancer Cells. *Rev. Bras. De Farmacogn.* 29(5): 591–596.

Abu-Reiadah, I.M., Jamous, R.M. and Ali-Shtayeh, M.S. 2014. Phytochemistry, Pharmacological Properties and Industrial Applications of *Rhus coriaria* L. (Sumac). *JJBS* 7(4): 233–244.

Ahangarpour, A., Heidari, H., Junghani, M.S., Absari, R., Khoogar, M. and Ghaedi, E. 2017. Effects of Hydroalcoholic Extract of *Rhus coriaria* Seed on Glucose and Insulin Related Biomarkers, Lipid Profile, and Hepatic Enzymes in Nicotinamide- Streptozotocin-Induced Type II Diabetic Male Mice. *Int. J. Med. Res. Pharm. Sci.* 12(5): 416–424.

Ahmadian-Attari, M.M., Amini, M., Farsam, H., Amin, G., Fazeli, M., Esfahani, H., Jamalifar, H. and Bairami, A. 2016. Isolation of Major Active Antibacterial Compounds of Sumac Fruit (*Rhus coriaria* L.). *Int. J. Enteric Pathog.* 4(4): 1–5.

Ali-Shtayeh, M.S., Al-Assali, A.A. and Jamous, R.M. 2013. Antimicrobial Activity of Palestinian Medicinal Plants against Acne-Inducing Bacteria. *Afr. J. Microbiol. Res.* 7(21): 2560–2573.

Alsamri, H., Athamneh, K., Pintus, G., Eid, A.H. and Iratni, R. 2021. Pharmacological and Antioxidant Activities of *Rhus coriaria* L. (Sumac). *Antioxidants (Basel)* 10(1): 73.

Anwar, M.A., Samaha, A.A., Baydoun, S., Iratni, R. and Eid, A.H. 2018. *Rhus coriaria* L. (Sumac) Evokes Endothelium-Dependent Vasorelaxation of Rat Aorta: Involvement of the CAMP and CGMP Pathways. *Front. Pharmacol.* 9: 688.

Ardakani, F., Reza, M., Vahidi, A.R., Karimi-Nazari, E., Dehghani, A. and Nadjarzadeh, A. 2016. Effect of *Rhus coriaria* L on Glycemic Control and Insulin Resistance in Patients with Type 2 Diabetes Mellitus. *Iran. J. Diabetes Obes.* 8: 172–178.

Ashoori, F., Fakhar, M., Goli, H.R., Mirzaee, F., Faridnia, R., Kalani, H. and Shahani, S. 2020. Antileishmanial and Antibacterial Activities of the Hydroalcoholic Extract of *Rhus coriaria* L. *Ann. Parasitol.* 62(2): 157–163.

Batiha, G.E.-S., Beshbishy, A.M., Adeyemi, O.S., Nadwa, E.H., Rashwan, E.K.M., Alkazmi, L.M., Elkelish, A.A. and Igarashi, I. 2020. Phytochemical Screening and Antiprotozoal Effects of the Methanolic *Berberis vulgaris* and Acetonic *Rhus coriaria* Extracts. *Molecules* 25(3): 550.

Baumgartner, R.R.,Steinmann, D., Heiss, E.H., Atanasov, A.G., Ganzera, M., Stuppner, H. and Dirsch, V.M. 2010. Bioactivity-Guided isolation of 1,2,3,4,6-Penta-O-galloyl-D-glucopyranose from *Paeonia lactiflora* roots as a PTP1B Inhibitor. *J. Nat. Prod.* 73(9): 1578–1581.

Beretta, G., Rossoni, G., Santagati, N.A. and Facino, R.M. 2009. Anti-ischemic Activity and Endothelium-Dependent Vasorelaxant Effect of Hydrolysable Tannins from the Leaves of *Rhus coriaria* (Sumac) in Isolated Rabbit Heart and Thoracic Aorta. *Planta Med. Med.* 75(14): 1482–1488.

Bozan, B., Koşar, M., Tunalıer, Z., Öztürk, N. and Başer, K.H.C. 2002. Antioxidant and Free Radical Scavenging Activities of *R. Coriaria* and *Cinnamomum cassia* Extracts. *Int. Conf. Exib. Nutraceut. Functional Foods*, 13–17 September. Houston, Texas.

Bursal, E. and Köksal, E. 2011. Evaluation of Reducing Power and Radical Scavenging Activities of Water and Ethanol Extracts from Sumac (*Rhus coriaria* L.). *Food Res. Int.* 44(7): 2217–2221.

Candan, F. 2003. Effect of *Rhus coriaria* L. (Anacardiaceae) on Superoxide Radical Scavenging and Xanthine Oxidase Activity. *J. Enzyme Inhib. Med. Chem.* 18(1): 59–62.

Dalar, A., Dogan, A., Bengu, A.S., Mukemre, M. and Celik, I. 2018. Screening In Vivo Antioxidant and Haematological Properties of Sumac and Acorn Bioactive Rich Extracts. *Ind. Crops Prod.* 124: 20–27.

Davis, P.H. 1997. *Flora of Turkey and East Aegein Islands.* Vol. 2. Edinburgh University Press, Edinburgh.

Dini, I. 2018. Spices and Herbs as Therapeutic Foods. In: Grumezescu, A.M., Holban, A.M. (Eds.). *Food Quality: Balancing Health and Disease.* Academic Press, 433–469.

Dogan, A. and Çelik, I. 2016. Healing Effects of Sumac (*Rhus coriaria*) in Streptozotocin-Induced Diabetic Rats. *Pharm. Biol.* 54(10): 2092–2102.

El Hasasna, H., Athamneh, K., Al Samri, H., Karuvantevida, N., Al Dhaheri, Y., Hisaindee, S., Ramadan, G., Al Tamimi, N., AbuQamar, S., Eid, A. and Iratni, R. 2015. *Rhus coriaria* Induces Senescence and Autophagic Cell Death in Breast Cancer Cells through a Mechanism Involving P38 and ERK1/2 Activation. *Sci. Rep.* 5: 13013.

El Hasasna, H., Saleh, A., Al Samri, H., Athamneh, K., Attoub, S., Arafat, K., Benhalilou, N., Alyan, S., Viallet, J., Al Dhaheri, Y., Eid, A. and Iratni, R. 2016. *Rhus coriaria* Suppresses Angiogenesis, Metastasis and Tumor Growth of Breast Cancer through Inhibition of STAT3, NF_B and Nitric Oxide Pathways. *Sci. Rep.* 6: 21144.

El-Sissi, H., Saleh, N. and Abd El Wahid, M. 1966. The Tannins of *Rhus coriaria* and *Mangifera indica*. *Planta Med.* 14(2): 222–231.

El-Sissi, H.I., Ishak, M.S. and Abd El Wahid, M.S. 1972. Polyphenolic Components of *Rhus coriaria* leaves. *Planta Med.* 21: 67–71.

El-Sissi, H.I., Ishak, M.S., El-Wahid, M.S. and El-Ansari, M.A. 1971. The Gallotannins of *Rhus coriaria* and *Mangifera indica*. *Planta Med.* 19(4): 342–351.

Ertürk, Ö. 2010. Antibacterial and Antifungal Effects of Alcoholic Extracts of 41 Medicinal Plants Growing in Turkey. *Czech J. Food Sci.* 28(1): 53–60.

Fazeli, M.R., Amin, G., Attari, M.M.A., Ashtiani, H., Jamalifar, H. and Samadi, N. 2007. Antimicrobial Activities of Iranian Sumac and Avishan-e Shirazi (*Zataria multiflora*) against Some Food-borne Bacteria. *Food Control* 18: 646–649.

Gabr, S.A. and Alghadir, A.H. 2019. Evaluation of the Biological Effects of Lyophilized Hydrophilic Extract of *Rhus coriaria* on Myeloperoxidase (MPO) Activity, Wound Healing, and Microbial Infections of Skin Wound Tissues. *Evid. Based Complement. Altern. Med.*: 1–14.

Giancarlo, S., Rosa, L.M., Nadjafi, F. and Francesco, M. 2006. Hypoglycaemic Activity of Two Spices Extracts: *Rhus coriaria* L. and *Bunium persicum Boiss*. *Nat. Prod. Res.* 20(9): 882–886.

Gottardi, D., Bukvicki, D., Prasad, S. and Tyagi, A.K. 2016. Beneficial Effects of Spices in Food Preservation and Safety. *Front. Microbiol.* 7: 1394.

Güner, A., Aslan, S., Ekim, T., Vural, M. and Babac, M.T. 2012. Türkiye Bitkileri Listesi (Damarli Bitkiler). Nezahat Gökyigit Botanik Bahçesi Yayinlari Flora Dizisi I, Istanbul. ISBN: 978-605-60425-7-7.

Hasanova, G., Öztürk, M. and Akçiçek, E. 2000. *Azerbaycan' Da Geleneksel Tedavide Kullanılan Bitkiler.* 13. BİHAT ÖzetKitabı, İstanbul.

Houssein, M., Abi Saab, W., Khalil, M., Khalife, H. and Fatfat, M. 2020. Cell Death by Gallotannin is Associated with Inhibition of JAK/STAT Pathway in Human Colon Cancer Cells. *Curr. Ther. Res.* 92: 100589.

Ibrahim, F.Y., EL-Khateeb, A.Y. and Mohamed, A.H. 2019. *Rhus* and *Safflower* Extracts as Potential Novel Food Antioxidant, Anticancer and Antimicrobial Agents Using Nanotechnology. *Foods* 8(4): 139.

Isik, S., Tayman, C., Cakir, U., Koyuncu, I., TaskinTurkmenoglu, T. and Cakir, E. 2019. Sumac (*Rhus coriaria*) for the Prevention and Treatment of Necrotizing Enterocolitis. *J. Food Biochem.* 43(12): e13068.

Ismail, M.I.M. 2020. Green Synthesis and Characterizations of Copper Nanoparticles. *Mater. Chem. Phys.* 240: 122283.

Janbaz, K.H., Shabbir, A., Mehmood, M.H. and Gilani, A.H. 2014. Pharmacological Basis Forthe Medicinal Use of *Rhus coriaria* in Hyperactive Gut Disorders. *Br. J. Pharm.* 9: 636–644.

Kacergius, T., Abu-Lafi, S., Kirkliauskiene, A., Gabe, V., Adawi, A., Rayan, M., Qutob, M., Stukas, R., Utkus, A., Zeidan, M. and Rayan, A. 2017. Inhibitory Capacity of *Rhus coriaria* L. Extract and its Major Component Methyl Gallate on Streptococcus mutans Biofilm Formation by Optical Profilometry: Potential Applications for Oral Health. *Mol. Med. Rep.* 16(1): 949–956.

Kadokura, K., Suruga, K., Tomita, T., Hiruma, W., Yamada, M., Kobayashi, A., Takatsuki, A., Nishio, T., Oku, T. and Sekino, Y. 2015. Novel Urushiols with Human Immunodeficiency Virus Type 1 Reverse Transcriptase Inhibitory Activity from the Leaves of *Rhus verniciflua*. *J. Nat. Med.* 69(1): 148–153.

Kang, S.Y., Kang, J.Y. and Oh, M.J. 2012. Antiviral Activities of Flavonoids Isolated from the Bark of *Rhus verniciflua* Stokes against Fish Pathogenic Viruses In Vitro. *J. Microbiol.* 50(2): 293–300.

Khalilpour, S., Sangiovanni, E., Piazza, S., Fumagalli, M., Beretta, G. and Dell'Agli, M. 2019. In Vitro Evidences of the Traditional Use of *Rhus coriaria* L. Fruits against Skin Inflammatory Conditions. *J. Ethnopharmacol.* 238: 111829.

Kizil, S. and Turk, M. 2010. Microelement Contents and Fatty Acid Compositions of *Rhus coriaria* L. and *Pistacia terebinthus L.* Fruits Spread Commonly in the South Eastern Anatolia Region of Turkey. *Nat. Prod. Res.* 24(1): 92–98.

Koşar, M., Bozan, B., Temelli, F. and Başer, K.H.C. 2007. Antioxidant activity and Phenolic Composition of Sumac (*Rhus coriaria* L.) Extracts. *Food Chem.* 103: 952–959.

Kossah, R., Nsabimana, C., Zhao, J., Chen, H., Tian, F., Zhang, H. and Chen, W. 2009. Comparative Study on the Chemical Composition of Syrian Sumac (*Rhus coriaria L.*) and Chinese Sumac (*Rhus typhina L.*) Fruits. *Pak. J. Nutr.* 8(10): 1570–1574.

Lin, Y.M., Flavin, M.T., Schure, R., Chen, F.C., Sidwell, R., Barnard, D.L., Huffman, J.H. and Kern, E.R. 1999. Antiviral Activities of Bioflavonoids. *Planta Med.* 65(2): 120–125.

Mamatkulova, N.M., Khidyrova, N.K., Mamadrakhimov, A.A. and Shakhidoyatov, K.M. 2014. Polyprenols from Leaves of *Rhus coriaria*. *Chem. Nat. Compd.* 50(5): 832–835.

Mamatkulova, N.M., Melikuziev, F.A., Mukarramov, N.I., Nishanbaev, S.Z., Khidyrova, N.K. and Shakhidoyatov, K.M. 2012. Polyisoprenoids from *Rhus coriaria*. *Chem. Nat. Compd.* 48(4): 661–662.

Mamıkoğlu, N.G. 2015. *Türkiye'nin Ağaçları ve Çalıları*. NTV Yayınları, İstanbul.

Mavlyanov, S.M., Islambekov, S.Y., Karimdzhanov, A.K. and Ismaikov, A.I. 1997. Anthocyanins and Organic Acids of the Fruits of Some Species of Sumac. *Chem. Nat. Compd.* 33(2): 209.

Mirian, M., Behrooeian, M., Ghanadian, M., Dana, N. and Sadeghi-Aliabadi, H. 2015. Cytotoxicity and Antiangiogenic Effects of *Rhus coriaria*, *Pistacia vera* and *Pistacia khinjuk* Oleoresin Methanol Extracts. *Res. Pharm. Sci.* 10(3): 233–240.

Najjar, F., Rizk, F., Carnac, G., Nassar, R., Jabak, S., Sobolev, A.P., BouSaada, Y., El Sabban, M. and Hamade, A. 2017. Protective Effect of *Rhus coriaria* Fruit Extracts against Hydrogen Peroxide-Induced Oxidative Stress in Muscle Progenitors and Zebrafish Embryos. *PeerJ* 5: e4144.

Nasar-Abbas, S.M., Halkman, A.K. and Al-Haq, M.I. 2004. Inhibition of Some Foodborne Bacteria by Alcohol Extract of Sumac (*Rhus coriaria* L.). *J. Food Saf.* 24(4): 257–267.

Onkar, S., Mohammed, A. and Nida, A. 2011. New Antifungal Aromatic Compounds from Seeds of *Rhus coriaria* L. *Inter. Res. J. Pharm.* 2: 188–194.

Özcan, M. 2003. Effect of Sumach (*Rhus coriaria* L.) Extracts on the Oxidative Stability of Peanut Oil. *J. Med. Food* 6(1): 63–66.

Özcan, M. 2004. Mineral Contents of Some Plants Used as Condiments in Turkey. *Food Chem.* 84(3): 37–440.

Özcan, M., Ünver, A., Arslan, D. and Koşar, M. 2007. *Değişik Yörelerden Sumak (Rhus coriaria L.) Meyvesinin Ayrıntılı Kimyasal Bileşimi and Oleorezin Üretiminde Kullanılması Üzerine Araştırma*. TÜBİTAK Projesi, Konya.

Parsania, M., Rezaee, M.B., Monavari, S.H., Jaimand, K., MousaviJazayeri, S.M., Razazian, M. and Nadjarha, M.H. 2017. Evaluation of Antiviral Effects of Sumac (*Rhus coriaria* L.) Fruit Extract on Acyclovir Resistant *Herpes simplex* Virus Type 1. *Med. Sci.* 27: 1–8.

Perkin, A.G. and Allen, G.Y. 1896. LXXIX—Colouring Matter of Sicilian Sumach, *Rhus coriaria*. *J. Chem. Soc. Trans.* 69: 1299–1303.

Radmehr, B. and Abdolrahimzade, M. 2009. Antimicrobial effects of Sumac (*Rhus coriaria* L.) Extract in Minced Meat. *Planta Med.* 75: 1068.

Rajabi, M. and Mousa, S. 2017. The Role of Angiogenesis in Cancer Treatment. *Biomedicines* 5: 1–12.

Raodah, M., Al-Ali, A.Z.H. and Faleeha, H.H. 2014. The Antioxidant and Antimicrobial of Syrian Sumac (*Rhus coriaria*) Fruit Extracts. *J. Nat. Sci. Res.* 4: 36–40.

Raut, J.S. and Karuppayil, S.M. 2014. A Status Review on the Medicinal Properties of Essential Oils. *Ind. Crops Prod.* 62: 250–264.

Regazzoni, L., Arlandini, E., Garzon, D., Santagati, N.A., Beretta, G. and MaffeiFacino, R. 2013. A Rapid Profiling of Gallotannins and Flavonoids of the Aqueous Extract of *Rhus coriaria* L. by Flow Injection Analysis with High-Resolution Mass Spectrometry Assisted with Database Searching. *J. Pharm. Biomed. Anal.* 72: 202–207.

Reichling, J., Neuner, A., Sharaf, M., Harkenthal, M. and Schnitzler, P. 2009. Antiviral Activity of *Rhus aromatica* (Fragrant Sumac) Extract against Two Types of Herpes Simplex Viruses in Cell Culture. *Pharmazie* 64(8): 538–541.

Romeo, F.V., Ballistreri, G., Fabroni, S., Pangallo, S., Nicosia, M.G.L.D., Schena, L. and Rapisarda, P. 2015. Chemical Characterization of Different Sumac and Pomegranate Extracts Effective against Botrytis cinerea Rots. *Molecules* 20(7): 11941–11958.

Roshankhah, S., Gholami, M.R. and Salahshoor, M.R. 2020. Evaluation of Male Infertilitytreatment Following *Rhus coriaria* Extract Administration on Rats Exposed Tomorphine. *Mol. Biol. Rep.* 47(8): 6073–6081.

Sabzghabaee, A.M., Kelishadi, R., Golshiri, K., Ghannadi, A. and Badri, S. 2014. Clinical Effects of *Rhus coriaria* Fruits on Dyslipidemia in Adolescents: A Triple-Blinded Randomized Placebo-Controlled Trial. *Med. Arch.* 68(5): 308–312.

Savarino, G., Corsello, A. and Corsello, G. 2021. Macronutrient Balance and Micronutrient Amounts through Growth and Development. *Ital. J. Pediatr.* 47(1): 109.

Sayed-Ahmed, M.Z., Ahdy, A.M., Younis, E.E., El-Khodery, S.A. and Baraka, H.N. 2019. Comparative Effectiveness of Sumaq and Neem Extract Cream, Eniloconazole and Glycerine Iodine on Dermatophytosis in Arabian Horses: A Randomized Clinical Trial. *Trop. Anim. Health Prod.* 51(4): 905–910.

Setorki, M., Rafieian, M., Heidarian, E., Ghatreh, K., Shahinfard, N., Ansari, R. and Forouzandeh, Z. 2012. Effect of *Rhus Coriaria* Consumption with High Cholesterol Food on Some Atherosclerosis Risk Factors in Rabbit. *J. Babol Univ. Med. Sci.* 14: 38–45.

Shabana, M.M., El Sayed, A.M., Yousif, M.F., El Sayed, A.M. and Sleem, A.A. 2011. Bioactive Constituents from *Harpephyllum caffrum Bernh.* and *Rhus coriaria* L. *Pharm. Mag.* 7: 298–306.

Shidfar, F., Rahideh, S.T., Rajab, A., Khandozi, N., Hosseini, S., Shidfar, S. and Mojab, F. 2014. The Effect of Sumac (*Rhus coriaria* L.) Powder on Serum Glycemic Status, ApoB, ApoA-I and Total Antioxidant Capacity in Type 2 Diabetic Patients. *Iran. J. Pharm. Sci.* 13(4): 1249–1255.

Süleyman, H., Gül, V. and Hirik, E. 2018. İskemi-reperfüzyonhasarı. *Erzincan Tıp Derg.* 1: 51–54.

Tanker, N., Coşkun, M. and Koyuncu, M. 2007. *Farmasötik Botanik.* Ankara Üniversitesi Yayınları, Ankara.

Tiryaki, G.Y. 2010. Kahramanmaras Province Produced in a Symbolic Traditional Product: Concentrated Citrus Tasting Water Extract Of Sumac (Rhus Coriaria L.) Fruit. *Gıda Mühendisliği Dergisi*, 31. 55–58.

Tuzlacı, E. 2006. *Şifa Niyetine Türkiye'nin Bitkisel Halk İlaçları.* Alfa Yayıncılık, İstanbul.

Tuzlacı, E. 2011. *Türkiye Bitkileri Sözlüğ ü.* Alfa Yayıncılık, İstanbul.

Zargham, H. and Zargham, R. 2008. Tannin Extracted from Sumac Inhibits Vascular Smooth Muscle Cell Migration. *McGill J. Med.* 11(2): 119.

7 Isolation of Bioactive Compounds From Medicinal Plants to Treat Neurological Disease

Mohammed Sohail Akhtar[1], Shyam Sundar Arputhanantham[1], and Wegdan Aly Shehata

CONTENTS

7.1 INTRODUCTION

The central nervous system (CNS) plays a pivotal role in carrying out key functions of the human body. Neurodegenerative disorders occur when there are anatomical problems in brain or spinal cord function. The increase in life expectancy has increased the prevalence of common neurodegenerative disorders and hence neuroprotection strategies are widely sought. A recent publication from the World Health Organization (WHO) estimates that almost one billion people are currently suffering from neurodegenerative disorders, and nearly 6.8 million individuals succumb to them every year, with such disorders being more prevalent in developing than in developed countries (Upadhyay, 2014; Kundap et al., 2017).

Mother Nature has generously provided numerous herbal remedies with limited or no adverse side effects for the treatment of many diseases. The increasing popularity of complementary and alternative medicine (CAM) in the past four decades has provided research teams across the globe with the focus to explore and validate the potency of herbal medicines for the treatment of various disorders (Lakea, 2015; Ahmed and Azam, 2014; Kantati et al., 2016).

The published literature already indicates that many herbal remedies are potent and safe for use in treating serious, medical conditions, including neurodegenerative disorders. These reports further suggest that 70% of new, active molecules are already derived from natural resources and that careful documentation of ethnobotanical information and further validation of such material has become an important strategy for health care researchers (Kumar et al., 2012; Newman and Cragg, 2012; Fabricant and Farnsworth, 2001).

DOI: 10.1201/9781003226925-7

The use of the phrase "reverse pharmacology," meaning drug discovery from traditional knowledge is now gaining considerable usage in Western societies, where many consumers now prefer herbal medicines over synthetic ones. Furthermore, due to the lower development cost and shorter development time associated with reverse pharmacology compared with conventional pharmacology, the exploration of herbal substances for their medicinal value is becoming increasingly important (Adams et al., 2007; Kumar, 2006).

Herbal practitioners often treat CNS disorders in an empirical way (through experience and occasional experimentation). The current interest and efforts of both public and health care researchers in this space also signals that traditional, conventional medicines will continue to play an important role in the treatment of nervous system disorders. The current challenge when using herbal medicines, however, is the lack of documented, precise information on the plant species, their extraction, and data on their safety profile (Abena et al., 1993; Bisong et al., 2010; Tabuti et al., 2010).

However, medicinal plants often contain a constellation of active compounds, such as various glycosides, flavonoids, alkaloids, etc., all of which may have been reported to be effective in the treatment of neurodegenerative disorders and offer potential for medical treatments if developed appropriately (Suk, 2005).

7.2 EPIDEMIOLOGY OF NEUROLOGICAL DISEASES

There are about 600 known neurological diseases, with only a few being very common, whereas most are rarely encountered (Hussain et al., 2017). Globally, a direct relationship has been noted between patient age and the prevalence of neurological diseases, with the prevalence of neurological diseases increasing in older patients (Erkkinen et al., 2018). Neurological diseases are the leading cause of disability and the second leading cause of mortality worldwide (Feigin et al., 2019). The two most common neurological diseases are Alzheimer's disease (AD), which affects approximately 6–10% of individuals above 65 years of age, and Parkinson's disease (PD), which affects about 1% of individuals above 60 years of age (Akkol et al., 2021). Epilepsy is another common, chronic, neurological disorder, prevalent in the younger population at around 0.5% and present in older population at 1.5% (Dumurgier and Tzourio, 2020). However, globally, the incidence and point prevalence of neurological disorders are increasing in developing countries due to many factors including malnutrition and poverty (Pandey, 2012). The annual incidence and prevalence of some common neurological diseases in developing countries with incidence above 1/100,000 in a year are presented in Table 7.1 (Shorvon, 2016).

7.3 HISTORY OF TRADITIONAL USES OF PLANTS TO TREAT NEUROLOGICAL DISEASE

Since antiquity, people have traditionally relied on natural sources to treat various health conditions, especially because they generally have fewer side effects when compared to synthetic treatments

TABLE 7.1
Incidence and Prevalence of Some Neurological Diseases in Developing Countries (Shorvon, 2016)

Condition	Incidence (per 100,000 persons/year)	Point prevalence (per 100,000 persons)
Migraine	370	12,100
Dementia	50	250
Parkinson's disease (PD)	20	200
Epilepsy	50	710
Multiple sclerosis (MS)	4	90

(Willcox and Gilbert, 2009). Historically, many plants were also used to treat neurological diseases (Beldar and Sidat, 2020). For instance, the paste formed from the roots of the plant *Aegle marmelos*, commonly known as bel or bael, has been traditionally used to treat migraine by applying the paste to the eyes at night for a duration of 4 to 5 days (Patil, 2012). The dried roots of *Cassia tora* have been used to treat epilepsy by mixing 100 g of the dried extract with a cup of milk which the patient consumes twice daily (Patil and Patil, 2012). In India, the plant *Sida cordifolia* is still used in traditional medicines to treat Parkinson's disease (Galal et al., 2015).

7.4 ACTIVE CONSTITUENTS FROM PLANTS AND THEIR ACTIVITY IN NEUROLOGICAL DISEASES

Active or natural constituents extracted from plants are mainly secondary metabolites (Sharifi-Rad et al., 2020). Secondary metabolites are small, organic compounds not directly essential for plant growth and development (Wink, 2010). However, secondary metabolites play a significant role in the adaptive processes key to plant survival (Salehi et al., 2019). Secondary metabolites may include alkaloids, flavonoids, phenolic acids, and terpenes (Buchanan et al., 2000). Numerous plants contain many active constituents that have promising, pharmacological activity for the treatment of neurological diseases, and some of the most bioactive plant compounds are reviewed in this article:

7.4.1 Berberine

Berberine is an isoquinoline alkaloid with a molecular formula ($C_{20}H_{19}NO_5$) and a molecular weight of 353.36 g/mol, Upadhyay (2014) with the structure shown in Figure 7.1 (Jiang et al., 2015). It is bitter in taste and its yellow color can be found in many plants, such as *Berberis vulgaris* (barberry), *Berberis aristata* (tree turmeric), *Coptis chinensis* (Chinese gold thread), and *Hydrastis canadensis* (goldenseal) (Cai et al., 2016). Berberine is effective against the neurodegenerative disease Huntington's disease (HD), which is caused by the accumulation of misfolded proteins in the brain that results in neurodegeneration (Landles and Bates, 2004). Transgenic mice given berberine orally exhibited significantly lower accumulation of the mutant Huntingtin protein as a result of degradation and autophagy processes, leading to reduced motor dysfunction, extending the rate of survival, and achieving protection against the mutation (Jiang et al., 2015). Due to its antioxidant properties, berberine was also beneficial in the treatment of Alzheimer's disease and Parkinson's disease (PD) (Ahmed et al., 2015).

7.4.2 Piperine

Piperine is the main alkaloid extracted from black pepper (*Piper nigrum*) and long pepper (*Piper longum*) (Ren and Zuo, 2019). Piperine is a secondary metabolite where its precursor in biosynthesis

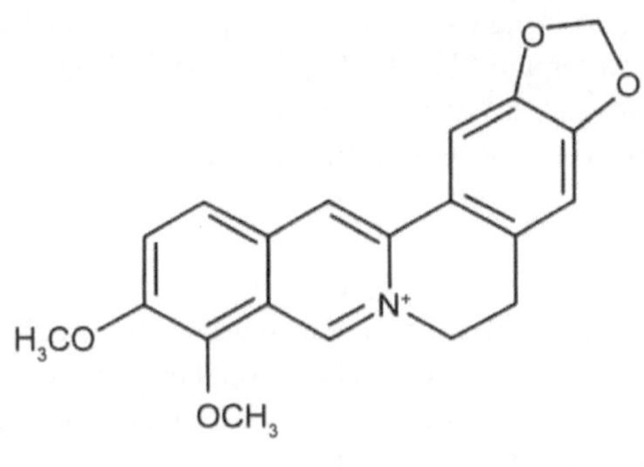

FIGURE 7.1 Berberine.

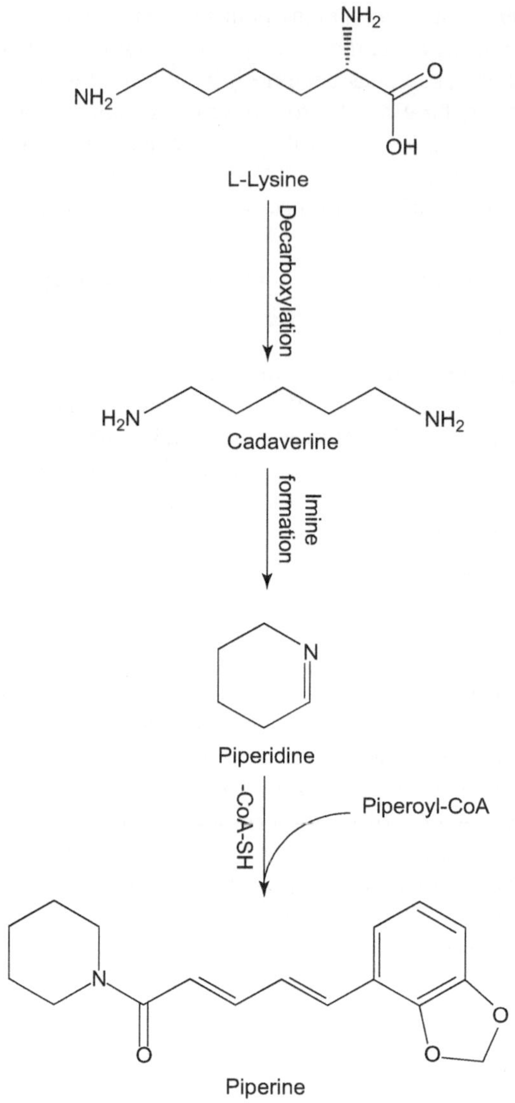

FIGURE 7.2 Piperine biosynthesis pathway.

is L-lysine (Figure 7.2) (Azam et al., 2022). A study was conducted to create an Alzheimer's disease (AD) model in mice by infusing streptozotocin (STZ) into the intracerebroventricular region on days 1 and 3 (1.5 mg/kg/day), creating oxidative stress; piperine was then administered as an intraperitoneal injection from day 8 for a total of 15 days. The study concluded that piperine had a promising effect in reducing oxidative stress and improving cognitive function in AD (Wang et al., 2019). Piperine was also effective in treating epilepsy, PD, and HD (Tripathi et al., 2022; Yang et al., 2015).

7.4.3 Quercetin

Quercetin is a polyphenolic flavonoid abundant in many plants including *Ruta graveolens* (rue), *Fagopyrum esculentum* (buckwheat), (Sofic et al., 2010). The structure of the flavonoid quercetin is shown in Figure 7.3 (Islam et al., 2021). Lipopolysaccharide (LPS) was found to stimulate the process of neuroinflammation which results in neurodegeneration, and this process contributes to the pathophysiology of many neurological diseases, such as AD, PD, HD, and multiple sclerosis (MS). It

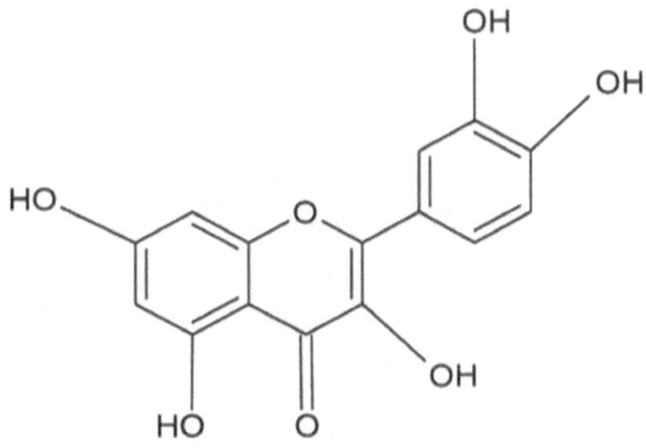

FIGURE 7.3 Quercetin.

was found that quercetin had a neuroprotective effect by reversing the LPS action, Khan et al. (2018) thus making quercetin highly sought after as a component in treatments for neurological conditions.

7.4.4 Kaempferol

Kaempferol (KPF) is a flavonoid found in many plants, such as broccoli, strawberry, beans, tomato, grapes, and in medicinal plants such as *Ginkgo biloba* and *Moringa oleifera* (Calderon-Montano et al., 2011). The chemical name of kaempferol is 3,5,7-trihydroxy-2-(4-hydroxyphenyl)-4H-1-ben zopyran-4-one; it is normally found in plants in the form of glycosides, such as kaempferitrin (KPF-3,7-dirhamnoside), and where it can undergo acylation with phenolic acids such as caffeic, sinapic, ferulic, and *p*-coumaric acids (Figure 7.4) (Silva dos Santos et al., 2021). MMP-3 is an endopeptidase which is well established to be a major factor in the inflammatory processes involved in multiple sclerosis (MS), which affects both the brain and the spinal cord. Kaempferol was found to effectively inhibit the expression of MMP-3 by its anti-neuroinflammatory action (Park et al., 2011). Kaempferol was also shown to have a neuroprotective effect against the cognitive dysfunction associated with AD (Kouhestani et al., 2018).

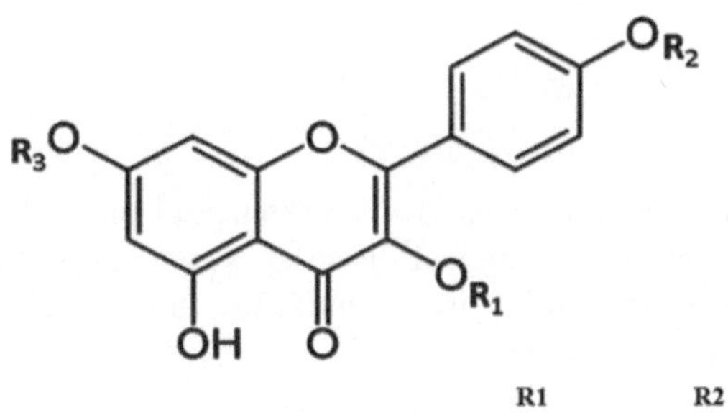

	R1	R2	R3
Kaempferol	H	H	H
kaempferol-3-O-glucoside	Glu	H	H
kaempferol-7-O-glucoside	H	H	Glu
kaempferol-3-O-rutinoside	Rut	H	H
kaempferol-3,7-dirhamnoside	Rha	H	Rha
kaempferol-3-O-caffeoyl-diglucoside-7-O-glucoside	Caff-Glu-Rha	H	Glu
kaempferol-3-O-sinapoyl-diglucoside-7-O-glucoside	Sin-Glu-Rha	H	Glu
kaempferol-3-O-feruloyl-diglucoside-7-O-glucoside	Fer-Glu-Rha	H	Glu
kaempferol-3-O-p-coumaroyl-diglucoside-7-O-glucoside	Cou-Glu-Rha	H	Glu
Kaempferol 4'-methyl-ether	H	Met	H

FIGURE 7.4 Kaempferol structure and its derivatives.

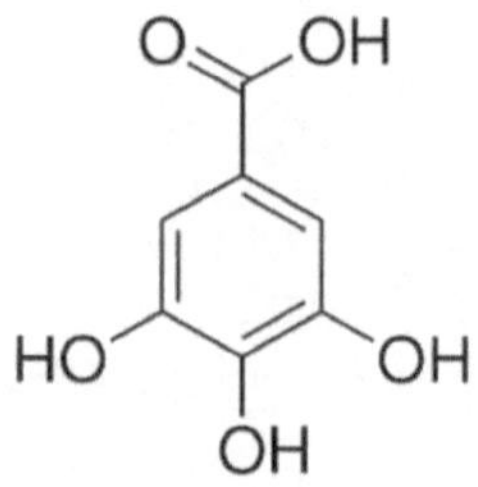

FIGURE 7.5 Gallic acid.

7.4.5 Gallic Acid

Gallic acid is a well-known antioxidant phenolic acid, having the chemical name 3,4,5-trihydroxy benzoic acid (Figure 7.5) (Shabani et al., 2020). Unlike other polyphenols, it can cross the blood–brain barrier (BBB) and be easily absorbed (Blas-Valdivia et al., 2021). It is found in many plants such as *Vitis vinifera* (grape seeds), *Camellia sinensis* (green tea), *Arctostaphylos uva-ursi* (common bearberry leaves), and *Corylus avellana* (hazelnuts) (Karamac et al., 2006). One of the main pathogenic factors in AD is the aggregation of beta-amyloid (Aβ) plaques and it was recently demonstrated that gallic acid had potent activity to prevent oligomerization of Aβ and hence possessed antioxidant activity which helped in treatment of AD and in prevention of dementia (Hajipour et al., 2016).

7.4.6 Curcumin

Curcumin is a well-known, polyphenolic compound extracted as a yellow dye from the rhizomes of turmeric (*Curcuma longa*) (Maiti and Dunbar, 2018). The percentage of curcumin derivatives that can be extracted as curcumin from turmeric is between 60 and 80%, with other forms extractable at different rates,- including bisdemethoxycurcumin (5–15%) and demethoxycurcumin (15–27%) (Goschorska et al., 2020). The antioxidant and analgesic activity of curcumin for the treatment of migraine was induced in rats by nitroglycerin, where it was found that curcumin action was more powerful than those of propranolol and indomethacin (Bulboacă et al., 2017). Due to its ability for inhibition of protein aggregation, free radical scavenging, and anti-inflammatory activities, curcumin has demonstrated pharmacological action against AD and PD (Figure 7.6) (Darvesh et al., 2012).

7.5 CONCLUSION

Bioactive compounds extracted and isolated from medicinal plants have a major role to play in the prevention and treatment of neurological diseases. Researchers should investigate the role of such bioactive compounds further to inspire potential use in the treatment of further neurological

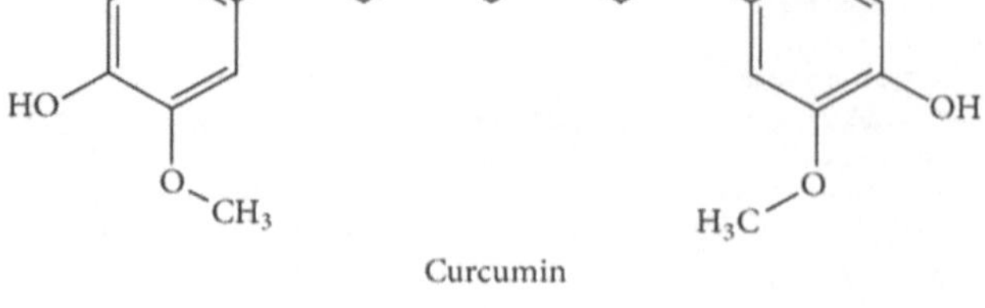

FIGURE 7.6 Curcumin.

diseases, especially those for which there are no effective curative medicines or as alternatives to existing medicines but with reduced side effects.

REFERENCES

Abena, A.A., Kintsangoula-Mbaya, G.S., Diantama, J. and Bioka, D., 1993. Analgesic effects of *Ageratum conyzoides* extract in the rat. *L'Encéphale*, 19(4): 329–332.

Adams, M., Gmünder, F. and Hamburger, M., 2007. Plants traditionally used in age related brain disorders—A survey of ethnobotanical literature. *Journal of Ethno Pharmacology*, 113(3): 363–381. https://www.sciencedirect.com/science/article/abs/pii/S037887410700356X.

Ahmed, T., Abdollahi, M., Daglia, M., Nabavi, S.F. and, Nabavi, S.M., 2015. Berberine and neurodegeneration: A review of literature. *Pharmacological Reports*, 67(5): 970–979. https://www.researchgate.net/publication/277592426_Berberine_and_neurodegeneration_A_review_of_literature.

Ahmed, M.N. and Azam, M.N.K., 2014. Traditional knowledge and formulations of medicinal plants used by the traditional medical practitioners of Bangladesh to treat schizophrenia like psychosis. *Schizophrenia and Research Treatment*, 2014: 679810. https://pubmed.ncbi.nlm.nih.gov/25101175/.

Akkol, E.K., Çankaya, I.T., Karatoprak, G.Ş., Carpar, E., Sobarzo-Sánchez, E. and Capasso, R., 2021. Natural compounds as medical strategies in the prevention and treatment of psychiatric disorders seen in neurological diseases. *Frontiers in Pharmacology*, 12. https://www.frontiersin.org/articles/10.3389/fphar.2021.669638/full.

Azam, S., Park, J.Y., Kim, I.S. and Choi, D.K., 2022. Piperine and its metabolite's pharmacology in neurodegenerative and neurological diseases. *Biomedicines*, 10(1): 1–16. https://www.researchgate.net/publication/357788271_Piperine_and_Its_Metabolite's_Pharmacology_in_Neurodegenerative_and_Neurological_Diseases.

Beldar, V.G. and Sidat, P.S., 2020. Ethnomedicinal plants used in Maharashtra for treatment of various diseases and disorders. *Malaysian Journal of Pharmaceutical Sciences*, 6: 12–36. https://ejournal.manipal.edu/mjps/docs/Vol6_Issue2/PDF/2.%20MJPS_6.2_Sep_2020_Review%20article.pdf.

Bisong, S.A., Brown, R. and Osim, E.E., 2010. Comparative effects of *Rauwolfia vomitoria* and chlorpromazine on locomotor behavior and anxiety in mice. *Journal of Ethnopharmacology*, 132(1): 334–339. https://pubmed.ncbi.nlm.nih.gov/20800668/.

Blas-Valdivia, V., Franco-Colín, M., Rojas-Franco, P., Chao-Vazquez, A. and Cano-Europa, E., 2021. Gallic acid prevents the oxidative and endoplasmic reticulum stresses in the hippocampus of adult-onset hypothyroid rats. *Frontiers in Pharmacology*, 12: 1645. https://www.frontiersin.org/articles/10.3389/fphar.2021.671614/full.

Buchanan, B.B., Gruissem, W. and Jones, R.L., 2000. Biochemistry & molecular biology of plants. *American Society of Plant Physiologists*, 1: 1250–1318. https://instruct.uwo.ca/biology/407b/restricted/pdf/Chpt24.pdf.

Bulboacă, A.E., Bolboacă, S.D., Stănescu, I.C., Sfrângeu, C.A. and Bulboacă, A.C., 2017. Preemptive analgesic and antioxidative effect of curcumin for experimental migraine. *BioMed Research International*, 1: 1–8. https://www.ncbi.nlm.nih.gov/pmc/articles/PMC5674483/.

Cai, Z., Wang, C. and Yang, W., 2016. Role of berberine in Alzheimer's disease. *Neuropsychiatric Disease and Treatment*, 12: 2509–2520. https://www.dovepress.com/role-of-berberine-in-alzheimerrsquos-disease-peer-reviewed-fulltext-article-NDT.

Calderon-Montano, J.M., Burgos-Morón, E., Pérez-Guerrero, C. and López-Lázaro, M., 2011. A review on the dietary flavonoid kaempferol. *Mini-Reviews in Medicinal Chemistry*, 11(4): 298–344. https://pubmed.ncbi.nlm.nih.gov/21428901/#:~:text=Kaempferol%20(3%2C5%2C7,Ginkgo%20biloba%2C%20Tilia%20spp%2C%20Equisetum.

Darvesh, A.S., Carroll, R.T., Bishayee, A., Novotny, N.A., Geldenhuys, W.J. and Van der Schyf, C.J., 2012. Curcumin and neurodegenerative diseases: A perspective. *Expert Opinion on Investigational Drugs*, 21(8): 1123–1140. https://pubmed.ncbi.nlm.nih.gov/22668065/#:~:text=Curcumin%20has%20shown%20therapeutic%20potential,its%20pharmacodynamic%20and%20pharmacokinetic%20profile.

Dumurgier, J. and Tzourio, C., 2020. Epidemiology of neurological diseases in older adults. *Revue neurologique*, 176(9): 642–648. https://www.sciencedirect.com/science/article/abs/pii/S0035378720303908.

Erkkinen, M.G., Kim, M.O. and Geschwind, M.D., 2018. Clinical neurology and epidemiology of the major neurodegenerative diseases. *Cold Spring Harbor Perspectives in Biology*, 10(4): a033118. https://www.ncbi.nlm.nih.gov/pmc/articles/PMC5880171/.

Fabricant, D.S. and Farnsworth, N.R., 2001. The value of plants used in traditional medicine for drug discovery. *Environmental Health Perspectives*, 109: 69–75.

Feigin, V.L., Nichols, E., Alam, T., Bannick, M.S., Beghi, E., Blake, N., Culpepper, W.J., Dorsey, E.R., Elbaz, A., Ellenbogen, R.G. and Fisher, J.L., 2019. Global, regional, and national burden of neurological disorders, 1990–2016: A systematic analysis for the global burden of disease study 2016. *The Lancet Neurology*, 18(5): 459–480. https://www.ncbi.nlm.nih.gov/pmc/articles/PMC6459001/.

Galal, A., Raman, V. and Khan, I.A., 2015. Sida cordifolia, a traditional herb in modern perspective–a review. *Current Traditional Medicine*, 1(1): 5–17. https://www.researchgate.net/publication/281611837_Sida_cordifolia_a_Traditional_Herb_in_Modern_Perspective_-_A_Review.

Goschorska, M., Gutowska, I., Baranowska-Bosiacka, I., Barczak, K. and Chlubek, D., 2020. The use of antioxidants in the treatment of migraine. *Antioxidants*, 9(2): 116. https://www.mdpi.com/2076-3921/9/2/116/htm.

Hajipour, S., Sarkaki, A., Farbood, Y., Eidi, A., Mortazavi, P. and Valizadeh, Z., 2016. Effect of gallic acid on dementia type of Alzheimer disease in rats: Electrophysiological and histological studies. *Basic and Clinical Neuroscience*, 7(2): 97–106. https://www.ncbi.nlm.nih.gov/pmc/articles/PMC4892325/.

Hussain, G., Rasul, A., Anwar, H., Sohail, M.U., Kamran, S.K.S., Baig, S.M. and Shabbir, A., 2017. Epidemiological data of neurological disorders in Pakistan and neighboring countries: A review. *Pakistan Journal of Neurological Sciences*, 12: 52–70. https://ecommons.aku.edu/cgi/viewcontent.cgi?article=1216&context=pjns.

Islam, M.S., Quispe, C., Hossain, R., Islam, M.T., Al-Harrasi, A., Al-Rawahi, A., Martorell, M., Mamurova, A., Seilkhan, A., Altybaeva, N., Abdullayeva, B., Docea, A.O., Calina, D. and Sharifi-Rad, J., 2021. Neuropharmacological effects of quercetin: A literature-based review. *Frontiers in Pharmacology*, 12: 1–16. https://www.frontiersin.org/articles/10.3389/fphar.2021.665031/full.

Jiang, W., Li, S. and Li, X., 2015. Therapeutic potential of berberine against neurodegenerative diseases. *Science in China (Life Sciences)*, 58(6): 564–569. https://www.ncbi.nlm.nih.gov/pmc/articles/PMC5823536/.

Jiang, W., Wei, W., Gaertig, M.A., Li, S. and Li, X.J., 2015. Therapeutic effect of berberine on Huntington's disease transgenic mouse model. *PLOS ONE*, 10(7): 1–16. https://pubmed.ncbi.nlm.nih.gov/26225560/#:~:text=We%20found%20that%20BBR%20can,transgenic%20N171%2D82Q%20HD%20mice.

Kantati, Y.T., Kodjo, K.M., Dogbeavou, K.S., Vaudry, D., Leprince, J. and Gbeassor, M., 2016. Ethnopharmacological survey of plant species used in folk medicine against central nervous system disorders in Togo. *Ethnopharmacology*, 181: 214–220. https://scholar.google.com/citations?view_op=view_citation&hl=fr&user=92yuQKkAAAAJ&citation_for_view=92yuQKkAAAAJ:u-x6o8ySG0sC.

Karamac, M., Kosiñska, A. and Pegg, R.B., 2006. Content of gallic acid in selected plant extracts. *Polish Journal of Food and Nutrition Sciences*, 15: 55–58. https://www.researchgate.net/publication/285726711_Content_of_Gallic_acid_in_selected_plant_extracts.

Khan, A., Ali, T., Rehman, S.U., Khan, M.S., Alam, S.I., Ikram, M., Muhammad, T., Saeed, K., Badshah, H. and Kim, M.O., 2018. Neuroprotective effect of quercetin against the detrimental effects of LPS in the adult mouse brain. *Frontiers in Pharmacology*, 9: 1–16. https://www.frontiersin.org/articles/10.3389/fphar.2018.01383/full#:~:text=It%20has%20been%20reported%20that,%3B%20Liao%20and%20Lin%2C%202015.

Kouhestani, S., Jafari, A. and Babaei, P., 2018. Kaempferol attenuates cognitive deficit via regulating oxidative stress and neuroinflammation in an ovariectomized rat model of sporadic dementia. *Neural Regeneration Research*, 13(10): 1827–1832. https://www.nrronline.org/article.asp?issn=1673-5374;year=2018;volume=13;issue=10;spage=1827;epage=1832;aulast=Kouhestani.

Kumar, V., 2006. Potential medicinal plants for CNS disorders: An overview. *Phytotherapy Research*, 20(12): 1023–1035. https://onlinelibrary.wiley.com/doi/abs/10.1002/ptr.1970.

Kumar, H., More, S.V., Han, S.D., Choi, J.Y. and Choi, D.K., 2012. Promising therapeutics with natural bioactive compounds for improving learning and memory—A review of randomized trials. *Molecules*, 17(9): 10503–10539. https://www.mdpi.com/1420-3049/17/9/10503.

Kundap, U.P., Bhuvanendran, S., Kumari, Y., Othman, I. and Shaikh, M.F., 2017. Plant derived phytocompound, embelin in CNS disorders: A systematic review. *Frontiers in Pharmacology*, 8: 76. https://doi.org/10.3389/fphar.2017.00076.

Lakea, S., 2015. Evidence–based medicine and the growing popularity of complementary and alternative treatments. *University of British Columbia Medical Journal*, 7: 40–41. https://ubcmj.med.ubc.ca.

Landles, C. and Bates, G.P., 2004. Huntingtin and the molecular pathogenesis of Huntington's disease: Fourth in molecular medicine review series. *EMBO Reports*, 5(10): 958–963. https://pubmed.ncbi.nlm.nih.gov/15459747/.

Maiti, P. and Dunbar, G.L., 2018. Use of curcumin, a natural polyphenol for targeting molecular pathways in treating age-related neurodegenerative diseases. *International Journal of Molecular Sciences*, 19(6): 1–42. https://www.mdpi.com/1422-0067/19/6/1637/htm.

Newman, D.J. and Cragg, D.J., 2012. Natural products as sources of new drugs over the 30 years from 1981 to 2010. *Journal of Natural Products*, 75(3): 311–335. https://pubmed.ncbi.nlm.nih.gov/22316239/.

Pandey, S., 2012. Challenges in neurological practice in developing countries. *Indian Journal of Public Health*, 56(3): 227–230. https://pubmed.ncbi.nlm.nih.gov/23229216/.

Park, S.E., Sapkota, K., Kim, S., Kim, H. and Kim, S.J., 2011. Kaempferol acts through mitogen-activated protein kinases and protein kinase B/AKT to elicit protection in a model of neuroinflammation in BV2 microglial cells. *British Journal of Pharmacology*, 164(3): 1008–1025. https://www.ncbi.nlm.nih.gov/pmc/articles/PMC3195922/.

Patil, H.M., 2012. Ethnobotanical notes on Satpura hills of Nandurbar district, Maharashtra, India. *Research Journal of Recent Sciences*, 1: 326–328. http://www.isca.in/rjrs/archive/v1/iISC-2011/53.ISCA-ISC-2011-3BS-41.pdf.

Patil, S.J. and Patil, H.M., 2012. Ethnomedicinal herbal recipes from Satpura Hill ranges of Shirpur Tahsil, Dhule, Maharashtra, India. *Research Journal of Recent Sciences*, 1: 333–336. http://www.isca.in/rjrs/archive/v1/iISC-2011/55.ISCA-ISC-2011-3BS-84.pdf.

Ren, T. and Zuo, Z., 2019. Role of piperine in CNS diseases: Pharmacodynamics, pharmacokinetics and drug interactions. *Expert Opinion on Drug Metabolism and Toxicology*, 15(10): 849–867. https://www.tandfonline.com/doi/abs/10.1080/17425255.2019.1672658.

Salehi, B., Selamoglu, Z., Sener, B., Kilic, M., Kumar Jugran, A., de Tommasi, N., Sinisgalli, C., Milella, L., Rajkovic, J., Morais-Braga, F.B., Bezerra, F. C., D M Coutinho, H., Oluwaseun Ademiluyi, A., Khan Shinwari, Z., Ahmad Jan, S., Erol, E., Ali, Z., Adrian Ostrander, E., Sharifi-Rad, J., de la Luz Cádiz-Gurrea, M., Taheri, Y., Martorell, M., Segura-Carretero, A. and Cho, W.C., 2019. Berberis plants—Drifting from farm to food applications, phytotherapy, and phytopharmacology. *Foods*, 8(10): 1–27. https://www.mdpi.com/2304-8158/8/10/522/htm.

Shabani, S., Rabiei, Z. and Amini-Khoei, H., 2020. Exploring the multifaceted neuroprotective actions of gallic acid: A review. *International Journal of Food Properties*, 23: 736–752. https://www.tandfonline.com/doi/full/10.1080/10942912.2020.1753769#:~:text=During%20the%20past%20years%2C%20it,valuable%20agent%20for%20nutraceutical%20interventions.

Sharifi-Rad, M., Lankatillake, C., Dias, D.A., Docea, A.O., Mahomoodally, M.F., Lobine, D., Chazot, P.L., Kurt, B., Boyunegmez Tumer, T., Catarina Moreira, A., Sharopov, F., Martorell, M., Martins, N., Cho, W.C., Calina, D. and Sharifi-Rad, J., 2020. Impact of natural compounds on neurodegenerative disorders: From preclinical to pharmacotherapeutics. *Journal of Clinical Medicine*, 9(4): 1–19. https://www.ncbi.nlm.nih.gov/pmc/articles/PMC7231062/#B10-jcm-09-01061.

Shorvon, S., 2016. Neurology worldwide: The epidemiology and burden of neurological disease. *Neurology: A Queen Square Textbook*, 1: 1–10. https://catalogimages.wiley.com/images/db/pdf/9781118486177.excerpt.pdf.

Silva dos Santos, J., GoncalvesCirino, J.P., de Oliveira Carvalho, P. and Ortega, M.M., 2021. The pharmacological action of kaempferol in central nervous system diseases: A review. *Frontiers in Pharmacology*, 11: 1–15. https://www.frontiersin.org/articles/10.3389/fphar.2020.565700/full.

Sofic, E., Copra-Janicijevic, A., Salihovic, M., Tahirovic, I. and Kroyer, G., 2010. Screening of medicinal plant extracts for quercetin-3-rutinoside (rutin) in Bosnia and Herzegovina. *Medicinal Plants – International Journal of Phytomedicines and Related Industries*, 2(2): 97–102. https://www.researchgate.net/publication/270191969_Screening_of_medicinal_plant_extracts_for_quercetin-3rutinoside_rutin_in_Bosnia_and_Herzegovina.

Suk, K., 2005. Regulation of neuroinflammation by herbal medicine and its implications for neurodegenerative diseases. *Neurosignals*, 14(1–2): 23–33. https://pubmed.ncbi.nlm.nih.gov/15956812/.

Tabuti, J.R.S., Kukunda, C.B. and Waako, P.J., 2010. Medicinal plants used by traditional medicine practitioners in the treatment of tuberculosis and related ailments in Uganda. *Journal of Ethnopharmacology*, 127(1): 130–136. https://pubmed.ncbi.nlm.nih.gov/19799983/.

Tripathi, A.K., Ray, A.K. and Mishra, S.K., 2022. Molecular and pharmacological aspects of piperine as a potential molecule for disease prevention and management: Evidence from clinical trials. *Beni-Suef University Journal of Basic and Applied Sciences*, 11(1): 1–24. https://bjbas.springeropen.com/articles/10.1186/s43088-022-00196-1.

Upadhyay, R.K., 2014. Drug delivery systems, CNS protection, and the blood brain barrier biomedical research. *International*, 1: 37. https://www.hindawi.com/journals/bmri/2014/869269/.

Wang, C., Cai, Z., Wang, W., Wei, M., Kou, D., Li, T., Yang, Z., Guo, H., Le, W. and Li, S., 2019. Piperine attenuates cognitive impairment in an experimental mouse model of sporadic Alzheimer's disease. *The Journal of Nutritional Biochemistry*, 70: 147–155. https://pubmed.ncbi.nlm.nih.gov/31207354/#:~

:text=In%20summary%2C%20our%20findings%20predict,restore%20neurotransmission%20and%20reduce%20neuroinflammation.

Willcox, M.L. and Gilbert, B., 2009. Traditional medicinal plants for the treatment and prevention of human parasitic diseases. *Ethnopharmacology*, 1: 1–10. https://www.eolss.net/Sample-Chapters/C03/E6-79-12.pdf.

Wink, M., 2010. Introduction: Biochemistry, physiology and ecological functions of secondary metabolites. *Annual Plant Reviews*, 40: 1–19. https://onlinelibrary.wiley.com/doi/pdf/10.1002/9781444320503.ch1.

Yang, W., Chen, Y.H., Liu, H. and Qu, H.D., 2015. Neuroprotective effects of piperine on the 1-methyl-4-phenyl-1, 2, 3, 6-tetrahydropyridine-induced Parkinson's disease mouse model. *International Journal of Molecular Medicine*, 36: 1369–1376. https://www.spandidos-publications.com/10.3892/ijmm.2015.2356.

8 Medicinal and Aromatic Plants against Paratuberculosis

Varsha Srivastava, Manthena Navabharath, Rabea Parveen, Shoor Vir Singh, and Sayeed Ahmad

CONTENTS

8.1 INTRODUCTION

Paratuberculosis, commonly referred to as para-TB, is a chronic enteric inflammatory condition. The infection occurs predominantly in domestic ruminant populations. The infected animals show various symptoms of chronic enteritis, such as weight loss and decreased productivity which may or may not be accompanied by diarrhea. Due to its sub-clinical nature, this chronic infection is often under-reported (Fecteau, 2018).

Mycobacterium avium paratuberculosis (MAP) is the apparent causative agent of para-TB and Johne's disease in ruminants and is endemic worldwide (Yayo Ayele et al., 2001). These bacilli are transmitted to the environment (as excreta) and in milk produced. Many studies conducted at the global level have revealed the presence of these bacilli in raw as well as processed milk products. Bio-load was 9.0–60.0%, 10.7–42.1%, and 21.4–71.4% in liquid milk, flavored milk, and milk powder, respectively, sold in local markets by leading Indian brands (Chaubey et al., 2017; Singh et al., 2018, 2019). Commercial milk products also contain a high bio-load (Chaubey et al., 2017). Through milk, the bacilli enter into human beings giving rise to Crohn's disease (CD) -like inflammatory disorders (Singh et al., 2018). It has been also associated with a number of incurable disorders of an autoimmune nature such as ulcerative colitis, rheumatoid arthritis, thyroiditis, type 1, type 2, and juvenile diabetes, and Alzheimer's and Parkinson's diseases (Sechi and Dow, 2015). CD is a major health problem arising in people due to the use of milk and products made from pasteurized milk.

DOI: 10.1201/9781003226925-8

In developed countries, patients suffering from CD undergo surgical removal of infected intestines, though, in many instances, even after removal of two-thirds of the intestines, CD relapses (Borody et al., 2002). Generally, people prefer surgery over therapy, and Ayurvedic science finds few adoptees in Western countries. In India, on the other hand, where surgery is kept as the last option, people prefer alternative medicines like ayurveda and homeopathy, since they are comparatively safe, cheaper, and efficient. Plants have been acknowledged in various literature for their antimycobacterial properties (Green et al., 2010; McGaw et al., 2008; Mohamad et al., 2011; Zheng et al., 2014). Integration of herbal plant-based therapy with anti-mycobacterial activity may help to combat infection in CD-like disorders and to overcome drug-resistant pathogens. The infection is chronic, and the conventional treatment schedule is long, making the overall cost of treatment high, which is also a major discouragement for the use of allopathic medicines in the treatment of MAP infection in animals and in human beings. Hence, there is a need to develop an alternative therapy for the management of such infections. This chapter identifies potential anti-MAP plants based on their traditional mention as exhibiting anti-mycobacterial, anti-inflammatory, and immunomodulatory properties in the AYUSH (Ayurveda, Yoga and Naturopathy, Unani, Siddha, and Homeopathy) system of medicine. Such plants can be explored to work on a therapy model using the Ayurvedic system of medicine against MAP infection.

8.2 TRAITS OF MAP

8.2.1 Genetic Makeup

Genetically, MAP is intracellular in nature, enclosed by a lipid-enriched cell wall. *M. tuberculosis paratuberculosis* exhibits DNA homology, including 16S rRNA similarities, with MAA (*Mycobacterium avium avium*). Due to this genetic parallelism, the former has been categorized into *Mycobacterium avium* subspecies as MAP (Stevenson, 2015). They can be distinguished on the basis of some phenotypic characteristics, such as pathogenicity and growth rate. Also, the insertion sequence IS 900, being a major genetic characteristic, acts as a distinguishing factor for MAP (Acharya et al., 2017).

8.2.2 *Effect of Heat*

MAP exhibits considerable heat tolerance. It has higher heat tolerance than other mycobacteria (*Mycobacterium phlei*, *M. avium*, *Mycobacterium chelonae*, *Mycobacterium xenopi*, and *Mycobacterium scrofulaceum*) (Clark et al., 2006). This trait explains the lack of effect of pasteurization against these bacilli, making the consumers of contaminated milk vulnerable to these bacilli (Singh and Vihan, 2004).

8.2.3 *Effect of Ultraviolet Light*

These sturdy, slow-growing bacilli are as sensitive to UV light as other bacilli. However, research conducted by Whittington and co-workers in Australia described only a notional influence of UV light on MAP (Manning, 2001; Whittington et al., 2004).

8.2.4 *Vulnerability to MAP*

Young animals (those under the age of six months) are thought to be most vulnerable to infection from intimate contact with infected feces. This theory is based on cattle data (Kennedy and Benedictus, 2001; Larsen et al., 1975). The biological mechanism for this age-related vulnerability is unknown (Larsen et al., 1975). Susceptibility to infection is likely linked to cellular immune system maturation and the inadequate ability of very young animals to cope with intracellular infections like *M. tuberculosis paratuberculosis*. Adult animals can become infected if they are exposed to a large and frequent enough dose of the pathogen (Mortier et al., 2015).

8.3 PATHOGENESIS OF MAP

MAP infection follows the pattern exhibited by *Mycobacterium bovis, Mycobacterium leprae*, or *M. tuberculosis*. The primary THI reaction is referred to as tuberculoid. The reaction is characterized by tissue infiltration in the presence of lymphocytes. MAP bacilli enter into the system through the lumen of the small intestine across M cells inside the lymphoid system. The bacilli further lead to the activation of epithelioid macrophages, eliciting activation of T cells and clonal multiplication (Stevenson, 2015). Further, helper T cells (TH1 and TH2) trigger the immune response. TH1 further produces various cytokine gamma interferons (IFN-γ), tumor necrosis factor-alpha (TNF-α), and interleukin 2 (IL-2). The cytokines produced are responsible for coordinating cell-mediated immune response which leads to the development of an intracellular infection (Mortier et al., 2013). Predominant helper T cell (TH1) activity appears at early sub-clinical levels which lasts for a few months to years, since MAP remains confined in microscopic granulomas and macrophages. T cell memory and reaction continuously maintains granulomas, controls bacillus spread, minimizing damage to tissues (Valentin-Weigand and Goethe, 1999). In the initial stage, the infected animals are characterized by loss of weight accompanied with or without diarrhea. In the lepromatous stage of infection, the infected animals enter helper T cell TH2, which is triggered through unknown causes. Although a humoral immune response is activated, composed of cytokines, including IL-4, IL-5, IL-6, and IL-10 stimulated by TH2 cells, it is incapable of blocking further progression of infection. When these inflammatory mediators penetrate the intestinal wall, it thickens, leading to its malfunctioning and leading to the disruption of nutrient absorption, causing intestinal disorders. Additionally, TNF-α escalation may contribute to tissue catabolism. During this period, the mobility of MAP is exhibited in the gastrointestinal tract (Brady et al., 2008). The affected animals yield infection within a few weeks. The pathogenesis of MAP is illustrated in Figure 8.1.

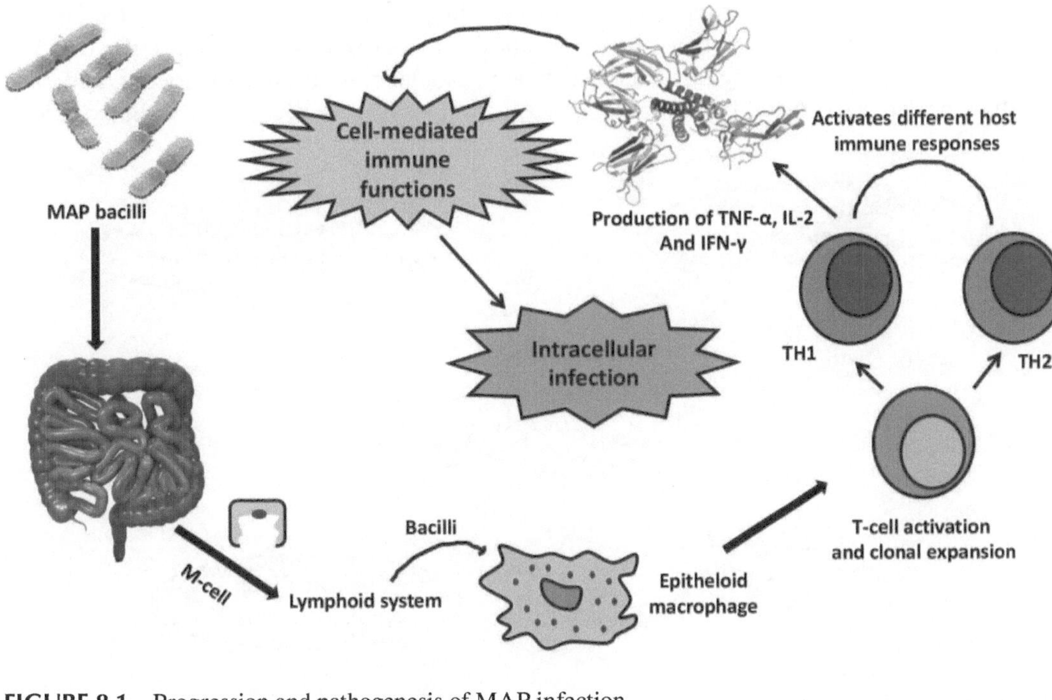

FIGURE 8.1 Progression and pathogenesis of MAP infection.

8.4 TRANSMISSION OF MAP

There are several ways to develop infection from the bacterium. Nursing from an infected dam (direct shedding of the organism into the milk/colostrum) or intake of fecally contaminated solid feed and water are the most common ways to contract the disease (Pierce, 2009). During the widespread phases of the infection, infected cows, and presumably other species, excrete *M. tuberculosis paratuberculosis* straight into milk. Given the volume of milk consumed and the young age of the animals consuming the milk, this is probably the most common infection pathway (Benedictus et al., 2008). *In-utero* infection has been recorded in infected cattle, both clinically and sub-clinically, and it may possibly occur in other species. When the dam is in the late stages of clinical illness, the risk of transplacental exposure is highest (Sweeney, 2011). Whereas *M. tuberculosis paratuberculosis* has been isolated from semen, no investigations have been conducted to determine the likelihood of infecting the dam or fetus by contaminated semen. This route is thought to have a very low possibility of transmission (Ayele et al., 2004). Different sources of transmission are illustrated in Figure 8.2.

8.5 POTENTIAL ANTI-MAP PLANTS

The potential use of plants against paratuberculosis remains unexplored. Based on the anti-inflammatory, anti-tubercular, and immunomodulatory activity, some potentially therapeutic anti-MAP plants and related possible para-TB bioactive metabolites are listed in Table 8.1.

8.5.1 Anti-Inflammatory Activity

MAP is expected to be a potential cause of Crohn's disease-like inflammatory disorders is human beings (Naser et al., 2004). Since the progression of this disease results in inflammation, plants with anti-inflammatory properties can be a suitable choice of treatment for such an infection. Inflammation is a dynamic process that is triggered by burns, mechanical injuries, microbial infections, or other unpleasant stimuli that may endanger the host's health (Abdulkhaleq et al., 2018). Changes in blood flow, increased vascular permeability, tissue destruction via activation, and migration of leucocytes, accompanied by the synthesis of reactive oxygen species (ROS), and the synthesis of local inflammatory mediators, such as prostaglandins (PGs), leukotrienes, and platelet-activating factors induced by phospholipase A2, cyclooxygenases (COXs), and lipoxygenases

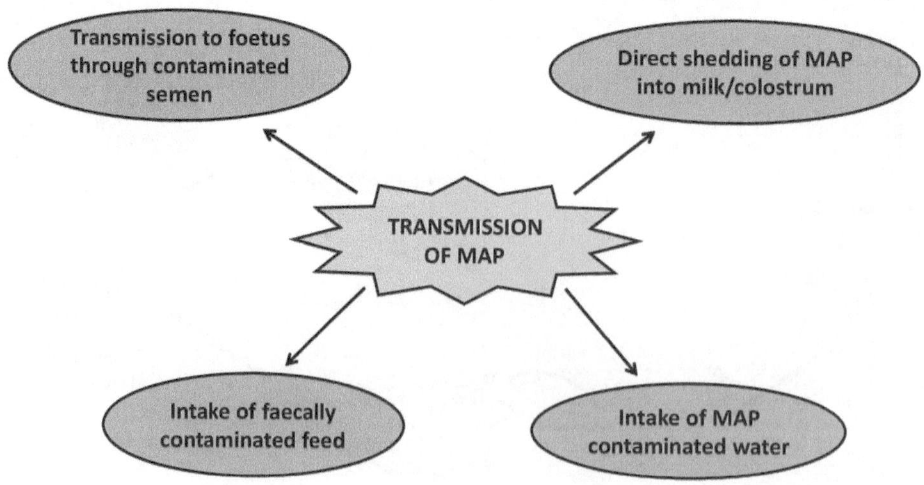

FIGURE 8.2 Sources of MAP transmission.

TABLE 8.1
Potential Anti-MAP Therapeutic Plants and Possible Active Para-TB Metabolites

Botanical Source	Family	Extract Type	Responsible Metabolites	References
Abrusprecatorius L.	Fabaceae	Methanolic	Abruquinone B	Limmatvapirat et al., 2004
Abutilon indicum (L.) Sweet	Malvaceae	Methanolic	β-Amyrin 3-palmitate	Macabeo and Lee, 2014
Allium sativum L.	Amaryllidaceae	Ethanolic	Allicin, ajoene	Viswanathan, et al., 2014; Hannan et al., 2011
Alpinia purpurata (Vieill.) K. Schum.	Zingiberaceae	Methanolic	β-Sitosteryl-3-*O*-6'-palmityl-β-D-glucoside	Villaflores et al., 2010
Andrographis paniculata (Burm.f.) Nees	Acanthaceae	Aqueous	Andrographolide	Prabu et al., 2015; Rashid et al., 2018; Chao et al., 2010
Anisochilus harmandii Doan ex Suddee & A. J. Paton	Lamiaceae	n-Hexane, ethanolic	9α-13α-Epidioxyabiet-8(14)-en-18-oic acid	Lekphrom et al., 2010
Anomianthus dulcis (Dunal) J. Sinclair	Annonaceae	Ethanolic	Water/(2S)-5-hydroxy-6,7-dimethoxyf lavanone	Ubonopas et al., 2014
Artocarpus lacucha Buch.-Ham.	Moraceae	Ethanolic	Lakoochins B and A	Puntumchai et al., 2004
Artocarpus rigidus Blume	Moraceae	n-Hexane, chloroform and methanolic	flavonoid artonin F.	Namdaung et al., 2007
Casearia grewiifolia Vent	Salicaceae	Hexane and dichloromethane	*rel*-(2S,5R,6R,8S,9S,10R,18S, 19R)-18,19-diacetoxy-18,19-epo xy-6-methoxy-2-(2-methylbutano yloxy)cleroda-3,13(16)	Kanokmedhakul et al., 2005
Chromolaena odorata (L.) R. M. King & H. Rob.	Asteraceae	Chloroform	Isosakuranetin (5,7-dihydroxy-4′-methoxyflavanone)	Suksamrarn et al., 2004
Clausena excavata Burm.f.	Rutaceae	Chloroform	Clauszoline J	Ismail Adam Arbab ,2011
Clausena wallichii Oliv.	Rutaceae	Acetone	Heptazoline & Fluroclausine A	Auranwiwat et al., 2014
Combretum wallichii var. *griffithii* (Van Heurck & Müll. Arg.) M. G. Gangop. & Chakrab.	Combretaceae	Methanolic	1-(2-Hydroxy-4-methoxyphe nyl)-3-(4-hydroxy-3-methoxyphenyl)p ropane	Moosophon et al., 2011
Cordia globifera W.W.Sm.	Boraginaceae	Ethanolic	Globiferin and cordiachrome C	Kanokmedhakul et al., 2007
Croton kongensis Gagnep	Euphorbiaceae	Dichloromethane	Ent-8,9-seco-7α-hydroxy-11β-acetoxy kaura-8(14),16-dien-9,15-dione	Thongtan et al., 2003

(Continued)

TABLE 8.1 (CONTINUED)
Potential Anti-MAP Therapeutic Plants and Possible Active Para-TB Metabolites

Botanical Source	Family	Extract Type	Responsible Metabolites	References
Dalbergia parviflora Roxb.	Fabaceae	Methanol, hexane and ethyl acetate	Flavanones, dalparvinene and dalparvone	Songsiang et al., 2009
Dasymaschalon dasymaschalum (Blume) I. M.Turner	Annonaceae	Chloroform	3β-Hydroxy-24-methylenecycloartane	Prawat et al., 2013
Dendrolobium lanceolatum (Dunn) Schindl.	Fabaceae	Chloroform	Flavanones (1 and 2), dibenzocycloheptene derivative (4), flavan, 4′-hydroxy-2′, 2′ dimethyl-pyranoflavan, flavan (3)	Kanokmedhakul et al., 2004
Derris indica (Lam.) Bennet	Fabaceae	Dichloromethane	Novel 2′-methoxy-4′, 5′-methylenedioxyfurano(7,8:4′′, 5′′)-flavone and karanjin	Koysomboon et al., 2006
Diospyros decandra Lour.	Ebenaceae	Dichloromethane	Novel 2-oxo-3β, 19α-dihydroxy-24-nor-urs-12-en-28-oic acid and betulinic acid	Nareeboon et al., 2006
Diospyros ehretioides Wall. ex G.Don	Ebenaceae	Dichloromethane	Palmarumycin JC2	Prajoubklang et al., 2005
Embelia ribes Burm.f.	Primulaceae	Aqueous	Embelin	Bhandari and Ansari, 2008
Eriosema chinense Vogel	Fabaceae	Methanol, hexane and dichloromethane	Khonklonginols A–H, lupinifolinol, yangambin, flemichin D, lupinifolin, eriosemaone A.	Thongnest et al., 2013
Etlingera pavieana (Pierre ex Gagnep.) R.M.Sm.	Zingiberaceae	Dichloromethane	(E)-((E)-3-(4-Methoxyphenyl)allyl)3-(4-hydroxyphenyl)acrylate	Tachai and Nuntawong, 2016
Friesodielsia discolor (Craib) D.Das	Annonaceae	Ethyl acetate and dichloromethane	3′-Formyl-2′,4′-dihydroxy-6′-methoxychalcone	Prawat et al., 2012
Goniothalamus laoticus (Finet&Gagnep.) Bân	Annonaceae	Ethyl acetate and n-hexane	Goniotriol, howiinin A, (+)-altholactone, (−)-nordicentrine	Lekphrom et al., 2009
Hedychium ellipticum Buch.-Ham. ex Sm	Zingiberaceae	Dichloromethane and n-hexane	11,13-Trien-15,16-olide, coronarin E	Songsri and Nuntawong, 2016

(Continued)

TABLE 8.1 (CONTINUED)
Potential Anti-MAP Therapeutic Plants and Possible Active Para-TB Metabolites

Botanical Source	Family	Extract Type	Responsible Metabolites	References
Jatropha integerrima Jacq.	Euphorbiaceae	Eth-8anol & Dichloromethane	Caniojane	Mongkolvisut et al., 2006
Koyamasia calcarea (Kitam.) H. Rob.	Asteraceae	Methanolic	5-Epiisogoyazensolides	Vongvanich et al., 2006
Lavandula angustifolia Mill.	Lamiaceae	Hydroalcoholic extract, polyphenolic fraction	Lavender essential oil	Garnier and Shahidi, 2021; Hajhashemi, Ghannadi, and Sharif, 2003
Momordica charantia L.	Cucurbitaceae	Ethanol	2,4-bis(2-phenylpropan-2-yl)phenol	Panlilio et al., 2012
Ocimumtenuiflorum L.	Lamiaceae	Hydro- distillation	Essential oil	Jayapal et al., 2021
Piper longum L.	Piperaceae	Methanolic	Piperlongumine	Sharma et al., 2014
Plectranthus scutellarioides (L.) R.Br.	Lamiaceae	Chloroform	2′,5′-Dimethyl benzopelargonolactone	Ahmad and Massi, 2014
Premna odorata Blanco	Lamiaceae	Dichloromethane and methanol	1-Heneicosyl formate	Lirio et al., 2014
Solanum virginianum L.	Solanaceae	-	Stigmasterol, carpesterol, diosgenin	Anwikar and Bhitre, 2010; Singh and Singh, 2010
Vitex negundo L.	Lamiaceae	Dichloromethane	Negundoins A–E, a 9,10-*seco*-abietane diterpene, negundoin F, a sandaracopimara-7,15-diene diterpene, negundoin G	Zheng et al., 2010

are major markers of inflammation. Arachidonic acid is a crucial biological intermediary that is transformed into a slew of eicosanoids with diverse biological functions (Vane and Botting, 1998). Herbs such as *Arnica montana*, belonging to the Asteraceae family (Iannitti et al., 2016; Bonamin et al., 2011; Alfredo et al., 2009; da Silva Prade et al., 2020), *Glycyrrhiza glabra*, belonging to the Fabaceae (Tanaka et al., 2008), *Matricaria chamomilla*, belonging to the Asteraceae (Bigagli et al., 2017), and *Salix alba*, belonging to the Salicaceae (Shrivastava et al., 2006) are some folk herbs utilized as anti-inflammatory drugs since ancient times.

A. montana (arnica) has been used in Western medicine to treat bruises and swelling because of its ability to produce sesquiterpene lactones like helenalin and dihydrohelenalin, which are thought to inhibit the activation of the transcription factor nuclear factor (NF-κB), which is responsible for the transcription of genes that encode inflammatory mediators (Bonamin et al., 2011; Lass et al., 2008).

The roots of *G. glabra* (liquorice) were known to Roman medics as Radixdulcis and to Arab physicians as a cough treatment, and the plant has been grown in Europe for its distinctive flavor since the eighteenth century. Glycyrrhizin (at a concentration of 6–13%) and glycyrrhizic acid, both found in liquorice, have anti-inflammatory properties (Račková et al., 2007; Yang et al., 2017). After hydrolysis of glycyrrhizic acid into glycyrrhetic acid, it acts primarily as a powerful inhibitor of 11-hydroxysteroid dehydrogenase, an enzyme that catalyzes the conversion of cortisol to cortisone, resulting in mineralocorticoid activity (Li et al., 2014).

The dried capitula of *Matricaria chamomilla* have been used as an anti-inflammatory and anti-spasmodic since ancient times due to the presence of bisabolol oxides, the efficacy of which has been experimentally evaluated (Harbourne et al., 2009).

Salicin, a glycoside found in *Salix alba* (white willow), is mostly responsible for the anti-inflammatory action of this plant. Salicin has antipyretic and analgesic activity and has traditionally been used to treat rheumatic fever. *Salix fragilis* and *Salix purpurea* (Salicaceae) are the most common, commercial sources of salicin (Schink et al., 2018). Other anti-inflammatory herbs include *Ananas comosus* (Kargutkar and Brijesh, 2018), *Echinacea angustifolia* (Rondanelli et al., 2017) and *Hamamelis virginiana* (Cheesman et al., 2021). Some potent, anti-inflammatory plants are listed in Table 8.1.

8.5.2 Anti-Tubercular Properties

Tuberculosis (TB) is a widespread and ancient disease caused by infection by the *Mycobacterium tuberculosis* complex (Koch and Mizrahi 2018). Because of their incredible chemical variety and millennia of evolution, natural products are critical sources of novel antimicrobials. Several plants from the Lauraceae (Salleh et al., 2015), Magnoliaceae (Cazzaniga et al., 2021), and Piperaceae (Bernuci et al., 2016) families have been proven to have anti-tubercular properties. Plants such as *Ocotea macrophylla* (Shinde et al., 2021) (Lauraceae), *Dugandiodendron argyrotrichum* (Shinde et al. 2021) (Magnoliaceae), *Piper hispidum*, and *Piper eriopodon*, belonging to the Piperaceae family, have exhibited growth suppression against *M. tuberculosis* and *M. bovis*, which are strongly linked to MAP (Bernuci et al., 2016). The herbs inhibit the peptidoglycan biosynthesis pathway by acting on MurE ligase (Shinde et al., 2021). Due to the close association of *M. tuberculosis* with MAP, these herbs can be an acceptable choice for treatment of MAP and related disorders. Examples of anti-tubercular plants are listed in Table 8.1.

8.5.3 *Immunomodulatory Properties*

The immunomodulatory properties of plants have been explored intensively in recent years, owing to a rising understanding of immune system regulation and the desire to achieve certain disease preventive outcomes. Several indigenous herbal treatments have anti-infective effects not only by directly impacting the pathogen, but also by activating the host's innate and adaptive defensive systems (Wen aet al., 2012). As a result of enhanced immune response, plant-based medicines have become more versatile. Modulation of the immune response to ameliorate illness has been practiced in ancient medical systems such as Ayurveda and Unani (Ram Mohan et al., 2019). Plants have also been utilized extensively as a source of medicine in these systems to improve health and decrease the body's susceptibility to illnesses. Nontoxic dosages of the galactoside-specific mistletoe lectin, as well as the edible fungi *Flammulina velutipes* (Wu et al., 2014), *Volvariella volvacea* (Li et al., 2021) and *Gonoderma lucidium* (Li et al., 2020) have been found to cause a cascade of events leading to cell activation, proliferation, and increased lymphokine production. Upon further investigation it was revealed that phytohemagglutinin (PHA), concanavalin A, and some fungal immunomodulatory proteins were responsible for such effects. Plant secondary metabolites are the components primarily responsible for immunomodulatory therapeutic actions. In animal studies, *Asparagus racemosus* (Pise et al., 2015) and *Tinospora cordifolia* (Singh et al., 2004) have been observed to induce substantial leukocytosis and predominant neutrophils, indicating powerful immunomodulatory effect.

The immunomodulatory effects of *Pisum sativum* agglutinin (PSA) have been discovered *in vivo* by its property of activating spleen cells (Chudzik-Kozłowska et al., 2020). The action of *Panax ginseng* on the immune system was discovered to be due to greatly enhanced intracellular killing (Kim et al., 2009). The response of the human immune system to application of aqueous extracts of *Phyllanthus debelis* (Seyed, 2019), *Melothria maderaspatana* (Choudhary et al., 2013), or *Osbeckia aspera* (Nicholl et al., 2001), which demonstrated a substantial anti-complement impact on both the classical system as well as alternative systems.

MAP is associated with the occurrence of autoimmune disorders like Crohn's disease in humans and Johne's disease in animals. Hence, immune-modulating plants are a prospective way for the management of the bacilli concerned and for treatment of associated disorders. Some potential anti-MAP therapeutic immunomodulatory plants are listed in Table 8.1.

8.6 MECHANISM OF ACTION OF VARIOUS PHYTOCHEMICALS

8.6.1 Anti-Inflammatory Phytochemicals

The pathways followed by various anti-inflammatory plants are illustrated in Figure 8.3.

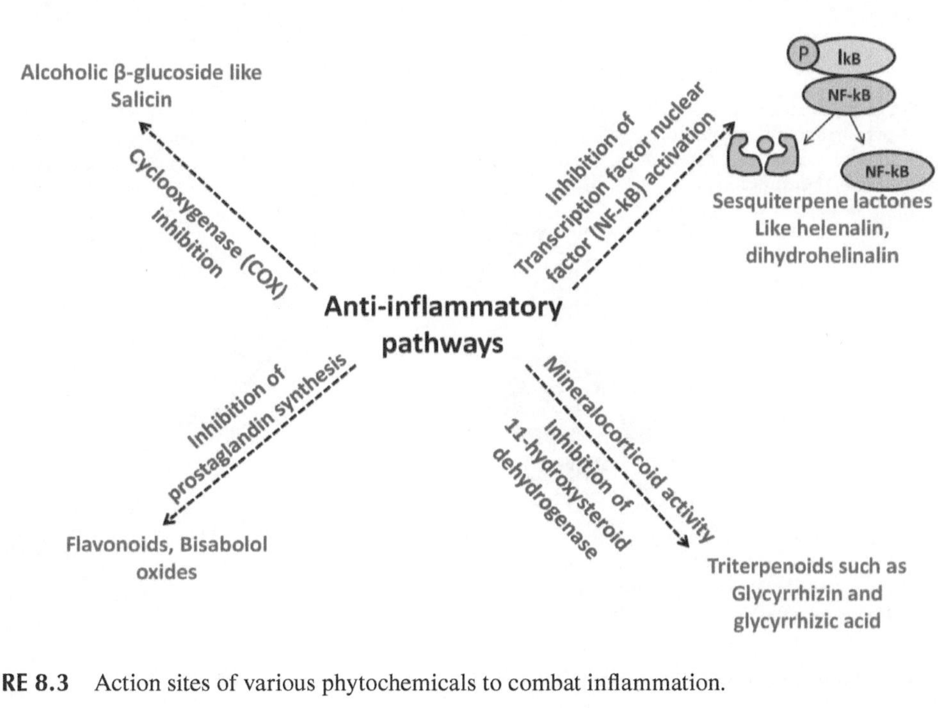

FIGURE 8.3 Action sites of various phytochemicals to combat inflammation.

8.6.2 Anti-Tubercular Phytochemicals

Anti-tubercular drugs act on Mur E ligase, inhibiting peptidoglycan cell wall synthesis, and thereby causing the death of the bacilli. Sites of action of anti-tubercular phytochemicals are illustrated in Figure 8.4.

8.6.3 Immunomodulatory Phytochemicals

The pathways followed by phytochemicals of various immunomodulatory plants is illustrated in Figure 8.5.

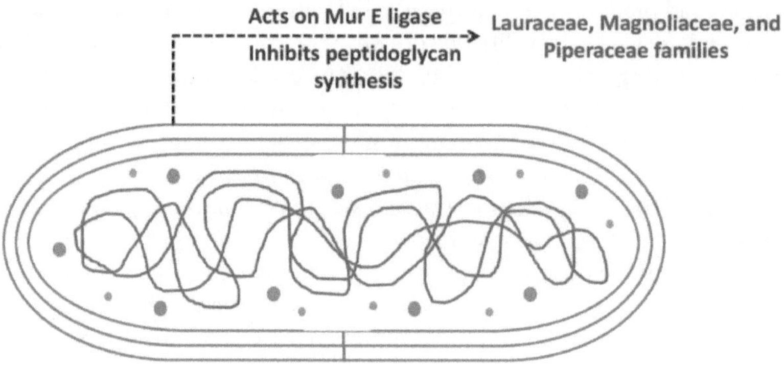

FIGURE 8.4 Mode of action of anti-tubercular plants.

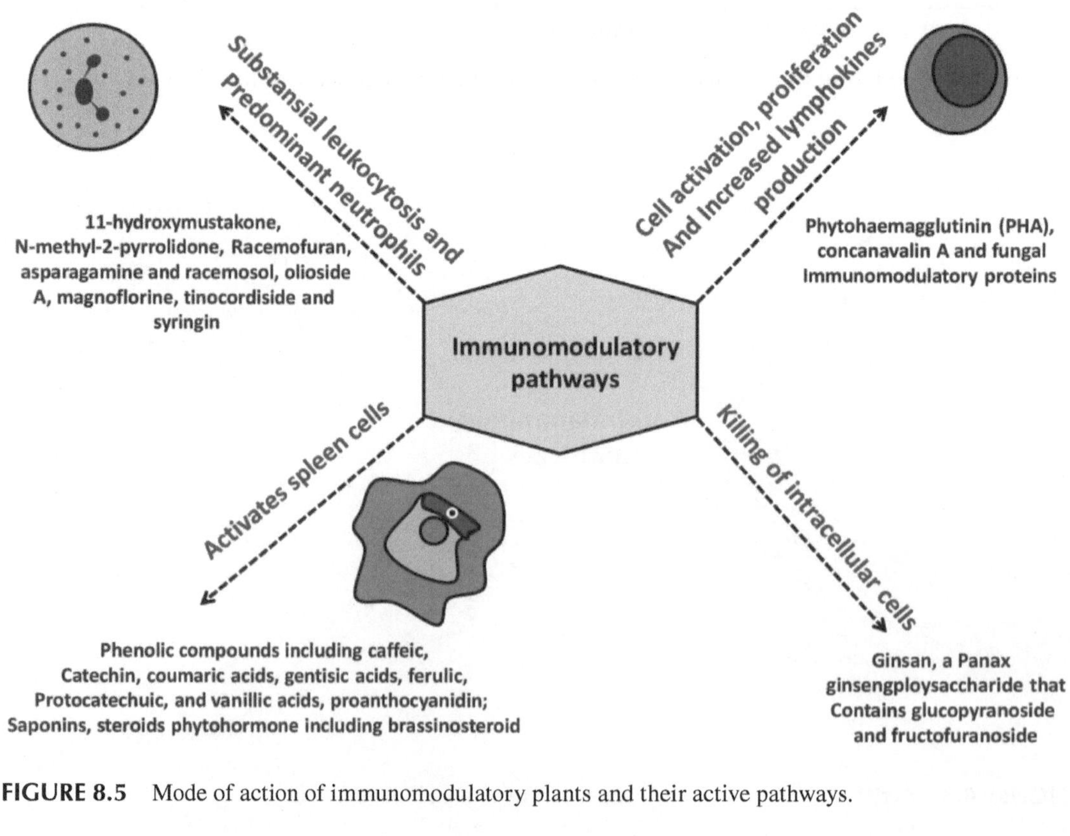

FIGURE 8.5 Mode of action of immunomodulatory plants and their active pathways.

8.7 CONCLUSION AND FUTURE PROSPECTS

Based on the literature, it can be concluded that MAP appears to be an emerging pathogen broadening from the dairy cattle population to human beings. Due to its delayed sub-clinical phase, it remains under-explored. These pathogenic bacilli have evolved as a potential cause behind various autoimmune disorders concerned with inflammation, such as Johne's disease (JD) in animals, particularly ruminants, and Crohn's disease (CD) -like disorders in humans. Although MAP remains under-explored, some countries have taken initiatives for its detection in various dairy samples as well as broader, environmental samples. However, a lacuna exists when it comes to its therapy. Hence, the current chapter deals with some potential anti-MAP plants which can be explored further at the experimental level, to provide a ray of hope for treatment strategies for such a chronic infection.

REFERENCES

Abdulkhaleq, L. A., M. A. Assi, Rasedee Abdullah, M. Zamri-Saad, Y. H. Taufiq-Yap, and M. N. M. Hezmee. 2018. "The Crucial Roles of Inflammatory Mediators in Inflammation: A Review." *Veterinary World*. https://doi.org/10.14202/vetworld.2018.627-635.

Acharya, Kamal R., Navneet K. Dhand, Richard J. Whittington, and Karren M. Plain. 2017. "Detection of Mycobacterium avium subspecies paratuberculosis in Powdered Infant Formula Using IS900 Quantitative PCR and Liquid Culture Media." *International Journal of Food Microbiology*. https://doi.org/10.1016/j.ijfoodmicro.2017.06.005.

Ahmad, Ahyar, and Muh Nasrum Massi. 2014. "The Antituberculosis Drug Rifampicin is Activated by 2', 5'-Dimethyl Benzopelargonolactone from the Leaf of Coleus atropurpureus L. Benth." *International Journal of Pharmacy and Biological Sciences*.

Alfredo, Patrícia P., Carlos A. Anaruma, Antônio C. S. Pião, Silvia M. A. João, and Raquel A. Casarotto. 2009. "Effects of Phonophoresis with Arnica montana onto Acute Inflammatory Process in Rat Skeletal Muscles: An Experimental Study." *Ultrasonics*. https://doi.org/10.1016/j.ultras.2008.12.002.

Anwikar, Shraddha, and Milind Bhitre. 2010. "Study of the Synergistic Anti-Inflammatory Activity of Solanum Xanthocarpum Schrad and Wendl and Cassia Fistula Linn." *International Journal of Ayurveda Research*. https://doi.org/10.4103/0974-7788.72489.

Arbab, Ismail Adam. 2011. "Clausena excavata Burm. f. (Rutaceae): A Review of its Traditional Uses, Pharmacological and Phytochemical Properties." *Journal of Medicinal Plants Research*. https://doi.org/10.5897/jmprx11.013.

Auranwiwat, Chiramet, Surat Laphookhieo, Kongkiat Trisuwan, Stephen G. Pyne, and Thunwadee Ritthiwigrom. 2014. "Carbazole Alkaloids and Coumarins from the Roots of Clausena Guillauminii." *Phytochemistry Letters*. https://doi.org/10.1016/j.phytol.2014.05.003.

Ayele, W. Y., M. Bartos, P. Svastova, and I. Pavlik. 2004. "Distribution of Mycobacterium avium Subsp. paratuberculosis in Organs of Naturally Infected Bull-Calves and Breeding Bulls." *Veterinary Microbiology*. https://doi.org/10.1016/j.vetmic.2004.07.011.

Benedictus, A., R. M. Mitchell, M. Linde-Widmann, R. Sweeney, T. Fyock, Y. H. Schukken, and R. H. Whitlock. 2008. "Transmission Parameters of Mycobacterium avium subspecies paratuberculosis Infections in a Dairy Herd Going through a Control Program." *Preventive Veterinary Medicine*. https://doi.org/10.1016/j.prevetmed.2007.07.008.

Bernuci, Karine Zanoli, Camila Cristina Iwanaga, Carla Maria Mariano Fernandez-Andrade, Fabiana Brusco Lorenzetti, Eduardo Caio Torres-Santos, Viviane Dos Santos Faiões, et al. 2016. "Evaluation of Chemical Composition and Antileishmanial and Antituberculosis Activities of Essential Oils of Piper species." *Molecules*. https://doi.org/10.3390/molecules21121698.

Bhandari, Uma, and M. Ansari. 2008. "Protective Effect of Aqueous Extract of Embelia Ribes Burm Fruits in Middle Cerebral Artery Occlusion-Induced Focal Cerebral Ischemia in Rats." *Indian Journal of Pharmacology*. https://doi.org/10.4103/0253-7613.44153.

Bigagli, Elisabetta, Lorenzo Cinci, Mario D'Ambrosio, and Cristina Luceri. 2017. "Pharmacological Activities of an Eye Drop Containing Matricaria Chamomilla and Euphrasia officinalis Extracts in UVB-Induced Oxidative Stress and Inflammation of Human Corneal Cells." *Journal of Photochemistry and Photobiology, Part B: Biology*. https://doi.org/10.1016/j.jphotobiol.2017.06.031.

Bonamin, Leoni Villano, Ana Paula Kawakami, Cesar Sato, and Thayna Neves Cardoso. 2011. "Inflammatory Process Modulation by Homeopathic Arnica Montana 6CH: The Role of Individual Variation." *Evidence-Based Complementary and Alternative Medicine: eCAM*. https://doi.org/10.1155/2011/917541.

Borody, T. J., S. Leis, E. F. Warren, and R. Surace. 2002. "Treatment of Severe Crohn's Disease Using Antimycobacterial Triple Therapy - Approaching a Cure?" *Digestive and Liver Disease*. https://doi.org/10.1016/S1590-8658(02)80056-1.

Brady, C., D. O'Grady, F. O'Meara, J. Egan, and H. Bassett. 2008. "Relationships between Clinical Signs, Pathological Changes and Tissue Distribution of Mycobacterium avium subspecies paratuberculosis in 21 Cows from Herds Affected by Johne's Disease." *Veterinary Record*. https://doi.org/10.1136/vr.162.5.147.

Cazzaniga, Giulia, Matteo Mori, Laurent Roberto Chiarelli, Arianna Gelain, Fiorella Meneghetti, and Stefania Villa. 2021. "Natural Products against Key Mycobacterium tuberculosis Enzymatic Targets: Emerging Opportunities for Drug Discovery." *European Journal of Medicinal Chemistry*. https://doi.org/10.1016/j.ejmech.2021.113732.

Chao, Wen Wan, Yueh Hsiung Kuo, and B. I.Fong Lin. 2010. "Anti-Inflammatory Activity of New Compounds from Andrographis paniculata by Nf-Kb Transactivation Inhibition." *Journal of Agricultural and Food Chemistry*. https://doi.org/10.1021/jf903629j.

Chaubey, Kundan Kumar, Shoor Vir Singh, Saurabh Gupta, Manju Singh, Jagdip Singh Sohal, Naveen Kumar, Manoj Kumar Singh, Ashok Kumar Bhatia, and Kuldeep Dhama. 2017. "Mycobacterium avium subspecies paratuberculosis – An Important Food Borne Pathogen of High Public Health Significance with Special Reference to India: An Update." *Veterinary Quarterly*. https://doi.org/10.1080/01652176.2017.1397301.

Cheesman, Matthew J., Sean Alcorn, Vishal Verma, and Ian E. Cock. 2021. "An Assessment of the Growth Inhibition Profiles of Hamamelis virginiana L. Extracts against Streptococcus and Staphylococcus spp." *Journal of Traditional and Complementary Medicine*. https://doi.org/10.1016/j.jtcme.2021.03.002.

Choudhary, Soniya, Babeet Singh Tanwer, Tribhuwan Singh, and Rekha Vijayvergia. 2013. "Total Phenolic, Total Flavonoid Content and the DPPH Free Radical Scavenging Activity of Melothria Maderaspatana (Linn.) Cogn." *International Journal of Pharmacy and Pharmaceutical Sciences*.

Chudzik-Kozłowska, Justyna, Ewa Wasilewska, and Dagmara Złotkowska. 2020. "Evaluation of Immunoreactivity of Pea (Pisum sativum) Albumins in BALB/C and C57BL/6 Mice." *Journal of Agricultural and Food Chemistry*. https://doi.org/10.1021/acs.jafc.0c00297.

Clark, Dorn L., Jennifer L. Anderson, Jeff J. Koziczkowski, and Jay L. E. Ellingson. 2006. "Detection of Mycobacterium avium subspecies paratuberculosis Genetic Components in Retail Cheese Curds Purchased in Wisconsin and Minnesota by PCR." *Molecular and Cellular Probes*. https://doi.org/10.1016/j.mcp.2005.12.006.

da Silva Prade, Josiéle, Eveline Costeira Bálsamo, Franciele Romero Machado, Márcia Rósula Poetini, Vandreza Cardoso Bortolotto, Stífani Machado Araújo, Larissa Londero, et al. 2020. "Anti-Inflammatory Effect of Arnica montana in a UVB Radiation-Induced Skin-Burn Model in Mice." *Cutaneous and Ocular Toxicology*. https://doi.org/10.1080/15569527.2020.1743998.

Fecteau, Marie Eve. 2018. "Paratuberculosis in Cattle." *Veterinary Clinics of North America. Food Animal Practice*. https://doi.org/10.1016/j.cvfa.2017.10.011.

Garnier, Andrew, and Fereidoon Shahidi. 2021. "Spices and Herbs as Immune Enhancers and Anti-Inflammatory Agents: A Review." *Journal of Food Bioactives*. https://doi.org/10.31665/jfb.2021.14266.

Green, Ezekiel, Amidou Samie, Chikwelu L. Obi, Pascal O. Bessong, and Roland N. Ndip. 2010. "Inhibitory Properties of Selected South African Medicinal Plants against Mycobacterium tuberculosis." *Journal of Ethnopharmacology*. https://doi.org/10.1016/j.jep.2010.04.033.

Hajhashemi, Valiollah, Alireza Ghannadi, and Badie Sharif. 2003. "Anti-Inflammatory and Analgesic Properties of the Leaf Extracts and Essential Oil of Lavandula angustifolia Mill." *Journal of Ethnopharmacology*. https://doi.org/10.1016/S0378-8741(03)00234-4.

Hannan, Abdul, Muhammad Ikram Ullah, Muhammad Usman, Shahid Hussain, Muhammad Absar, and Khursheed Javed. 2011. "Anti-Mycobacterial Activity of Garlic (Allium sativum) against Multi-Drug Resistant and Non-Multi-Drug Resistant Mycobacterium tuberculosis." *Pakistan Journal of Pharmaceutical Sciences*.

Harbourne, Niamh, Jean Christophe Jacquier, and Dolores O'Riordan. 2009. "Optimisation of the Extraction and Processing Conditions of Chamomile (Matricaria Chamomilla L.) for Incorporation into a Beverage." *Food Chemistry*. https://doi.org/10.1016/j.foodchem.2008.11.044.

Iannitti, Tommaso, Julio Cesar Morales-Medina, Paolo Bellavite, Valentina Rottigni, and Beniamino Palmieri. 2016. "Effectiveness and Safety of Arnica montana in Post-surgical Setting, Pain and Inflammation." *American Journal of Therapeutics*. https://doi.org/10.1097/MJT.0000000000000036.

Jayapal, Venugopal, C. K. Vidya Raj, Muthuraj Muthaiah, Vineet Kumar Chadha, Usharani Brammacharry, Stephen Selvaraj, and Joshy M. Easow. 2021. "In-Vitro Anti-mycobacterium Tuberculosis Effect of Essential Oil of Ocimum Sanctum L. (Tulsi/Basil) Leaves." *Indian Journal of Tuberculosis*. https://doi.org/10.1016/j.ijtb.2021.02.009.

Kanokmedhakul, Somdej, Kwanjai Kanokmedhakul, and Mongkol Buayairaksa. 2007. "Cytotoxic Clerodane Diterpenoids from Fruits of Casearia Grewiifolia." *Journal of Natural Products*. https://doi.org/10.1021/np070083y.

Kanokmedhakul, Somdej, Kwanjai Kanokmedhakul, Tanika Kanarsa, and Mongkol Buayairaksa. 2005. "New Bioactive Clerodane Diterpenoids from the Bark of Casearia Grewiifolia." *Journal of Natural Products*. https://doi.org/10.1021/np049757k.

Kanokmedhakul, Somdej, Kwanjai Kanokmedhakul, Komkrich Nambuddee, and Palangpon Kongsaeree. 2004. "New Bioactive Prenylflavonoids and Dibenzocycloheptene Derivative from Roots of Dendrolobium lanceolatum." *Journal of Natural Products*. https://doi.org/10.1021/np030519j.

Kargutkar, Samira, and S. Brijesh. 2018. "Anti-Inflammatory Evaluation and Characterization of Leaf Extract of Ananas comosus." *Inflammopharmacology*. https://doi.org/10.1007/s10787-017-0379-3.

Kennedy, D. J., and G. Benedictus. 2001. "Control of Mycobacterium avium Subsp. paratuberculosis Infection in Agricultural Species." *OIE Revue Scientifique et Technique*. https://doi.org/10.20506/rst.20.1.1274.

Kim, Mi Hyoung, Yun Young Byon, Eun Ju Ko, Jie Young Song, Yeon Sook Yun, Taekyun Shin, and Hong Gu Joo. 2009. "Immunomodulatory Activity of Ginsan, a Polysaccharide of Panax Ginseng, on Dendritic Cells." *Korean Journal of Physiology and Pharmacology*. https://doi.org/10.4196/kjpp.2009.13.3.169.

Koch, Anastasia, and Valerie Mizrahi. 2018. "Mycobacterium tuberculosis." *Trends in Microbiology*. https://doi.org/10.1016/j.tim.2018.02.012.

Koysomboon, Sorwaporn, Ian van Altena, Shigeru Kato, and Kan Chantrapromma. 2006. "Antimycobacterial Flavonoids from Derris indica." *Phytochemistry*. https://doi.org/10.1016/j.phytochem.2006.03.019.

Larsen, A. B., R. S. Merkal, and R. C. Cutlip. 1975. "Age of Cattle as Related to Resistance to Infection with Mycobacterium paratuberculosis." *American Journal of Veterinary Research*.

Lass, Christian, Marc Vocanson, Steffen Wagner, Christoph M. Schempp, Jean Francois Nicolas, Irmgard Merfort, and Stefan F. Martin. 2008. "Anti-Inflammatory and Immune-Regulatory Mechanisms Prevent Contact Hypersensitivity to Arnica montana L." *Experimental Dermatology*. https://doi.org/10.1111/j.1600-0625.2008.00717.x.

Lekphrom, Ratsami, Somdej Kanokmedhakul, and Kwanjai Kanokmedhakul. 2009. "Bioactive Styryllactones and Alkaloid from Flowers of Goniothalamus laoticus." *Journal of Ethnopharmacology*. https://doi.org/10.1016/j.jep.2009.06.023.

Lekphrom, Ratsami, Somdej Kanokmedhakul, and Kwanjai Kanokmedhakul. 2010. "Bioactive Diterpenes from the Aerial Parts of Anisochilus harmandii." *Planta Medica*. https://doi.org/10.1055/s-0029-1240656.

Li, Chao, Zhuping Dong, Bin Zhang, Qiang Huang, Guang Liu, and Xiong Fu. 2020. "Structural Characterization and Immune Enhancement Activity of a Novel Polysaccharide from Moringa oleifera Leaves." *Carbohydrate Polymers*. https://doi.org/10.1016/j.carbpol.2020.115897.

Li, Jian Yuan, Hong Yan Cao, Ping Liu, Gen Hong Cheng, and Ming Yu Sun. 2014. "Glycyrrhizic Acid in the Treatment of Liver Diseases: Literature Review." *BioMed Research International*. https://doi.org/10.1155/2014/872139.

Li, Ju Pi, Yi Pang Lee, Jung Chein Ma, Betty Revon Liu, Nien Tsu Hsieh, Dz. Chi Chen, Ching Liang Chu, and Ren In You. 2021. "The Enhancing Effect of Fungal Immunomodulatory Protein-Volvariella volvacea (Fip-Vvo) on Maturation and Function of Mouse Dendritic Cells." *Life*. https://doi.org/10.3390/life11060471.

Limmatvapirat, Chutima, Sinenart Sirisopanaporn, and Prasat Kittakoop. 2004. "Antitubercular and Antiplasmodial Constituents of Abrus Precatorius." *Planta Medica*. https://doi.org/10.1055/s-2004-818924.

Lirio, Stephen B., Allan Patrick G. Macabeo, Erickson M. Paragas, Matthias Knorn, Paul Kohls, Scott G. Franzblau, Yuehong Wang, and Ma Alicia M. Aguinaldo. 2014. "Antitubercular Constituents from Premna odorata Blanco." *Journal of Ethnopharmacology*. https://doi.org/10.1016/j.jep.2014.04.015.

Macabeo, G. Allan Patrick, and Christopher A. Lee. 2014. "Sterols and Triterpenes from the Non-polar Antitubercular Fraction of Abutilon indicum." *Pharmacognosy Journal*. https://doi.org/10.5530/pj.2014.4.7.

Manning, Elizabeth J. B. 2001. "Mycobacterium avium subspecies paratuberculosis: A Review of Current Knowledge." *Journal of Zoo and Wildlife Medicine*. https://doi.org/10.1638/1042-7260(2001)032[0293:maspar]2.0.co;2.

McGaw, L. J., N. Lall, J. J. M. Meyer, and J. N. Eloff. 2008. "The Potential of South African Plants against Mycobacterium Infections." *Journal of Ethnopharmacology*. https://doi.org/10.1016/j.jep.2008.08.022.

Mohamad, Suriyati, Nabihah Mohd Zin, Habibah A. Wahab, Pazilah Ibrahim, Shaida Fariza Sulaiman, Anis Safirah Mohd Zahariluddin, and Siti Suraiya Md. Noor. 2011. "Antituberculosis Potential of Some Ethnobotanically Selected Malaysian Plants." *Journal of Ethnopharmacology*. https://doi.org/10.1016/j.jep.2010.11.037.

Mongkolvisut, Wantana, Somyote Sutthivaiyakit, Heiko Leutbecher, Sabine Mika, Iris Klaiber, Wolfgang Möller, Harald Rösner, Uwe Beifuss, and Jürgen Conrad. 2006. "Integerrimides A and B, Cyclic Heptapeptides from the Latex of Jatropha Integerrima." *Journal of Natural Products*. https://doi.org/10.1021/np0602012.

Moosophon, Panawan, Somdej Kanokmedhakul, and Kwanjai Kanokmedhakul. 2011. "Diarylpropanes and an Arylpropyl Quinone from Combretum griffithii." *Journal of Natural Products*. https://doi.org/10.1021/np200593d.

Mortier, Rienske A. R., Herman W. Barkema, Janet M. Bystrom, Oscar Illanes, Karin Orsel, Robert Wolf, Gordon Atkins, and Jeroen De Buck. 2013. "Evaluation of Age-Dependent Susceptibility in Calves Infected with Two Doses of Mycobacterium avium subspecies paratuberculosis Using Pathology and Tissue Culture." *Veterinary Research*. https://doi.org/10.1186/1297-9716-44-94.

Mortier, Rienske A. R., Herman W. Barkema, and Jeroen De Buck. 2015. "Susceptibility to and Diagnosis of Mycobacterium avium subspecies paratuberculosis Infection in Dairy Calves: A Review." *Preventive Veterinary Medicine*. https://doi.org/10.1016/j.prevetmed.2015.08.011.

Namdaung, Umalee, Nuntana Aroonrerk, Sunit Suksamrarn, Kannawat Danwisetkanjana, Janya Saenboonrueng, Wisit Arjchomphu, and Apichart Suksamrarn. 2007. "Bioactive Constituents of the Root Bark of Artocarpus Rigidus subsp. Rigidus." *ChemInform*. https://doi.org/10.1002/chin.200714222.

Nareeboon, Parichat, Wolfgang Kraus, Uwe Beifuss, Juergen Conrad, Iris Klaiber, and Somyote Sutthivaiyakit. 2006. "Novel 24-Nor-, 24-nor-2,3-Seco-, and 3,24-Dinor-2,4-Seco-Ursane Triterpenes from Diospyros decandra: Evidences for Ring A Biosynthetic Transformations." *Tetrahedron*. https://doi.org/10.1016/j.tet.2006.03.034.

Naser, Saleh A., George Ghobrial, Claudia Romero, and John F. Valentine. 2004. "Culture of Mycobacterium avium subspecies paratuberculosis from the Blood of Patients with Crohn's Disease." *Lancet*. https://doi.org/10.1016/S0140-6736(04)17058-X.

Nicholl, Dawn S., Helena M. Daniels, M. Ira Thabrew, Renée J. Grayer, Monique S. J. Simmonds, and Robin D. Hughes. 2001. "In Vitro Studies on the Immunomodulatory Effects of Extracts of Osbeckia aspera." *Journal of Ethnopharmacology*. https://doi.org/10.1016/S0378-8741(01)00319-1.

Panlilio, Bella G., Allan Patrick G. MacAbeo, Matthias Knorn, Paul Kohls, Pascal Richomme, Simeon Fogue Kouam, Dietmar Gehle, et al. 2012. "A Lanostane Aldehyde from Momordica charantia." *Phytochemistry Letters*. https://doi.org/10.1016/j.phytol.2012.07.006.

Pierce, Ellen S. 2009. "Possible Transmission of Mycobacterium avium subspecies paratuberculosis through Potable Water: Lessons from an Urban Cluster of Crohn's Disease." *Gut Pathogens*. https://doi.org/10.1186/1757-4749-1-17.

Pise, Mashitha Vinod, Jaishree Amal Rudra, and Avinash Upadhyay. 2015. "Immunomodulatory Potential of Shatavarins Produced from Asparagus racemosus Tissue Cultures." *Journal of Natural Science, Biology and Medicine*. https://doi.org/10.4103/0976-9668.160025.

Prabu, Amudha, Sameer Hassan, A. S. Prabuseenivasan, L. E. Hanna Shainaba, and Vanaja Kumar. 2015. "Andrographolide: A Potent Antituberculosis Compound that Targets Aminoglycoside 2′-N-Acetyltransferase in Mycobacterium tuberculosis." *Journal of Molecular Graphics and Modelling*. https://doi.org/10.1016/j.jmgm.2015.07.001.

Prajoubklang, Areerat, Busaban Sirithunyalug, Panarat Charoenchai, Rapheephat Suvannakad, Nongluksna Sriubolmas, Sirivipa Piyamongkol, Palangpon Kongsaeree, and Prasat Kittakoop. 2005. "Bioactive Deoxypreussomerins and Dimeric Naphthoquinones from Diospyros Ehretioides Fruits: Deoxypreussomerins May Not Be Plant Metabolites but May Be from Fungal Epiphytes or Endophytes." *Chemistry and Biodiversity*. https://doi.org/10.1002/cbdv.200590108.

Prawat, Uma, Orapan Chairerk, Ratchanee Lenthas, Abdul Wahab Salae, and Pittaya Tuntiwachwuttikul. 2013. "Two New Cycloartane-Type Triterpenoids and One New Flavanone from the Leaves of Dasymaschalon Dasymaschalum and their Biological Activity." *Phytochemistry Letters*. https://doi.org/10.1016/j.phytol.2013.03.002.

Prawat, Uma, Duangraudee Phupornprasert, Akkarapol Butsuri, Abdul Wahab Salae, Sompong Boonsri, and Pittaya Tuntiwachwuttikul. 2012. "Flavonoids from Friesodielsia discolor." *Phytochemistry Letters*. https://doi.org/10.1016/j.phytol.2012.09.007.

Puntumchai, Apirak, Prasat Kittakoop, Shuleewan Rajviroongit, Saovaluk Vimuttipong, Kittisak Likhitwitayawuid, and Yodhathai Thebtaranonth. 2004. "Lakoochins A and B, New Antimycobacterial Stilbene Derivatives from Artocarpus Lakoocha." *Journal of Natural Products*. https://doi.org/10.1021/np030429e.

Račková, Lucia, Viera Jančinová, Margita Petríková, Katarína Drábiková, Radomír Nosáľ, Milan Stefek, Daniela Košťálová, Naďa Prónayová, and Mária Kováčová. 2007. "Mechanism of Anti-Inflammatory Action of Liquorice Extract and Glycyrrhizin." *Natural Product Research*. https://doi.org/10.1080/14786410701371280.

Ram Mohan, Manda, Garige Baba Shankar Rao, Boggula Narender, Chettupalli Ananda Kumar, Pragada Venkateswara Rao, and Vasudha Bakshi. 2019. "Indian Medicinal Plants Used as Immunomodulatory Agents: A Review." *International Journal of Green Pharmacy*.

Rashid, Parisa Tamannur, Muniruddin Ahmed, Md. Mizanur Rahaman, and Md. Abdul Muhit. 2018. "14-Deoxyandrographolide Isolated from Andrographis paniculata (Burm. f) Nees Growing in Bangladesh and its Antimicrobial Properties." *Dhaka University Journal of Pharmaceutical Sciences*. https://doi.org/10.3329/dujps.v17i2.39185.

Rondanelli, Mariangela, Antonella Riva, Paolo Morazzoni, Pietro Allegrini, Milena Anna Faliva, Maurizio Naso, Alessandra Miccono, Gabriella Peroni, Irene Degli Agosti, and Simone Perna. 2017. "The Effect and Safety of Highly Standardized Ginger (Zingiber officinale) and Echinacea (Echinacea angustifolia) Extract Supplementation on Inflammation and Chronic Pain in NSAIDs Poor Responders. A Pilot Study in Subjects with Knee Arthrosis." *Natural Product Research*. https://doi.org/10.1080/14786419.2016.1236097.

Salleh, Wan Mohd Nuzul Hakimi Wan, Farediah Ahmad, Khong Heng Yen, and Razauden Mohamed Zulkifli. 2015. "A Review on Chemical Constituents and Biological Activities of the Genus Beilschmiedia (Lauraceae)." *Tropical Journal of Pharmaceutical Research*. https://doi.org/10.4314/tjpr.v14i11.28.

Schink, Anne, Jan Neumann, Anna Lena Leifke, Kira Ziegler, Janine Fröhlich-Nowoisky, Christoph Cremer, Eckhard Thines, et al. 2018. "Screening of Herbal Extracts for TLR2-And TLR4-Dependent Anti-Inflammatory Effects." *PLOS ONE*. https://doi.org/10.1371/journal.pone.0203907.

Sechi, Leonardo A., and Coad Thomas Dow. 2015. "Mycobacterium avium Ss. Paratuberculosis Zoonosis - The Hundred Year War - Beyond Crohn's Disease." *Frontiers in Immunology*. https://doi.org/10.3389/fimmu.2015.00096.

Seyed, Mohamed Ali. 2019. "A Comprehensive Review on Phyllanthus Derived Natural Products as Potential Chemotherapeutic and Immunomodulators for a Wide Range of Human Diseases." *Biocatalysis and Agricultural Biotechnology*. https://doi.org/10.1016/j.bcab.2019.01.008.

Sharma, Sandeep, Nitin Pal Kalia, Pankaj Suden, Prashant Singh Chauhan, Manoj Kumar, Anshu Beulah Ram, Anamika Khajuria, Sarang Bani, and Inshad Ali Khan. 2014. "Protective Efficacy of Piperine against Mycobacterium tuberculosis." *Tuberculosis*. https://doi.org/10.1016/j.tube.2014.04.007.

Shinde, Yashodeep, Iqrar Ahmad, Sanjay Surana, and Harun Patel. 2021. "The Mur Enzymes Chink in the Armour of Mycobacterium tuberculosis Cell Wall." *European Journal of Medicinal Chemistry*. https://doi.org/10.1016/j.ejmech.2021.113568.

Shrivastava, R., J. C. Pechadre, and G. W. John. 2006. "Tanacetum parthenium and Salix Alba (Mig-RL®) Combination in Migraine Prophylaxis: A Prospective, Open-Label Study." *Clinical Drug Investigation*. https://doi.org/10.2165/00044011-200626050-00006.

Singh, Manju, Saurabh Gupta, Kundan Kumar Chaubey, Shoor Vir Singh, and Jagdip Singh Sohal. 2019. "Profiling of Mycobacterium avium subspecies paratuberculosis in the Milk of Lactating Goats Using Antigen-Antibody Based Assays." *Comparative Immunology, Microbiology and Infectious Diseases*. https://doi.org/10.1016/j.cimid.2019.02.006.

Singh, Manju, Shoor Vir Singh, Saurabh Gupt, Kundan Kumar Chaubey, Jagdip Singh Sohal, and Kuldeep Dhama. 2018. "Bio-incidence Of Mycobacterium avium subspecies paratuberculosis in the Pasteurized Liquid Milk, Flavoured Milk and Milk Powder Commercially Sold by Leading Market Brands in India." *Journal of Experimental Biology and Agricultural Sciences*. https://doi.org/10.18006/2018.6(1).188.203.

Singh, Nisha, Sukh Mahendra Singh, and Pratima Shrivastava. 2004. "Immunomodulatory and Antitumor Actions of Medicinal Plant Tinospora cordifolia Are Mediated Through Activation of Tumor Associated Macrophages." *Immunopharmacology and Immunotoxicology*. https://doi.org/10.1081/IPH-120029952.

Singh, Okram Mukherjee, and Thokchom Prasanta Singh. 2010. "Phytochemistry of Solanum Xanthocarpum: An Amazing Traditional Healer." *Journal of Scientific and Industrial Research*.

Singh, S. V., and V. S. Vihan. 2004. "Detection of Mycobacterium avium subspecies paratuberculosis in Goat Milk." *Small Ruminant Research*. https://doi.org/10.1016/j.smallrumres.2003.12.002.

Songsiang, Uraiwan, Suchana Wanich, Siripit Pitchuanchom, Siwames Netsopa, Komphet Uanporn, and Chavi Yenjai. 2009. "Bioactive Constituents from the Stems of Dalbergia parviflora." *Fitoterapia*. https://doi.org/10.1016/j.fitote.2009.06.002.

Songsri, Sineenard, and Nuchnipa Nuntawong. 2016. "Cytotoxic Labdane Diterpenes from Hedychium ellipticum Buch.-Ham. Ex Sm." *Molecules*. https://doi.org/10.3390/molecules21060749.

Stevenson, Karen. 2015. "Genetic Diversity of Mycobacterium avium subspecies paratuberculosis and the Influence of Strain Type on Infection and Pathogenesis: A Review." *Veterinary Research*. https://doi.org/10.1186/s13567-015-0203-2.

Suksamrarn, Apichart, Apinya Chotipong, Tananit Suavansri, Somnuk Boongird, Puntip Timsuksai, Saovaluk Vimuttipong, and Aporn Chuaynugul. 2004. "Antimycobacterial Activity and Cytotoxicity of Flavonoids from the Flowers of Chromolaena odorata." *Archives of Pharmacal Research*. https://doi.org/10.1007/BF02980123.

Sweeney, Raymond W. 2011. "Pathogenesis of Paratuberculosis." *Veterinary Clinics of North America: Food Animal Practice*. https://doi.org/10.1016/j.cvfa.2011.07.001.

Tachai, Sirichan, and Nuchnipa Nuntawong. 2016. "Uncommon Secondary Metabolites from Etlingera Pavieana Rhizomes." *Natural Product Research*. https://doi.org/10.1080/14786419.2016.1146884.

Tanaka, Aki, Masahiro Horiuchi, Katsummi Umano, and Takayuki Shibamoto. 2008. "Antioxidant and Anti-Inflammatory Activities of Water Distillate and its Dichloromethane Extract from Licorice Root (Glycyrrhiza uralensis) and Chemical Composition of Dichloromethane Extract." *Journal of the Science of Food and Agriculture*. https://doi.org/10.1002/jsfa.3191.

Thongnest, Sanit, Thitima Lhinhatrakool, Nuanchawee Wetprasit, Pakawadee Sutthivaiyakit, and Somyote Sutthivaiyakit. 2013. "Eriosema chinense: A Rich Source of Antimicrobial and Antioxidant Flavonoids." *Phytochemistry*. https://doi.org/10.1016/j.phytochem.2013.06.004.

Thongtan, Jirabhorn, Prasat Kittakoop, Nijsiri Ruangrungsi, Janya Saenboonrueng, and Yodhathai Thebtaranonth. 2003. "New Antimycobacterial and Antimalarial 8,9-Secokaurane Diterpenes from Croton Kongensis." *Journal of Natural Products*. https://doi.org/10.1021/np030067a.

Ubonopas, L., P. Wongsinkongman, W. Chuakul, K. Suwanborirux, K. H. Lee, and N. Soonthornchareonnon. 2014. "Bioactive Flavonoids and Alkaloids from Anomianthus dulcis (Dunal) J. Sinclair Stem Bark." *Mahidol University Journal of Pharmaceutical Sciences*.

Valentin-Weigand, Peter, and Ralph Goethe. 1999. "Pathogenesis of Mycobacterium avium subspecies paratuberculosis Infections in Ruminants: Still More Questions than Answers." *Microbes and Infection/Institut Pasteur*. https://doi.org/10.1016/S1286-4579(99)00203-8.

Vane, J. R., and R. M. Botting. 1998. "Anti-Inflammatory Drugs and their Mechanism of Action." *Inflammation Research*. https://doi.org/10.1007/s000110050284.

Villaflores, Oliver B., Allan Patrick G. MacAbeo, Dietmar Gehle, Karsten Krohn, Scott G. Franzblau, and Alicia M. Aguinaldo. 2010. "Phytoconstituents from Alpinia purpurata and their In Vitro Inhibitory Activity against Mycobacterium tuberculosis." *Pharmacognosy Magazine*. https://doi.org/10.4103/0973-1296.71785.

Viswanathan, V., A. Phadatare, and Alka Mukne. 2014. "Antimycobacterial and Antibacterial Activity of Allium sativum Bulbs." *Indian Journal of Pharmaceutical Sciences*.

Vongvanich, Namphung, Prasat Kittakoop, Panarat Charoenchai, Sutichai Intamas, Kanlayanee Sriklung, and Yodhathai Thebtaranonth. 2006. "Antiplasmodial, Antimycobacterial, and Cytotoxic Principles from Camchaya calcarea." *Planta Medica*. https://doi.org/10.1055/s-2006-951711.

Wen, Chih Chun, Hui Ming Chen, and Ning Sun Yang. 2012. "Developing Phytocompounds from Medicinal Plants as Immunomodulators." *Advances in Botanical Research*. https://doi.org/10.1016/B978-0-12-394591-4.00004-0.

Whittington, Richard J., D. Jeff Marshall, Paul J. Nicholls, Ian B. Marsh, and Leslie A. Reddacliff. 2004. "Survival and Dormancy of Mycobacterium avium Subsp. paratuberculosis in the Environment." *Applied and Environmental Microbiology*. https://doi.org/10.1128/AEM.70.5.2989-3004.2004.

Wu, Ming, Xia Luo, Xiaoyan Xu, Wei Wei, Yu Mengyao, Nan Jiang, Liming Ye, Zhirong Yang, and Xiaofan Fei. 2014. "Antioxidant and Immunomodulatory Activities of a Polysaccharide from Flammulina velutipes." *Journal of Traditional Chinese Medicine*. https://doi.org/10.1016/s0254-6272(15)30089-3.

Yang, Rui, Bo Chuan Yuan, Yong Sheng Ma, Shan Zhou, and Ying Liu. 2017. "The Anti-Inflammatory Activity of Licorice, a Widely Used Chinese Herb." *Pharmaceutical Biology*. https://doi.org/10.1080/13880209.2016.1225775.

Yayo Ayele, W., M. Macháčková, and I. Pavlík. 2001. "The Transmission and Impact of Paratuberculosis Infection in Domestic and Wild Ruminants." *Veterinarni Medicina*. https://doi.org/10.17221/7878-vetmed.

Zheng, Cheng Jian, Bao Kang Huang, Yang Wang, Qi Ye, Ting Han, Qiao Yan Zhang, Hong Zhang, and Lu Ping Qin. 2010. "Anti-Inflammatory Diterpenes from the Seeds of Vitex negundo." *Bioorganic and Medicinal Chemistry*. https://doi.org/10.1016/j.bmc.2009.11.004.

Zheng, Yuejuan, Xin Jiang, Feng Gao, Junxiang Song, Jinxia Sun, Lixin Wang, Xiaoxia Sun, Zhenhui Lu, and Huiyong Zhang. 2014. "Identification of Plant-Derived Natural Products as Potential Inhibitors of the Mycobacterium tuberculosis Proteasome." *BMC Complementary and Alternative Medicine*. https://doi.org/10.1186/1472-6882-14-400.

9 Standardization and Evaluation of Medicinal Properties of *Cleome ambylocarpa* Barr. and Murb. (*Cleome africana*)

Mohammad Kamil

CONTENTS

9.1 INTRODUCTION

In recent years, with the ever-growing commercialization of herbal medicines, there has been a consequential increased demand for quality control procedures to ensure standardization. These studies have been concerned with the identity, purity, and quality of the genuine material and checking for adulteration. This is considered to be essential as a set of standards would be a significant deterrent to substitution and adulteration and would be a considerable aid for drug enforcement.

Cleome is a genus of flowering plants in the family Cleomaceae, commonly known as spider flowers, spider plants, spider weeds, or bee plants (Pagani, 1970; Ahmed, 2001). Previously, it had been classified into the family Capparaceae, until DNA studies found the Cleomaceae genera to be more closely related to the Brassicaceae than to the Capparaceae. *Cleome* and clammyweed (*Polanisia dodecandra*) can also sometimes be confused. The simplest way to differentiate the two is to compare the seed pods which project out or down on *Cleome* and up on clammy weed.

DOI: 10.1201/9781003226925-9

FIGURE 9.1 A general view of *Cleome ambylocarpa* in nature.

Cleome amblyocarpa Barr. & Murb. (syn *Cleome africana*), commonly known as spider flower, is usually found on the Arabian Peninsula, from Egypt to Morocco. The plant is also common in the United Arab Emirates, in the sandy desert of the northern emirates. It is an erect perennial glandular, somewhat fetid herb, 50 cm high, sometimes sub-frutescent at the base. Leaves are tri-foliolate, with elliptical-oblanceolate leaflets 4–25 mm long, 2–5 mm wide, one petiole about as long as the blade. The uppermost leaves are often simple and smaller. Flowers are pedicelled, with a rather obscure corolla; petals are white, broadly veined yellow, with purple-veined tips. Fruit is a compressed, glandular, two-valved capsule, 55 mm long, 3–10 mm wide, mucronate at the apex, on pedicels 3–10 mm long. Seeds are densely covered with white hairs (Mandaville, 1990; Jongbloed, 2003).

Normally, it is the leaves and aerial parts which are used medicinally. Infusions of leaves have an almost immediate effect on abdominal and rheumatic pains (Loutfy Boulos, 1983). The dried leaf is pale green in color and lanceolate, but the dry leaf curls inwards, leaving the dorsal part to the outside. It is easily broken. At the top of the green plant, a number of leaves (3–4 leaves) are found in a group (Figure 9.1).

A surface view of the leaf lamina shows an intricate network of venation, and, in addition to the entire margin of the leaf, both surfaces bear many glandular trichomes. These trichomes are short and have multicellular, rounded heads and multiseriate stalks. Both epidermizes consist of small, polygonal cells. The upper epidermis bears many small, rounded to oval a nomocytic a normocytic. A transverse section of the leaf shows that each epidermis consists of one layer. The leaf is dorsiventral and one palisade layer exists below the upper epidermis, consisting of small, compactly packed, cylindrical cells with rounded tips, containing chloroplasts. The spongy mesophyll is differentiated into elongated cells with irregular shapes and thin cell walls, and they also contain chloroplasts. The vascular tissues are embedded in the spongy mesophyll, and the vessels are annularly thickened. The stem is pale yellow in color. It is almost quadrangular in cross-section, with prominent, longitudinal striations with slight protrusions. When held between fingers, the stem surface gives a coarse feeling. It branches at the internodes and, when broken, it shows white internal tissues.

A transverse section through the stem shows an almost circular or semi-oblong cross-section. From the periphery inwards, the section shows the following: an epidermis consisting of polygonal cells, which exhibit small, arch-like protrusions, and also bear a number of glandular trichomes with rounded heads and multiseriate stalks, in addition to oval stomata; the light- yellow-colored, multi-layered cortex the outer layers of which consist of collenchyma cells whereas the inner layers consist of polygonal cells interrupted with groups of long lignified fibers; a few layers of compressed parenchyma cells surrounding a circle of heavily lignified, vascular tissues with annularly and spirally thickened vessels; and other layers of compressed parenchyma cells that surround the pith and then the pith that consists of many, greenish-yellow cells.

The herbal medicinal material consists of a dark-greenish-yellow, homogeneous powder. It has a light odor reminiscent of dried straw and a slightly, somewhat astringent, acrid taste. Microscopically,

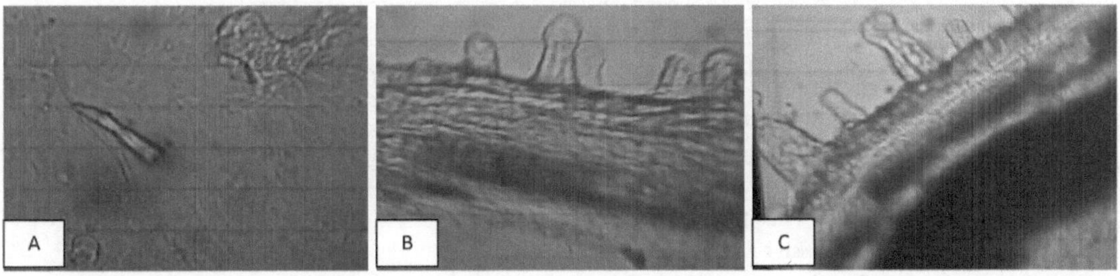

FIGURE 9.2 A, B, C. Microscopical Evaluation of *Cleome ambylocarpa.*

the powder shows longitudinal fragments of the stems and branches that contain the gray-colored, fibro-vascular tissues. The vessels of the xylem are compactly arranged, and they are spirally and annularly thickened. It also shows fragments of colorless polygonal, epidermal cells and the almost rounded, colorless pith cells with their obvious cell walls (Figure 9.2A, B, C).

A. A surface view of the leaf showing its small, polygonal cells and a glandular trichome, which is short with a multicellular, rounded head and a multiseriate stalk, quite characteristic of this plant.
B. A surface view of the stem shows the epidermal polygonal cells to have small, arch-like protrusions, oval stomata, and a number of glandular trichomes similar to those of the leaf.
C. A transverse section through a portion of the stem exhibiting the epidermal layer that bears a number of glandular trichomes, the light-yellow outer layer of cortex, which consists of collenchyma cells, the inner layer of cortex consisting of parenchyma cells interrupted with groups of long, lignified fibers, layers of compressed parenchyma cells, heavily lignified vascular tissues, compressed parenchyma layers and then the rounded cells of the pith.

9.2 CHEMICAL CONSTITUENTS

The aerial parts of *C. amblyocarpa* yield dammarane-type triterpenes. Four new dammarane triterpenes, amblyone, cleomblynol A, isocleomblynol A, and cleomblynol B, and two known compounds, cleocarpanol and cabraleahydroxy lactone, were isolated from the aerial parts of this species (Harraz, 1995). Previous studies had shown that *Cleome* species contained ten methylated flavonoids and eight flavonol glycosides (Sharaf, 1997). The dichloromethane-methanol extraction of the aerial parts of *C. amblyocarpa* led to the isolation of a new dammarane triterpenoid, 15 alpha-acetoxycleomblynol A (Ahmed, 2001).

The following chemical studies have previously been carried out (Quality Control Methods, 1998; Evans, 1996) on *C. amblyocarpa* (Tables 9.1, 9.2, and 9.3; Figure 9.3).

Thin-layer chromatography (TLC) fingerprint of petroleum ether (60–80°C) (track 1) and methanolic extract (track 2)

Mobile phase	Figure 9.4A, B: Toluene: ethyl acetate (93:7)
	2-C: Ethyl acetate: methanol: water (100:13.5:10)
	2-D: Toluene: ethyl formate: formic acid (5:4:1)
Detection:	UV 254 nm (A)
Derivatization,	2-B, C and D: vanillin-sulfuric acid-visible

9.3 PHARMACOLOGICAL AND TOXICOLOGICAL STUDIES

Although antioxidant activity has been reported from this plant (Selloum, 1997), it also has diuretic and laxative activity (Bose, 2009). *Cleome gynandropsis* also has antibacterial activity (Perumal,

TABLE 9.1
Physicochemical Studies of *Cleome amblyocarpa*

Physicochemical Constants (%)	
Loss of weight on drying at 105°C	9.60
Absolute alcohol solubility	8.00
Water solubility	28.00
Successive Extractives (%)	
Petroleum ether (60–80°C)	4.76
Chloroform	2.15
Absolute alcohol	9.07
Ash Values (%)	
Total ash	13.50
Water-soluble ash	5.33
Acid insoluble ash (10% HCl)	0.67
pH Values (Aqueous Solution)	
pH of 1% solution	5.599–5.601
pH of 10% solution	5.013–5.020

TABLE 9.2
Elemental Analysis

	Ash values (British Herbal Pharmacopeia; BHMA, 1996)				
	Assay and identification of element (AOAC International)				
Apparatus	(AA-6800 Shimadzu-Flame method)				
Element	Std. conc. µg/ml (ppm)	Sample conc. mg/ml	Sample absorbance	Actual conc. mg/ml	Actual conc. (%)
Cr	1, 2, 4	20.014	0.0016	0.01045	0.001045
Zn	0.5, 1, 2	20.014	0.0878	0.0420750	0.0042075
Cu	0.5, 1, 2	20.014	0.0138	0.006585	0.0006585
Fe	1, 2, 4	20.014	0.2330	0.446655	0.0446655
K	1, 2, 4	20.014	2.2704	1.2847600	0.128476
Pb	1, 2, 4	20.014	0.0000	0.0000	0.0000
Cd	0.25, 0.5, 1	20.014	0.0003	0.0000500	0.000005

1ppm conc. = 1µg/ml; Actual conc. (%) = Actual conc. (ppm) x 0.0001 [1ppm=0.0001%]

TABLE 9.3
UV Spectral Studies:

Sample Conc. mg/ml	Solvent Used	Max	Min	Remark
5.02 mg/ml	Intestinal fluid simulated without pancreatic (pH = 7.5)	-ve	-ve	225 to 400
1.02 mg/ml	Gastric Fluid simulated without pepsin (pH = 1.2)	319	309	225 to 400

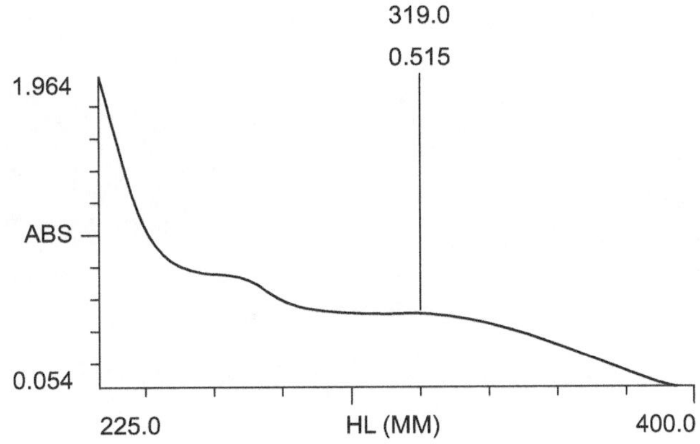

FIGURE 9.3 A graphical representation of gastric fluid simulated without pepsin, pH= 1.2 ± 0.1.

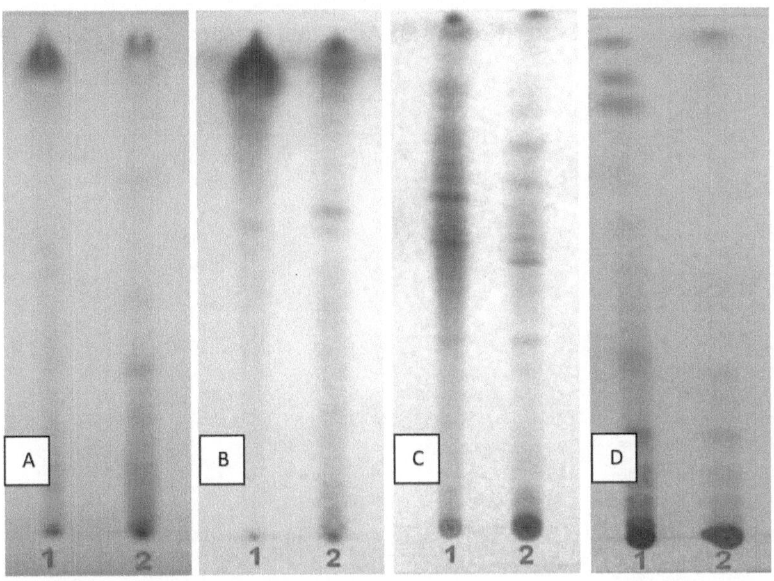

FIGURE 9.4 A general view of the thin-layer chromatography (TLC) (Wagner, 1996).

1999). The anti-inflammatory activity of *Cleome arabica* leaf extract has been studied both *in vivo* and in vitro. The *in vivo* studies examined the extract's anti-inflammatory activity using the carrageenan-induced rat paw edema model of acute inflammation.

Non-parenchymal cells of *Cleome* might play an important role in the modulation of xenobiotic metabolism in the liver and its pharmacological and toxicological consequences. Therefore, the role of cell-to-cell interactions in herbal-induced liver toxicity was investigated in monocultures of cells from the human hepatocyte cell line (HepG2) and in co-cultures of cells from the HepG2 cell line and cells from the human monocyte cell line (THP1). Cells were treated with various concentrations (1–500 µl/ml) of extracts of *Pistacia palaestina*, *Juglans regia*, and *Quercus ithaburensis* for 24 h. Extracts from *Cleome droserifolia*, a known toxic plant, were regarded as a positive control. In the co-culture system, toxic effects were observed after exposure to extracts of *P. palaestina* and *C. droserifolia*. The extract significantly reduced cell viability as measured by the MTT test and the LDH assay (Saad, 2006).

Pharmacological and safety evaluation of the aqueous extract of the plant *C.amb lyocarpa* was carried out (Table 9.4).

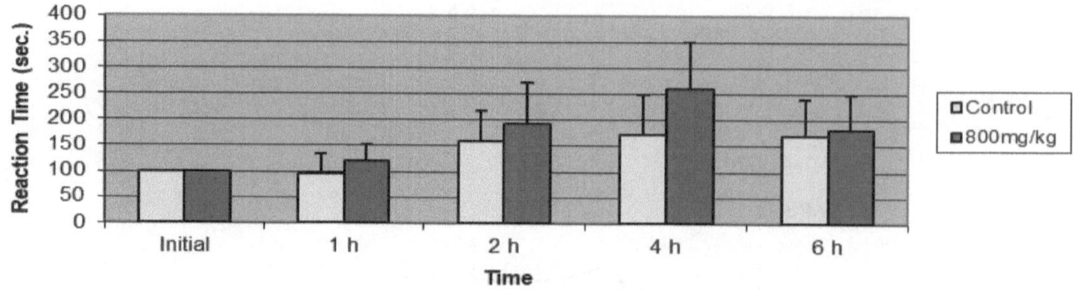

FIGURE 9.5 Anti-nociceptive activity of *Cleome* aqueous extract (tail flick method; p.o., n=7).

TABLE 9.4
Pharmacological and Safety Evaluation of an Aqueous Extract of the Plant *Cleome amblyocarpa*

Activity	Results			
	Strong	Moderate	Mild	Negative
Analgesic			√	
Anti-inflammatory				√
Cardiovascular		√		
Effect on isolated rat fundus	√			
Sexual effects				√
Effect on rat fundus	√			
Effect on guinea pig tracheal chain			√	
Antithrombotic effect	√			
Studies on biochemical parameters			√	
Studies on hematological parameters		√		
Locomotor				√
Motor co-ordination (grip strength and motor activity)				√
Rectal temperature				√
Body weight				
Vital organs				
Lethality				√

9.4 ANTINOCICEPTIVE ACTIVITY

9.4.1 Acetic Acid-Induced Writhing Method

The number of writhings in 25 min after acetic acid injection was recorded **at intervals** of 5 minutes. The ethanolic extract of *C. amblyocarpa* at a dose of 800 mg/kg., p.o. (by mouth), did not show analgesic activity in the acetic acid writhing experiment. The onset of writhing was reduced and the total number of writhings was increased in the treated group compared to the control group, although these changes were not statistically significant.

An aqueous extract of *C. amblyocarpa* had a significant antinociceptive effect, which increased the systolic and diastolic blood pressure as well as the heart rate at both doses studied.

C. amblocarpa aqueous extract produced a significant increase in resting tension of isolated rat fundus strip. The extract produced mild relaxation (~11%) in the histamine pre-contracted tracheal chain. The extract had no effect on the copulatory behavior of male rats. The extract appears to be non-toxic at the doses studied with an oral, half-maximal lethal dose (LD_{50}) of 4.067 g/kg.

9.5 EFFECTS ON THE CARDIOVASCULAR SYSTEM

9.5.1 *Effect on Blood Pressure and Heart Rate* (Figure 9.6, Figure 9.7, and Figure 9.12)

The extract (400 or 800 mg/kg, i.p.) or normal saline was administered intraperitoneally in different animals and blood pressures were recorded at different time intervals following administration. Aqueous extracts of *C. amblyocarpa* at 400 mg/kg, i.p., increased the systolic blood pressure in anesthetized rats at 5-, 10-, 15-, and 30-min observation periods, which were significant at 5, 10, and 15 minutes. At 800 mg/kg, the systolic pressure was significantly increased at 5, 10, 15, and 30 minutes.

C. amblyocarpa extract increased diastolic blood pressure significantly (Figure 9.6) at the doses of 400 and 800 mg/kg at 5-, 10-, 15-, 30-, or 60-min observation points. At 90 and 120 min, there was also an increase in the diastolic pressure with both doses, although the increases were not significant.

Heart rate was significantly increased by *Cleome* extract at 400 mg/kg at 5, 10, and 15 min. At 800 mg/kg, *C. amblyocarpa* extract increased the heart rate significantly at 10 and 15 min. Aqueous extracts of *C. amblyocarpa* increased both heart rate and blood pressure in anesthetized rats. The possibility of the therapeutic dose affecting the cardiovascular parameters has to be ruled out before sending the preparation for clinical studies with an aqueous extract of *C. amblyocarpa*

9.6 EFFECT ON ISOLATED RAT FUNDUS

9.6.1 In Vitro Experiment

Rat fundus was set up in an organ bath and maintained at a resting tension of 1 g. After a 30-min stabilization period, the *Cleome* extract at concentrations of 0 (control), 0.1, 0.3, 0.5, 1.0, 3.0, or 5.0 mg/ml were added to the bath and responses recorded.

The *C. amblyocarpa* aqueous extract at concentrations of 0.1, 0.3, 0.5, 1.0, 3.0, or 5.0 mg/ml induced increases in the resting tension of isolated rat fundus strip which were statistically

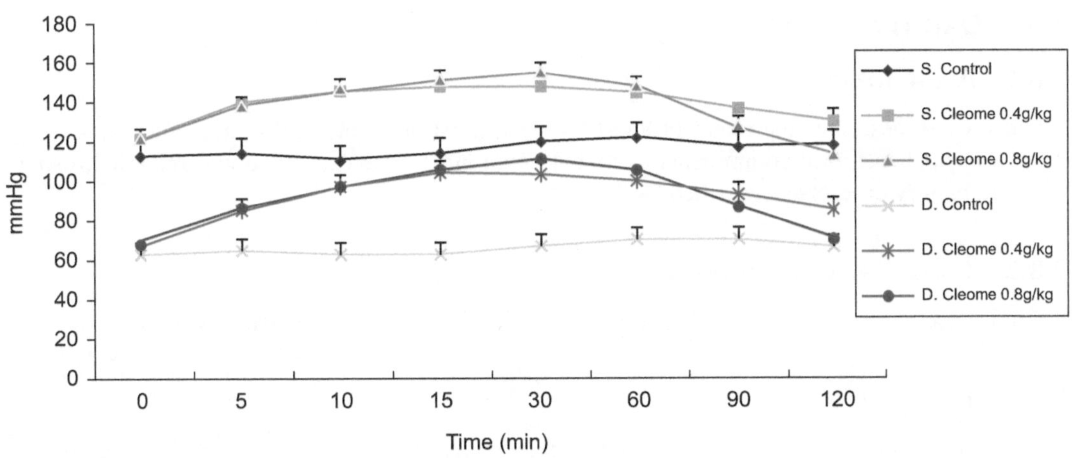

FIGURE 9.6 Effect of *Cleome* water extract on systolic and diastolic BP in rats.

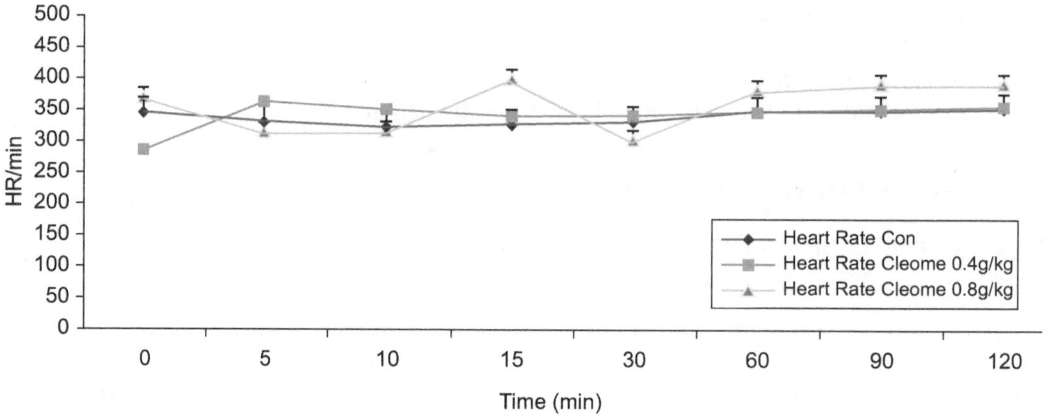

FIGURE 9.7 Effect of *Cleome* water extract on heart rate in rats.

significant. The maximum response (134% of resting tension) was induced by the 5.0 mg/ml concentration. The responses were dose dependent. The increase in tension by the *Cleome* extract was almost completely abolished by pre-treatment of the tissue with 100 μM atropine. The extract may thus be able to increase gastric motility acting through muscarinic mechanisms.

9.7 EVALUATION OF SEX-STIMULANT ACTIVITY

9.7.1 Copulatory Behavior in Mice

Female rats (used as the mating stimulus) were brought into estrus with a single subcutaneous dose of 600 μg of progesterone 72 h or 6 h before testing. The male rats were first placed in the observation arena for an adaptation period of 3–5 min and then a sexually receptive female rat was introduced. The Mount Latency, Intromission Latency, Ejaculation Latency, Mount Frequency, Intromission Frequency, Mean Intromission Interval, and Post-Ejaculatory Interval were recorded. Results showed that behavior of male rats treated with *C. amblyocarpa* aqueous extract produced no significant change in any of the sex parameters observed compared to the control

9.8 TOXICITY STUDIES

9.8.1 Acute Toxicity

Acute toxicity describes the result of short-term administration of a substance, via a single high dose. Various behavioral components as toxicity endpoints were examined and recorded, using a standard battery of tests as described below.

9.8.2 Gross Behavioral Studies

A single dose of the extract at doses of 0.8, 1.6, 3.2, and 6.4 mg/kg., p.o., was given to the mice.

9.8.3 *LD_{50} Studies*

This involves giving different doses of the plant extract to matched groups of animals, usually mice, in order to identify the dose causing 50% mortality. The acute group was given one of the

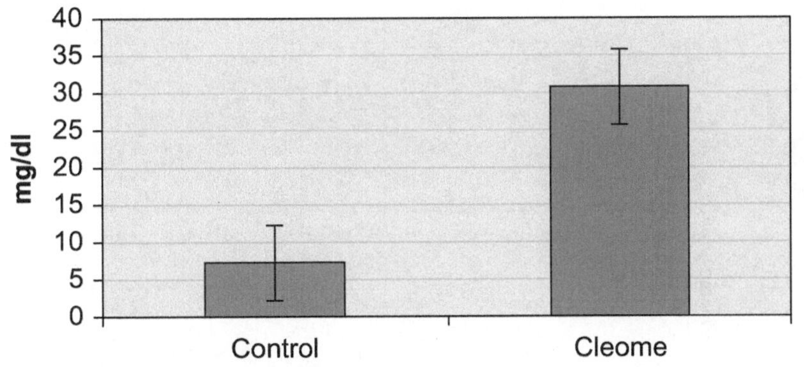

FIGURE 9.8 Effect of *Cleome* Aq. Ext. on febrinogen level.

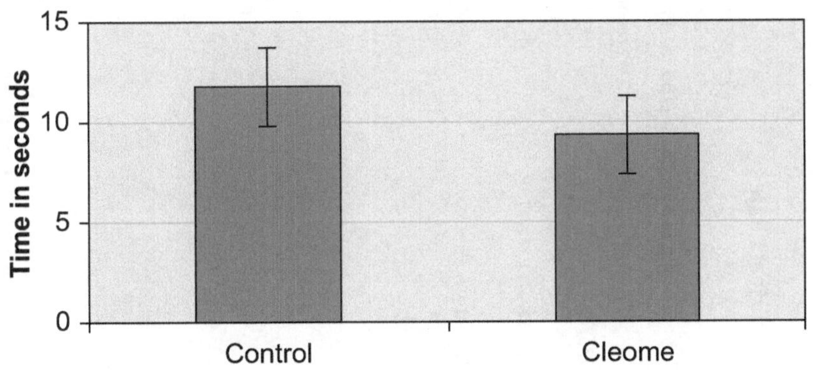

FIGURE 9.9 Effect of *Cleome* Aq. Ext. on prothrombin.

four increasing doses of the extract, i.e., 4.47, 4.869, 5.308, and 6.4 g/kg orally corresponding to a regular logarithmic thatrval. Mortality was recorded in the following 24 h. Doses that caused no death or 100% death were determined and the LD_{50} was calculated according to the Spearman-Karber method. No mortality was found during the observation period at the tested doses indicating that the extract produces no observable toxicity and that it may be used for further pharmacological experiments.

The aqueous extract of *C. amblyocarpa* induced signs of toxicity with mortality rates of 0%, 37.5%, 62.5%, and 87.5% respectively for doses of 4.47, 4.869, 5.308, and 6.4 g/kg, p.o., indicating that the extract appears to be toxic at the doses studied (Figure 9.8, Figure 9.9).

9.8.4 *Cleome* Antidepressant Activity

The Effects of *Cleome amblyocarpa* on Blood Pressure and Heart Rate in Rats (Table 9.5, Figure 9.10, Figure 9.11).

TABLE 9.5
Effects of *Cleome amblyocarpa* Extract on Blood Pressure and Heart Rate

	Control Group			*Cleome amblyocarpa* (100 mg/kg)			*Cleome amblyocarpa* (200 mg/kg)		
Time (min)	SBP	Fine V.	Percentage	SBP	Fine V.	Percentage	SBP	Fine V.	Percentage
Initial	115.67± 11.46		100.00± 0.00	135.00 ± 5.16		100.00± 0.00	129.17± 11.86		100.00± 0.00
5 min	112.50 ± 10.91		97.36± 0.86	129.17 ± 4.73		95.76± 1.49	115.00± 12.04		88.66± 2.69*
10 min	110.33 ± 11.32		95.15± 1.37	125.50 ± 6.18		92.97± 2.88	112.50± 11.81		86.97± 3.20*
15 min	108.67 ± 10.70		93.96± 1.65	122.17 ± 4.47		90.74± 2.91	108.00± 11.02		83.92± 4.09*
30 min	109.83± 9.04		95.85± 2.10	123.00 ± 4.83		91.31± 2.80	113.83± 10.72		88.91± 4.78
60 min	105.83± 10.30		91.74± 3.36	120.50 ± 5.41		89.43± 3.48	128.00± 9.93		99.90± 2.92
90 min	107.67 ± 10.39		93.55± 3.35	123.83± 5.89		91.96± 3.86	132.50± 9.90		103.76± 4.31
120 min	107.67 ± 10.28		93.58± 3.09	123.33± 7.26		91.67± 5.29	131.67± 10.54		102.73± 4.23
N	6		6	6		6	6		6

* *P*<0.05, ** *P*<0.01 vs Control Group. Animals: Wistar rats, both male and female.
Systolic Blood Pressure: mm Hg mean ±SE. Extract: *Cleome* (ethanol extract); dosage and route: 100–200 mg/kg, i.p.

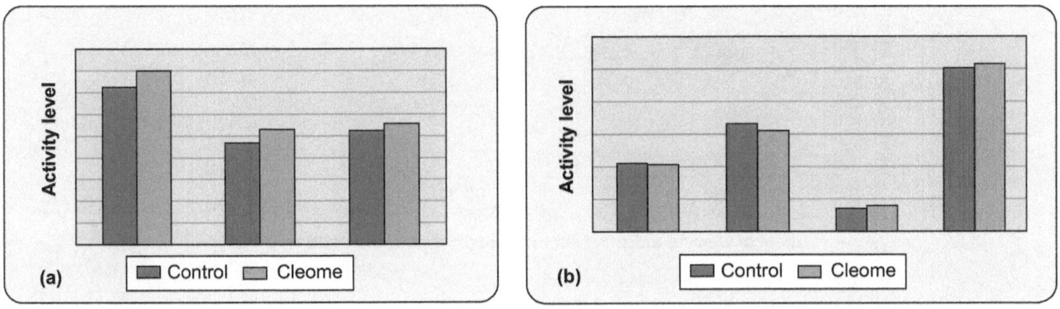

FIGURE 9.10 (a) Effect of *Cleome* on total, ambulatory and distance time activities. (b) Effect of *Cleome* on locomotor activities.

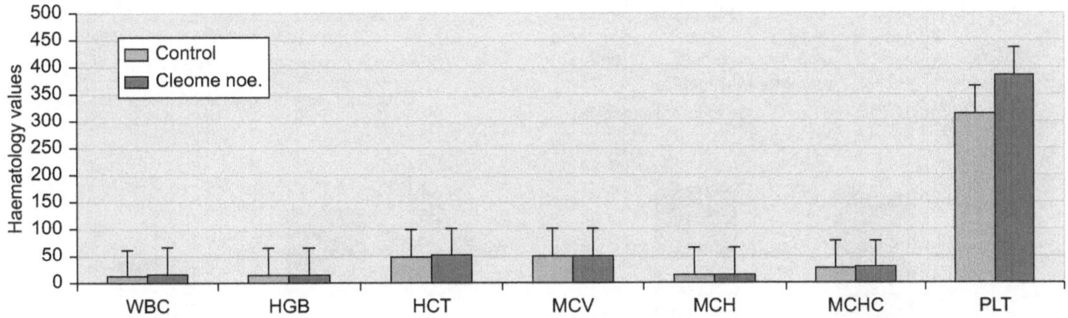

FIGURE 9.11 Haematological effect of *Cleome*.

9.8.4.1 Anti-Inflammatory Activity

9.8.4.1.1 Carrageenan-Induced Rat Paw Edema Method

Carrageenan was injected 30 min after the administration of the extract (400 or 800 mg/kg p.o. single dose). The volume of the inflamed paw was measured plethysmometrically before the injection of carrageenan and at 1, 2, 3, 5, and 24 h after the drug was administered. Two more groups were studied in parallel—one treated with a standard drug (diclofenac, 10 mg/kg, p.o.) and another treated with vehicle only (control) for comparison purposes.

9.8.4.1.2 Cotton Pellet Method (Sub-Acute Effect)

Acute oral administration of 70% ethanolic extract of *C. amblyocarpa* (400 or 800 mg/kg p.o., 8 days) was carried out in rats. A dose of 400 mg/kg did not cause any anti-inflammatory activity toward carrageenan-induced paw edema model at any of the observation points. However, *Cleome* extract at 800 mg/kg p.o. significantly reduced the carrageenan-induced edema relative to the vehicle-treated control group. Diclofenac sodium at a dose of 10 mg/kg, p.o, also reduce the paw edema significantly. At 24 h, only the *Cleome*-treated group (800 mg/kg) showed anti-inflammatory activity, with the *Cleome*-treated group (400 mg/kg) and the diclofenac group not exhibiting any anti-inflammatory effects at 24 h.

Subacute oral treatment of rats for 8 days using a cotton pellet method showed significant anti-inflammatory activity in the group treated with 400 mg/kg of the 70% ethanolic extract of *Cleome*. The anti-inflammatory effects of Cleome 70% ethanolic extract (800 mg/kg) were comparable to the effects of diclofenac sodium (10 mg/kg).

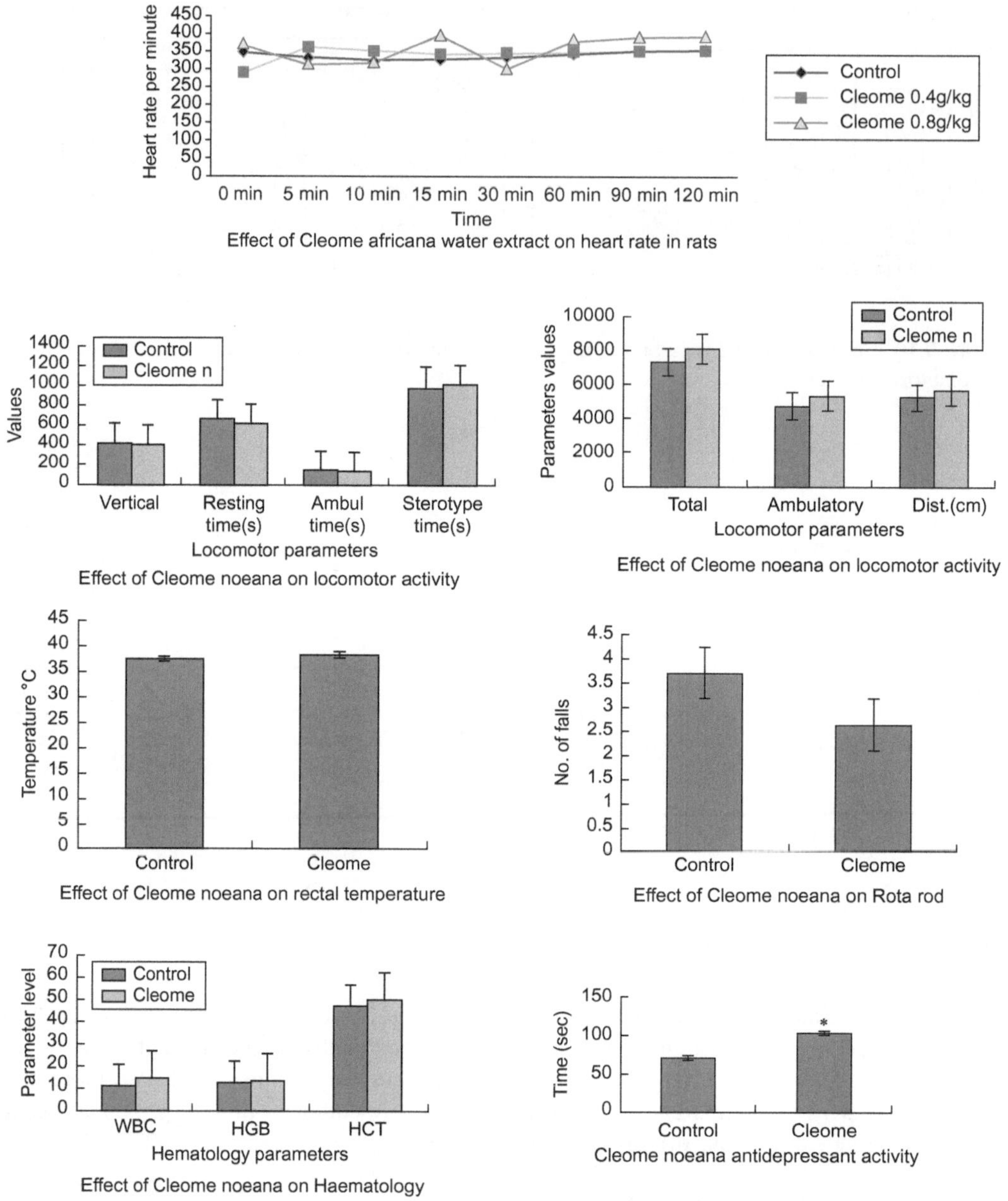

FIGURE 9.12 Effect of different *Cleome* species on different activities.

9.9 CONCLUSION

Quality control studies based on comprehensive laboratory studies on extracts of the aerial parts of *Cleome amblyocarpa* studied the botanical, physicochemical, macroscopic, microscopic, spectroscopic, chromatographic, phytochemical, pharmacological, and toxicological aspects in detail. While evaluating different pharmacological studies, it was observed that an aqueous extract of *Cleome* had significant antinociceptive effect, which increased the systolic and diastolic blood pressure as well as heart rate at both the doses studied. At 800 mg/kg p.o., *Cleome* extract significantly reduced the carrageenan-induced edema, compared to the vehicle-treated control group. Diclofenac sodium at 10 mg/kg p.o., also reduced the paw edema significantly. At 24 h, subacute oral treatment of rats for 8 days using the cotton pellet method showed significant anti-inflammatory activity in the group treated with 400 mg/kg of the 70% ethanolic extract of *Cleome*. The anti-inflammatory effects of *Cleome* 70% ethanolic extract (800 mg/kg) were comparable to the effects of diclofenac sodium (10 mg/kg).

Acknowledgment Thanks are due to Mr MT Abdulla and Mr F Ahmad for their technical support.

REFERENCES

Ahmed A. A., Kattab A. M., Bodige S. G., Mao Y., Minter D. E., Reinecke M. G., Watson W. H., Mabry T. J. 15alpha-Acetoxycleomblynol A from Cleome amblyocarpa. *J Nat Prod* 2001;64(1):106–7.

Bauer A. W., Kirby W. M., Sherris J. C., Turck M. Antibiotic susceptibility testing by a standardized single disk method. *Am J Clin Pathol* 1966;45(4):493–6.

Bose A., Mondal S., Gupta J. K., Dash G. K., Ghosh T., Si S. *Studies on diuretic* and laxative activity of ethanolic extract and its fractions of Cleome rutidosperma aerial parts. *Pharmacognosy Magzine Jadavpur University* 2009, Kolkata, 32.

Boulos L. *Medicinal Plants of North Africa*. Reference Publications Incorporated; 1983.

Bouriche H., Selloum L., Tigrine C., Boudoukha C. Effect of Cleome arabica Leaf extract on rat paw edema and human neutrophil migration. *Pharm Biol* 2003;41(1 February):10–5.

British Herbal Medicine Association. *British Herbal Pharmacopoeia*. 4th ed. BHMA; 1996.

Currniff P. *Official Methods of Analysis of AOAC International*. 16th ed. Vol. (I) and (II); 1999.

Das P. C., Patra A., Mandal S., Mallick B., Das A., Chatterjee A. Cleogynol, a novel dammarane triterpenoid from Cleome gynandra. *J Nat Prod* 1999;62(4):616–8.

Derelanko M. J., Hollinger M. A. *Handbook of Toxicology*. 2nd ed. Boca Raton: CRC Press; 2002.

Evans W. C. *Trease and Evans' Pharmacognosy*. 14th ed. London: Saunders, p. 105; 1996.

Fushia S., Kishi Y., Hattori K., Batkhuu J., Takano F., Singab A. N., Okuyama T. Flavonoids from Cleome droserifolia suppress NO production in activated macrophages in vitro. *Planta Med* 1999;65(5):404–7.

Han J., Hoosier G. L. V. J. *Handbook of Laboratory Science, Animal Models*. 2nd ed. Vol. (II). Boca Raton: CRC Press; 2003.

Harraz F. M., Sevil A. U., Ulubelen A., Nur T. Dammarane triterpenes from *Cleome amblyocarpa*. *Phytochemistry* 1995;39(1):175–8.

Hashem F. A., Wahba H. E. Isothiocyanates in myrosinase treated herb extract of Cleome chrysantha decne. and their antimicrobial activities. *Phytother Res* 2000;14(4):284–7.

Jongbloed M., Feulner G., Boer B., Western A. R. *The Comprehensive Guide to the Wildflowers of the United Arab Emirates*. Abu Dhabi, UAE: ERWDA; 2003.

Kärber G. Archiv f. experiment. *Pathol. u. Pharmakol* 1931;162(4):480–3. doi:10.1007/BF01863914.

Mandaville J. *Flora of Eastern Saudi Arabia*. London: Kegan Paul International Ltd; 1990.

Mothana R. A. A., Abdo S. A. A., Hasson S., Althawab F. M. N., Alaghbari S. A. Z., Lindquist U. Antimicrobial, antioxidant and cytotoxic activity and phytochemical screening of some Yemeni medicinal plants. *Evid Based Complement Alternat Med* 2008;7(3):323–30.

Nagaya H., Tobita Y., Nagae T., Itokawa H., Takeya K., Halim A. F., Abdel-Halim O. B. Cytotoxic triterpenes from Cleome africana. *Phytochemistry* 1997;44(6):1115–9.

Nicola W. G., Ibrahim K. M., Mikhail T. H., Girgis R. B., Khadr M. E. Role of the hypoglycemic plant extract cleome droserifolia in improving glucose and lipid metabolism and its relation to insulin resistance in fatty liver. *Boll Chim Farm* 1996;135(9):507–17.

Pagani F., Romussi G. On the constituents of Cleome pungens Willd seeds. *Farmaco Prat* 1967;22(10):553–71. Italian.

Pagani F., Romussi G. On the constituents of the seeds of Cleome pungens Willd. II. Isolation of feruloylcholine chloride. *Farmaco Sci* 1969;24(2):257–62. Italian.

Pagani F., Romussi G. Components of Cleome pungens Willd. 3. Analysis of the main chemical components of the plant and synthesis of related compounds of potential pharmacological interest. *Farmaco Sci* 1970;25(10):727–48. Italian.

Perumal Samy R., Ignacimuthu S., Raja D. P. Preliminary screening of ethnomedicinal plants from India. *J Ethnopharmacol* 1999;66(2):235–40.

Polasa K., Rukmini C. Mutagenicity tests of cashewnut shell liquid, rice-bran oil and other vegetable oils using the Salmonella typhimurium/microsome system. *Food Chem Toxicol* 1987;25(10):763.

Qin G. W., Hamed A. I., el-Emary N. A., Chen Y. G., Wang L. Q., Cheung K. K., Cheng K. F. A new trinor-triterpenoid from Cleome chrysantha. *Planta Med* 2000; 66(2):191–3.
Rukmini C. Chemical, nutritional and toxicological evaluation of the seed oil of Cleome viscosa. *Indian J Med Res* 1978;67:604–7.
Rukmini C., Deosthale Y. G. Nutritive value of defatted seed cake of Cleome viscosa. *J Am Oil Chem Soc* 1979;56(4):503–5.
Saad B., Dakwar S., Said O., Abu-Hijleh G., Al Battah F., Kmeel A., Aziazeh H. Evaluation of medicinal plant hepatotoxicity in co-cultures of hepatocytes and monocytes. *Evid Based Complement Alternat Med* 2006;3(1):93–8.
Saxena R. C., Dixit O. P., Sukumaran P. Laboratory assessment of indigenous plant extracts for anti-juvenile hormone activity in Culex quinquefasciatus. *Indian J Med Res* 1992;95:204–6.
Selloum L., Arrar L., Medani B., Khenchouche A., Bisker H. Effect of Cleome arabica leaves extract on inflammatory cells response in rat. *Biochem Soc Trans* 1995;23(4):609S.
Selloum L., Sebihi L., Mekhalfia A., Mahdavi R., Senator A. Antioxidant activity of Cleome arabica leaves extract. *Biochem Soc Trans* 1997;25(4):S608.
Sharaf M., El-Ansari M. A., Saleh N. A. M. Flavonoids of four *cleome* and three *Capparis* species. *Biochem Syst Ecol* 1997;25(2):161–6.
Utsalo S. J., Onoyom-Ita V., Ifeanyi-Chukwu M., Akpan J. O. Home medication and microbiological profile in chronic otitis media in some Nigerian children. *Cent Afr J Med* 1990;36(11):278–83.
Wagner H., Bladt S. *Plant Drug Analysis-A Thin Layer Chromatography Atlas.* 2nd ed. Berlin Heidelberg: Springer-Verlag; 1996.
World Health Organization. *Quality Control Methods for Medicinal Plant Materials.* Geneva: WHO; 1998.
Yaniv Z., Dafni A., Friedman J., Palevitch D. Plants used for the treatment of diabetes in Israel. *J Ethnopharmacol* 1987;19(2):145–51.

10 The Concept and Status of Medicinal and Aromatic Plants

History, Pharmacognosy, Ecology, and Conservation

Mohd Kafeel Ahmad Ansari, Muhammad Iqbal, Noureddine Chaachouay, Abdullah Adil Ansari, and Gary Owens

CONTENTS

10.1 INTRODUCTION

The term "medicinal plant" applies broadly to any type of plant commonly used as "herbal medicine" for treating a variety of human and animal diseases worldwide (Adaikan et al., 2001; Beigh et al., 2003a, 2003b; Rishton, 2008; Schmidt et al., 2008). Historically, the word 'herb' is derived from the Latin "*herba*" and French "*herbe*" (i.e., grass or green plant). In botanical science, the term "herb" applies specifically to non-woody plants, but, in terms of traditional medication, "herbal medicine" is often synonymous with "plant medicine," irrespective of whether it relates to herbs, shrubs, or trees (Schmidt et al., 2008; Parveen et al., 2020). Herbal medicine may consist of a whole plant or any part(s) of the plant, be it seeds, fruits, flowers, any specific floral part, stem, bark, leaf, or root, as well as the chemical materials derived from plants, and the products developed from these materials. Many of these medicinal plants also act as a source of food, perfumes, dyes, and other chemicals (Parveen et al., 2020).

As evident from the ancient Chinese, Egyptian, Mesopotamian, Indian, and other scriptures, Greek medicine men, Arab and Persian *Hakims*, Indian *Vaids*, and the physicians of European and Mediterranean regions had been using plants as medicine for over 4,000 years (Iqbal et al., 2018; Parveen et al., 2020). Many Indigenous cultures of Rome, Egypt, Iran, Africa, and America used herbs in their healing rituals, whereas experts in the traditional Unani, Ayurvedic, and Chinese medicines developed formal medical systems in which herbal therapies were prescribed and administered methodically and systematically (Adaikan et al., 2001; Bent, 2008; Parveen et al., 2020, 2022).

Even today, in many places all over the world, traditional systems of medicine continue to be widely practiced. This has received impetus from such factors as population rise, inadequate supply of conventional drugs, prohibitive cost of Western treatments, side effects of several synthetic

DOI: 10.1201/9781003226925-10

drugs, and the development of resistance to antibiotics currently in use against infectious diseases (Adaikan et al., 2001; Parveen et al., 2020). The amazing results from herbal medication in curing a range of ailments have strengthened the faith of modern-day people in traditional therapies. Diverse folklore with respect to ethnomedicines, that keep coming from various parts of the world even today, are consistently widening the horizon of traditional medicine, largely having a plant origin (Atique et al., 1984a, 1984b, 1985a, 1985b, 1993; Anis and Iqbal, 1986, 1994; Anis et al., 2000). Herbal medication is not only safe, with no or minimal side effects, but also seems to be independent of any age or gender specifications (Pieroni et al., 2004; Barnes and Prasain, 2005). All these factors have added to the trust and dependence of people on use of natural plant materials as a source of medication for their physical ailments (Beigh et al., 2002; Parveen et al., 2020, 2022).

Among the ancient civilizations, India is well known to have a rich repository of medicinal and aromatic plants, where these plants have long been collected as raw materials for the manufacture of drugs and perfumery products. For example, about 8,000 herbal remedies are claimed to have been codified in AYUSH (Ayurveda, Unani, Siddha, and Homeopathy) systems in India. Of these, Ayurveda and Unani are especially popular and widely practiced traditional medical systems in rural India (Vaidya and Devasagayam, 2007; Parveen et al., 2020, 2022). The WHO (World Health Organization) estimates that 80% of the global populations relies on herbal medicines for their primary health care needs and that around 21,000 plant species have so far shown potential for use as medicine (WHO, 2002).

In well-developed countries, such as the United States, plant-derived drugs constitute nearly 25% of the total drugs used, compared with 80% in rapidly developing countries, such as India and China, with a history of traditional herbal-medicine usage (Jamshidi-Kia et al., 2018; Parveen et al., 2020). In countries like India, development of the herbal medicine industry also has great economic importance. About two-thirds of the plants used in the modern systems of medicine come from developing countries; a large number of secondary metabolites (natural products) of therapeutic value used in modern medicine are obtained from these plants. Efforts are also in progress to increase the production of these metabolites by external manipulation; where several studies have found that stress stimulates the production of these secondary metabolites (Ali et al., 1998; Saba et al., 2000; Arshi et al., 2006; Ahmad et al., 2007; Ali et al., 2008; Beigh et al., 2008; Ansari et al., 2016; Ansari et al., 2021a). Recently, nanotechnological approaches have also been tried (Husen and Iqbal, 2019; Mishra et al., 2019). Medicinal plants are a valuable source of new drugs, and the use of medicinal plants in modern medicine is also growing rapidly throughout the world (Nalawade et al., 2003; Cole et al., 2007). There are over 1,300 medicinal plants commonly used in Europe, of which 90% are harvested from wild resources; in the United States, about 118 of the top 150 prescription drugs are based on natural sources (Hamilton, 2004; Balunas and Kinghorn, 2005).

Plants that contain or exude volatile compounds, such as essential oils, are known as aromatic plants. The medicinal and aromatic plants (MAPs) not only form a reliable source of essential medicines but also contribute to the economic development of rural populations (Olsen and Larsen, 2003; Shackleton and Shackleton, 2004). The term "essential oil" is concomitant with fragrance or perfume because these fragrances are oily in nature, representing the essence or active constituents of plants (Olsen and Larsen, 2003; Joy et al., 2008). They are also called volatile or ethereal oils, as they evaporate at ordinary temperatures. In reality, they are not simple single chemicals, but consist of hundreds of organic compounds including terpenoids, benzenoids, and organic sulfur and nitrogenous compounds, which all work together at different levels to deliver an overall effect. The average essential oil may contain anywhere from 80 to 400 distinct chemical constituents contributing to the oil's fragrance, flavor, or therapeutic effects. There are at least 300 different types of essential oils (Olsen and Larsen, 2003), with a dozen major crops contributing to 80% of the world market for essential oils (Joy et al., 2008).

The essential oils have a uniform flavor profile without imparting color to the final product. They are used in perfumes and cosmetics, for flavoring food and drink, and for scenting incense and household cleaning products (Boeckel, 2017). The alternative and less costly healthcare system of aromatherapy offers therapeutic effects leading to calmness, emotional balance, stress relief, and rejuvenation (Kauluni and Joshi, 2018). Due to their physiological properties, these oils are absorbed into the body. Some of them have antibiotic, antiseptic, and anti-inflammatory properties, and are thus pain relieving.

In this chapter, we examine the current status of MAPs with emphasis on their historical background, pharmacognosy, ecology, and conservation strategies.

10.2 CURRENT STATUS OF MEDICINAL AND AROMATIC PLANTS

Medicinal and aromatic plants are extensively used in traditional Chinese medicine and the Indian medicinal systems (Ayurveda, Unani, and Siddha). About 3,000 species of MAPs are traded internationally, among which 2,000, in particular, are traded in the European nations like France, Germany, and Switzerland. The international market for MAPs is dominated by China, France, Germany, Italy, Japan, Spain, the UK, and the USA (Schippmann et al., 2005, 2006; Kauluni and Joshi, 2018). Likewise, India has a very long history of a safe and continuous usage of herbal drugs as officially recognized alternative systems of therapy (Vaidya and Devasagayam, 2007).

Of the 26,092 Chinese higher plant species, 4,941 (18.9%) are medicinal plant species. India has 15,000 higher plant species, of which 3,000 (20%) are medicinal plants. Likewise, Indonesia has 22,500 higher plant species, of which 1000 (4.4%) are medicinal plants; Malaysia has 15,500 higher plant species, of which 1,200 (7.7%) have medicinal value; Nepal has 6,973 higher plant species, of which 700 (10%) are medicinal. Similarly, of the 4,950 higher plant species in Pakistan, 300 (6.1%) are medicinal plants; of the 8,931 higher plant species in Philippines, 850 (nearly 9.5%) are medicinal plants; of the 3,314 higher plant species in Sri Lanka, 550 (16.6%) are medicinal plants; of the 11,625 higher plant species in Thailand, 1,800 (15.5%) have medicinal value. Similarly, the USA has 21,641 higher plant species, and 2,564 (11.8%) of these are used as medicinal plants (Kauluni and Joshi, 2018). Let us first examine the history, pharmacognosy, ecology and conservation of medicinal plants.

10.2.1 History of Medicinal *Plants*

The art of healing is believed to date back to the fifth generation of Adam (the father of all mankind). Asclepius, a student of Hermes (Prophet Idrees), is said to have been the first to develop the preliminary knowledge of successfully treating sick humans. The oldest written evidence of medicinal plants for preparation of drugs was found on a Sumerian clay tablet from Nagpur, dated 5000 BC. It comprised 12 recipes for drug preparation referring to over 250 different plants, like henbane, mandrake, and poppy (Kelly, 2009). The Indian holy books, Vedas, also mentioned numerous plant treatments, of which many spice plants of Indian origin, such as nutmeg, pepper, and clove, are still globally today used (Tucakov, 1971; Petrovska, 2012).

The Chinese book on roots and grasses, "*Pen T'Sao*," written by Emperor Shen Nung circa 2500 BC, described 365 drugs (dried parts of medicinal plants), many of which, including Rhei rhizome (rhubarb rhizome), camphor, Theae folium (tea leaf), *Podophyllum*, the great yellow gentian, ginseng, jimson weed, cinnamon bark, and ephedra, are used even today (Bottcher, 1965; Tucakov, 1971; Petrovska, 2012). In Egypt, "Ebers Papyrus," written circa 1550 BC, is an ancient collection of 800 prescriptions referring to 700 plant species including aloe, castor oil plant, coriander, fig, garlic, juniper, onion, pomegranate, senna, and willow (Glesinger, 1954; Petrovska, 2012).

In Homer's epics, "The Iliad" and "The Odyssey," written circa 800 BC, 63 plant species from the Minoan, Mycenaean, and Egyptian civilizations were reported. Some of these plants were named after mythological characters from these epics; for instance, Elecampane (*Inula helenium*, Asteraceae) was named in honor of Elena, the central figure of the Trojan War. The genus *Artemisia* (Asteraceae), believed to restore strength and protect health, was derived from the Greek word *artemis*, meaning healthy (Galle, 2005). Herodotus (500 BC) referred to the castor oil plant,

Orpheus (an ancient Greek prophet) to the fragrant hellebore and garlic, and Pythagoras (570–495 BC) to the sea onion (*Scilla maritima*), mustard, and cabbage. The works of Hippocrates (459–370 BC) contained reference to 300 medicinal plants classified by physiological action: wormwood (*Artemisia absinthium*) and common centaury (*Centaurium umbellatum*) were used to treat fever; garlic (*Allium sativum*) against intestinal parasites; opium poppy (*Papaver somniferum*), henbane (*Hyoscyamus* spp.), deadly nightshade (*Atropa belladonna*), and mandrake (*Mandragora officinarum*) were used as narcotics; fragrant hellebore (*Helleborus odorus*) and hazelwort (*Asarum europaeum*) as emetics; climbing sea onion (*Bowiea volubilis*), celery (*Apium graveolens*), parsley (*Petroselinum crispum*), asparagus (*Asparagus officinalis*), and garlic (*Allium sativum*) as diuretics; and oak (*Quercus* spp.) and pomegranate (*Punica granatum*) as astringents (Bojadzievski, 1992; Gorunovic and Lukic, 2001).

Theophrastus (371–287 BC) founded botanical science with his two books "De Causis Plantarium"—Plant Etiology—and "De Historia Plantarium"—Plant History. In these books, he generated a classification of more than 500 of the medicinal plants known at that time. In his writings, he frequently referred to cinnamon, cardamom, false hellebore, fragrant hellebore, iris rhizome, mint, monkshood, and pomegranate, among others (Bazala, 1943; Nikolovski, 1995; Petrovska, 2012).

Aulus Cornelius Celsus (25 BC–AD 50), a renowned Roman medical writer, mentioned 250 medicinal plants in his work "*De Medicinaa.*" Pedanius Dioscorides (c. 30–90 AD), known as the father of pharmacognosy, studied medicinal plants wherever he traveled as a military physician and pharmacognosist in Nero's Roman Army (Petrovska, 2012), and published his work "*De Materia Medica*" around AD 77. This classical work of ancient history, translated many times since then, offers plenty of data on medicinal plants constituting the basic *Materia Medica* until the late Middle Ages and the Renaissance (Thorwald, 1991; Petrovska, 2012). Of the 944 drugs described, 657 were of plant origin, with descriptions of their outward appearance, mode of collection, preparation of drugs, and therapeutic effects. In addition to the plant description, their names in other languages, coupled with the localities where they grow, were provided.

The Greek philosopher Hippocrates (460–370 BC) was the first to transform the raw concepts of medication into logical and practical knowledge. He considered the human body to be an integrated system, which responds to and is affected by the surrounding environment (Kleisiaris, 2014; Parveen et al., 2022). The impact of Greek medicine reached the Arab lands around AD 600. Healing practices of that period (the seventh century AD) were greatly influenced by the teachings of the Prophet of Islam, Muhammad bin Abdullah (AD 570–632), who laid special emphasis on cleanliness of body and habitat, risks associated with prohibited food and drink, benefits of nasal, dental, and inter-finger cleaning and nail cutting, prevention from visiting infected places, and isolation of patients from healthy persons. He also emphasized the use of certain edibles and herbs like honey, vinegar, cucurbits, date fruit, garlic, onion, figs, *kasni* (chicory, and *kalonji* (nigella). For the next 600 years, covering the Umayyad period (AD 660–750) and the Abbasid period (AD 750–1258), medical knowledge prospered in that region. Emperor Harun Rasheed established a great academic center, called *Baital-Hikmah* (House of Wisdom) in Baghdad, which was joined by many foreigners including the Ayurveda experts from India, such as Manka-al-Hindi, Kanka-al-Hindi, and Judr-al-Hindi. Several ancient Sanskrit works on medicine were then translated into Arabic (Azmi, 1989; Nigrami, 2009; Qadeer, 2016).

The most eminent Arab physicians and herbalists of that era included Jabir bin Hayyan (AD 717–813), known as the father of chemistry and inventor of distillation apparatus; Ibne Raban Tabari (AD 810–895), an author of 14 books including the famous "*Firdous-ul-Hikmat*" (Paradise of Wisdom); Abu Bakar Zakariya Razi or Rhazes (AD 850–923), the proposer of the concept of acquired immunity, and the author of nearly 200 books, including the encyclopedic "*Kitab-al-Hawi fit-Tibb*" (in Latin, *Liber Continens*); Ibn El-Jazzar Al-Qayrawani ((AD 895–979), the author of many books, who described 272 drugs mainly of plant origin; Ali Ibn Abbas Majoosi or Haly Abbas (AD 930–994), the author of "*Kamilus-Sana'ah*" (in Latin, *Liber Regius*); Bu Ali Ibn Sina or Avicenna (AD

980–1037), the author of the world famous "*Al-Qanoon fit-Tibb*" (The Canon of Medicine); Ibn Rushd (AD 1126–1198), an author of over 20 books; Abdul-Latif Baghdadi (AD 1163–1231), a famous anatomist and author of over 57 books; and Ibne Baitar (AD 1197–1248), a botanist who authored the comprehensive "*Kitab Jameul-Mufradaat*" on nearly 2000 medicinal plants and their properties (Browne, 1921; El-Demerdash, 2001; Nigrami, 2009; Saad, 2014; Parveen et al., 2022). All this basic information paved the way for later modern research in European universities and elsewhere.

10.2.2 Pharmacognosy

Pharmacognosy is a term derived from the Greek words for drug (*pharmakon*) and knowledge (*gnosis*). It is a field of study within chemistry which is focused on natural products isolated from different sources and their biological activities. Research on natural products began more than one hundred years ago and continues today, with numerous research groups routinely discovering more and more novel active constituents everyday (Adaikan et al., 2001; Peixoto et al., 2012). Pharmacognosy initially included the study of crude drugs of plant or animal origin (in the form of tinctures, teas, poultices, powders, and other herbal formulations), and now incorporates authentication and quality control of such drugs, based on macroscopic, microscopic, and molecular examination of crude drugs. The term pharmacognosy was first introduced by the Austrian physician Schmidt in 1811 and then by Seydlerin in 1815 in a work titled "*Analecta Pharmacognostica*" (Marcolin, 2013). Now, pharmacognosy is defined as the science of biogenic or naturally derived drugs, pharmaceuticals, and poisons, and incorporates various modern analytical techniques to authenticate and conduct quality control on crude drugs as well as purified active extracts, fractions, and components, and even medicinal foods. The process of drug development from medicinal plants has extended from the formulation of crude drugs to the isolation, identification, and assessment of individual bioactive compounds (Adaikan et al., 2001; Rishton, 2008; Marcolin, 2013). The American Society of Pharmacognosy defines pharmacognosy as "the study of the physical, chemical, biochemical and biological properties of drugs, drug substances, or potential drugs or drug substances of natural origin as well as the search for new drugs from natural sources." Modern pharmacognosy involves a broad study of natural products from various sources including plants, bacteria, fungi, and marine organisms. Phytochemistry and phytochemical analysis from *in-vivo* as well as *in-vitro* studies have also now become integral parts of pharmacognosy (Mughal et al., 1999; Saba et al., 1999a, 1999b, 1999c; Siddiqi et al., 2003; Trag et al., 2005a, 2005b; Peixoto et al., 2012).

Molecular biology has also become essential for drug discovery through screening assays directed toward physiologically relevant molecular targets, so modern pharmacognosy encapsulates all these relevant disciplines. The emphasis and focus of research in pharmacognosy have changed significantly, from focusing on identification of drugs to the isolation of active principles, and, more recently, to the investigation of biological activity. The metabolomic and transcriptomic studies of medicinal plants now seem to have occupied the forefront of pharmacognostic research. Not only the extracts of single plants but also the polyherbal and polyherbo-mineral compound formulations of traditional medicines are being analyzed by a variety of modern techniques to identify the enormous gamut of metabolites present therein, as indicated by Bhattacharyya and Ghosh (2022) and Ahmad et al. (2022), while explaining the role of metabolomics and ethnopharmacology in the development of traditional herbal medicine. Many fine collections of articles are now available on the role of omics studies in understanding the potential of MAPs under different environmental conditions (Aftab and Hakeem, 2020; Husen and Iqbal, 2022).

Research in ethnobotany, ethnomedicine, and ethnopharmacology has also become an important element in pharmacognosy (Atique and Iqbal, 1992; Iqbal et al., 2018; Parveen et al., 2020). Pharmacognosy, with its pivotal role in the discovery and development of new drugs and therapies, formed the basis for the development of modern-day pharmacy (Rishton, 2008; Kaur et al., 2019). It has never been more important for today's pharmacists to have knowledge of traditional medicines.

A good example of the usefulness of not neglecting indigenous herbal remedies is *Artemisia annua* (commonly known in China as *qinghao*), which has long been used in China for the treatment of intermittent fever or malaria. Extracts from this plant have been studied in the west since the 1980s, resulting in the discovery of artemisinin (*qinghaosu*) and a new class of effective antimalarials, the synthetic artemisinins (Adaikan et al., 2001; Marcolin, 2013).

Each individual medicinal plant species carries a unique profile of secondary metabolites (SMs) that contain the medicinally active components. More than 80,000 SMs have been identified thus far. These phytochemicals may have diverse activities, like working as an antibacterial, antifungal, or antiparasite agent, inhibiting cell division, modifying DNA and RNA, or modulating the activity of specific enzymes. A single species often has a large number of these phytochemicals, e.g., *Hypericum perforatum* has more than 25 bioactive chemicals, while *Echinacea* has more than 20. It is likely that the therapeutic efficacy of such a plant is due to these different compounds working in a synergistic way. The situation becomes more complex when a drug formulation contains more than one plant species (Murch et al., 2001). Moreover, there may be many unidentified metabolites. For instance, the presence of melatonin, a neurotransmitter, was detected much later in the prophylactic migraine medication Tanacet 1125 ™ (McNeil Consumer Products Inc.), which had not been standardized for this compound (Murch et al., 1997).

Furthermore, the active ingredients of medicinal plants can also be affected by genetic variation within the species, wild populations typically having a wider genetic variability, and by the ontogenetic stages of the plant (Iqbal et al., 2011a, 2018). Growing conditions and management practices, including the post-harvest treatment and storage, as well as the availability of nutrients in the soil also influence the overall metabolite composition of a plant preparations (Ahmad et al., 2005; Umar et al., 2003, 2008). As the skeleton of an active secondry metabolite is based on the carbon assimilated via photosynthesis, temperature, photoperiod, and the intensity and spectrum of the available light also have an impact on the final product. Warm weather conditions favor biosynthesis, whereas a rainy climate can inhibit alkaloid production (Bernath, 1986). *Atropa belladonna* (deadly nightshade) accumulated 1.3% of active alkaloid when grown in the Crimean Peninsula, but only 0.3% when grown in Leningrad (Waller and Nowacki, 1978).

Problems also arise by misidentification of related species with similar morphology, which have different profiles of bioactive compounds, as in the case of *Hypericum* and *Echinacea* species, for example. In addition, contamination of plant material with bacteria, fungi, or even with heavy metals, may cause serious health risks. Furthermore, plant-based medicines normally prove effective in clinical trials, but sometimes they may have negative reaction with other pharmaceuticals; St. John's wort preparations could induce metabolic degradation of the HIV protease inhibitor (indinavir), the immunosuppressant (cyclosporin), and many oral contraceptives (Moore et al., 2000).

10.2.3 Ecology and Plant Diversity

More than one-tenth of all plant species (approximately 50,000 species) are used in drugs and health products. Their distribution is not uniform across the world because different species may require different environmental conditions to grow (Huang, 2011; Rafieian-Kopaei, 2013; Chen et al., 2016). China and India have the highest numbers of medicinal plants, with 11,146 and 7,500 species, respectively, followed by Colombia, South Africa, the United States, and another 16 countries, with percentages of total medicinal plant species ranging from 7% in Malaysia to 44% in India (Rafieian-Kopaei, 2013; Rao et al., 2012). Certain plant families not only have higher numbers of medicinal plants, but also have higher proportions of threatened species (Huang, 2011). In fact, only a relatively small proportion of medicinal plants that currently suffer from genetic erosion and resource destruction have been listed as threatened (Schippmann et al., 2005).

Since a large proportion of Himalayan flora possess medicinal value, the region is identified as the global center for medicinal plants (Hamilton and Radford, 2007; Lakey and Dorji, 2016). A number of diverse ecological conditions and a wide altitude range from 100 to 7,500 m above sea level favor the growth of diverse plant types with a range of medicinal properties (Hamilton and Radford, 2007). These medicinal plants also serve as a primary source of income for the local population (Rasul et al., 2012), thus contributing to the Gross National Happiness (GNH) (Wangchuk and Tobgay, 2015). Due to its richness in medicinal plants, Bhutan, a small nation in the Himalayan range, was known as *Menjong* (land of medicine/medicinal plants) (Wangchuk et al., 2010), and the medicinal plants used in Bhutanese traditional medicine (BTM) are known by different names based on the altitude of their habitat, like "*Sngonsman*" (higher-elevation medicinal plants) and "*Khrogsman*" (lower-elevation medicinal plants) (Wangchuk et al., 2011). The amounts of active ingredients even within the same species often varies with altitude (Beigh et al., 2008).

The increased international market for medicinal plants has also led to illegal harvesting, thereby reducing plant population in their natural habitats. However, much of the decline in Trans-Himalayan regions has been attributed to loss of natural habitat (Kumar et al., 2015).

Nowadays, increasing levels of environmental degradation and unprecedented global climate change are causing detrimental effects on plants, including those of medicinal importance (Yunus and Iqbal, 1996; Yunus et al., 1996; Iqbal et al., 2000; Ahmad et al., 2011; Ansari et al., 2020), thus disturbing their form, function, and productivity, and hence their utility and usefulness (Mahmooduzzafar et al., 1987; Ghouse et al., 1989; Verma et al., 2006; Wali et al., 2007; Iqbal et al., 2010a, 2010b, 2010c; Bashir et al., 2015; Husein et al., 2017). Different stresses, such as drought, heat, salinity, pollutants, and nutrient deficiency, affect even the biosynthesis of plant metabolites, which may receive a fillip under moderate stress conditions but ultimately decline under heavy stress (Ali et al., 2008; Nighat and Iqbal, 2008; Hakeem et al., 2012, 2013; Bagheri et al., 2015, 2017; Yousuf et al., 2016a, 2016b). Initially, plants develop tolerance to the abiotic stresses, by using mechanical means, activating antioxidant defence mechanism, and/or producing relevant secondary metabolites (Iqbal et al., 1996; Bashir and Iqbal, 2014; Bashir et al., 2007, 2014; Aref et al., 2016; Rahman et al., 2022). Sometimes, plants growing under stressful conditions exhibit changes in not only quantity but also quality of the metabolites produced (Iqbal et al., 2011a, 2011b, 2018; Yahyazadeh et al., 2022); which becomes a matter of special concern when it comes to medicinal plants, because many of their secondary metabolites are used for treating human diseases. A similar situation may arise when a medicinal plant is used for the purpose of phytoremediation of heavy metals (HMs) (Shmaefsky and Husen, 2022). In fact, several plants with considerable medicinal properties have proved to be hyperaccumulators or at least good accumulators of various HMs (Ansari 2013a, 2013b, 2015, 2018, 2021b; Iqbal et al., 2015). Such plants are likely to accumulate enough metallic ions, if they grow on a soil contaminated with HMs, to possibly cause toxicity when consumed by humans as a medicine. Care, therefore, needs to be taken with reference to cultivation of medicinal plants.

10.2.4 Conservation

A highly conservative estimate indicates that the current loss of plant species on Earth is 100–1000 times faster than the expected, natural extinction rate, and the Earth is losing at least one potential major drug every two years (Cole et al., 2007). According to the International Union for Conservation of Nature, and the World Wildlife Fund, the number of flowering plant species being used for medicinal purposes worldwide lies between 50,000 and 80,000. Among these, about 15,000 species are threatened with extinction due to overharvesting and habitat destruction, and 20% of wild resources have already been exhausted because of the increased human population and concomitant plant consumption (Ross, 2005; Bentley, 2018). This accelerated loss of species and habitat destruction worldwide has drastically increased the risk of extinction of medicinal plants, especially in China, India, Nepal, Kenya, Tanzania, and Uganda (Nalawade et al., 2003; Hamilton, 2008; Heywood and Iriondo, 2003; Zerabruk and Yirga, 2012). The International Union for Conservation of Nature and Natural Resources (IUCN) Red Data book has assigned "endangered" status to 34,000 plant species. Conservation of biodiversity has thus become one of the most pressing environmental issues worldwide. Species rarity is used to assess the extinction risk of medicinal plants, although there may be different reasons for the decline of rare species. It is difficult to explain the susceptibility of individual species or its resilience to harvest pressure. Multiple biological characters, such as habitat specificity, distribution range, population size, species diversity, growth rate, and reproductive system, correlate with extinction risk (Chen et al., 2016).

The Convention on International Trade of Endangered Species (CITES), which was enforced in 1975, monitors and restricts the trade of species threatened due to over-exploitation. The signatory nations are bound to promote its objectives. Likewise, the Convention on Biological Diversity, introduced in 1993, is an international legal instrument to effect biodiversity conservation and achieve sustainable use of its components, including medicinal plants. It emphasizes protection of habitats of the threatened species, integration of general protection measures for the natural and semi-natural habitats with land-use planning, with back-up of monitoring and control measures and aims to create incentives for sustainable use and disincentives for over-exploitation (Leaman, 2001).

The guidelines on medicinal plants, referred to as the Chiang Mai Declaration, identified the following priority actions: (a) Basic studies to identify medicinal plants and traditional knowledge of their uses, sustainable utilization through cultivation, controlled wild harvest, and reduction of waste; (b) Conservation both *in situ* as well as *ex situ* (for those species which cannot be conserved *in situ*, i.e. species that need to be stored in seed banks); and (c) Increased public support for conservation of medicinal plants, built through communication and cooperation (WHO, 1993). Some donor agencies have come forward to help implement these objectives. For instance, the International Development Research Centre (IDRC), Canada, helped in developing the regional research networks on sustainable use of medicinal plants in Central America, Africa, and Asia, while the World Bank also considered the conservation of medicinal plants to be a global development issue (Leaman, 2001).

In situ conservation of plant biodiversity, i.e., conservation within ecosystems and natural habitats, allows and enables the species concerned to be maintained and to flourish within the context of the ecosystem in which they occur naturally (Olorode, 2004). This is an important consideration because many wild plants behave differently during domestication outside their natural habitat, and the concentrations of their natural products change. The very survival of many plant species may be at stake, or the production of the desired metabolites of therapeutic value may alter considerably, as is the case for *Aconitum heterophyllum* and *Podophyllum hexandrum*, two endangered species endemic to the Northwest Himalayas (Beigh et al., 2005, 2011). Strategies for *in situ* conservation, like biosphere reserves, wildlife sanctuaries, national parks, and sacred groves, are therefore more successful than *ex situ* strategies, such as gene banks, field gene banks, cryobanks, and botanic gardens. The sensitivity of each species to habitat change is directly related to its adaptability or versatility; with the most versatile species being the least sensitive to habitat manipulation (McNeely, 1988).

In *ex situ* conservation, gene banks can be built to prioritize medicinal plants on the basis of threat assessment, ethno-medicinal, botanical, and ecological surveys, intraspecific variability assessment, species recovery programs, establishment of seed centers, and networking of forest reserves with the involvement of local stakeholders. Botanic gardens help conservation biology in terms of arboriculture, cultivar conservation, dendrology, introduction of new crop genetic resources and their maintenance, and research into habitat restoration (Jackson and Sutherland, 2000). This is not a new concept, as the first formal botanic gardens were developed in Italian universities around the middle of the sixteenth century. According to the Botanic Gardens Conservation International (BGCI) database, there are currently 1,775 botanic gardens and arboreta in 148 countries around the world (Parveen et al., 2020). Of late, large-scale cultivation of some important medicinal plants has also been put into practice on a commercial basis, as encouraged by the pharmaceutical industry. Scientifically approved fertilization strategies are important in making such programs a real success (Ahmad et al., 2005; Umar et al., 2003, 2008; Iqbal et al., 2019). Domestication and cultivation of medicinal plants ensure several advantages over the wild harvest, such as reliable botanical identification, steady supplies of raw materials, availability of standardized or improved genotypes, and controlled post-harvest handling (Leaman, 2001).

10.3 CONCLUSION

Since the dawn of civilization and during the subsequent development of human society, more and more plants have been identified as possessing healing and aromatic properties. The growing global interest in MAPs has culminated in the development of today's modern herbal medicinal treatments. However, the increasing demand for herbal medicines has caused a heavy pressure on the natural plant species from which these medicines are derived, thus threatening their existence and questioning the long-term viability and sustainability of the associated industries.

Despite the fact that the safety of traditional herbal drugs had been established by their continued usage over thousands of years, extensive research still needs to be undertaken in order to allay public concerns regarding their quality control and clinical validation. The growing and widespread use of herbal medicines in the form of complementary and alternative medicine is bound to increase the pressure for clinical evidence. Using the "reverse pharmacology" approach, further efforts may be made to confirm the potential health benefits of herbal drugs. Observational therapeutics is a prelude to the path of reverse pharmacology for the development of new natural drugs.

REFERENCES

Aftab, T. & Hakeem, K.R. (2020). *Medicinal and Aromatic Plants: Expanding their Horizons through Omics*. Academic Press. (an Imprint of Elsevier).

Adaikan, P.G., Gauthaman, K. & Prasad, R.N.V. (2001). History of herbal medicines with an insight on the pharmacological properties of *Tribulus terrestris*. *The Aging Male*, 4(3): 163–169.

Ahmad, A., Khan, I., Anjum, N.A., Diva, I., Abdin, M.Z. & Iqbal, M. (2005). Effect of timing of sulphur fertilizer application on growth and yield of rapeseed. *Journal of Plant Nutrition*, 28(6): 1049–1059.

Ahmad, A., Siddiqi, T.O. & Iqbal, M. (2011). *Medicinal Plants in Changing Environment*. Capital Publishing Company, New Delhi.

Ahmad, S., Jamal, A., Fazli, I.S., Alam, T., Khan, M.A., Kamaluddin, I.M. & Abdin, M.Z. (2007). Impact of sulphur and nitrogen application on seed xanthotoxin yield in *Ammi majus* L. Korean. *Journal of Crop Science*, 52: 153–161.

Ahmad, S., Katiyar, C.K., Ulrich-Merzenich, G.S. & Mukherjee, P.K. (2022). Editorial: Metabolomics and ethnopharmacology in the development of herbal and traditional medicine. *Frontiers in Pharmacology*, 13: 851023. https://doi.org/10.3389/fphar.2022.851023.

Ali, G., Iqbal, M. & Srivastava, P.S. (1998). Interactive effect of Cd and Zn in morphogenic potential of *Bacopa monniera* (L.) Wettst. *Plant Tissue Culture & Biotechnology*, 4: 159–164.

Ali, S.T., Mahmooduzzafar, A.M.Z. & Iqbal, M. (2008). Ontogenetic changes in foliar features and psoralen content of *Psoralea corylifolia* Linn. exposed to SO_2 stress. *Journal of Environmental Biology*, 29(5): 661–668.

Anis, M. & Iqbal, M. (1986). Antipyretic utility of some Indian plants in traditional medicine. *Fitoterapia*, 57: 52–55.

Anis, M. & Iqbal, M. (1994). Medicinal plantlore of Aligarh, India. *International Journal of Pharmacognosy*, 32(1): 59–64.

Anis, M., Sharma, M.P. & Iqbal, M. (2000). Herbal ethnomedicine of the Gwalior-forest division in Madhya Pradesh, India. *Pharmaceutical Biology*, 38(4): 241–253.

Ansari, M.K.A., Ahmad, A., Umar, S. & Iqbal, M. (2009). Mercury induced changes in growth variables and antioxidative enzyme activities in Indian mustard. *Journal of Plant Interecations*, 4(2): 131–136.

Ansari, M.K.A., Oztetik, E., Ahmad, A., Umar, S., Iqbal, M. & Owens, G. (2013a). Identification of the phytoremediation potential of Indian mustard genotypes for copper, evaluated from a hydroponic experiment. *CLEAN – Soil, Air, Water*, 41(8): 789–796.

Ansari, M.K.A., Shao, B.-H., Umar, S., Ahmad, A., Ansari, S.H., Iqbal, M. & Owens, G. (2013b). Screening Indian mustard genotypes for phytoremediating arsenic-contaminated soils. *CLEAN – Soil, Air, Water*, 41(2): 195–201.

Ansari, M.K.A., Ahmad, A., Umar, S., Zia, M.H., Iqbal, M. & Owens, G. (2015). Genotypic variation in phytoremediation potential of Indian mustard exposed to nickel stress: A hydroponic study. *International Journal of Phytoremediation*, 17(1–6): 135–144.

Ansari, M.K.A., Umar, M.K., Tasneem, F. & Owens, G. (2016). Evaluation of methyl red tolerant cyanobacteria for simultaneous laccase production and dye decolorization. *International Journal of Waste Resources*, 6(2). https://doi.org/10.4172/2252-5211.1000221.

Ansari, M.K.A., Ahmad, A., Aref, I.M., Owens, G. & Iqbal, M. (2018). Tolerance capacity of Turkish Genotypic of barley (*Hordeum vulagre* L.) for cadmium stress. *Journal of Environmental Biology*, 39(6): 1027–1035.

Ansari, M.K.A., Nasreen, I.K., Abdullah, A.A. & Gary, O. (2020). The world wide environmental encroachments of covid-19 outbreak. *Acta Scientific Microbiology*, 4: 114–221.

Ansari, M.K.A., Oksana, L., Muhammad, I., Abdullah, A.A., Tasneem, F., Susana, R. & Owens, G. (2021a). Lacacse-the wonderful enzyme for a variety of industries. *Acta Scientific Microbiology*, 4: 52–66.

Ansari, M.K.A., Ahmad, A., Umar, S., Iqbal, M., Zia, M.H., Husen, A. & Owens, G. (2021b). Suitability of Indian mustard genotypes for phytoremediation of mercury-contaminated sites. *South African Journal of Botany*, 142: 12–18.

Aref, I.M., Khan, P.R., Khan, S., El-Atta, H., Ahmed, A.I. & Iqbal, M. (2016). Modulation of antioxidant enzymes in *Juniperus procera* needles in relation to habitat environment and dieback incidence. *Trees - Structure & Function*, 30(5): 1669–1681.

Arshi, A., Abdin, M.Z. & Iqbal, M. (2006). Sennoside content and yield attributes of *Cassia angustifolia* Vahl. as affected by NaCl and $CaCl_2$. *Scientia Horticulturae*, 111(1): 84–90.

Atique, A., Iqbal, M. & Ghouse, A.K.M. (1984a). *Folklore Studies on an Indigenous Antifertility Drug*. Indian Botnical Contactor, IA, p. 117.

Atique, A., Iqbal, M. & Ghouse, A.K.M. (1984b). Medico-botany of *Jasminum officinale* var. *grandiflorum* Linn. (Oleaceae). *Indian Botanical Contactor IA*, 118.

Atique, A., Iqbal, M. & Ghouse, A.K.M. (1985a). Use of *Annona squamosa* and *Piper nigrum* against diabetes. *Fitoterapia*, 56: 190–192.

Atique, A., Iqbal, M. & Ghouse, A.K.M. (1985b). Ethnobotanical study of cluster fig (*Ficus racemosa*). *Fitoterapia*, 56: 236–240.

Atique, A. & Iqbal, M. (1992). Traditional single drug administration for the psychosomatic dermal diseases: Psoriasis. *Hamdard Medicus*, 35: 76–81.

Atique, A., Khair, S. & Iqbal, M. (1993). Traditional herbal contraceptives from north-west Uttar Pradesh. In: Govil, J.N., Singh, V.K. & Hashmi, S. (eds.), *Medicinal Plants: New Vistas of Research, Part I. Today & Tomorrow's Printers & Publishers*. New Delhi, pp. 41–47.

Azmi, W.A. (1989). *Baital-Hikmat ki TibbiKhidmaat. (Medical Contribution of the House of Wisdom)*. Mehboob Press, Deoband.

Balunas, M.J. & Kinghorn, A.D. (2005). Drug discovery from medicinal plants. *Life Sciences*, 78(5): 431–441.

Barnes, S. & Prasain, J. (2005). Current progress in the use of traditional medicines and nutraceuticals. *Current Opinion in Plant Biology*, 8(3): 324–328.

Bazala, V. (1943). *The Historical Development of Medicine in the Croatian Lands*. Croatian Publishing Bibliographic Institute, Zagreb, pp. 9–20.

Bent, S. (2008). Herbal medicine in the United States: Review of efficacy, safety and regulation. *International Journal of General Medicine*, 23: 854–859.

Bentley, R. (2018). *Medicinal Plants (Classic Reprint)*. Forgotten Books, London, pp. 23–46.

Boeckel, L. van (2017). *MAPs and Essential Oils From Nepal–Market Analysis and Market Entry Strategy in Khumaltar, Lalitpur: Deutsche Gesellschaft Fur Internationale Zusammenarbeit (GIZ) GmbH:Trade Promotion Programme*. National Trust for Nature Conservation, Kathmandu.

Bojadzievski, P. (1992). *The Health Services in Bitola through the Centuries*. Society of Science and Art, Bitola, pp. 15–27.

Bottcher, H. (1965). *Miracle Drugs*. Zora, Zagreb, pp. 23–139.

Bagheri, R., Bashir, H., Ahmad, J., Iqbal, M. & Qureshi, M.I. (2015). Spinach (*Spinacia oleracea* L.) modulates its proteome differentially in response to salinity, cadmium and their combination stress. *Plant Physiology & Biochemistry*, 97: 235–245.

Bagheri, R., Ahmad, J., Bashir, H., Iqbal, M. & Qureshi, M.I. (2017). Changes in RuBisCO, cysteine-rich proteins and antioxidant system of spinach (*Spinacia oleracea* L.) due to sulphur deficiency, cadmium stress and their combination. *Protoplasma*, 254(2): 1031–1043.

Bashir, F. & Iqbal, M. (2014). Lambda-cyhalothrin-induced oxidative stress alters ascorbate-glutathione cycle in *Glycine max* (L.) Merr. *Trends in Biosciences*, 7(9): 703–710.

Bashir, F., Mahmooduzzafar, Siddiqi, T.O. & Iqbal, M. (2007). The antioxidative response system in *Glycine max* (L.) Merr. exposed to Deltamethrin, a synthetic pyrethroid insecticide. *Environmental Pollution*, 147(1): 94–100.

Bashir, F., Mahmooduzzafar, Siddiqi, T.O. & Iqbal, M. (2014). Alphamethrin (a synthetic pyrethroid)-induced oxidative stress and antioxidant-defence mechanism in *Glycine max* (L.) Merr. Internationl. *Journal on Agricultural Sciences*, 5(1): 27–41.

Bashir, H., Qureshi, M.I., Ibrahim, A.M. & Iqbal, M. (2015). Chloroplast and photosystems: Impact of cadmium and iron deficiency. *Photosynthetica*, 53(3): 321–335.

Beigh, S.Y., Arshi, A., Nawchoo, I.A., Aref, I.M. & Iqbal, M. (2011). Conservation and reproductive biology of *Aconitum heterophyllum* and *Podophyllum hexandrum*. In: Ahmad, A., Siddiqi, T.O. & Iqbal, M. (eds.), *Medicinal Plants in Changing Environment*. Capital Publishing Company, New Delhi, pp. 140–164.

Beigh, S.Y., Nawchoo, I.A. & Iqbal, M. (2002). Herbal drugs in India: Past and present uses. *Journal of Tropical Medicinal Plants*, 3: 197–204.

Beigh, S.Y., Nawchoo, I.A. & Iqbal, M. (2003a). Plants in traditional medicine of Kashmir Himalayas. *Journal of Economic & Taxonomic Botany*, 27(1): 99–104.

Beigh, S.Y., Nawchoo, I.A. & Iqbal, M. (2003b). Traditional veterinary medicine among the tribes of Kashmir Himalaya. *Journal of Herbs, Spices & Medicinal Plants*, 10(4): 121–127.

Beigh, S.Y., Nawchoo, I.A. & Iqbal, M. (2005). Cultivation and conservation of *Aconitum heterophyllum*, a critically endangered medicinal herb of the north-west Himalayas. *Journal of Herbs, Spices & Medicinal Plants*, 11: 47–56.

Beigh, S.Y., Nawchoo, I.A. & Iqbal, M. (2008). Alkaloid contents of *Aconitum heterophyllum* growing at different altitudes of the Kashmir Himalaya. *Journal of Indian Botanical Society*, 87: 252–255.

Bernath, J. (1986). Production ecology of secondary plant products. *Herbs, Spices & Medicinal Plants*, 1: 185–234.

Bhattacharyya, P. & Ghosh, A. (2022). Medicinal plants metabolomics in response to abiotic stresses. In: Husen, A. & Iqbal, M. (eds.), *Medicinal Plants: Their Response to Abiotic Stress*. Springer Nature, Switzerland.

Browne, E.C. (1921). *Arabian Medicine*. Hijra International Publishers, Lahore.

Chen, S.L., Yu, H., Luo, H.M., Wu, Q., Li, C.F. & Steinmetz, A. (2016). Conservation and sustainable use of medicinal plants: Problems, progress, and prospects. *Chinese Medicine*, 11: Art. 37. https://doi.org/10.1186/s13020-016-0108-7.

Cole, I.B., Saxena, P.K. & Murch, S. (2007). Medicinal biotechnology in the genus Scutellaria. *In Vitro Cellular and Developmetal Biology-Plant*, 43(4): 318–327.

El-Demerdash, M. (2001). Medicinal plants of Egypt. In: Saxena, P.K. (ed.), *Development of Plant-Based Medicines: Conservation, Efficacy and Safety*. Kluwer Academic Publishers, Dordrecht, the Netherlands, pp. 69–94.

Galle, T.K. (2005). *Domestic Medicinal Plants*. Mozaic Book, Zagreb, pp. 60–61.

Ghouse, A.K.M., Mahmooduzzafar, Iqbal, M. & Dastgiri, M. (1989). Effect of coal-smoke pollution on the stem anatomy of *Cajanus cajan* (L.) Mill. *Indian Journal of Applied & Pure Biology*, 4: 147–149.

Glesinger, L. (1954). *Medicine through Centuries*. Zora, Zagreb, pp. 21–38.

Gorunovic, M. & Lukic, P. (2001). *Pharmacognosy*. Beograd, pp. 1–5.

Hakeem, K.R., Khan, F., Chandna, R., Siddiqui, T.O. & Iqbal, M. (2012). Genotypic variability among soybean genotypes under NaCl stress and proteome analysis of salt-tolerant genotype. *Applied Biochemistry & Biotechnology*, 168(8): 2309–2329.

Hakeem, K.R., Chandana, R., Ahmad, A. & Iqbal, M. (2013). Reactive nitrogen inflows and nitrogen use efficiency in agriculture: An environment perspective. In: Ahmad, P. & Prasad, M.N.V. (eds.), *Environmental Adaptations and Stress Tolerance of Plants in the Era of Climate Change*. Springer Science & Business Media LLC, New York, pp. 217–232.

Hamilton, A.C. (2004). Medicinal plants, conservation and livelihoods. *Biodiversity & Conservation*, 13(8): 1477–1517.

Hamilton, A.C. & Radford, E.A. (2007). *Identification and Conservation of Important Plant Areas for Medicinal Plants in the Himalaya*. Ethnobotanical Society of Nepal, Nepal; Plant Life International and Kathmandu, Salisbury.

Hamilton, A.C., ed. (2008). *Medicinal plants in conservation and development: Case studies and lessons learnt*. Plantlife International, Salisbury.

Heywood, V.H. & Iriondo, J.M. (2003). Plant conservation: Old problems, new perspectives. *Biological Conservation*, 113(3): 321–335.

Huang, H. (2011). Plant diversity and conservation in China: Planning a strategic bioresource for a sustainable future. *Botanical Journal of the Linnean Society*, 166(3): 282–300.

Husen, A. & Iqbal, M. (2019). Nanomaterials and plant potential: An overview. In: Husen, A. & Iqbal, M. (eds.), *Nanomaterials and Plant Potential*. Springer Nature, Switzerland, pp. 3–29.

Husen, A. & Iqbal, M. (2022). *Medicinal Plants: Their Response to Abiotic Stress*. Springer Nature, Switzerland.

Hussein, M., Embiale, A., Husen, A., Aref, I.M. & Iqbal, M. (2017). Salinity-induced modulation of plant growth and photosynthetic parameters in faba bean (*Vicia faba*) cultivars. *Pakistan Journal of Botany*, 49(3): 867–877.

Iqbal, M., Abdin, M.Z., Mahmooduzzafar, Yunus, M. & Agrawal, M. (1996). Resistance mechanism in plants against air pollution. In Yunus, M. & Iqbal, M. (ed.), *Plant Response to Air Pollution*. John Willey & Sons, Chichester, pp. 195–240.

Iqbal, M., Jacob, M., Mehta, S. & Mahmooduzzafar. (2000). Plant responses to particulate matters suspended in the air. In: Iqbal, M., Srivastava, P.S. & Siddiqi, T.O. (eds.), *Environmental Hazards: Plants and People*.CBS Publishers, New Delhi, pp. 80–112.

Iqbal, M., Beigh, S.Y. & Nawchoo, I.A. (2004). Variability in morphology and active constituents of *Podophyllum hexandrum*: A Himalayan herb known for its anti-cancer activity. *Journal of Tropical Medicinal Plants*, 5: 33–36.

Iqbal, M., Jura-Morawiec, J., Wloch, W. & Mahmooduzzafar. (2010a). Foliar characterstics, cambial activity and wood formation in *Azadirachta indica* A. Juss. as affected by coal-smoke pollution. *Flora*, 205(1): 61–71.

Iqbal, M., Mahmooduzzafar, Aref, I.M. & Khan, P.R. (2010b). Behavioural responses of leaves and vascular cambium of *Prosopis cineraria* (L.) Druce to different regimes of coal-smoke pollution. *Journal of Plant Interactions*, 5(2): 117–133.

Iqbal, M., Mahmooduzzafar, Nighat, F. & Khan, P.R. (2010c). Photosynthetic, metabolic and growth responses of *Triumfetta rhomboidea* to coal-smoke pollution at different stages of plant ontogeny. *Journal of Plant Interactions*, 5(1): 11–19.

Iqbal, M., Ahmad, A. & Siddiqi, T.O. (2011a). Characterization of controversial plant drugs and effect of changing environment on active ingredients. In: Ahmad, A., Siddiqi, T.O. & Iqbal, M. (eds.), *Medicinal Plants in Changing Environment*. Capital Publishing Company, New Delhi, pp. 1–10.

Iqbal, M., Mahmooduzzafar, Nighat, F. & Aref, I.M. (2011b). Composition of seed oils of *Peristrophe bicalyculata* and *Ruellia tuberosa* as affected by coal-smoke stress. *Journal of Food, Agriculture & Environment*, 9(3–4): 1101–1104.

Iqbal, M., Ahmad, A., Ansari, M.K.A., Qureshi, M.I., Aref, M.I., Khan, P.R., Hegazy, S.S., El-Atta, H., Husen, A. & Hakeem, K.R. (2015). Improving the phytoextraction capacity of plants to scavenge metal(loid)-contaminated sites. *Environmental Reviews*, 23(1): 44–65.

Iqbal, M., Parveen, R., Parveen, A., Parveen, B. & Aref, I.M. (2018). Establishing the botanical identity of plant drugs based on their active ingredients under diverse growth conditions. *Journal of Environmental Biology*, 39(1): 123–136.

Iqbal, M., Umar, S. & Mahmooduzzafar. (2019). Nano-fertilization to enhance nutrient use efficiency and productivity of crop plants. In: Husen, A. & Iqbal, M. (eds.), *Nanomaterials and Plant Potential*. Springer Nature, Switzerland, pp. 473–505.

Jackson, W.P.S. & Sutherland, L.A. (2000). *International Agenda for Botanic Gardens in Conservation, Botanic Gardens*. Conservation International, Kew.

Jamshidi-Kia, F., Lorigooini, Z. & Amini-Khoei, H. (2018). Medicinal plants: Past history and future perspective. *Journal of HerbMed Pharmacology*, 7(1): 1–7.

Joy, P.P., Thomas, J., Mathew, S., Jose, G. & Joseph, J. (2008). Aromatic plants. In: Bose, T.K., Kabir, J., Das, P. & Joy, P.P.J. (eds.), *Tropical Horticulture vol. 2*. Naya Prokash, Calcutta, pp. 633–733.

Kauluni, D. & Joshi, A. (2018). Status of medicinal and aromatic plant (MAPs) and socio-economic influence in Nepalese livelihood - A review research. *Acta Scientific Agriculture*, 2(9): 125–130.

Kaur, N., Kumar, V., Rishi, P., Sehrawat, N., Dilawari, R., Kumar, P. & Aggarwal, N.K. (2019). Phytomedicine: History, scope and future prospects. In: Yadav, M., Kumar, V. & Sehrawat, N. (eds.), *Industrial Biotechnology: Plant Systems, Resources and Products*. De Gruyter, Berlin, pp. 105–120.

Kelly, K. (2009). *History of Medicine*. Facts on File, New York, pp. 29–50.

Kleisiaris, C.F., Sfakianakis, C. & Papathanasiou, I.V. (2014). Health care practices in ancient Greece: The Hippocratic ideal. *Journal of Medical Ethics & History of Medicine*, 7: 1–5.

Kumar, G.P., Kumar, R., Chaurasia, O. & Singh, S.B. (2015). Current status and potential prospects of medicinal plant sector in trans-Himalayan Ladakh. *Journal of Medicinal Plants Research*, 5: 2929–2940.

Lakey & Dorji, K. (2016). Ecological status of high-altitude medicinal plants and their sustainability: Lingshi, Bhutan. *BMC Ecology*, 16(1): 45. https://doi.org/10.1186/s12898-016-0100-1.

Leaman, D.J. (2001). Conservation, trade, sustainability and exploitation of medicinal plant species. In: Saxena, P.K. (ed.), *Development of Plant-Based Medicines: Conservation, Efficacy and Safety*. Kluwer Academic Publishers, Dordrecht, the Netherlands.

Mahmooduzzafar, Iqbal, M. & Ghouse, A.K.M. (1987). Impact of air pollution on the anatomy of *Cassia occidentalis* L. and *Cassia tora* L. *Indian Journal of Applied & Pure Biology*, 2: 45–47.

Marcolin, N. (2013). A glória do botânico. *Pesquisa FAPESP*, 210: 88–89.

McNeely, J.A. (1988). *Economics and Biodiversity: Developing and Using Economic Incentives to Conserve Biological Resources*. IUCN, Gland.

Mishra, V.K., Husen, A., Rahman, Q.I., Iqbal, M., Sohrab, S.S. & Yassin, M.O. (2019). Plant-based fabrication of silver nanoparticles and their application. In: Husen, A. & Iqbal, M. (eds.), *Nanomaterials and Plant Potential*. Springer Nature, Switzerland, pp. 135–175.

Moore, L.B., Goodwin, B., Jones, S.A., Wisely, G.B., Serabjit-Singh, C.J., Wilson, T.M., Collins, J.L. & Kliewer, S.A. (2000). St. John's wort induces hepatic drug metabolism through activation of the pregnane X receptor. *Proceedings of the National Academy of Sciences of the United States of America*, 97(13): 7500–7502.

Mughal, M.H., Saba, Ali, G., Srivastava, P.S. & Iqbal, M. (1999). Pharmaceutical potentials of *Butea monosperma* (Lamk.) Kuntze. *Pakistan Journal of Plant Sciences*, 5: 147–158.

Murch, S.J., Simmons, C.B. & Saxena, P.K. (1997). Melatonin in feverfew and other medicinal plants. *The Lancet*, 350(9091): 1598–1599.

Murch, S.J., Choffe, K.L. & Saxena, P.K. (2001). The unique challenges of plant-based medicines. In: Saxena, P.K. (ed.), *Development of Plant-Based Medicines: Conservation, Efficacy and Safety*. Kluwer Academic Publishers, Dordrecht, the Netherlands.

Nalawade, S.M., Sagare, A.P., Lee, C.Y., Kuo, C.L. & Tsay, H.S. (2003). Studies on tissue culture of Chinese medicinal plant resources in Taiwan and their sustainable utilization. *Botanical Bulletin of Academia Sinica*, 44: 79–98.

Nikolovski, B. (1995). *Essays on the History of Health Culture in Macedonia*. Macedonian Pharmaceutical Association, Skopje, pp. 17–27.

Nighat, F., Mahmooduzzafar & Iqbal, M. (2008). Coal-smoke pollution modifies physio-chemical characterstics of tissues during the ontogeny of *Peristrophe bycalyculata*. *Biologia*, 63(6): 1128–1134.

Nigrami, S.M.H. (2009). *Tarikhe-Tibb (History of Medicine)*, 5th ed. NCPUL, New Delhi.

Olorode, O. (2004). Conservation of plant genetic resources. *African Journal of Traditional, Complementary & Alternative Medicines*, 1(1): 4–14.

Olsen, C.S. & Larsen, H.O. (2003). Alpine medicinal plant trade and Himalayan mountain livelihood strategies. *Geographical Journal*, 169(3): 243–254.

Parveen, A., Parveen, R., Akhatar, A., Parveen, B., Siddiqui, K.M. & Iqbal, M. (2020). Concepts and quality considerations in Unani system of medicine. *Journal of AOAC International*, 103(3): 609–633.

Parveen, B., Parveen, A., Parveen, R., Ahmad, S., Ahmad, M. & Iqbal, M. (2020). Challenges and opportunities for traditional herbal medicine today, with special reference to its status in India. *Annals of Phytomedicine*, 9(2): 97–112.

Parveen, A., Ahmad, M., Parveen, B., Parveen, R. & Iqbal, M. (2022). Unani system of medicine, its origin, evolution and Indianization: A critical appraisal. *Indian Journal of Traditional Knowledge*, 21(3): (in press).

Peixoto, A.L., Guedes-Bruni, R.R., Haverroth, M. & Silva, I.M. (2012). Saberes e práticas sobre plantas: A contribuição de Barbosa Rodrigues. *Revista brasileira de história da ciência*, 5: 22–30.

Petrovska, B.B. (2012). Historical review of medicinal plants' usage. *Pharmacognosy Reviews*, 6(11): 1–5.

Pieroni, A., Howard, P., Volpato, G. & Santoro, R.F. (2004). Natural remedies and nutraceuticals used in ethnoveterinary practices in inland southern Italy. *Veterinary Research Communications*, 28(1): 55–80.

Qadeer, A. (1997). *Tibb-e-Unani: Challenges*. (Unani Medicine: Challenges). Nayi Awaz (Maktaba Jamia Limited), New Delhi.

Rahman, S., Iqbal, M. & Husen, A. (2022). Medicinal plants and abiotic stress: An overview. In: Husen, H. & Iqbal, M. (eds.), *Medicinal Plants: Responses to Abiotic Stress*. Springer Nature, Switzerland (in press).

Rafieian-Kopaei, M. (2013). Medicinal plants and the human needs. *Journal of Herbmed Pharmacology*, 1(1): 1–2.

Rao, R.V., Descamps, O., John, V. & Bredesen, D.E. (2012). Ayurvedic medicinal plants for Alzheimer's disease: A review. *Alzheimer's Research & Therapy*, 4(3): 22.

Rasul, G., Choudhary, D., Pandit, B.H. & Kollmair, M. (2012). Poverty and livelihood impacts of a medicinal and aromatic plant project in India and Nepal. *Mountain Research & Development*, 32(2): 137–148.

Rishton, G.M. (2008). Natural products as a robust source of new drugs and drug leads: Past successes and present-day issues. *American Journal of Cardiology*, 101(10A): 43D–49D.

Ross, I.A. (2005). *Medicinal Plants of the World: Chemical Constituents, Traditional and Modern Medicinal Uses*. Vol. 3. Humana Press, Totowa.

Saad, B. (2014). Greco-Arab and islamic herbal medicine: A review. *European Journal of Medicinal Plants*, 4(3): 249–258.

Saba, Iqbal, M. & Srivastava, P.S. (1999a). Three new phytoconstituents from *Lepidium sativum*. *Die Pharmazie*, 51: 851–853.

Saba, Iqbal, M. & Srivastava, P.S. (1999b). Detection of hyoscyamine in *Lepidium sativum*. *Hamdard Medicus*, 42: 72–75.

Saba, Mughal, M.H., Ali, M., Iqbal, M. & Srivastava, P.S. (1999c). A steryl ester from *Lepidium sativum*. *Phytochemistry*, 50: 1375–1377.

Saba, Pande, D., Iqbal, M., & Srivastava, P.S. (2000). Effect of $ZnSO_4$ and $CuSO_4$ on regeneration and lepidine content in *Lepidium sativum* L. *Biologia Plantarum*, 43(2): 253–256.

Shackleton, C.M. & Shackleton, S.E. (2004). The importance of non-timber forest products in rural livelihood security and as safety nets: A review of evidence from South Africa. *South African Journal of Science*, 1: 658–664.

Shmaefsky, B.R. & Husen, A. (2022). Phytoremediation potential of medicinal plants. In: Husen, H. & Iqbal, M. (eds.), *Medicinal Plants: Responses to Abiotic Stress*. Springer Nature, Switzerland (in press).

Siddiqi, T.O., Trag, A.R. & Iqbal, M. (2003). Pharmacognosy of the fruits of *Pimpinella anisum* L. *Hamdard Medicus*, 45: 46–48.

Schippmann, U., Leaman, D.J., Cunningham, A.B. & Walter, S. (2005). Impact of cultivation and collection on the conservation of medicinal plants: Global trends and issues. In: Jatisatienr, A., Paratasilpin, T., Elliott, S., et al. (eds.), *Conservation, Cultivation and Sustainable Use of MAPs: Proceedings of IIIrd World Congress on Medicinal and Aromatic Plants*, Chiang Mai, Thailand (February 3–7, 2003). Acta Horticulturae Art. 676, pp. 31–44.

Schippmann, U., Leaman, D.J. & Cunningham, A.B. (2006). A comparison of cultivation and wild collection of medicinal and aromatic plants under sustainability aspects. In: Bogers, R.J., Craker, L.E. & Lange, D. (eds.), *Medicinal and Aromatic Plants*, pp. 75–95. Springer, the Netherlands, pp. 75–95.

Schmidt, B., Ribnicky, D., Poulev, A., Logendra, S., Cefalu, W. & Raskin, I. (2008). A natural history of botanical therapeutics. *Metabolism*, 57: S3–S9.

Trag, A.R., Siddiqi, T.O., Ali, M., Mahmooduzzafar & Iqbal, M. (2005a). Phytochemical investigation of *Azadirachta indica* stem bark. *Journal of Saudi Chemical Society*, 9: 143–150.

Trag, A.R., Siddiqi, T.O., Ali, M., Mahmooduzzafar & Iqbal, M. (2005b). New aliphatic esters and pyran derivative isolated from stem bark of *Dalbergia sissoo* Roxb. *Journal of Saudi Chemical Society*, 9: 341–346.

Thorwald, J. (1991). *Power and Knowledge of Ancient Physicians*. August Cesarec, Zagreb, pp. 10–255.

Tucakov, J. (1971). *Healing with Plants – Phytotherapy*. Culture, Beograd, pp. 180–190.

Umar, S., Anjana & Iqbal, M. (2003). Potassium fertilization: A substitute to pesticide application in field crops. In: Prakash, S. (ed.), *Proceedings of National Symposium on Biochemical Sciences, Health and Environmental Aspects*. Allied Publishers Pvt. Ltd., New Delhi, pp. 494–496.

Umar, S., Diva, I., Anjum, N.A. & Iqbal, M. (2008). Potassium nutrition reduces cadmium accumulation and oxidative burst in mustard (*Brassica campestris* L.). *Electronic International Fertilizer Correspondent*, 16: 6–9.

Vaidya, A.D.B. & Devasagayam, T.P.A. (2007). Current status of herbal drugs in India: An overview. *Clinical Biochemistry*, 41(1): 1–11.

Verma, R.B., Mahmooduzzafar, Siddiqi, T.O. & Iqbal, M. (2006). Foliar response of *Ipomea pes-tigridis* L. to coal-smoke pollution. *Turkish Journal of Botany*, 30: 413–417.

Wali, B., Iqbal, M. & Mahmooduzzafar. (2007). Anatomical and functional responses of *Calendula officinalis* L to SO_2 stress as observed at different stages of plant development. *Flora*, 202(4): 268–280.

Waller, G.R. & Nowacki, E.W. (1978). *Alkaloid Biology and Metabolism in Plants*. Plenum Press, New York.

Wangchuk, P., Bremner, J.B., Samten, Rattanajak, R. & Kamchongpaisan, S. (2010). Anti-plasmodial agents from the Bhutanese medicinal plant Corydalis calliantha. *Phytotherapy Research*, 24(4): 481–485.

Wangchuk, P., Keller, P.A., Pyne, S.G., Taweechotipatr, M., Tonsomboon, A., Rattanajak, R. & Kamchonwongpaisan, S. (2011). Evaluation of an ethnopharmacologically selected Bhutanese medicinal plants for their major classes of phytochemicals and biological activities. *Journal of Ethnopharmacology*, 137(1): 730–742.

Wangchuk, P. & Tobgay, T. (2015). Contributions of medicinal plants to the gross national happiness and biodiscovery in Bhutan. *Journal of Ethnobiology & Ethnomedicine*, 11: 48.

WHO. (1993). *Guidelines on the Conservation of Medicinal Plants*. ICUN, The World Conservation Union, Gland, Switzerland.

WHO. (2002). *WHO Traditional Medicine Strategy 2002–2005*. World Health Organization, Geneva, Switzerland.

Yahyazadeh, M., Abouzeid, S., Lewerenz, L., Tahani Hijazin, T. & Selmar, D. (2022). Impact of aridity on specialized metabolism: Concentration of natural products in plants. In: Husen, H. & Iqbal, M. (eds.), *Medicinal Plants: Responses to Abiotic Stress*. Springer Nature, Switzerland (in press).

Yousuf, P.Y., Ahmad, A., Aref, I.M., Ozturk, M., Hemant, Ganie, A.H. & Iqbal, M. (2016a). Salt-stress-responsive chloroplast proteins in *Brassica juncea* genotypes with contrasting salt tolerance and their quantitative PCR analysis. *Protoplasma*, 253(6): 1565–1575.

Yousuf, P.Y., Ganie, A.H., Khan, I., Qureshi, M.I., Ibrahim, M.M., Sarwat, M., Iqbal, M. & Ahmad, A. (2016b). Nitrogen-efficient and nitrogen-inefficient Indian mustard showed differential expression pattern in response to elevated CO_2 and low nitrogen. *Frontiers in Plant Science*, 7: Art. 1074. https://doi.org/10.3389/fpls.2016.01074.

Yunus, M. & Iqbal, M. (1996). Major highlights of research on plant response to air pollution. In: Yunus, M. & Iqbal, M. (eds.), *Plant Response to Air Pollution*. John Wiley & Sons, Chichester, pp. 489–493.

Yunus, M., Singh, N. & Iqbal, M. (1996). Global status of air pollution: An overview. In: Yunus, M. & Iqbal, M. (eds.), *Plant Response to Air Pollution*. John Wiley & Sons, Chichester, pp. 1–34.

Zerabruk, S. & Yirga, G. (2012). Traditional knowledge of medicinal plants in Gindeberet district, Western Ethiopia. *South African Journal of Botany*, 78: 165–169.

11 Effect of Abiotic Stress on Production of Secondary Metabolites in Plants

Monalisha Samal, Sageer Abass, Rabea Parveen, Sayeed Ahmad, and Muhammad Iqbal

CONTENTS

11.1 INTRODUCTION

Plants are vulnerable to a variety of environmental stressors, both biotic and abiotic. They respond to stressful circumstances by altering their metabolism at the genetic level, which has an impact on metabolite synthesis. Plants are sessile creatures that respond to stress by altering the expression levels of genes involved in the synthesis of metabolites that are released in response to various stresses. Many metabolic pathways create distinct types of metabolites, which are classed as primary or secondary, depending on their biosynthetic pathway and their importance to plant survival, although a firm delineation between them is difficult to make (Rejeb et al., 2014). The primary metabolites (PMs) are necessary for cellular development and reproduction, whereas the secondary metabolites (SMs) are the molecules created in reaction to any biotic or abiotic stress, which can be external or endogenous to the cell (Isah, 2019). Thus, these metabolites are basically an outcome of the plant's defence strategy and provide protection to the plant. Additionally, many of these are also beneficial for humans, as they normally possess properties of therapeutic significance and are, therefore, used in treating a variety of human disorders and diseases (Iqbal et al., 2018; Parveen et al., 2020a, 2020b). This chapter focuses only on abiotic stresses and their impact on SM biosynthesis. There are around 214,000 different types of plant secondary metabolites, which are categorized based on their structure, function, and metabolic process. Terpenes, phenolic compounds, alkaloids, and sulfur-containing compounds are the major secondary metabolites found in plants (Wink 1999). It is the presence of these metabolites that so many plant species, believed to possess

DOI: 10.1201/9781003226925-11

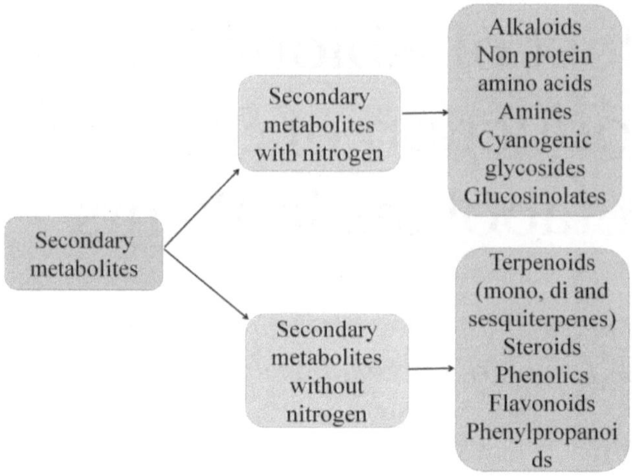

FIGURE 11.1 Types of secondary metabolites.

medicinal properties, gave rise to numerous folklores about their therapeutic efficacy in different tribal and under-developed parts of various countries, including the Indian sub-continent (Anis et al., 2000; Atique and Iqbal, 1992; Atique et al., 1985a, 1985b, 1993; Beigh et al., 2003), and have long been providing a form of ethnomedicine or herbal drugs, which are rapidly gaining ground all over the globe (Parveen et al., 2022).

Based on the presence or absence of nitrogen in the chemical structure, plant secondary metabolites can be classified into two major groups, as shown in Figure 11.1:

Secondary metabolites in plants are the derivatives of primary metabolites generated by plants because of various physiological changes. The SMs support plant growth and survival under various environmental conditions (Zandalinas et al., 2018). Plants exhibit diversity in the production and accumulation of SMs in response to abiotic stress that influences SM biosynthesis. The concentration of SMs may differ in plants of the same species growing under different environmental circumstances (Iqbal et al., 2018; Radušienė et al., 2012; Zhi-lin et al., 2007). Plants have evolved several complex and tightly controlled mechanisms to defend themselves against a variety of stressors. These efficient mechanisms, effecting adjustments in photosynthetic rates, stomatal conductance, membrane remodeling, transpiration, cell cycle and cell division, have an impact on the overall plant performance, and fine-tune the physiology and metabolism of bioactive compounds (Guo et al., 2018). Environmental stressors like pollutants, drought, excessive salinity, and extreme temperatures have a significant impact on plant development and output. As the stresses often occur simultaneously in the field, it is important to equip crops with multi-stress tolerance in order to reduce the burden of environmental impacts. Metabolite biosynthesis is influenced by the physiology and the developmental stage of the plant and by the environmental factors that affect the pathways involved in their biosynthesis. Many studies have shown that the secondary metabolism in plants serves several ecological purposes, ranging from stress tolerance to nutrient uptake. Metabolic activities in plants are enhanced under mild stress. This has led to the use of many biotic and abiotic elicitors for enhancing metabolite production in plant cells through *in-vivo* and *in-vitro* manipulation of growth conditions (Ali et al., 1999a, 1999b, 2000; Kim et al., 2010; Saba et al., 2000; Zhao et al., 2005).

11.2 ABIOTIC STRESS AND THEIR EFFECTS ON BIOSYNTHETIC PATHWAYS

Since the beginning of time, plants have existed in an intrinsically hostile environment. Abiotic stresses, such as drought, flooding, temperature extremes, salinity, toxic gases, and heavy metals, are the key factors restricting agricultural growth and sustainability across the world, although

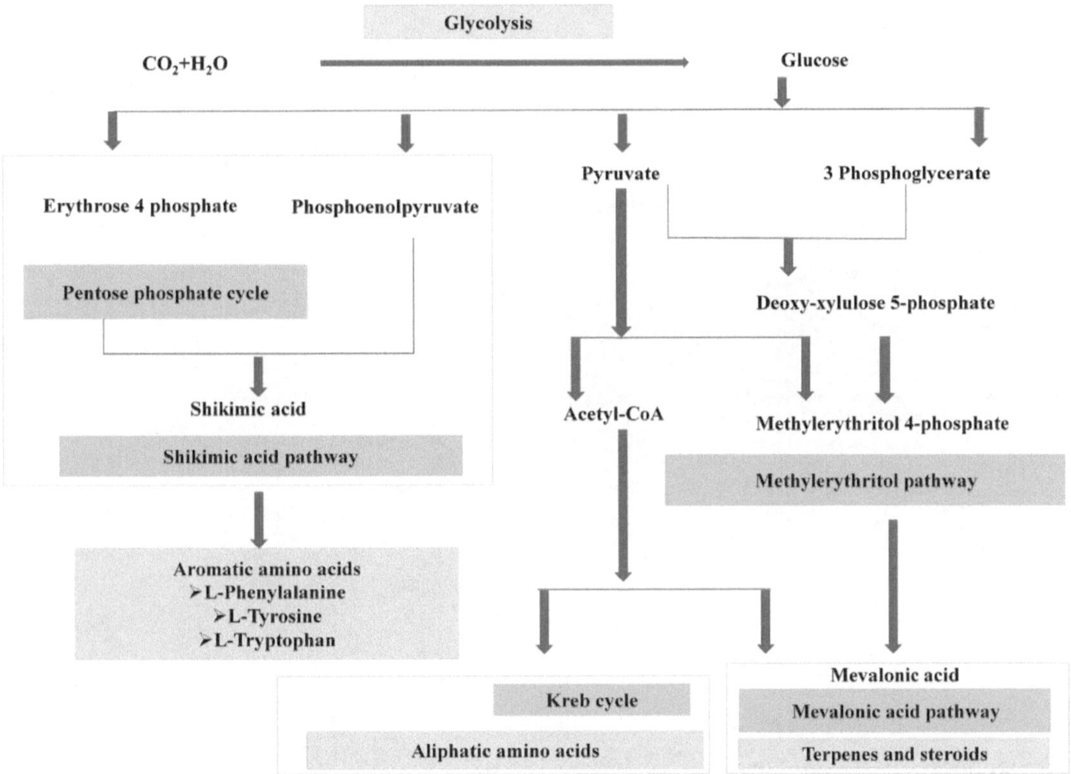

FIGURE 11.2 Overall scheme of metabolic pathways in plants.

plants adopt several means to defend themselves (Iqbal et al., 1996). These factors pose a danger to agriculture and ecosystems, accounting for significant losses to agricultural output and leading to reduced production of plant SMs (Wang et al., 2003; Zhao et al., 2005). On the other hand, low levels of these abiotic stressors, as well as certain nutrients, can act as potent elicitors and stimulate the plant growth processes as well as the metabolic routes for SM formation (Ahmad et al., 2007; Aref et al., 2016a; Karuppusamy, 2009). The SMs are biosynthesized mainly by three basic pathways: (a) shikimic acid pathway, (b) acetate pathway, and (c) amino acid pathway. The overall scheme of metabolic pathways in plants is shown in Figure 11.2.

11.2.1 Shikimic Acid Pathway

Shikimic acid, formed through the shikimic acid pathway, is a product of carbohydrate metabolism and acts as a key intermediate in the biosynthesis of C6-C3 units called phenylpropane derivatives. It also serves as a precursor for the biosynthesis of aromatic amino acids (phenylalanine, tyrosine, and tryptophan), and acts as an intermediate in the genesis of tannins, flavones, coumarins, and vanillin (Haslam, 2014).

The shikimic acid pathway is used mainly by bacteria, fungi, algae, some protozoans, and plants for the biosynthesis of folates and aromatic amino acids. Animals (including humans) do not have this pathway, and therefore they obtain these amino acids from their diet. This can be through direct consumption of microorganisms and plants by humans or via their indirect consumption through animals. Phenylalanine and tryptophan represent essential amino acids, which can be obtained from the animal's diet. Animals can synthesize tyrosine from phenylalanine, which is not an essential amino acid except for those individuals which are unable to hydroxylate phenylalanine to tyrosine (Maeda and Dudareva, 2012) (Figure 11.3).

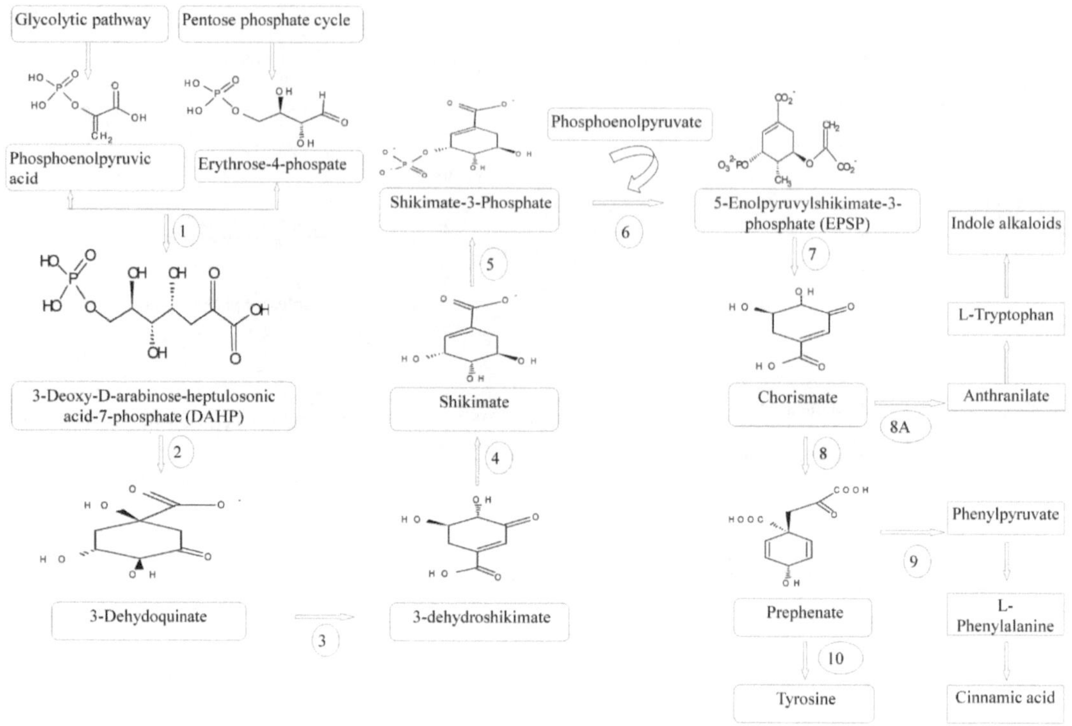

FIGURE 11.3 Biosynthesis of aromatic compounds via the shikimic acid pathway by incorporating subsequent enzyme systems such as (1) DAHP synthase, (2) 3-dehydroquinase, (3) 3-dehydroshikimase, (4) shikimate oxidoreductase, (5) shikimate kinase, (6) EPSP synthase, (7) chorismate synthase, (8) chorismate mutasc, (8A) anthranilate synthase, (9) phenylpyruvate synthase, and (10) prephenate dehydrogenase responsible for the synthesis of shikimic acid (key product) and the final products of the pathway.

11.2.2 Acetate Pathway

In the acetate pathway, also known as the polyketide pathway, the key product is acetyl coenzyme A (acetyl CoA), a thioester which is produced from primary carbon metabolism such as the glycolytic pathway via pyruvic acid, fatty acid, and amino acid oxidation. It acts as a metabolic pool for the formation of a large class of secondary metabolites, such as aromatic compounds, anthraquinones, fatty acids, flavonoids, isoprenoids, stilbenoids, macrolides, polyketides, and steroids (Perez de Souza et al., 2020). The formation of acetyl CoA and further biosynthesis of secondary metabolites are shown in Figure 11.4.

11.2.3 Amino Acid Pathway

The enzyme-driven set of biochemical reactions by which amino acids are produced is called the amino acid pathway. It is an essential pathway required for the regular functioning of the life of organisms.

Amino acids, the building blocks of proteins, are organic compounds containing one or more amino groups (-NH_2) and one or more carboxyl group (-COOH) groups, along with a side chain group (-R) specific to each amino acid. Amino acids are mainly classified as α, β, γ, δ, etc., according to the relative positions of their two functional groups. Specific amino acids contain only carbon, hydrogen, nitrogen, and oxygen, but other atoms may also be present, e.g., sulfur in cystine, and iodine in thyroxin.

Every organism is unable to synthesize all the amino acids. Therefore, some amino acids are taken from food; these amino acids are known as essential amino acids, whereas those produced

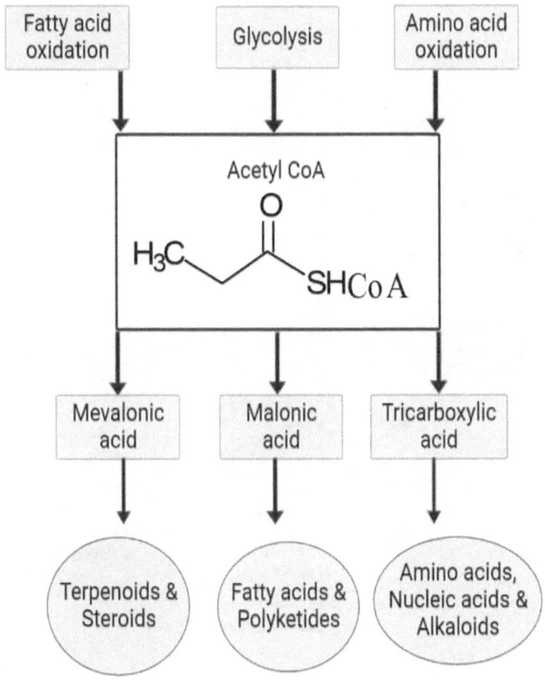

FIGURE 11.4 Origin of biosynthesis of predominant secondary metabolites via acetyl CoA as an intermediate.

within the body are called non-essential amino acids. There are twenty amino acids, of which eight cannot be synthesized by humans. Furthermore, arginine, cysteine, glycine, glutamine, histidine, proline, serine, and tyrosine are considered to be conditionally essential amino acids, as they are not normally required for the human diet, but need to be supplied exogenously to particular populations that do not synthesize them in adequate amounts. The pathways for the synthesis of non-essential amino acids are quite simple but those of essential amino acids are much more complex (Figure 11.5).

11.3 EFFECT OF ABIOTIC STRESS ON THE PRODUCTION OF SECONDARY METABOLITES

The biosynthetic pathways play a key role in SM formation. Efficiency of SM biosynthesis in plants is subject to abiotic stress.

11.3.1 Water Stress and Plant Secondary Metabolites

Drought is one of the most damaging abiotic stressors, that exerts a huge impact on agriculture (Verma and Deepti 2016), causing a 50–70% decrease in agricultural output. It affects 40% of the world's population; it is likely to force 700 million people to flee their homes by 2030 as reported by WHO. Drought stress stimulates the production of reactive oxygen species (ROS), which are extremely toxic and reactive, disrupting the cellular equilibrium by destroying proteins, carbohydrates, and DNA. Plant height, canopy size, root growth, and leaf area index are affected due to scarcity of water. Plant physiology, such as osmotic potential, stomatal conductance, photosynthetic rate, pressure potential, and transpiration rates, are also adversely affected by drought (Davis et al. 2014).In response to drought stress, plants undergo several morphological, physiological, and biochemical modifications (Anjum et al., 2008; Aref et al., 2013a, 2013b; Husen et al., 2014, 2017), which lead to alterations to the transcriptome, proteome, and metabolome of plants, and affect the

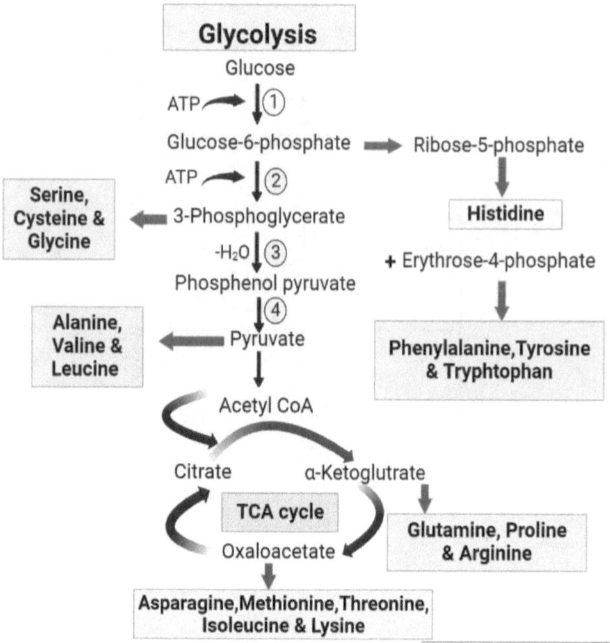

FIGURE 11.5 Biosynthesis of different amino acids at various levels of glycolysis and TCA cycle with the help of primary enzymes such as hexokinase, phosphoglycerate kinase, enolase, and pyruvate kinase.

cellular biosynthesis and breakdown activities. As a result, plants employ a variety of tactics to sustain growth and productivity under drought conditions, including increased synthesis of SMs, phytohormones, and ROS signaling (Khan et al., 2015; Ullah et al., 2018; Yadav et al., 2021). Plants produce a huge number of SMs that are necessary for survival in harsh environments. A balanced production of SMs is necessary for optimal plant development under changing environmental circumstances (Fàbregas and Fernie, 2019). Since water is a basic requirement for plant metabolism and physiology and the movement of metabolites and nutrients throughout the plant, a high transpiration rate and a low water content pave the way for drought stress in plants, resulting in reduced water potential/turgor pressure. Consequently, changes occur in various physiological functions, including the biosynthesis of SMs (Hildebrandt, 2018). This altered SM synthesis helps plants in inducing drought tolerance (Verma and Shukla, 2015).

Plants grown under drought stress have a higher concentration and a greater number of endogenous secondary metabolites than plants grown in well-watered conditions (Kleinwächter and Selmar 2015). *Trachyspermum ammi*, for example, showed an increase in phenolic compounds when subjected to drought stress (Azhar et al., 2011). SMs such as betulinic acid, rutin, quercetin, and artemisinin were found in high concentrations in *Hypericum brasiliense* and *Artemisia annua* (Chen et al. 2011). However, in a study of *Salvia sclarea*, Mahajan et al. (2020) found a decrease in plant biomass, SM content, yield and yield components under drought stress. The concentration of proline and essential oils in *Ocimum americanum* and *Ocimum basilicum* were altered due to water scarcity (Zhang et al., 2012). *Sesamum indicum* showed a reduction in sesamin, quercetin, and oil content when subjected to drought stress, but the overall content of flavonoids, phenolics, and different polyphenolics increased (Kermani et al., 2019). Naringin content in leaves of *Citrus unshiu* (Malik et al., 2015) and hesperidin content in *Mentha piperita* (Gholamreza et al., 2019) increased significantly under water deficit, although the isoflavone content of soybeans (Lim et al., 2021) and saponin content of *Chenopodium quinoa* (Solíz-Guerrero et al., 2002) were reduced. Drought frequently produces oxidative stress, which is often linked to the activation of the antioxidant defence system (Aref et al., 2016b), or even to increases in the levels of flavonoids and phenolic acids, as in willow leaves (Olajire and Azeez, 2011). *Glycyrrhiza glabra* produced enhanced glycyrrhizin synthesis with an expansion in the glycyrrhizin biosynthesis pathway under normal as well as water-deficit conditions. However, under extreme drought, the amount of glycyrrhizin declined, as compared with mild and moderate drought conditions (Hosseini et al., 2018). Drought stress reduced the overall saponin, phenolic, and flavonoid contents in *Mentha piperita* and *Catharanthus roseus*, but increased the levels of other SMs, such as tannins and alkaloids, in both plants (Alhaithloul et al., 2019).

11.3.2 Influence of Light on Secondary Metabolites

Plants need a certain amount of light for photosynthesis, which later affects the quality and concentration of SMs. Photoperiod (time), intensity (amount), and quality are all important aspects of light radiation (frequency or wavelength). Plants may adapt to changing conditions by releasing and accumulating a variety of SMs, such as phenolic compounds, triterpenoids, alkaloids, and flavonoids in response to light radiation (Shohael et al., 2006a; Zoratti et al., 2014). The output of these compounds is usually increased under mild stress or when the environment is variable. *In-vitro* approaches may be useful for scaling up the creation of these phytochemicals. Many abiotic elicitors are utilized to enhance the synthesis of SMs, and light is a powerful elicitor that influences plant growth, development, and morphogenesis (Adil et al., 2019; Tariq et al., 2014). It has been used to boost SM synthesis in several *in-vitro* cultures of medicinally significant plant species (Fazal et al., 2016; Shohael et al., 2006a). Light elicitation has increased the generation of a variety of useful secondary metabolites, opening a new avenue of study that might have considerable commercial implications for the pharmaceutical and nutraceutical industries. Various light sources, including ultraviolet (UV), light-emitting diodes (LED), and fluorescent lights, have been demonstrated to be effective elicitors for pharmacologically significant secondary metabolites (Dou et al., 2019; Kim et al., 2014).

It has been shown that there is a favorable relationship between increasing light intensity and phenolic levels. UV-A and UV-B have been found to induce biosynthesis of some flavonoids in *Ligustrum vulgare* leaves (Agati et al., 2011). In *Mikania glomerata*, accumulation of coumarins is promoted by sunlight; it was seen that a shorter photoperiod reduces coumarin levels, whereas a longer one increases them in the leaves and stem of this plant. The photoperiod had a considerable impact on the coumarin concentration. Foliar tannin and phenolic glycosides were found to decrease in shaded willow leaves (de Castro et al., 2007; Kolehmainen et al., 1995). When UV light from 280 to 320 nm was coupled with red light, it stimulated anthocyanin production synergistically in apples (Arakawa et al., 1985). UV-B has been shown to increase flavonoid content in barley (Liu et al. 1995), specifically flavonol content in Norway spruce (Fischbach et al., 1999), and polyamine content in cucumber (Kramer et al., 1991). When exposed to UV-B light, *Catharanthus roseus* plants produces greater amounts of vinblastine and vincristine (Binder et al. 2009). UV-B radiation stimulates the production of dimeric terpenoids and strictosidine synthase enzyme involved in the synthesis of indole alkaloids in *C. roseus* (Ramani and Jayabaskaran, 2008). Similarly, UV-B application enhanced flavonoids content in pea plant roots (Ramani and Jayabaskaran 2008), flavonoids and phenolic acids concentrations in chrysanthemum (Ma et al., 2016), epicatechin, quercetin, phloroglucinol, chlorogenic acid, and ferulic acid contents in *Nymphoides humboldtiana* (Nocchi et al., 2020), and the nataloin, barbaloin, and homonataloin contents in *Aloë mutabilis* grown under shade (Chauser-Volfson and Gutterman, 1998). However, in *Ocimum basilicum*, the concentration of linalool and eugenol is dramatically reduced when the plant is shaded (Chang et al., 2008).

A study of *Ocimum basilicum* seeds showed an increased production of phenolics after three days of UV-B exposure (Mosadegh et al., 2018). Likewise, 48 hours after UV-C treatment, the induced callus culture of *Vitis vinifera* produced 8–10 times more resveratrol than in the first 24 hours (Keskin and Kunter, 2010). *Stevia rebaudiana* callus cultures produced more secondary metabolites and accumulated more biomass when exposed to various fluorescent spectrum lights.

The maximum potential to increase the total phenolic content and total flavonoids content appeared under blue light (Ahmad et al., 2016).

11.3.3 Salinity Stress and Plant Secondary Metabolites

Salinity is defined as the amount of dissolved salt content in water. Salts are substances that dissolve into ions, such as sodium chloride, magnesium sulfate, potassium nitrate, and sodium carbonate. Salinity stress is one of the most damaging abiotic stressors that affect plant growth and development, plant yield and physiology, including the antioxidant defence system (Ali et al., 1999b; Arshi et al., 2002, 2004, 2012; Hakeem et al., 2012). Salinity inhibits plant development by reducing the leaf water potential, triggering morpho-physiological changes, producing the reactive oxygen species (ROS), increasing the osmotic stress and ion toxicity, and modifying metabolic processes (Bagheri et al., 2015; Khan et al., 2014; Qureshi et al., 2013; Yousuf et al., 2017). Sodium chloride is one of the major causes of salinity. In leaves, salinity increases the concentration of Na^+ and Cl^- ions while diminishing the concentrations of Mg^{+2}, K^+, and Ca^{+2} (Khalid and Cai, 2011; Li et al., 2006). Increasing salinity and falling crop production are the common concerns of all agroecosystems throughout the world these days. In the absence of adequate management of soil salinity, the loss of yield and quality has increased dramatically. Increased salt levels not only degrade the soil structure, but also limit plant development by creating hazardous imbalances and metabolic disorders (Husen et al., 2016; Hussein et al., 2017). Soil salinity is a serious global concern that is reducing the agricultural output and compromising the soil productivity capacity (Arora et al., 2018). It is projected that the rising levels of soil salinization will cause more than one-third of irrigated land to become barren. Increased salinity in the soil causes hyperosmotic stress and nutritional imbalances in plants, leading to reduced photosynthesis, retarded plant growth, and altered accumulation of secondary metabolites in plant tissues (Ali et al., 1999b; Arshi et al., 2006; Hussein et al., 2017). Human-caused activities that promote soil salinization are consistently increasing the proportion of the world's salinized land mass, threatening the medicinal plants' survival as well as the availability of bioactive compounds they generate.

Significant progress has been achieved in the discovery and characterization of several salt-stress-induced reactions related to SM formation in medicinal and crop plants (Gupta and Huang, 2014; Sytar et al., 2018).A wide range of plant SMs, including terpenoids and steroids, phenolics, flavonoids, and alkaloids, are known to be generated, implicated, or activated in response to salinity stress (Bourgaud et al., 2001). Safflower (*Carthamus tinctorius*) increases the production of secondary metabolites when exposed to salt stress (Gengmao et al., 2015). However, the concentration of naringenin is reduced in the nodules of chickpea (*Cicer arietinum*) plants under high sodium concentration (Garg and Singla, 2016).The concentration of alkaloids in *Datura inoxia* plants also shows an increasing trend (Selmar, 2008). Another study showed enhancement of diterpene glycoside content under low salt stress, but the same metabolite decreased under high concentrations of NaCl in *Stevia rebaudiana* (Shahverdi et al., 2019). The concentrations of carotenoids, phenolics, and flavonoids were significantly increased in tomato and mung bean plants subjected to varying degrees of salt stress (Iqbal et al., 2020; Mahmood et al., 2016). Also, different alkaloids, such as reserpine and vincristine, accumulated in *Rauwolfia tetraphylla* and *Catharanthus roseus,* respectively (Ahl and Omer, 2011). *Medicago arborea* exhibited alterations in the concentration of secondary metabolites in roots but not in shoots, when exposed to salt stress (Sarri et al., 2021).

11.3.4 Heavy Metal Stress and Secondary Metabolites

Heavy metal (HM) toxicity is a crucial abiotic stress, which influences all aspects of a plant's life: structural, functional, as well as chemical (Ali et al., 1999a; Qureshi et al., 2006; Umar et al., 2005). It alters the physiological and metabolic processes in plants by activating harmful ROS generation that causes ion toxicity and retards growth phenomena (Ansari et al., 2009; Bagheri et al., 2015;

Qureshi et al., 2007). Metal stress results in increased biosynthesis of polyphenols that protect plants from oxidative stress. However, the response of the shikimate pathway to different metals varies; copper stress caused an increased enzymatic activity of dehydrogenase and peroxidase, whereas nickel stress resulted in increased phenylalanine ammonia-lyase (PAL) activity (Díaz et al. 2001). It is, therefore, safe to say that plant responses to HM stress are metal specific, a phenomenon that could be associated with the physical properties of metals, and results in different stimulation of metabolic pathways within phenolic acid synthesis (Kováčik and Klejdus, 2008).Certain flavonoids exhibit the ability to enhance metal chelation process to reduce the harmful hydroxyl radical level in plant cells, thus enabling defense against heavy metal toxicity in an effective form (Diwan et al., 2010; Villiers et al., 2011). The severity of the effect of toxic or nutritive substances on plants depends on the uptake rate of the respective ions by plants, and this is significantly affected by the type and efficiency of the transporter system involved (Ahmad et al., 2005). Studies have shown that, under aluminum toxicity, accumulation of flavonoids such as anthocyanin sand flavonols is enhanced, fortifying the plant defense mechanism. Thus, the phenylpropanoid pathway in plants is activated by undergoing upregulation of specific key enzymes like chalcone synthase, cinnamyl alcohol dehydrogenase, phenylalanine ammonia-lyase, polyphenol oxidase, and shikimate dehydrogenase. This mechanistic hypothesis was confirmed through increased concentration of catechin and quercetin in roots of *Zea mays* grown on soil with high aluminum ion content (Kidd et al., 2001).

Cadmium (Cd) and copper (Cu) stress stimulated accumulation of phenolic compounds such as *p*-coumaric acid and ferulic acid, respectively, along with the accumulation of lignins (Sarker and Oba, 2018). Flavonols are appropriate mediators in a plant's responses to toxic heavy metals, as seen in increased rutin content in *Erica andevalensis* grown under Cd stress (Márquez-García et al., 2012). Similarly, enhanced accumulation of quercetin was reported in roots of maize after exposure to aluminum toxicity (Kidd et al., 2001). In *Fagopyrum esculentum*, increased total phenolic, flavonoid, and anthocyanin contents were observed owing to enhanced activity of the PAL enzyme post triggering under aluminum toxicity(Smirnov et al., 2015).Applications of metals in nanoform also cause explicit impacts on plant SM production (Husen and Iqbal, 2019). The derivatives of quercetin and kaempferol in *Vitis vinifera* increased in amount after treatment with titanium nanoparticles (Ben-Abdallah et al., 2019). In another study on *Nigella arvensis*, nanoparticles of NiO, TiO_2, and Al_2O_3 increased the quercetin content and reduced the kaempferol content (Martinez et al., 2016).

11.3.5 Temperature Level and Secondary Metabolites

Temperature, being one of the major weather variables, can markedly affect the various morpho-physiological traits of plants, including the composition of SMs, which helps to identify heat-/cold-tolerant genotypes or hybrids (Shohael et al., 2006b; Noor et al., 2019). An increased level of flavonols has been recorded in tomatoes (*Solanum lycopersicon* cv. Boludo) and *Hypericum Brasiliense* in response to high temperatures (de Abreu and Mazzafera, 2005). Too high a temperature may reduce the concentration of secondary metabolites in plants, as observed in soybean wherein high temperature decreased isoflavone content at late reproductive stages. On the other hand, studies have revealed a conflicting response of phenols and flavonoids toward low temperatures. A study reported an increased concentration of two superior isoflavones, calycosin and its 7-*O*-D-glucoside, in seedlings of *Astragalus membranaceus* var. *mongholicus* in response to chilling conditions (Kumar and Sharma, 2018). A phenolic antioxidant, *viz*., catechin, accumulated at chilling temperatures in different plants such as date palm (Wang et al., 2017), *Quercus suber* (Djoukeng et al., 2008), and Arabidopsis (Meng et al., 2015). On the contrary, in *Prunus persica*, low temperature increased the amount of flavonols (Wang et al., 2019). The influence of heat and cold stresses on the production rate and amount of SMs helped in establishing guidelines for the commercial exploitation of plants. For instance, positive correlations with temperature change were seen in the composition of phenolic compounds, delphinidin-3-*O*-glucoside, delphinidin-3-*O*-rutinoside, and myricetin-3-*O*-glucoside in three cultivars of *Ribes nigrum* (Zheng et al., 2012). On similar grounds, prolonged chilling induced stimulation of anthocyanin biosynthetic pathway genes that resulted in increased anthocyanin content in maize (*Zea mays*) seedlings (Yüzbaşıoğlu et al., 2017). Similarly, in leaves and stems of *Arabidopsis thaliana*, the cold-stress-mediated phenylpropanoid biosynthetic pathway facilitated the increased accumulation of anthocyanins (Jaini et al., 2017). On the other hand, heat-stress-induced inhibition of anthocyanin biosynthetic gene transcription followed by anthocyanin degradation resulted in a reduced total anthocyanin content in *Vitis vinifera* (Borsani et al., 2010). However, in some plants such as *Petunia* × *hybrida* (Andersson, 2011) and *Citrus sinensis* (Vince-Prue and Canham, 1983), anthocyanin accumulation was induced by low temperature and inhibited by high temperature. Stiles and co-workers investigated the influence of hot, cold, and room-temperature conditions on anthocyanins originating from precursor compounds, cyanidin and delphinidin in an anthocyanin biosynthetic pathway that includes eight cyanidin-based, six petunidin-based, two delphinidin-based, and one peonidin-based anthocyanin from the floral extracts of *Plantago lanceolata* (Stiles et al., 2007). Thirteen individual anthocyanidin concentrations increased at low temperatures, whereas the content of four petunidin-based anthocyanins were unaffected. Cold conditions have a marked influence on alkaloid biosynthesis, as in *Papaver somniferum*, where concentrations of morphinan, phthalisoquinoline, and benzylisoquinoline were restricted at low temperature (Matyášová et al., 2011). In contrast, increased total phenolics, like genistein, daidzein, and genistein, were recorded in roots of *Glycine max* under the influence of low temperature for 24 h, compared with the control. The highest rise, of about 310%, was observed in genistein after treatment at 10°C for 24 h (Posmyk et al., 2005). Similarly, in *Lupinus angustifolius*, a high alkaloid content was recorded for four consecutive years, under high ambient temperature (Otterbach et al. 2019). Investigation of 60-day-old *Catharanthus roseus* seedlings has shown that, under temporary heat shock, the alkaloid contents of vindoline, catharanthine, and vinblastine were higher at 40°C than at 30°C. In fact, catharanthine content was enhanced by 40% after incubation at 40°C for 2 h and attained the highest value at 6 h. Furthermore, in a long-term experiment at 35°C, a spiky increase was recorded in the concentrations of monomeric alkaloids, such as catharanthine and vindoline (Guo et al., 2007). On the contrary, *C. roseus* leaves showed a nearly 2-fold and 2- to 4-fold decline in levels of catharanthine and vindoline, respectively, under the impact of low temperature (Mustafa and Verpoorte, 2007).The correlation between varying temperature and simultaneous terpenoid accumulation was observed in *Quercus rubra* and *Quercus alba* under warm as well as cold conditions and showed a two-fold rise in terpenoid content with high temperature than with low temperature (Hanson and Sharkey, 2001). Experimental analysis showed that, under high temperature, terpene content of *Daucus carota* root was directly correlated to the temperature, except for terpinolene, which showed a significant reduction with increase in temperature (Simon et al. 2021).

11.3.6 Secondary Metabolites in Relation to Other Abiotic Stresses

Several anthropogenic abiotic stresses, caused by toxic gases, pesticides (insecticides, herbicides, and fungicides), and other pollutants, are also known to affect plant performance by altering the structural, chemical and functional features of plants, including their metabolic activities, growth, and yield (Iqbal et al., 2000, 2005; Verma et al., 2006; Wali et al., 2007; Yunus and Iqbal, 1996; Yunus et al., 1996). Plants increase the biosynthesis of secondary metabolites to protect themselves from the phytotoxic effects of such stresses (Ali et al., 2008; Sharma et al., 2019; Singh et al., 2000; Świeca, 2015). A large number of secondary metabolites are activated in plants under uncontrolled pesticidal application. Pesticides may influence the production rate of secondary metabolites like hydroxycinnamic acids, anthocyanins, tropane alkaloids, and terpenoids by unknown mechanisms or by interfering with key biosynthetic steps. Increased phenolic biosynthesis leads to the accumulation of polyphenols in plants (Mahdavi et al., 2015; Sharma et al., 2016a). The volatility of toxic or beneficial essential oils is also altered by a change in their chemical composition, as pesticide residues and atmospheric gases are soluble in volatile oils (Karimi and Meiners, 2021; Spinelli et al., 2011). Not only the quantity but also the composition (ratio of the constituent fatty acids) of

TABLE 11.1
Impact of Different Abiotic Stresses on the Production of Secondary Metabolites in Plants

S. No	Plant name	Effect of stress on secondary metabolites	References
Drought stress			
1.	*Brassica napus* L.	Increased total phenols, flavonoids, and flavanols	Raza et al., 2021
2.	*Nicotiana tabacum* L.	Increased kaempferol and quercetin contents	García-Calderón et al., 2015
3.	*Camellia sinensis* L.	Increased epicatechin and quinine	Hernández et al., 2006
4.	*Labisia pumila* Blume Mez	Increased total phenolic compounds	Jaafar et al., 2012
5.	*Gossypium hirsutum* L.	Increased ascorbic acid, glutathione, and α-tocopherol concentrations	Hussien et al., 2015
6.	*Pinus sylvestris* L. and *Picea abies* L.	Increased concentration of terpenes	Turtola et al., 2003
7.	*Crataegus laevigata* (Poir.) DC and *Crataegus monogyna* Jacq.	Increased flavonol content	Kirakosyan et al., 2003
8.	*Triticum aestivum* L.	Flavonoids and phenolic acids synthesized in a large amount	Ma et al., 2014
9.	*Chenopodium quinoa* L.	Decreased saponin content	Solíz-Guerrero et al., 2002
10.	*Sesamum indicum* L.	Decreased the sesamin, quercetin, and oil content	Kermani et al., 2019
11.	*Nicotiana tabacum* L.	Increased flavonoid accumulation	Meng et al., 2015
12.	*Achillea pachycephala* Rech. f.	Decreased apigenin-7-*O*-glucoside, and increased luteolin-7-*O*-glucoside	Gharibi et al., 2019
13.	*Scutellaria baicalensis* Georgi	Increased baicalin content under drought stress	Zhang et al., 2020
14.	*Hypericum brasiliense* Choisy	Increased rutin and quercetin contents	de Abreu and Mazzafera, 2005
15.	*Salvia officinalis* L.	Increased monoterpene concentrations	Bettaieb et al., 2009
16.	*Thymus vulgaris* L.	Increased rosmarinic acid content	Trócsányi et al., 2015
17.	*Papaver somniferum* L.	Increased alkaloid content	Wang et al., 2010
18.	*Ocimum basilicum* L.	Increased anthocyanins	Hassani and Omidbaigi, 2006
19.	*Ocimum tenuiflorum* L.	Decreased concentration of eugenol	Rastogi et al., 2019
Radiation stress			
20.	*Vanilla planifolia* Jacks. ex Andrews	Increased vanillin content under blue light stress	Havkin-Frenkel et al., 1996
21.	*Astragalus compactus* Lam.	Increased phenolic content under UV radiation	Naghiloo et al., 2012
22.	*Withania somnifera* Dunal	UV solar radiation increased the production of alkaloids, anthocyanins, carotenoids, flavonoids, lignin, phytosterols, saponins and tannins	Takshak and Agrawal, 2014
23.	*Asparagus officinalis* L.	Increased flavonol, quercetin-4′-*O*-monoglucoside content in response to increased UV-B dose	Eichholz et al., 2012
24.	*Coleus forskohlii* Willd.	Increased phenolics and flavonoids	Takshak and Agrawal, 2015

(*Continued*)

TABLE 11.1 (CONTINUED)
Impact of Different Abiotic Stresses on the Production of Secondary Metabolites in Plants

S. No	Plant name	Effect of stress on secondary metabolites	References
25.	*Prunella vulgaris* L.	UV-B increased rosmarinic acid and caffeic acid contents	Chen et al., 2018
26.	*Kalanchoe pinnata* Lam.	Combination of UV-B and white light increased the quercetin content	dos Santos Nascimento et al., 2015
27.	*Mahonia breviracema* Y.S. Wang & P.G. Xiao	Increased alkaloid content	Li et al., 2018
28.	*Mahonia bodinieri* Gagnep.	Increased alkaloid content	Kong et al., 2016
29.	*Glechoma longituba* Kuprian.	Decreased ursolic and oleanolic acids	Zhang et al., 2015
30.	*Centella asiatica* L.	Increased chlorogenic acid content	Alqahtani et al., 2015
31.	*Capsicum annum* L.	UV-A/B increased cynaroside content	Ellenberger et al., 2020
32.	*Ginkgo biloba* L.	UV-B increased quercetin, kaempferol	Zhao et al., 2020
33.	*Vitis vinifera* L.	UV-C increased catechin and chlorogenic acid	Cetin, 2014
34.	*Lepidium sativum* L.	UV-C increased kaempferol and quercetin	Ullah et al., 2019
35.	*Clematis terniflora* DC.	High UV radiation increased alkaloid biosynthesis	Gao et al., 2016
Salinity stress			
36.	*Brassica rapa* L.	Increased flavonoid content	Abdel-Farid et al., 2009
37.	*Plantago ovata* Forssk.	Increased phenolics and flavonoids	Haghighi et al., 2012
38.	*Oryza sativa* L.	Increased phenolics and flavonoids contents	Hussain et al., 2013
39.	*Acalypha wilkesiana* Mull. Arg.	Increased saponin content	Odjegba and Alokolaro, 2013
40.	*Nigella sativa* L.	Increased apigenin	Bourgou et al., 2012
41.	*Trachyspermum ammi* L.	Decreased essential oil yield	Ashraf and Orooj, 2006
42.	*Mentha piperita* L.	Decreased essential oil yield	Tabatabaie and Nazari, 2007
43.	*Thymus maroccanus* Ball.	Decreased essential oil yield	Belaqziz et al., 2009
44.	*Carthamus tinctorius* L.	Increased flavonoids under different NaCl concentrations	Belaqziz et al., 2009
45.	*Pilocarpus jaborandi* Holmes	Increased pilocarpine	De Abreu et al., 2005
46.	*Catharanthus roseus* L.	Increased reservine production	Misra and Gupta, 2006
47.	*Artemisia annua* L.	Increased artemisinin production	Vashisth et al., 2018
48.	*Arabidopsis thaliana* L.	Decreased anthocyanin production	Van Oosten et al. 2013
49.	*Solanum nigrum* L.	Increased production of solasodine and proline	Šutković et al., 2011
50.	*Sesuviumportulacastrum*L.	Increased polyphenol production	Slama et al., 2017
51.	*Olea europaea* L.	Increased oleuropein and phenolic production	Petridis et al., 2012
Heavy metal stress			
52.	*Brassica juncea* L.	Increased phenolic and flavonoid contents under Ba stress	Bouslimi et al., 2021
53.	*Kandelia obovate* H.Y. Liu and J. Yong	Increased total phenolics due to Cd and Zn stress	Chen et al., 2019
54.	*Zea mays* L.	Increased contents of total phenols and some polyphenols like chlorogenic and vanillic acids due to Cu, Pb, and Cd stress	Veeraraghavan et al., 2021
55.	*Withania somnifera* L. Dunal	Increased total contents of flavonoids and phenolics due to Cd stress	Dobrikova et al., 2021

(Continued)

TABLE 11.1 (CONTINUED)
Impact of Different Abiotic Stresses on the Production of Secondary Metabolites in Plants

S. No	Plant name	Effect of stress on secondary metabolites	References
56.	*Belamcanda chinensis* L. DC.	Increased total flavonoids (iristectorigenin A and tectorigenin) due to Cu stress	Zhu et al., 2020
57.	*Solanum lycopersicum* L.	Increased phenolic compounds and anthocyanins due to Ni stress	Jahan et al., 2020
58.	*Salix purpurea* × *viminalis* L.	Increased production of salicylic acid and phenolic compounds under Cu and Ni stress	Drzewiecka et al., 2017
59.	*Oryza sativa* L.	Increased production of protocatechuic, *p*-hydroxybenzoic, vanillic, *p*-coumaric, caffeic, and gallic acids under Cr stress	Dubey et al., 2018
60.	*Catharanthus roseus* L.	Increased total alkaloids, phenolic compounds, and flavonoid contents under Cd stress	Abnosi et al., 2015
Temperature stress			
61.	*Quercus rubra* L. and *Quercus alba* L.	Increased isoprene concentration under high temperature	Hanson and Sharkey, 2001
62.	*Daucus carota* L.	Increased α-caryophyllene, decreased α-terpinolene, and increased α-farnesene concentrations	Yang et al., 2018
63.	*Betula pendula* L.	Increased (*E*)-4,8–dimethyl–1,3,7-nonatriene (DMNT) concentration under night-time warming	Ibrahim et al., 2010
64.	*Glycine max* L.	Increased genistein and daidzein contents at 10$^{o'}$ C and 11°C, respectively, for 24 h	Janas et al., 2002
65.	*Zea mays* L.	Increased pelargonidin content at low temperature	Pant et al., 2021
66.	*Catharanthus roseus* L.	Increased catharanthine content with prolonged heat stress, and decreased vindoline content at low temperature	Liu et al., 2021
67.	*Papaver somniferum* L.	Decreased morphine content at low temperature	Matyášová et al., 2011
68.	*Curcuminus acuminata*L.	Increased 10-hydroxycamptothecin at 40°C for 2 h	Zu et al., 2003
69.	Somatic embryos of *Eleutherococcus senticosus* (Rupr. & Maxim.) Maxim.	Decreased total phenolics, flavonoids and total eleutheroside accumulation, but increased eleutheroside E at low temperature	Shohael et al., 2006
70.	*Rhodiola crenulate* (Hook.f. & Thomson) H.Ohba	Increased melatonin concentration at low temperature	Zhao et al., 2010
71.	*Panax ginseng* C.A. Mey, *Panax quinquefolius* L.	Increased ginsenoside at high temperature	Hou et al., 2021
72.	*Hypericum perforatum* L.	Increased hypericin and hyperforin contents at high temperature	Yao et al., 2019
73.	*Pinus pinaster* Aiton	Increased endogenous jasmonates at low temperature	Pedranzani et al., 2007
Pesticide stress			
74.	*Polyalthia bullata* King	Herbicide 2,4-dichlorophenoxyacetic acid (2,4-D) increased alkaloid production	Zaman et al., 2021
75.	*Solanum lyratum* Thunb.	Herbicide (2,4-D) and naphthaleneacetic acid increased the steroidal alkaloid production	Kuo et al., 2012

(*Continued*)

TABLE 11.1 (CONTINUED)
Impact of Different Abiotic Stresses on the Production of Secondary Metabolites in Plants

S. No	Plant name	Effect of stress on secondary metabolites	References
76.	*Catharanthus roseus* L.	Herbicide (2,4-D) increased the vincristine, vinblastine, and ajmalicine contents	Verma et al., 2012
77.	*Centella asiatica* L.	Methyl-jasmonic acid increased alkaloid, saponin and terpenoid production	Luthra et al., 2022
78.	*Moringa oleifera* Lam	Phenylalanine increased the niazirin and vincosamide contents	Mahood et al., 2018
79.	*Erythroxylum coca* Lam.	Herbicide glyphosate increased the cocaine content	Casale and Lydon, 2007
80.	*Datura stramonium* L.	Herbicides chlorsulfuron and glyphosate decreased the overall tropane alkaloid biosynthesis	Deng, 2005
81.	*Catharanthus roseus* L.	Herbicide dimethylarsinatem decreased alkaloid production	Cullen and Hettipathirana, 1993
82.	*Solanum aviculare* G. Forst and *Solanum laciniatum* Aiton	Herbicide barban 4-chlorobut-2-ynyl-3-chlo rophenylcarba-mate decreased the solasodine content	Mann and Lancaster, 1976
83.	*Brassica juncea* L.	Insecticide imidacloprid increased polyphenol and total phenol content	Sharma et al., 2016b
84.	*Brassica juncea* L.	Insecticide imidacloprid increased the carotenoid, anthocyanin, and xanthophyll contents	Sharma et al., 2016a
85.	*Oryza sativa* L.	Insecticide diazinon caused variation in amounts of sugars, amino acids, organic acids, and phenylpropanoids	Mahdavi et al., 2015
86.	*Brassica juncea* L.	Insecticide imidacloprid increased the total phenol and anthocyanin contents	Sharma et al., 2018
87.	*Oryza sativa* L.	Insecticide fipronil, omethoate, and buprofezin altered the metabolic responses	Zhou et al., 2012

some seed oils were found to change under the impact of coal-smoke pollution (Iqbal et al., 2011); this change in the relative proportions of fatty acids is likely to affect even the quality of the oils concerned. Insecticide applications may increase the accumulation of flavonoids, such as anthocyanins, in plant leaves; this helps in improving plant photosynthetic efficiency (Shahzad et al., 2018). However, several herbicides, such as alachlor and glyphosate, have been reported to reduce the levels of bioactives by non-specific mechanisms. Whereas alachlor reduces flavonoid biosynthesis by inhibiting a crucial yet unknown step in the biosynthetic pathway, glyphosate blocks the synthesis of all cinnamate-derived phenolic compounds by inhibiting 5-enolpyruvyl shikimate-3-phosphate (EPSP) synthase. This inhibition of EPSP synthase leads to increased accumulation of important metabolites, such as shikimate, benzoic acids, and benzoic acid derivatives (Lydon and Duke, 1989). Other herbicides, such as sulfonylureas and *p*-nitro-substituted diphenylether (DPE) have been reported to increase the level of cinnamate derivatives and terpenoids, respectively (Zheljazkov et al., 2006). Certain fungicides are believed to act by influencing the plant's capacity to generate phytoalexins and other useful secondary metabolites (Majid et al., 2013). These data suggest that sublethal effects of pesticides (insecticides, herbicides, and fungicides) on targeted and non-targeted plants can significantly affect plant secondary metabolism by altering the biosynthetic pathways of secondary metabolites having therapeutic significance (Singh et al., 2020) (Table 11.1).

11.4 SUMMARY AND CONCLUSIONS

Environmental stress factors such as salinity, drought or flood, temperature extremes (freezing, cold, or hot), heavy metals, nutrient stress, air pollution, pesticides, UV light, other radiations, etc. cause severe problems for plant growth and development. High stress conditions often inflict grave damage to the production of plant biomass. Higher plants show case great stress-tolerance responses and exhibit adaptations to maintain optimal growth. Stress induces the cell-signaling process, resulting in transcriptional upregulation of majorly interlinked biosynthetic pathways, such as the shikimic acid and acetate pathways, due to improved stress tolerance of plants. By activating the key enzymes (like phenylalanine ammonia-lyase, chalcone synthase, shikimate dehydrogenase, chorismate synthase, cinnamyl alcohol dehydrogenase, polyphenol oxidase, flavonol synthase, and chalcone isomerase) involved in the production of secondary metabolites (such as phenols, flavonoids, alkaloids, terpenes, tannins, glycosides, saponins, and steroids), several environmental stressors increase the ROS-scavenging capacity of plants, as shown in Figure 11.6. In addition, the increase in plant tolerance to stress is correlated with polyphenol functions, which are primarily related to antioxidant ability, and/or to the capacity of flavonoid- and anthocyanin-like bioactive compounds that protect plants from external pressures.

Secondary metabolism regulates plant growth and development and serves as a reservoir of essential phytochemicals that protect plants from environmental stresses. Many plant SMs have antioxidant properties and function as the first line of defense against oxidative damage. Plant-environment interactions rely heavily on the SM status of plants, as the SMs play a crucial role in plant adaptation to environmental restrictions. Interestingly, individual environmental stress can increase the content of many SMs in plants. The ability of plant SMs to protect plants from oxidative damage is a topic that deserves deeper investigation. Future studies must focus on controlling the

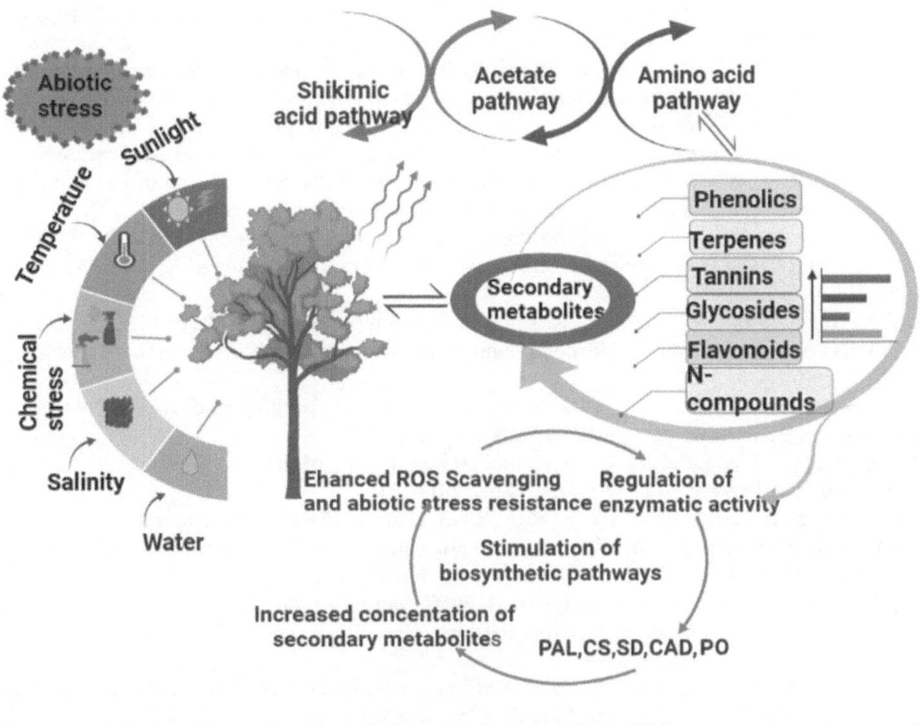

FIGURE 11.6 Diagrammatic explanation of the role of abiotic stress (light, temperature, water, salinity, or air pollution) on production of secondary metabolites. ROS (reactive oxygen species); PAL (phenylalanine ammonia-lyase); CS (chalcone synthase); SD (shikimate dehydrogenase); CAD (cinnamyl alcohol dehydrogenase); PO (polyphenol oxidase).

regulatory proteins and genes involved in SM biosynthesis so as to ensure an increased production of bioactive substances in plants.

REFERENCES

Abdel-Farid, I., Jahangir, M., van den Hondel, C., Kim, H., Choi, Y., & Verpoorte, R. 2009. Fungal infection-induced metabolites in *Brassica rapa*. *Plant Science*, 176(5): 608–615.

Abnosi, M., Amirjani, M., Mahdiyeh, M., & Moradipoor, H. 2015. Biochemical and cellular response of *Catharanthus roseus* callus cells to cadmium toxicity. *Journal of Genetic Resources*, 1(2): 101–114.

Adil, M., Ren, X., & Jeong, B. R. 2019. Light elicited growth, antioxidant enzymes activities and production of medicinal compounds in callus culture of *Cnidium officinale* Makino. *Journal of Photochemistry and Photobiology, Part B: Biology*, 196: 111509.

Agati, G., Biricolti, S., Guidi, L., Ferrini, F., Fini, A., & Tattini, M. 2011. The biosynthesis of flavonoids is enhanced similarly by UV radiation and root zone salinity in L. *vulgare* leaves. *Journal of Plant Physiology*, 168(3): 204–212.

Ahl, S.-A., & Omer, E. 2011. Medicinal and aromatic plants production under salt stress: A review. *Herba Polonica*, 57(1): 72–87.

Ahmad, A., Khan, I., Abrol, Y. P., & Iqbal, M. 2005. Role of sulphate transporter system in sulphur efficiency of mustard genotypes. *Plant Science*, 169(5): 842–846.

Ahmad, N., Rab, A., & Ahmad, N. 2016. Light-induced biochemical variations in secondary metabolite production and antioxidant activity in callus cultures of *Stevia rebaudiana* (Bert). *Journal of Photochemistry and Photobiology, Part B: Biology*, 154: 51–56.

Ahmad, S., Jamal, A., Fazli, I. S., Alam, T., Khan, M. A., Kamaluddin, M., Iqbal, M., & Abdin, M. Z. 2007. Impact of sulphur and nitrogen application on seed xanthotoxin yield in *Ammi majus* L. *Korean Journal of Crop Science*, 52: 153–161.

Alhaithloul, H. A., Soliman, M. H., Ameta, K. L., El-Esawi, M. A., & Elkelish, A. 2019. Changes in ecophysiology, osmolytes, and secondary metabolites of the medicinal plants of *Mentha piperita* and *Catharanthus roseus* subjected to drought and heat stress. *Biomolecules*, 10(1): E43. https://doi.org/10.3390/biom10010043.

Ali, G., Srivastava, P. S., & Iqbal, M. 1999a. Some physiochemical responses of *Bacopa monniera* (L.) Wettst. cultures to copper toxicity. *Indian Journal of Experimental Biology*, 37: 1144–1147.

Ali, G., Srivastava, P. S., & Iqbal, M. 1999b. Proline accumulation, protein pattern and photosynthesis in *Bacopa monniera* regenerants grown under NaCl stress. *Biologia Plantarum*, 42(1): 89–95.

Ali, G., Srivastava, P. S., & Iqbal, M. 2000. Influence of cadmium and zinc on growth and photosynthesis of *Bacopa monniera* cultivated in vitro. *Biologia Plantarum*, 43(4): 599–601.

Ali, S. T., Mahmooduzzafar, Abdin, M. Z., & Iqbal, M. 2008. Ontogenetic changes in foliar features and psoralen content of *Psoralea corylifolia* Linn. exposed to SO_2 stress. *Journal of Environmental Biology*, 29(5): 661–668.

Alqahtani, A., Tongkao-on, W., Li, K. M., Razmovski-Naumovski, V., Chan, K., & Li, G. Q. 2015. Seasonal variation of triterpenes and phenolic compounds in Australian *Centella asiatica* (L.) Urb. *Phytochemical Analysis*, 26(6): 436–443.

Andersson, N. 2011. The influence of water stress and air velocity on growth of *Impatiens walleriana* and Petunia×hybrid. *Scientia Horticulturae*, 128(2): 146–151.

Anis, M., Sharma, M. P., & Iqbal, M. 2000. Herbal ethnomedicine of the Gwalior-forest division in Madhya Pradesh, India. *Pharmaceutical Biology*, 38(4): 241–253.

Anjum, N. A., Umar, S., Iqbal, M., & Khan, N. A. 2008. Growth characteristics and antioxidant metabolism of mungbean genotypes differing in photosynthetic capacity subjected to water deficit stress. *Journal of Plant Interactions*, 3(2): 127–136.

Ansari, M. K. A., Ahmad, A., Umar, S., & Iqbal, M. 2009. Mercury-induced changes in growth variables and antioxidative enzymes activities in Indian mustard (*Brassica juncea* L.). *Journal of Plant Interactions*, 4(2): 131–136.

Arakawa, O., Hori, Y., & Ogata, R. 1985. Relative effectiveness and interaction of ultraviolet-B, red and blue light in anthocyanin synthesis of apple fruit. *Physiologia Plantarum*, 64(3): 323–327.

Aref, M. I., Ahmed, A. I., Khan, P. R., El-Atta, H., & Iqbal, M. 2013a. Drought-induced adaptive changes in the seedling anatomy of *Acacia ehrenbergiana* and *Acacia tortilis* subsp. *raddiana*. *Trees – Structure and Function*, 27(4): 959–971.

Aref, M. I., El-Atta, H., El-Obeid, M., Ahmed, A., Khan, P. R., & Iqbal, M. 2013b. Effect of water stress on relative water and chlorophyll contents of *Juniperus procera* Hochst. ex Endlicher in Saudi Arabia. *Life Science Journal*, 10(4): 681–685.

Aref, I. M., Khan, P. R., Al-Sahli, A. A., Husen, A., Ansari, M. K. A., Mahmooduzzafar, & Iqbal, M. 2016a. Response of *Datura innoxia* Linn. to gamma rays and its impact on plant growth and productivity. *Proceedings of the National Academy of Sciences, India (Biological Sciences)*, 86(3): 623–629.

Aref, I. M., Khan, P. R., Khan, S., El-Atta, H., Ahmed, A. I., & Iqbal, M. 2016b. Modulation of antioxidant enzymes in *Juniperus procera* needles in relation to habitat environment and dieback incidence. *Trees - Structure and Function*, 30(5): 1669–1681.

Arora, N. K., Fatima, T., Mishra, I., Verma, M., Mishra, J., & Mishra, V. 2018. Environmental sustainability: Challenges and viable solutions. *Environmental Sustainability*, 1(4): 309–340.

Arshi, A., Abdin, M. Z., & Iqbal, M. 2002. Growth and metabolism of Senna as affected by salt stress. *Biologia Plantarum*, 45(1): 295–298.

Arshi, A., Abdin, M. Z., & Iqbal, M. 2004. Changes in biochemical status and growth performance of Senna (*Cassia angustifolia* Vahl.) grown under salt stress. *Phytomorphology*, 54: 109–124.

Arshi, A., Abdin, M. Z., & Iqbal, M. 2006. Sennoside content and yield attributes of *Cassia angustifolia* Vahl. as affected by NaCl and $CaCl_2$. *Scientia Horticulturae*, 111(1): 84–90.

Arshi, A., Ahmad, A., Aref, I. M., & Iqbal, M. 2012. Comparative studies on antioxidant enzyme action and ion accumulation in soybean cultivars under salinity stress. *Journal of Environmental Biology*, 33(1): 9–20.

Ashraf, M., & Orooj, A. 2006. Salt stress effects on growth, ion accumulation and seed oil concentration in an arid zone traditional medicinal plant ajwain (Trachyspermum ammi [L.] Sprague). *Journal of Arid Environments*, 64(2): 209–220.

Atique, A., & Iqbal, M. 1992. Traditional single drug administration for the psychosomatic dermal diseases: Psoriasis. *Hamdard Medicus*, 35: 76–81.

Atique, A., Iqbal, M., & Ghouse, A. K. M. 1985a. Use of *Annona squamosa* and *Piper nigrum* against diabetes. *Fitoterapia*, 56: 190–192.

Atique, A., Iqbal, M., & Ghouse, A. K. M. 1985b. Ethnobotanical study of cluster fig (*Ficus racemosa*). *Fitoterapia*, 56: 236–240.

Atique, A., Khair, S., & Iqbal, M. 1993. Traditional herbal contraceptives from north-west Uttar Pradesh. In: Govil, J. N., Singh, V. K., & Hashmi, S. (eds.), *Medicinal Plants: New Vistas of Research, Part I*. Today & Tomorrow's Printers & Publishers, New Delhi, pp. 41–47.

Azhar, N., Hussain, B., Ashraf, M. Y., & Abbasi, K. Y. 2011. Water stress mediated changes in growth, physiology and secondary metabolites of desi ajwain (*Trachyspermum ammi* L.). *Pakistan Journal of Botany*, 43(1): 15–19.

Bagheri, R., Bashir, H., Ahmad, J., Iqbal, M., & Qureshi, M. I. 2015. Spinach (*Spinacia oleracea* L.) modulates its proteome differentially in response to salinity, cadmium and their combination stress. *Plant Physiology and Biochemistry*, 97: 235–245.

Beigh, S. Y., Nawchoo, I. A., & Iqbal, M. 2003. Plants in traditional medicine of Kashmir Himalayas. *Journal of Economic and Taxonomic Botany*, 27(1): 99–104.

Belaqziz, R., Romane, A., & Abbad, A. 2009. Salt stress effects on germination, growth and essential oil content of an endemic thyme species in Morocco (*Thymus maroccanus* Ball.). *Journal of Applied Sciences Research*, 5(7): 858–863.

Ben-Abdallah, S., Zorrig, W., Amyot, L., Renaud, J., Hannoufa, A., Lachâal, M., & Karray-Bouraoui, N. 2019. Potential production of polyphenols, carotenoids and glycoalkaloids in *Solanum villosum* Mill. under salt stress. *Biologia*, 74(3): 309–324.

Bettaieb, I., Zakhama, N., Wannes, W. A., Kchouk, M., & Marzouk, B. 2009. Water deficit effects on *Salvia officinalis* fatty acids and essential oils composition. *Scientia Horticulturae*, 120(2): 271–275.

Binder, B. Y., Peebles, C. A., Shanks, J. V., & San, K. Y. 2009. The effects of UV-B stress on the production of terpenoid indole alkaloids in *Catharanthus roseus* hairy roots. *Biotechnology Progress*, 25(3): 861–865.

Borsani, O., Gonzalez-Neves, G., Ferrer, M., & Monza, J. 2010. Anthocyanins accumulation and genes-related expression in berries of cv. Tannat (*Vitis vinifera* L.). *Journal of Applied Horticulture*, 12(1): 3–9.

Bourgaud, F., Gravot, A., Milesi, S., & Gontier, E. 2001. Production of plant secondary metabolites: A historical perspective. *Plant Science*, 161(5): 839–851.

Bourgou, S., Pichette, A., Marzouk, B., & Legault, J. 2012. Antioxidant, anti-inflammatory, anticancer and antibacterial activities of extracts from *Nigella sativa* (black cumin) plant parts. *Journal of Food Biochemistry*, 36(5): 539–546.

Bouslimi, H., Ferreira, R., Dridi, N., Brito, P., Martins-Dias, S., Caçador, I., & Sleimi, N. 2021. Effects of barium stress in *Brassica juncea* and *Cakile maritima*: The indicator role of some antioxidant enzymes and secondary metabolites. *Phyton*, 90(1): 145.

Casale, J., & Lydon, J. 2007. Apparent effects of glyphosate on alkaloid production in coca plants grown in Colombia. *Journal of Forensic Sciences*, 52(3): 573–578.

Cetin, E. S. 2014. Induction of secondary metabolite production by UV-C radiation in *Vitis vinifera* L. Öküzgözü callus cultures. *Biological Research*, 47(1): 37. https://doi.org/10.1186/0717-6287-47-37.

Chang, X., Alderson, P. G., & Wright, C. J. 2008. Solar irradiance level alters the growth of basil (*Ocimum basilicum* L.) and its content of volatile oils. *Environmental and Experimental Botany*, 63(1–3): 216–223.

Chauser-Volfson, E., & Gutterman, Y. 1998. Content and distribution of anthrone C-glycosides in the South African arid plant species *Aloe mutabilis* growing in direct sunlight and in shade in the Negev Desert of Israel. *Journal of Arid Environments*, 40(4): 441–451.

Chen, S., Wang, Q., Lu, H., Li, J., Yang, D., Liu, J., & Yan, C. 2019. Phenolic metabolism and related heavy metal tolerance mechanism in *Kandelia obovata* under Cd and Zn stress. *Ecotoxicology and Environmental Safety*, 169: 134–143.

Chen, Y., Guo, Q., Liu, L., Liao, L., & Zhu, Z. 2011. Influence of fertilization and drought stress on the growth and production of secondary metabolites in *Prunella vulgaris* L. *Journal of Medicinal Plants Research*, 5(9): 1749–1755.

Chen, Y., Zhang, X., Guo, Q., Liu, L., Li, C., Cao, L., Qin, Q., Zhao, J., & Wang, W. 2018. Effects of UV-B radiation on the content of bioactive components and the antioxidant activity of *Prunella vulgaris* L. spica during development. *Molecules*, 23(5): 989.

Cullen, W. R., & Hettipathirana, D. I. 1993. The effect of arsenicals on alkaloid production by cell suspension cultures of *Catharanthus roseus*. *Applied Organometallic Chemistry*, 7(7): 477–486.

Davis, R., Earl, H., & Timper, P. 2014. Effect of simultaneous water deficit stress and *Meloidogyne incognita* infection on cotton yield and fiber quality. *Journal of Nematology*, 46(2): 108.

de Abreu, I. N., & Mazzafera, P. 2005. Effect of water and temperature stress on the content of active constituents of *Hypericum brasiliense* Choisy. *Plant Physiology and Biochemistry*, 43(3): 241–248.

De Abreu, I. N., Sawaya, A. C. H., Eberlin, M. N., & Mazzafera, P. 2005. Production of pilocarpine in callus of jaborandi (*Pilocarpus microphyllus* Stapf). *In Vitro Cellular and Developmental Biology – Plant*, 41(6): 806–811.

de Castro, E. M., Pinto, J., Bertolucci, S. K., Malta, M. R., Cardoso, M. D. G., & de Silva, F. A. M. 2007. Coumarin contents in young *Mikania glomerata* plants (Guaco) under different radiation levels and photoperiod. *Acta Farmaceutica Bonaerense*, 25(3): 387.

Deng, F. 2005. Effects of glyphosate, chlorsulfuron, and methyl jasmonate on growth and alkaloid biosynthesis of jimsonweed (*Datura stramonium* L.). *Pesticide Biochemistry and Physiology*, 82(1): 16–26.

Díaz, J., Bernal, A., Pomar, F., & Merino, F. 2001. Induction of shikimate dehydrogenase and peroxidase in pepper (*Capsicum annuum* L.) seedlings in response to copper stress and its relation to lignification. *Plant Science*, 161(1): 179–188.

Diwan, H., Khan, I., Ahmad, A., & Iqbal, M. 2010. Induction of phytochelatins and antioxidant defence system in *Brassica juncea* and *Vigna radiata* in response to chromium treatments. *Plant Growth Regulation*, 61: 97–107.

Djoukeng, J. D., Arbona, V., Argamasilla, R., & Gomez-Cadenas, A. 2008. Flavonoid profiling in leaves of citrus genotypes under different environmental situations. *Journal of Agricultural and Food Chemistry*, 56(23): 11087–11097.

Dobrikova, A. G., Apostolova, E. L., Hanć, A., Yotsova, E., Borisova, P., Sperdouli, I., Adamakis, I. S., & Moustakas, M. 2021. Cadmium toxicity in *Salvia sclarea* L.: An integrative response of element uptake, oxidative stress markers, leaf structure and photosynthesis. *Ecotoxicology and Environmental Safety*, 209: 111851.

dos Santos Nascimento, L. B., Leal-Costa, M. V., Menezes, E. A., Lopes, V. R., Muzitano, M. F., Costa, S. S., & Tavares, E. S. 2015. Ultraviolet-B radiation effects on phenolic profile and flavonoid content of *Kalanchoe pinnata*. *Journal of Photochemistry and Photobiology, Part B: Biology*, 148: 73–81.

Dou, H., Niu, G., & Gu, M. 2019. Pre-harvest UV-B radiation and photosynthetic photon flux density interactively affect plant photosynthesis, growth, and secondary metabolites accumulation in basil (*Ocimum basilicum*) plants. *Agronomy*, 9(8): 434. https://doi.org/10.3390/agronomy9080434.

Drzewiecka, K., Mleczek, M., Gąsecka, M., Magdziak, Z., Budka, A., Chadzinikolau, T., Kaczmarek, Z., & Goliński, P. 2017. Copper and nickel co-treatment alters metal uptake and stress parameters of *Salix purpurea* × *viminalis*. *Journal of Plant Physiology*, 216: 125–134.

Dubey, S., Gupta, A., Khare, A., Jain, G., Bose, S., & Rani, V. 2018. Long-and short-term protective responses of rice seedling to combat Cr (VI) toxicity. *Environmental Science and Pollution Research International*, 25(36): 36163–36172.

Eichholz, I., Rohn, S., Gamm, A., Beesk, N., Herppich, W. B., Kroh, L. W., Ulrichs, C., & Huyskens-Keil, S. 2012. UV-B-mediated flavonoid synthesis in white asparagus (*Asparagus officinalis* L.). *Food Research International*, 48(1): 196–201.

Ellenberger, J., Siefen, N., Krefting, P., Schulze Lutum, J.-B., Pfarr, D., Remmel, M., Schröder, L., & Röhlen-Schmittgen, S. 2020. Effect of UV radiation and salt stress on the accumulation of economically relevant secondary metabolites in bell pepper plants. *Agronomy*, 10(1): 142. https://doi.org/10.3390/agronomy10010142.

Fàbregas, N., & Fernie, A. R. 2019. The metabolic response to drought. *Journal of Experimental Botany*, 70(4): 1077–1085.

Fazal, H., Abbasi, B. H., Ahmad, N., Ali, S. S., Akbar, F., & Kanwal, F. 2016. Correlation of different spectral lights with biomass accumulation and production of antioxidant secondary metabolites in callus cultures of medicinally important *Prunella vulgaris* L. *Journal of Photochemistry and Photobiology, Part B: Biology*, 159: 1–7.

Fischbach, R., Kossmann, B., Panten, H., Steinbrecher, R., Heller, W., Seidlitz, H., Sandermann, H., Hertkorn, N., & Schnitzler, J. P. 1999. Seasonal accumulation of ultraviolet-B screening pigments in needles of Norway spruce (*Picea abies* (L.) Karst.). *Plant, Cell and Environment*, 22(1): 27–37.

Gao, C., Yang, B., Zhang, D., Chen, M., & Tian, J. 2016. Enhanced metabolic process to indole alkaloids in *Clematis terniflora* DC. after exposure to high level of UV-B irradiation followed by the dark. *BMC Plant Biology*, 16(1): 1–15.

García-Calderón, M., Pons-Ferrer, T., Mrazova, A., Pal'ove-Balang, P., Vilkova, M., Pérez-Delgado, C. M., Vega, J. M., Eliášová, A., Repčák, M., Márquez, A. J., & Betti, M. 2015. Modulation of phenolic metabolism under stress conditions in a *Lotus japonicus* mutant lacking plastidic glutamine synthetase. *Frontiers in Plant Science*, 6: 760.

Garg, N., & Singla, P. 2016. Stimulation of nitrogen fixation and trehalose biosynthesis by naringenin (Nar) and arbuscular mycorrhiza (AM) in chickpea under salinity stress. *Plant Growth Regulation*, 80(1): 5–22.

Gengmao, Z., Yu, H., Xing, S., Shihui, L., Quanmei, S., & Changhai, W. 2015. Salinity stress increases secondary metabolites and enzyme activity in safflower. *Industrial Crops and Products*, 64: 175–181.

Gharibi, S., Tabatabaei, B. E. S., Saeidi, G., Talebi, M., & Matkowski, A. 2019. The effect of drought stress on polyphenolic compounds and expression of flavonoid biosynthesis related genes in *Achillea pachycephala* Rech. f. *Phytochemistry*, 162: 90–98.

Gholamreza, A., Shokrpour, M., Karami, L., & Salami, S. A. 2019. Prolonged water deficit stress and methyl jasmonate-mediated changes in metabolite profile, flavonoid concentrations and antioxidant activity in peppermint (*Mentha × piperita* L.). *Notulae Botanicae Horti Agrobotanici Cluj-Napoca*, 47(1): 70–80.

Guo, R., Qiao, H., Zhao, J., Wang, X., Tu, M., Guo, C., Wan, R., Li, Z., & Wang, X. 2018. The grape VIWRKY3 gene promotes abiotic and biotic stress tolerance in transgenic *Arabidopsisthaliana*. *Frontiers in Plant Science*, 9: 545. https://doi.org/10.3389/fpls.2018.00545.

Guo, X.-R., Yang, L., Yu, J.-H., Tang, Z.-H., & Zu, Y.-G. 2007. Alkaloid variations in Catharanthus roseus seedlings treated by different temperatures in short term and long term. *Journal of Forestry Research*, 18(4): 313–315.

Gupta, B., & Huang, B. 2014. Mechanism of salinity tolerance in plants: Physiological, biochemical, and molecular characterization. *International Journal of Genomics*, 2014: Art. 701596. https://doi.org/10.1155/2014/701596.

Haghighi, Z., Modarresi, M., & Mollayi, S. 2012. Enhancement of compatible solute and secondary metabolites production in *Plantago ovata* Forsk. by salinity stress. *Journal of Medicinal Plants Research*, 6(18): 3495–3500.

Hakeem, K. R., Khan, F., Chandna, R., Siddiqui, T. O., & Iqbal, M. 2012. Genotypic variability among soybean genotypes under NaCl stress and proteome analysis of salt-tolerant genotype. *Applied Biochemistry and Biotechnology*, 168(8): 2309–2329.

Hanson, D., & Sharkey, T. D. 2001. Effect of growth conditions on isoprene emission and other thermotolerance-enhancing compounds. *Plant, Cell and Environment*, 24(9): 929–936.

Haslam, E. 2014. *The Shikimate Pathway: Biosynthesis of Natural Products Series*. Elsevier Science, Amsterdam.

Hassani, A., & Omidbaigi, R. 2006. Morphological and biochemical characteristics of purple basil (*Ocimum basilicum*). *Journal of Biological Sciences*, 6(4): 763–767.

Havkin-Frenkel, D., Podstolski, A., & Knorr, D. 1996. Effect of light on vanillin precursors formation by in vitro cultures of *Vanilla planifolia*. *Plant Cell, Tissue and Organ Culture*, 45(2): 133–136.

Hernández, I., Alegre, L., & Munné-Bosch, S. 2006. Enhanced oxidation of flavan-3-ols and proanthocyanidin accumulation in water-stressed tea plants. *Phytochemistry*, 67(11): 1120–1126.

Hildebrandt, T. M. 2018. Synthesis versus degradation: Directions of amino acid metabolism during Arabidopsis abiotic stress response. *Plant Molecular Biology*, 98(1): 121–135.

Hosseini, M. S., Samsampour, D., Ebrahimi, M., Abadía, J., & Khanahmadi, M. 2018. Effect of drought stress on growth parameters, osmolyte contents, antioxidant enzymes and glycyrrhizin synthesis in licorice (*Glycyrrhiza glabra* L.) grown in the field. *Phytochemistry*, 156: 124–134.

Hou, M., Wang, R., Zhao, S., & Wang, Z. 2021. Ginsenosides in *Panax* genus and their biosynthesis. *Acta Pharmaceutica Sinica B*, 11(7): 1813–1834.

Husen, A., & Iqbal, M. 2019. Nanomaterials and plant potential: An overview. In: Husen, A., & Iqbal, M. (eds.), *Nanomaterials and Plant Potential*. Springer Nature, Cham, Switzerland, pp. 3–29.

Husen, A., Iqbal, M., & Aref, I. M. 2014. Growth, water status and leaf characteristics of *Brassica carinata* under drought and rehydration conditions. *Brazilian Journal of Botany*, 37(3): 217–227.

Husen, A., Iqbal, M., & Aref, I. M. 2016. IAA-induced alteration in growth and photosynthesis of pea (*Pisum sativum* L.) plants grown under salt stress. *Journal of Environmental Biology*, 37(3): 421–429.

Husen, A., Iqbal, M., & Aref, I. M. 2017. Plant growth and foliar characteristics of faba bean (*Vicia faba* L.) as affected by indole-acetic acid under water-sufficient and water-deficient conditions. *Journal of Environmental Biology*, 38(2): 179–186.

Hussain, M., Park, H.-W., Farooq, M., Jabran, K., & Lee, D.-J. 2013. Morphological and physiological basis of salt resistance in different rice genotypes. *International Journal of Agriculture and Biology*, 15(1): 113–118.

Hussein, M., Embiale, A., Husen, A., Aref, I. M., & Iqbal, M. 2017. Salinity-induced modulation of plant growth and photosynthetic parameters in faba bean (*Vicia faba*) cultivars. *Pakistan Journal of Botany*, 49(3): 867–877.

Hussien, H.-A., Salem, H., & Mekki, B. E.-D. 2015. Ascorbate-glutathione-α-tocopherol triad enhances antioxidant systems in cotton plants grown under drought stress. *International Journal of ChemTech Research*, 8: 1463–1472.

Ibrahim, M. A., Mäenpää, M., Hassinen, V., Kontunen-Soppela, S., Malec, L., Rousi, M., Pietikäinen, L., Tervahauta, A., Kärenlampi, S., Holopainen, J. K., & Oksanen, E. J. 2010. Elevation of night-time temperature increases terpenoid emissions from *Betula pendula* and *Populus tremula*. *Journal of Experimental Botany*, 61(6): 1583–1595.

Iqbal, M., Abdin, M. Z., Mahmooduzzafar, Yunus, M., & Agrawal, M. 1996. Resistance mechanism in plants against air pollution. In: Yunus, M., & Iqbal, M. (eds.), *Plant Response to Air Pollution*. John Willey & Sons, Chichester, pp. 195–240.

Iqbal, M., Bano, R., & Wali, B. 2005. Plant growth responses to air pollution. In: Chaturvedi, S. N., & Singh, K. P. (eds.), *Plant Biodiversity, Microbial Interaction and Environmental Biology*. Avishkar Publishers, Jaipur, pp. 166–188.

Iqbal, M., Mahmooduzzafar, & Abdin, M. Z. 2000. *Studies on Anatomical, Physiological and Biochemical Response of Trees to Coal-Smoke Pollution around a Thermal Power Plant*. Research Project 14/62/89-MAB/Re, Ministry of Environment & Forests (Govt. of India), New Delhi, p. 335.

Iqbal, M., Mahmooduzzafar, Nighat, F., & Aref, I. M. 2011. Composition of seed oils of *Peristrophe bicalyculata* and *Ruellia tuberosa* as affected by coal-smoke stress. *Journal of Food, Agriculture and Environment*, 9(3–4): 1101–1104.

Iqbal, M., Parveen, R., Parveen, A., Parveen, B., & Aref, I. M. 2018. Establishing the botanical identity of plant drugs based on their active ingredients under diverse growth conditions. *Journal of Environmental Biology*, 39(1): 123–136.

Iqbal, S., Hussain, S., Qayyaum, M. A., & Ashraf, M. 2020. The response of maize physiology under salinity stress and its coping strategies. In: Hossain, A. (ed.), *Plant Stress Physiology*. IntechOpen, London, pp. 1–25.

Isah, T. 2019. Stress and defense responses in plant secondary metabolites production. *Biological Research*, 52(1): 39. https://doi.org/10.1186/s40659-019-0246-3.

Jaafar, H. Z., Ibrahim, M. H., & Mohamad Fakri, N. F. 2012. Impact of soil field water capacity on secondary metabolites, phenylalanine ammonia-lyase (PAL), maliondialdehyde (MDA) and photosynthetic responses of Malaysian Kacip Fatimah (*Labisia pumila* Benth). *Molecules*, 17(6): 7305–7322.

Jahan, M. S., Guo, S., Baloch, A. R., Sun, J., Shu, S., Wang, Y., Ahammed, G. J., Kabir, K., & Roy, R. 2020. Melatonin alleviates nickel phytotoxicity by improving photosynthesis, secondary metabolism and oxidative stress tolerance in tomato seedlings. *Ecotoxicology and Environmental Safety*, 197: 110593. https://doi.org/10.1016/j.ecoenv.2020.110593.

Jaini, R., Wang, P., Dudareva, N., Chapple, C., & Morgan, J. A. 2017. Targeted metabolomics of the phenylpropanoid pathway in *Arabidopsis thaliana* using reversed phase liquid chromatography coupled with tandem mass spectrometry. *Phytochemical Analysis : PCA*, 28(4): 267–276.

Janas, K. M., Cvikrová, M., Pałagiewicz, A., Szafranska, K., & Posmyk, M. M. 2002. Constitutive elevated accumulation of phenylpropanoids in soybean roots at low temperature. *Plant Science*, 163(2): 369–373.

Karimi, A., & Meiners, T. 2021. Antifungal activity of *Zataria multiflora* Boiss. essential oils and changes in volatile compound composition under abiotic stress conditions. *Industrial Crops and Products*, 171: 113888. https://doi.org/10.1016/j.indcrop.2021.113888.

Karuppusamy, S. 2009. A review on trends in production of secondary metabolites from higher plants by in vitro tissue, organ and cell cultures. *Journal of Medicinal Plants Research*, 3(13): 1222–1239.

Kermani, S. G., Saeidi, G., Sabzalian, M. R., & Gianinetti, A. 2019. Drought stress influenced sesamin and sesamolin content and polyphenolic components in sesame (*Sesamum indicum* L.) populations with contrasting seed coat colors. *Food Chemistry*, 289: 360–368.

Keskin, N., & Kunter, B. 2010. Production of trans-resveratrol in callus tissue of Öküzgözü (*Vitis vinifera* L.) in response to ultraviolet-C irradiation. *Journal of Animal and Plant Sciences*, 20(3): 197–200.

Khalid, K. A., & Cai, W. 2011. The effects of mannitol and salinity stresses on growth and biochemical accumulations in lemon balm. *Acta Ecologica Sinica*, 31(2): 112–120.

Khan, M. I. R., Asgher, M., Fatma, M., Per, T. S., & Khan, N. A. 2015. Drought stress vis a vis plant functions in the era of climate change. *Climate Change and Environmental Sustainability*, 3(1): 13–25.

Khan, N. A., Khan, M. I. R., Asgher, M., Fatma, M., Masood, A., & Syeed, S. 2014. Salinity tolerance in plants: Revisiting the role of sulfur metabolites. *Journal of Plant Biochemistry and Physiology*, 2(1): 1000120. https://doi.org/10.4172/2329-9029.1000120.

Kidd, P., Llugany, M., Poschenrieder, C., Gunse, B., & Barcelo, J. 2001. The role of root exudates in aluminium resistance and silicon-induced amelioration of aluminium toxicity in three varieties of maize (*Zea mays* L.). *Journal of Experimental Botany*, 52(359): 1339–1352.

Kim, E.-Y., Park, S., Park, B.-J., Lee, Y., & Oh, M.-M. 2014. Growth and antioxidant phenolic compounds in cherry tomato seedlings grown under monochromatic light-emitting diodes. *Horticulture, Environment, and Biotechnology*, 55(6): 506–513.

Kim, Y.-S., Choi, Y.-E., & Sano, H. 2010. Plant vaccination: Stimulation of defense system by caffeine production in planta. *Plant Signaling and Behavior*, 5(5): 489–493.

Kirakosyan, A., Seymour, E., Kaufman, P. B., Warber, S., Bolling, S., & Chang, S. C. 2003. Antioxidant capacity of polyphenolic extracts from leaves of *Crataegus laevigata* and *Crataegus monogyna* (Hawthorn) subjected to drought and cold stress. *Journal of Agricultural and Food Chemistry*, 51(14): 3973–3976.

Kleinwächter, M., & Selmar, D. 2015. New insights explain that drought stress enhances the quality of spice and medicinal plants: Potential applications. *Agronomy for Sustainable Development*, 35(1): 121–131.

Kolehmainen, J., Julkunen-Tiitto, R., Roininen, H., & Tahvanainen, J. 1995. Phenolic glucosides as feeding cues for willow-feeding leaf beetles. *Entomologia Experimentalis et Applicata*, 74(3): 235–243.

Kong, D.-X., Li, Y.-Q., Wang, M.-L., Bai, M., Zou, R., Tang, H., & Wu, H. 2016. Effects of light intensity on leaf photosynthetic characteristics, chloroplast structure, and alkaloid content of *Mahonia bodinieri* (Gagnep.) Laferr. *Acta Physiologiae Plantarum*, 38(5): 1–15.

Kováčik, J., & Klejdus, B. 2008. Dynamics of phenolic acids and lignin accumulation in metal-treated *Matricaria chamomilla* roots. *Plant Cell Reports*, 27(3): 605–615.

Kramer, G. F., Norman, H. A., Krizek, D. T., & Mirecki, R. M. 1991. Influence of UV-B radiation on polyamines, lipid peroxidation and membrane lipids in cucumber. *Phytochemistry*, 30(7): 2101–2108.

Kumar, I., & Sharma, R. K. 2018. Production of secondary metabolites in plants under abiotic stress: An overview. *Significances of Bioengineering & Biosciences*, 2(4): 196–200.

Kuo, C. I., Chao, C. H., & Lu, M. K. 2012. Effects of auxins on the production of steroidal alkaloids in rapidly proliferating tissue and cell cultures of *Solanum lyratum*. *Phytochemical Analysis: PCA*, 23(4): 400–404.

Li, X.-G., Li, F.-M., Ma, Q.-F., & Cui, Z.-J. 2006. Interactions of NaCl and Na_2SO_4 on soil organic C mineralization after addition of maize straws. *Soil Biology and Biochemistry*, 38(8): 2328–2335.

Li, Y., Kong, D., Liang, H., & Wu, H. 2018. Alkaloid content and essential oil composition of *Mahonia breviracema* cultivated under different light environments. *Journal of Applied Botany and Food Quality*, 91: 171–179.

Lim, J.-S., Kim, S.-Y., & Kim, Y.-H. 2021. Differential expression of isoflavone biosynthetic genes in soybean during germination. *Korean Journal of Crop Science*, 66(4): 365–374.

Liu, L., Gitz III, D. C., & McClure, J. W. 1995. Effects of UV-B on flavonoids, ferulic acid, growth and photosynthesis in barley primary leaves. *Physiologia Plantarum*, 93(4): 725–733.

Liu, Y., Patra, B., Singh, S. K., Paul, P., Zhou, Y., Li, Y., Wang, Y., Pattanaik, S., & Yuan, L. 2021. Terpenoid indole alkaloid biosynthesis in *Catharanthus roseus*: Effects and prospects of environmental factors in metabolic engineering. *Biotechnology Letters*, 43(11): 2085–2103.

Luthra, R., Satija, G., & Roy, A. 2022. Centellosides: Pharmaceutical applications and production enhancement strategies. *Plant Cell, Tissue and Organ Culture*, 149(1–2): 25–39.

Lydon, J., & Duke, S. O. 1989. Pesticide effects on secondary metabolism of higher plants. *Pesticide Science*, 25(4): 361–373.

Ma, C. H., Chu, J. Z., Shi, X. F., Liu, C. Q., & Yao, X. Q. 2016. Effects of enhanced UV-B radiation on the nutritional and active ingredient contents during the floral development of medicinal chrysanthemum. *Journal of Photochemistry and Photobiology, Part B: Biology*, 158: 228–234.

Ma, D., Sun, D., Wang, C., Li, Y., & Guo, T. 2014. Expression of flavonoid biosynthesis genes and accumulation of flavonoid in wheat leaves in response to drought stress. *Plant Physiology and Biochemistry: PPB*, 80: 60–66.

Maeda, H., & Dudareva, N. 2012. The shikimate pathway and aromatic amino acid biosynthesis in plants. *Annual Review of Plant Biology*, 63: 73–105.

Mahajan, M., Kuiry, R., & Pal, P. K. 2020. Understanding the consequence of environmental stress for accumulation of secondary metabolites in medicinal and aromatic plants. *Journal of Applied Research on Medicinal and Aromatic Plants*, 18: 100255. https://doi.org/10.1016/j.jarmap.2020.100255.

Mahdavi, V., Farimani, M. M., Fathi, F., & Ghassempour, A. 2015. A targeted metabolomics approach toward understanding metabolic variations in rice under pesticide stress. *Analytical Biochemistry*, 478: 65–72.

Mahmood, S., Daur, I., Al-Solaimani, S. G., Ahmad, S., Madkour, M. H., Yasir, M., Hirt, H., Ali, S., & Ali, Z. 2016. Plant growth promoting rhizobacteria and silicon synergistically enhance salinity tolerance of mung bean. *Frontiers in Plant Science*, 7: 876. https://doi.org/10.3389/fpls.2016.00876.

Mahood, H. E., Alwash, B. M., & Ibrahim, K. M. 2018. Improvement of alkaloids yield using phenylalanine as a precursor supplemented to *Morina oleifera* L. callus cultures. *Biochemical and Cellular Archives*, 18(1): 913–919.

Majid, U., Mahmooduzzafar, Siddiqi, T. O., Aref, I. M., & Iqbal, M. 2013. Quantitative changes in proteins, pigments and sennosides of *Cassia angustifolia* Vahl. treated with Mancozeb. *Pakistan Journal of Botany*, 45(5): 1509–1514.

Malik, N., Perez, J., Kunta, M., & Olanya, M. 2015. Changes in polyphenol levels in Satsuma (*Citrus unshiu*) leaves in response to Asian *Citrus* psyllidinfestation and water stress. *Open Agriculture Journal*, 9(1): 1–5.

Mann, J., & Lancaster, J. 1976. Effects of herbicides and growth regulators on solasodine production by *Solanum aviculare* and *S. laciniatum* Ait. *New Zealand Journal of Experimental Agriculture*, 4(3): 347–350.

Márquez-García, B., Fernández-Recamales, M., & Córdoba, F. 2012. Effects of cadmium on phenolic composition and antioxidant activities of *Erica andevalensis*. *Journal of Botany*, 2012: 324–329.

Martinez, V., Mestre, T. C., Rubio, F., Girones-Vilaplana, A., Moreno, D. A., Mittler, R., & Rivero, R. M. 2016. Accumulation of flavonols over hydroxycinnamic acids favors oxidative damage protection under abiotic stress. *Frontiers in Plant Science*, 7: 838. https://doi.org/10.3389/fpls.2016.00838.

Matyášová, E., Novak, J., Stranska, I., Hejtmankova, A., Skalický, M., Hejtmankova, K., & Hejnak, V. 2011. Production of morphine and variability of significant characters of *Papaver somniferum* L. *Plant, Soil and Environment*, 57(9): 423–428.

Meng, C., Zhang, S., Deng, Y.-S., Wang, G.-D., & Kong, F.-Y. 2015. Overexpression of a tomato flavanone 3-hydroxylase-like protein gene improves chilling tolerance in tobacco. *Plant Physiology and Biochemistry*, 96: 388–400.

Misra, N., & Gupta, A. K. 2006. Effect of salinity and different nitrogen sources on the activity of antioxidant enzymes and indole alkaloid content in *Catharanthus roseus* seedlings. *Journal of Plant Physiology*, 163(1): 11–18.

Mosadegh, H., Trivellini, A., Ferrante, A., Lucchesini, M., Vernieri, P., & Mensuali, A. 2018. Applications of UV-B lighting to enhance phenolic accumulation of sweet basil. *Scientia Horticulturae*, 229: 107–116.

Mustafa, N. R., & Verpoorte, R. 2007. Phenolic compounds in *Catharanthus roseus*. *Phytochemistry Reviews*, 6(2): 243–258.

Naghiloo, S., Movafeghi, A., Delazar, A., Nazemiyeh, H., Asnaashari, S., & Dadpour, M. R. 2012. Ontogenetic variation of total phenolics and antioxidant activity in roots, leaves and flowers of *Astragalus compactus* Lam. (Fabaceae). *BioImpacts*, 2(2): 105–109.

Nocchi, N., Duarte, H. M., Pereira, R. C., Konno, T. U. P., & Soares, A. R. 2020. Effects of UV-B radiation on secondary metabolite production, antioxidant activity, photosynthesis and herbivory interactions in *Nymphoides humboldtiana* (Menyanthaceae). *Journal of Photochemistry and Photobiology, Part B: Biology*, 212: 112021. https://doi.org/10.1016/j.jphotobiol.2020.112021.

Noor, J. J., Vinayan, M. T., Umar, S., Devi, P., Iqbal, M., Seetharam, K., & Zaidi, P. H. 2019. Morpho-physiological traits associated with heat stress tolerance in tropical maize (*Zea mays* L.) at the reproductive stage. *Australian Journal of Crop Science*, 13(4): 536–545.

Odjegba, V. J., & Alokolaro, A. A. 2013. Simulated drought and salinity modulates the production of phytochemicals in *Acalypha wilkesiana*. *Journal of Plant Studies*, 2(2): 105–112.

Olajire, A., & Azeez, L. 2011. The antioxidants of higher plants phytochemistry. *African Journal of Food Science and Technology*, 2(2): 22–29.

Otterbach, S. L., Yang, T., Kato, L., Janfelt, C., & Geu-Flores, F. 2019. Quinolizidine alkaloids are transported to seeds of bitter narrow-leafed lupin. *Journal of Experimental Botany*, 70(20): 5799–5808.

Pant, P., Pandey, S., & Dall'Acqua, S. 2021. The influence of environmental conditions on secondary metabolites in medicinal plants: A literature review. *Chemistry and Biodiversity*, 18(11): e2100345. https://doi.org/10.1002/cbdv.202100345.

Parveen, A., Parveen, R., Akhatar, A., Parveen, B., Siddiqui, K. M., & Iqbal, M. 2020a. Concepts and quality considerations in Unani system of medicine. *Journal of AOAC International*, 103(3): 609–633.

Parveen, B., Parveen, A., Parveen, R., Ahmad, S., Ahmad, M., & Iqbal, M. 2020b. Challenges and opportunities for traditional herbal medicine today, with special reference to its status in India. *Annals of Phytomedicine*, 9(2): 97–112.

Parveen, A., Ahmad, M., Parveen, B., Parveen, R., & Iqbal, M. 2022.The traditional system of Unani medicine, its origin, evolution and Indianization: A critical appraisal. *Indian Journal of Traditional Knowledge*, 21(3): 511–521.

Pedranzani, H., Sierra-de-Grado, R., Vigliocco, A., Miersch, O., & Abdala, G. 2007. Cold and water stresses produce changes in endogenous jasmonates in two populations of *Pinus pinaster* Ait. *Plant Growth Regulation*, 52(2): 111–116.

Perez de Souza, L., Garbowicz, K., Brotman, Y., Tohge, T., & Fernie, A. R. 2020. The acetate pathway supports flavonoid and lipid biosynthesis in Arabidopsis. *Plant Physiology*, 182(2): 857–869.

Petridis, A., Therios, I., Samouris, G., & Tananaki, C. 2012. Salinity-induced changes in phenolic compounds in leaves and roots of four olive cultivars (*Olea europaea* L.) and their relationship to antioxidant activity. *Environmental and Experimental Botany*, 79: 37–43.

Posmyk, M. M., Bailly, C., Szafrańska, K., Janas, K. M., & Corbineau, F. 2005. Antioxidant enzymes and isoflavonoids in chilled soybean (*Glycine max* (L.) Merr.) seedlings. *Journal of Plant Physiology*, 162(4): 403–412.

Qureshi, M. I., Abdin, M. Z., Ahmad, J., & Iqbal, M. 2013. Effect of long-term salinity on cellular antioxidants, compatible solute and fatty acid profile of Sweet annie (*Artemisia annua* L.). *Phytochemistry*, 95: 215–223.

Qureshi, M. I., Abdin, M. Z., Qadir, S., Kamaluddin, M., & Iqbal, M. 2006. Responses of some medicinal plants to heavy metal, salinity and oxidative stress. In: Abdin, M. Z., & Abrol, Y. P. (eds.), *Traditional Systems of Medicine*. Narosa Publishing House, New Delhi, pp. 522–533.

Qureshi, M. I., Abdin, M. Z., Qadir, S., & Iqbal, M. 2007. Lead-induced oxidative stress and metabolic alterations in *Cassia angustifolia* Vahl. *Biologia Plantarum*, 51(1): 121–128.

Radušienė, J., Karpavičienė, B., & Stanius, Ž. 2012. Effect of external and internal factors on secondary metabolites accumulation in St. John's worth. *Botanica Lithuanica*, 18(2): 101–108.

Ramani, S., & Jayabaskaran, C. 2008. Enhanced catharanthine and vindoline production in suspension cultures of *Catharanthus roseus* by ultraviolet-B light. *Journal of Molecular Signaling*, 3(1): 1–6.

Rastogi, S., Shah, S., Kumar, R., Vashisth, D., Akhtar, M. Q., Kumar, A., Dwivedi, U. N., & Shasany, A. K. 2019. Ocimum metabolomics in response to abiotic stresses: Cold, flood, drought and salinity. *PLOS ONE*, 14(2): e0210903. https://doi.org/10.1371/journal.pone.0210903.

Raza, A., Razzaq, A., Mehmood, S. S., Hussain, M. A., Wei, S., He, H., Zaman, Q. U., Xuekun, Z., & Hasanuzzaman, M. 2021. Omics: The way forward to enhance abiotic stress tolerance in *Brassica napus* L. *GM Crops and Food*, 12(1): 251–281.

Rejeb, I. B., Pastor, V., & Mauch-Mani, B. 2014. Plant responses to simultaneous biotic and abiotic stress: Molecular mechanisms. *Plants*, 3(4): 458–475.

Saba, Pande, D., Iqbal, M., & Srivastava, P. S. 2000. Effect of $ZnSO_4$ and $CuSO_4$ on regeneration and lepidine content in *Lepidium sativum* L. *Biologia Plantarum*, 43(2): 253–256.

Sarker, U., & Oba, S. 2018. Augmentation of leaf color parameters, pigments, vitamins, phenolic acids, flavonoids and antioxidant activity in selected *Amaranthus tricolor* under salinity stress. *Scientific Reports*, 8(1): 1–9.

Sarri, E., Termentzi, A., Abraham, E. M., Papadopoulos, G. K., Baira, E., Machera, K., Loukas, V., Komaitis, F., & Tani, E. 2021. Salinity stress alters the secondary metabolic profile of *M. sativa*, *M. arborea* and their hybrid (Alborea). *International Journal of Molecular Sciences*, 22(9): 4882. https://doi.org/10.3390/ijms22094882.

Selmar, D. 2008. Potential of salt and drought stress to increase pharmaceutical significant secondary compounds in plants. *Landbauforschung Volkenrode*, 58(1): 139–144.

Shahverdi, M. A., Omidi, H., & Tabatabaei, S. J. 2019. Stevia (Stevia rebaudiana Bertoni) responses to NaCl stress: Growth, photosynthetic pigments, diterpene glycosides and ion content in root and shoot. *Journal of the Saudi Society of Agricultural Sciences*, 18(4): 355–360.

Shahzad, B., Tanveer, M., Che, Z., Rehman, A., Cheema, S. A., Sharma, A., Song, H., Rahman, S., & Zhaorong, D. 2018. Role of 24-epibrassinolide (EBL) in mediating heavy metal and pesticide induced oxidative stress in plants: A review. *Ecotoxicology and Environmental Safety*, 147: 935–944.

Sharma, A., Kumar, V., Thukral, A. K., & Bhardwaj, R. 2016a. Epibrassinolide-imidacloprid interaction enhances non-enzymatic antioxidants in *Brassica juncea* L. *Indian Journal of Plant Physiology*, 21(1): 70–75.

Sharma, A., Kumar, V., Yuan, H., Kanwar, M. K., Bhardwaj, R., Thukral, A. K., & Zheng, B. 2018. Jasmonic acid seed treatment stimulates insecticide detoxification in *Brassica juncea* L. *Frontiers in Plant Science*, 9: 1609. https://doi.org/10.3389/fpls.2018.01609.

Sharma, A., Thakur, S., Kumar, V., Kanwar, M. K., Kesavan, A. K., Thukral, A. K., Bhardwaj, R., Alam, P., & Ahmad, P. 2016b. Pre-sowing seed treatment with 24-epibrassinolide ameliorates pesticide stress in *Brassica juncea* L. through the modulation of stress markers. *Frontiers in Plant Science*, 7: 1569. https://doi.org/10.3389/fpls.2016.01569.

Sharma, A., Yuan, H., Kumar, V., Ramakrishnan, M., Kohli, S. K., Kaur, R., Thukral, A. K., Bhardwaj, R., & Zheng, B. 2019. Castasterone attenuates insecticide-induced phytotoxicity in mustard. *Ecotoxicology and Environmental Safety*, 179: 50–61.

Shohael, A. M., Ali, M. B., Yu, K.-W., Hahn, E.-J., Islam, R., & Paek, K. Y. 2006a. Effect of light on oxidative stress, secondary metabolites and induction of antioxidant enzymes in Eleutherococcus senticosus somatic embryos in bioreactor. *Process Biochemistry*, 41(5): 1179–1185.

Shohael, A. M., Ali, M. B., Yu, K.-W., Hahn, E.-J., & Paek, K.-Y. 2006b. Effect of temperature on secondary metabolites production and antioxidant enzyme activities in Eleutherococcus senticosus somatic embryos. *Plant Cell, Tissue and Organ Culture*, 85(2): 219–228.

Simon, P. W., Rolling, W. R., Senalik, D., Bolton, A. L., Rahim, M. A., Mannan, A. T. M., Islam, F., Ali, A., Nijabat, A., Naveed, N. H., Hussain, R., & Shah, A. I. 2021. Wild carrot diversity for new sources of abiotic stress tolerance to strengthen vegetable breeding in Bangladesh and Pakistan. *Crop Science*, 61(1): 163–176.

Singh, N., Ali, G., Soh, W. Y., & Iqbal, M. 2000. Growth responses and hyoscyamine content of *Datura innoxia* under the influence of coal-smoke pollution. *Journal of Plant Biology*, 43(2): 69–75.

Singh, S., Kumar, V., Gill, J. P. K., Datta, S., Singh, S., Dhaka, V., Kapoor, D., Wani, A. B., Dhanjal, D. S., Kumar, M., Harikumar, S. L., & Singh, J. 2020. Herbicide glyphosate: Toxicity and microbial degradation. *International Journal of Environmental Research and Public Health*, 17(20): 7519.

Slama, I., M'Rabet, R., Ksouri, R., Talbi, O., Debez, A., & Abdelly, C. 2017. Effects of salt treatment on growth, lipid membrane peroxidation, polyphenol content, and antioxidant activities in leaves of *Sesuvium portulacastrum* L. *Arid Land Research and Management*, 31(4): 404–417.

Smirnov, O., Kosyan, A., Kosyk, O., & Taran, N. Y. 2015. Response of phenolic metabolism induced by aluminium toxicity in *Fagopyrum esculentum* Moench. plants. *The Ukrainian Biochemical Journal*, 87(6): 129–135.

Solíz-Guerrero, J. B., De Rodriguez, D. J., Rodríguez-García, R., Angulo-Sánchez, J. L., & Méndez-Padilla, G. 2002. *Quinoa Saponins: Concentration and Composition Analysis*. ASHS Press, Alexandria, Egypt, pp. 110–114.

Spinelli, F., Cellini, A., Marchetti, L., Mudigere, K., & Piovene, C. 2011. Emission and function of volatile organic compounds in response to abiotic stress. In: Shanker, A. K., & Venkateswarlu, B. (eds.), *Abiotic Stress in Plants–Mechanisms and Adaptations*. InTech Open, London, pp. 367–394.

Stiles, E. A., Cech, N. B., Dee, S. M., & Lacey, E. P. 2007. Temperature-sensitive anthocyanin production in flowers of *Plantago lanceolata*. *Physiologia Plantarum*, 129(4): 756–765.

Šutković, J., Ler, D., Ragab, M., & Gawwad, A. 2011. In vitro production of solasodine alkaloid in *Solanum nigrum* under salinity stress. *Journal of Phytology*, 3(1): 43–49.

Świeca, M. 2015. Elicitation with abiotic stresses improves pro-health constituents, antioxidant potential and nutritional quality of lentil sprouts. *Saudi Journal of Biological Sciences*, 22(4): 409–416.

Sytar, O., Mbarki, S., Zivcak, M., & Brestic, M. 2018. The involvement of different secondary metabolites in salinity tolerance of crops. In: Kumar, V., Wani, S. H., Suprasanna, P., & Tran, L.-S. P. (eds.), *Salinity Responses and Tolerance in Plants, Volume 2*. Springer, Cham, Switzerland, pp. 21–48.

Tabatabaie, S. J., & Nazari, J. 2007. Influence of nutrient concentrations and NaCl salinity on the growth, photosynthesis, and essential oil content of peppermint and lemon verbena. *Turkish Journal of Agriculture and Forestry*, 31(4): 245–253.

Takshak, S., & Agrawal, S. 2015. Defence strategies adopted by the medicinal plant Coleus forskohlii against supplemental ultraviolet-B radiation: Augmentation of secondary metabolites and antioxidants. *Plant Physiology and Biochemistry : PPB*, 97: 124–138.

Takshak, S., & Agrawal, S. Á. 2014. Secondary metabolites and phenylpropanoid pathway enzymes as influenced under supplemental ultraviolet-B radiation in *Withania somnifera* Dunal, an indigenous medicinal plant. *Journal of Photochemistry and Photobiology, Part B: Biology*, 140: 332–343.

Tariq, U., Ali, M., & Abbasi, B. H. 2014. Morphogenic and biochemical variations under different spectral lights in callus cultures of *Artemisia absinthium* L. *Journal of Photochemistry and Photobiology, Part B: Biology*, 130: 264–271.

Trócsányi, E., György, Z., Inotai, K., Szabó, K., Pluhár, Z., Radácsi, P., Malekzadeh, M., & Németh-Zámboriné, E. 2015. Enhanced rosmarinic acid accumulation and rosmarinic acid synthase gene expression under drought stress in thyme (*Thymus vulgaris*). *Planta Medica*, 81(16). https://doi.org/10.1055/s-0035-1565623.

Turtola, S., Manninen, A.-M., Rikala, R., & Kainulainen, P. 2003. Drought stress alters the concentration of wood terpenoids in Scots pine and Norway spruce seedlings. *Journal of Chemical Ecology*, 29(9): 1981–1995.

Ullah, A., Manghwar, H., Shaban, M., Khan, A. H., Akbar, A., Ali, U., Ali, E., & Fahad, S. 2018. Phytohormones enhanced drought tolerance in plants: A coping strategy. *Environmental Science and Pollution Research International*, 25(33): 33103–33118.

Ullah, M. A., Tungmunnithum, D., Garros, L., Drouet, S., Hano, C., & Abbasi, B. H. 2019. Effect of ultraviolet-C radiation and melatonin stress on biosynthesis of antioxidant and antidiabetic metabolites produced in in vitro callus cultures of *Lepidium sativum* L. *International Journal of Molecular Sciences*, 20(7): 1787. https://doi.org/10.3390/ijms20071787.

Umar, S., Moinuddin, & Iqbal, M. 2005. Heavy metals: Availability, accumulation and toxicity in plants. In: Dwivedi, P., & Dwivedi, R. S. (eds.), *Physiology of Abiotic Stress in Plants*. Agrobios (India), Jodhpur, pp. 325–348.

Van Oosten, M. J., Sharkhuu, A., Batelli, G., Bressan, R. A., & Maggio, A. 2013. The *Arabidopsis thaliana* mutant air1 implicates SOS3 in the regulation of anthocyanins under salt stress. *Plant Molecular Biology*, 83(4): 405–415.

Vashisth, D., Kumar, R., Rastogi, S., Patel, V. K., Kalra, A., Gupta, M. M., Gupta, A. K., & Shasany, A. K. 2018. Transcriptome changes induced by abiotic stresses in *Artemisia annua*. *Scientific Reports*, 8(1): 1–14.

Veeraraghavan, V. P., Hussain, S., Papayya Balakrishna, J., Dhawale, L., Kullappan, M., Mallavarapu Ambrose, J., & Krishna Mohan, S. 2021. A comprehensive and critical review on ethnopharmacological importance of desert truffles: *Terfezia claveryi*, *Terfezia boudieri*, and *Tirmania nivea*. *Food Reviews International*, 37: 1–20 (DOI:10.1080/87559129.2021.1889581).

Verma, A., & Deepti, S. 2016. Abiotic stress and crop improvement: Current scenario. *Adv. Plants Agric. Res*, 4(4): 345–346.

Verma, A. K., Singh, R., & Singh, S. 2012. Improved alkaloid content in callus culture of *Catharanthus roseus*. *Botanica Serbica*, 36(2): 123–130.

Verma, N., & Shukla, S. 2015. Impact of various factors responsible for fluctuation in plant secondary metabolites. *Journal of Applied Research on Medicinal and Aromatic Plants*, 2(4): 105–113.

Verma, R. B., Mahmooduzzafar, Siddiqi, T. O., & Iqbal, M. 2006. Foliar response of *Ipomea pes-tigridis* L. to coal-smoke pollution. *Turkish Journal of Botany*, 30: 413–417.

Villiers, F., Ducruix, C., Hugouvieux, V., Jarno, N., Ezan, E., Garin, J., Junot, C., & Bourguignon, J. 2011. Investigating the plant response to cadmium exposure by proteomic and metabolomic approaches. *Proteomics*, 11(9): 1650–1663.

Vince-Prue, D., & Canham, A. 1983. Horticultural significance of photomorphogenesis. In: Shropshire Jr., W., & Mohr, H. (eds.), *Photomorphogenesis*. Springer-Verlag,Berlin, pp. 518–544.

Wang, D., Du, F., Liu, H., & Liang, Z. 2010. Drought stress increases iridoid glycosides biosynthesis in the roots of Scrophularia ningpoensis seedlings. *Journal of Medicinal Plants Research*, 4(24): 2691–2699.

Wang, L., Shan, T., Xie, B., Ling, C., Shao, S., Jin, P., & Zheng, Y. 2019. Glycine betaine reduces chilling injury in peach fruit by enhancing phenolic and sugar metabolisms. *Food Chemistry*, 272: 530–538.

Wang, W., Vinocur, B., & Altman, A. 2003. Plant responses to drought, salinity and extreme temperatures: Towards genetic engineering for stress tolerance. *Planta*, 218(1): 1–14.

Wang, Y., Liu, Y., Liu, J., Zhang, Z., & Tang, Z. 2017. Effect of light intensity on the contents of main secondary metabolites in Astragalus. *Chinese Journal of Applied and Environmental Biology*, 23(5): 928–933.

Wink, M. 1999. *Functions of Plant Secondary Metabolites and their Exploitation in Biotechnology, Vol. 3*. Sheffield Academic Press/CRC Press.

Yadav, B., Jogawat, A., Lal, S. K., Lakra, N., Mehta, S., Shabek, N., & Narayan, O. P. 2021. Plant mineral transport systems and the potential for crop improvement. *Planta*, 253(2): 1–30.

Yang, L., Wen, K.-S., Ruan, X., Zhao, Y.-X., Wei, F., & Wang, Q. 2018. Response of plant secondary metabolites to environmental factors. *Molecules*, 23(4): 762. https://doi.org/10.3390/molecules23040762.

Yao, Y., Kang, T., Jin, L., Liu, Z., Zhang, Z., Xing, H., Sun, P., & Li, M. 2019. Temperature-dependent growth and hypericin biosynthesis in *Hypericum perforatum. Plant Physiology and Biochemistry : PPB*, 139: 613–619.

Yousuf, P. Y., Ahmad, A., Ganie, A. H., Sareer, O., Krishnapriya, V., Aref, I. M., & Iqbal, M. 2017. Antioxidant response and proteomic modulations in Indian mustard grown under salt stress. *Plant Growth Regulation*, 81(1): 31–50.

Yunus, M., & Iqbal, M. 1996. Major highlights of research on plant response to air pollution. In: Yunus, M., & Iqbal, M. (eds.), *Plant Response to Air Pollution*. John Wiley & Sons, Chichester, UK, pp. 489–493.

Yunus, M., Singh, N., & Iqbal, M. 1996. Global status of air pollution: An overview. In: Yunus, M., & Iqbal, M. (eds.), *Plant Response to Air Pollution*. John Wiley & Sons, Chichester, UK, pp. 1–34.

Yüzbaşıoğlu, E., Dalyan, E., & Akpınar, I. 2017. Changes in photosynthetic pigments, anthocyanin content and antioxidant enzyme activities of maize (*Zea mays* L.) seedlings under high temperature stress conditions. *Trakya University Journal of Natural Sciences*, 18(2): 97–104.

Zaman, M. A. K., Azzeme, A. M., Ramle, I. K., Normanshah, N., Shaharuddin, N. A., Ahmad, S., & Abdullah, S. N. A. 2021. Prolonged incubation of callus on auxin herbicide 2, 4-D displayed significant effect on alkaloid production in callus of the woody medicinal plant *Polyalthia bullata. In Vitro Cellular and Developmental Biology – Plant*, 57(5): 749–759.

Zandalinas, S. I., Mittler, R., Balfagón, D., Arbona, V., & Gómez-Cadenas, A. 2018. Plant adaptations to the combination of drought and high temperatures. *Physiologia Plantarum*, 162(1): 2–12.

Zhang, L., Guo, Q., Chang, Q., Zhu, Z., Liu, L., & Chen, Y. 2015. Chloroplast ultrastructure, photosynthesis and accumulation of secondary metabolites in *Glechoma longituba* in response to irradiance. *Photosynthetica*, 53(1): 144–153.

Zhang, N., Wen, Q., Feng, H., Cao, R., Zhou, X., Tagn, J., & Wu, N. 2012. Effects of water stress and nitrogen nutrition on regulation of *Catharanthus roseus* alkaloids metabolism. *China Journal of Chinese Materia Medica*, 37(10): 1346–1352.

Zhang, Z., Zhong, M., Wang, J., Xia, D., & Bao, J. 2020. Synthesis of baicalein modified cerium oxide nanoparticles for inhibitory activation of NF-κB and mitogen-activated protein kinase signals in rotenone-induced Parkinsonian rats. *Science of Advanced Materials*, 12(1): 93–100.

Zhao, B., Wang, L., Pang, S., Jia, Z., Wang, L., Li, W., & Jin, B. 2020. UV-B promotes flavonoid synthesis in *Ginkgo biloba* leaves. *Industrial Crops and Products*, 151: 112483. https://doi.org/10.1016/j.indcrop.2020.112483.

Zhao, J.-L., Zhou, L.-G., & Wu, J.-Y. 2010. Promotion of Salvia miltiorrhiza hairy root growth and tanshinone production by polysaccharide–protein fractions of plant growth-promoting rhizobacterium *Bacillus cereus. Process Biochemistry*, 45(9): 1517–1522.

Zhao, J., Davis, L. C., & Verpoorte, R. 2005. Elicitor signal transduction leading to production of plant secondary metabolites. *Biotechnology Advances*, 23(4): 283–333.

Zheljazkov, V. D., Zhalnov, I., & Nedkov, N. K. 2006. Herbicides for weed control in blessed thistle (Silybum marianum). *Weed Technology*, 20(4): 1030–1034.

Zheng, J., Yang, B., Ruusunen, V., Laaksonen, O., Tahvonen, R., Hellsten, J., & Kallio, H. 2012. Compositional differences of phenolic compounds between black currant (*Ribes nigrum* L.) cultivars and their response to latitude and weather conditions. *Journal of Agricultural and Food Chemistry*, 60(26): 6581–6593.

Zhi-Lin, Y., Chuan-Chao, D., Lian-Qing, C., & Key, J. 2007. Regulation and accumulation of secondary metabolites in plant-fungus symbiotic system. *African Journal of Biotechnology*, 6(11): 57436. https://doi.org/10.4314/AJB.V6I11.57436.

Zhou, J., Zhang, L. P., Chang, Y., Lu, X., Zhu, Z., & Xu, G. 2012. Alteration of leaf metabolism in Bt-transgenic rice (*Oryza sativa* L.) and its wild type under insecticide stress. *Journal of Proteome Research*, 11(8): 4351–4360.

Zhu, Y., Chen, Y., Zhang, X., Xie, G., & Qin, M. 2020. Copper stress-induced changes in biomass accumulation, antioxidant activity and flavonoid contents in *Belamcanda chinensis* calli. *Plant Cell, Tissue and Organ Culture*, 142(2): 299–311.

Zoratti, L., Karppinen, K., Luengo Escobar, A., Häggman, H., & Jaakola, L. 2014. Light-controlled flavonoid biosynthesis in fruits. *Frontiers in Plant Science*, 5: 534. https://doi.org/10.3389/fpls.2014.00534.

Zu, Y.-G., Zohanghua, T., Yu, J.-H., Liu, S.-G., Wang, W., & Guo, X.-R. 2003. Different responses of camptothecin and 10-hydroxycamptothecin to heat shock in *Camptotheca acuminata* seedlings. *Acta Botanica Sinica*, 45(7): 809–814.

12 Opportunities and Challenges for Medicinal and Aromatic Plants in Japan

Maiko Inoue and Shinichiro Hayashi

CONTENTS

12.1 INTRODUCTION

In recent years, herbal medicine has been successful in attracting the attention of people all over the world and being identified as traditional, complementary, or alternative medicine. For instance, fully established traditional systems of medicine, based primarily on plants, are in practice in India, China, and Tibet (Parveen et al., 2020a, 2022). More and more therapeutic properties of plants keep coming to light through the folk claims collected from various tribal and rural people (Anis and Iqbal, 1986; Anis et al., 2000; Atique et al., 1985a, 1985b). These claims are then tested by laboratory experiments and clinical trials (Iqbal et al., 2018; Parveen et al., 2020b) and, if the claims are proved to be genuine, efforts are made to enhance the efficacy of the plants concerned by promoting the production of their active ingredients (Ahmad et al., 2005, 2006a, 2006b; Ali et al., 1997b, 2000; Saba et al., 2000a, 2000b).

Japan is a long island extending to the north and south. It includes subarctic and tropical areas and has a humid climate with four distinct seasons. Therefore, many endemic medicinal and aromatic plants (MAPs) are found there and have long been in use as medicine. They also provide basic materials for clothing, food, and housing (NMJH, 2006). Recently, many of these plant species have been faced with the danger of rapid loss and possible extinction, leading ultimately to the

DOI: 10.1201/9781003226925-12

loss of accumulated knowledge regarding their utility. However, people's interest is increasing in Kampo (traditional Japanese medicine) and, consequently, Kampo has been incorporated into integrative medicine, but simultaneously, many effective traditional folk medicines are being ignored. Therefore, many professionals as well as companies have started focusing on the functionality and characteristics of native MAPs and on the popularization of their use and product development (Nishizawa, 2020). The movement is gaining popularity among patients who are suffering from ailments that are difficult to resolve with modern drug therapy or among those who are fond of nature and safety. Recent research has focused on natural remedies, and the healing effects of tree-based essential oils and other plant-based formulations are being recorded (Ikei et al., 2015, 2016). The Japanese government is promoting over-the-counter use of preventative medicine as it is cost effective.

12.2 STATUS OF MEDICINAL AND AROMATIC PLANTS IN JAPAN

12.2.1 History

The oldest Japanese medicine, called "*waho*," was the main medicine until the introduction of traditional Chinese medicine. The oldest Japanese medical book, "*Daidouruijuho*," in which various *waho* remedies were compiled in the year 808, was written in compliance with the dying injunctions of Emperor *Kanmu*, who was apprehensive of the abolition of *waho* because of the influx of traditional Chinese medicine. Unfortunately, *waho* was unsystematic, compared with traditional Chinese medicine. Consequently, *waho* survives merely as folk medicine in the present time. The traditional Chinese medicine, introduced to Japan in the seventh century, became the primary medicine in Japan due to an overwhelming variety of crude drugs and systematized remedies. The oldest record of the use of MAPs in Japan appears in the historical book "*Nihon shoki*," written in 720 (Table 12.1). It describes an event showing that the empress *Suiko* went to *Udano* (Nara prefecture) to harvest medicinal and aromatic plants in the year 611.

In 1635, a national isolation began in Japan, and most of the crude drugs were no longer imported, except via a few ports. Therefore, traditional Chinese medicine continued to develop in Japan. Kampo (traditional Japanese medicine) was established using the native crude drugs. The first two medicinal botanical gardens, *Azabu* and *Otsuka*, were established in Edo in 1638. Natural history flourished among academicians and physicians. Later, the first Japanese herbal book, "*Yamatohonzo*," was compiled by *Ekiken Kaibara* (1630–1714). After a while, the government started facing financial difficulties and the domestic production of medicinal plants, extracts, and essential oils was affected. However, without relying on imports, full-scale distillation of essential oils was started at Dejima Island in Nagasaki Harbor in 1672 (Yoshitake, 2012).

At the same time, Shogun *Yosimune Tokugawa* (1716–1745) directed the herbalist *Masakatsu Uemura* (1695–1777) to investigate and collect medicinal plants from all over the country. In 1804, physician *Seishu Hanaoka* (1760–1835) performed the first breast cancer surgery using general anesthesia based on a Kampo formula. Thereafter, modern Western medicine was introduced into Japan. In 1874, the Japanese government decided that Japanese medicine should follow German medicine. Thus, the use of Kampo medicine gradually declined for about a hundred years. Immediately after World War II, only folk medicine based mainly on medicinal plants was available due to shortage of supplies. Afterwards, the modern Western medicine became the primary medicine once again. At present, Kampo and folk medicine are used by those exhibiting pre-symptomatic states or the side effects of modern medicine. From 1967, Kampo medicine was covered by health insurance due to the great demand. Additionally, Kampo medicine has become part of the core curriculum in the modern medical education system since 2001. In 2005, the rate of prescribing Kampo medicines by medical doctors was 78%, showing a significant increase over the 70% figure of 1999 (Fujiwara et al., 2011).

TABLE 12.1
Historical Events Related to Medicinal and Aromatic Plants (MAPs) in Japan.

Timeline	Historical Events
611	Medicinal plant picking by Empress *Suiko*.
600–894	Traditional Chinese medicine was introduced by mission to Sui and Tang.
712	The oldest history book "*Kojiki*" was published.
720	A Japanese chronicle, "*Nihon shoki*," described the first driftage of incense wood.
756	Treasures of relics, including MAPs, were dedicated to *Shosoin*.
808	"*Daidouruijuho*," which compiled various *waho* remedies, was written.
927	A list of 207 medicines was published in *Engi-shiki*.
984	*Ishinhō*, the oldest surviving Japanese medical text, was completed by *Yasuyori Tanba* (912–995).
1498	*Sanki Tashiro* (1465–1537) returned after studying in China and introduced Lee Zhu Medicine.
1556	Western medicine was introduced by *Luís de Almeida* (1525–1583).
1590	Beginning of apothecary in Edo.
1607	Chinese Compendium of Materia Medica was dedicated to Shogunate *Ieyasu Tokugawa* (1543–1616).
1635	National isolation started.
1638	*Azabu* and *Otsuka* medicinal botanical gardens were opened in Edo.
1645	Manufacturing of various folk medicine started.
1672	The first full-scale distillation of essential oil was started.
1684	*Azabu* medicinal botanical garden changed to *Koishikawa* medicinal botanical garden.
1709	First Japanese herbal book, *Yamatohonzo*, was compiled by *Ekiken Kaibara* (1630–1714).
1712	A doctor, *Ryōan Terajima* (1654–unknown), published "*Wakansansaizue*," which was an illustrated Japanese encyclopedia. *Engelbert Kaempfer* published *Amoenitatum exoticarum*, which was the first extensive description of Japanese plants (Flora Japonica).
1684–1751	Shogun *Yosimune Tokugawa* (1716–1745) directed the herbalist *Masakatsu Uemura* (1695–1777) to investigate and collect medicinal plants all over the country.
1722	Edo Shogunate established an institution for quality control of Japanese medicine.
1729	*Tosuke Morino* (1690–1767) went on the investigation and collection trip with *Masakatsu Uemura* and opened *Morino* medicinal botanical garden.
1802	*Kinzan Murai* (1733–1815) wrote a book, "*Wahoichimanhou*" which introduced folk medicine.
1804	*Seishu Hanaoka* (1760–1835) performed the first breast cancer surgery using general anesthesia.
1823	A German physician, botanist, and traveler, *Philipp Franz Balthasar von Siebold* (1796–1866), came to Japan and introduced the modern Western medicine.
1835	*Youan Udagawa* wrote "*Shokugakukeigen*" which introduced Western botany.
1854	National isolation was finished.
1861	Shogunate established a school of modern Western medicine.
1874	Japanese government decided that Japanese medicine would follow German medicine.
1885	Ephedrine was first isolated from *Ephedra* by *Nagayoshi Nagai* (1845–1929).
1886	The first edition of the Japanese Pharmacopeia was released.
1915	A Japanese medical scientist, *Yu Fujikawa*," wrote a book, "*Folk Medicine*," including 5,000 kinds of folk medicine.
1967	Kampo medicine was covered by health insurance.
2001	Kampo medicine became part of the core curriculum in the medical education system.

12.2.2 Conventional Use of Medicinal and Aromatic Plants in Japan

Folk medicine using MAPs in difficult times is almost a forgotten story in the present-day Japan because markedly effective modern medicines are now easily available everywhere. However, many MAPs that are supposed to be quite effective are listed in Japanese Pharmacopeia, and some phytochemicals derived from the MAPs are reborn as pharmaceuticals (Kotaro Pharmaceutical, 2022).

Geranium nepalense subsp. *thunbergii* — *Houttuynia cordata* — *Swertia japonica*

FIGURE 12.1 Japan's three major medicinal plants.

In particular, *Geranium nepalense* subsp. *thunbergii*, *Houttuynia cordata*, and *Swertia japonica* are said to be Japan's three major medicinal plants and have been in use continuously since ancient times (Figure 12.1). *G. nepalense thunbergii* is a wonderful medicinal plant as it is used for treating both diarrhea and constipation. The active ingredient of therapeutic value is a tannin named geraniin, which produces ellagic acid by hydrolysis and shows intestinal regulation and antiviral effects. *H. cordata* is used in Kampo medicine by the name "*jyuyaku*" to manage diuresis, bowel movement, parasitic intestinal worms, hypertension, and for topical application to suppurated areas and wounds. *S. japonica* is mainly used as a bitter stomachic. The bittering component is swertiamarin, a type of secoiridoid glycoside. Extract of *S. japonica*is has the effect of improving blood circulation in the skin and it is also used as a hair restorer (Funayama, 2015; Sashida and Kihara, 2013).

Kampo, based on traditional Chinese medicine, uniquely developed after the national isolation began in 1635. It is a typical, complementary and alternative medicine (CAM) in Japan, quite effective in treating pre-symptomatic states and chronic symptoms. Basically, Kampo formulas blend several kinds of crude drugs, some including minerals and animals. Most crude drugs come from China, but some of them, such as *Bupleurum falcatum, Coptis japonica*, and *Magnolia hypoleuca* also grow in Japan. In early days, Kampo was too expensive to use for the common masses, hence they used folk medicine quite often. *Lithospermum erythrorhizon* produces naphthoquinones, such as shikonin and acetyl shikonin. The roots were used for dyeing, called "Edo purple" which was very popular early in the eighteenth century. *L. erythrorhizon* is also used in Kampo as an anti-inflammatory and antipyretic drug, and for external application on burns, eczema, and hemorrhoids. The aroma of Yuzu (*Citrus junos*), an important condiment in Japanese cuisine, is becoming very popular all over the globe. According to Japanese custom, taking a yuzu bath on the winter solstice is said to be a fine way to avoid catching cold.

12.2.3 Current Status of Medicinal and Aromatic Plants in Japan

In 2013, the Japanese Ministry of Agriculture, Forestry and Fisheries (MAFF) promoted an alliance between the medical, welfare, food, and agricultural bodies (including MAPs). Projects have been started in universities, in the community, and in various organizations (MAFF, 2013; NU, 2014; TCDA, 2014) for branding indigenous medicinal plants and creating a campaign to increase public awareness and domestic cultivation of native aromatic plants.

A survey report by Aroma Environment Association of Japan (AEAJ) on the aromatics market in Japan in 2018 revealed that the size of the aromatics market had expanded (107%), compared with the last survey in 2015 (AEAJ, 2019). The aromatics market consists of aromatic products, containing essential oils and aromatherapy; products here refer to cosmetics, clothes detergents, fragrances, and bath additives. The aromatherapy includes products, education, and aromatherapy services. According to the survey results, insect repellent spray showed a 151% growing, and cosmetics were popular among both females and males. On the other hand, growth of the aromatherapy market was slow. Using services to give character to certain spaces using aromas grew by 135%.

In 2010, the Japanese Ministry of Health, Labour, and Welfare (MHLW) established the integrative medicine project team and generated evidence-based information about medicinal plants, aromatherapy, Ayurveda, and homeopathy, and so on (MHLW, 2014). The degree of spread of integrative medicine in Japan is gradually rising, especially combined with Kampo medicine (Fujiwara et al., 2011). The possible reasons were 1) education on Kampo is done in all medical schools, 2) Kampo medicine is covered by health insurance, and 3) the scientific evidence is gradually increasing. Almost no combination of herbs and aromatherapy as integrative medicine is seen. One of the reasons for this situation is probably that Japanese medical doctors don't really feel the need for herbal and aroma therapy because Kampo medicine is widespread (Imanishi, 2017). Although the functionality of herbs attracts worldwide attention, and knowledge from leading-edge research is accumulated, most of the educators of medicine, pharmacy and healthcare professionals in Japan haven't caught up with that information. Therefore, incorrect or outdated information has become widespread (Hayashi, 2017).

12.3 MULTIFOLD USES OF NATIVE MEDICINAL AND AROMATIC PLANTS

12.3.1 Native Medicinal Plants

Since ancient times, native medicinal plants, including condiments, crude drugs, and cosmetics, have been part of our daily life. Recent studies have revealed the significance and functionality of native medicinal plants (Table 12.2) (Cheng et al., 2020; Nakamura et al., 2017; Perera et al., 2015).

Cultivation and processing of native MAPs is a traditional handiwork in Japan. For example, *Crocus sativus* is cultivated in Oita prefecture inside buildings and *Echinacea purpurea* is cultivated in Saitama prefecture using methodology practiced for green tea production (TAY, 2022). These native medicinal plants products are not only useful for self-care, but also provide local employment. A recent issue is the end of production due to the aging of producers and a shortage of successors.

12.3.1.1 Hokkaido Siberian ginseng (*Eleutherococcus senticosus*)

Eleutherococcus senticosus, also called Siberian ginseng or *Ci wujia* in Chinese, inhabits the Russian Far East, Northern China, and parts of Hokkaido. In China, *E. senticosus* is used as an energizing and anti-aging herb, which has increased Qi for 2000 years. It is considered to be an adaptogen herb, together with Korean ginseng and American ginseng (Bleakney, 2008). Adaptogen is defined as a substance that can improve adaptability to stress, remains available for a long-time use, and is harmless. In the former Soviet Union, *E. senticosus* was used for enhancing the performance of military personnel, cosmonauts, and Olympic athletes facing stressful environments (Winston and Maimes, 2011). The European Scientific Cooperative on Phytotherapy (ESCOP) monograph described that application of *E. senticosus* is effective for treating decreased vitality of mind and body, such as frailty, fatigue, languor, poor concentration, and convalescent period after illness (ESCOP 2009). In the 16th edition of the Japanese Pharmacopeia, *E. senticosus* is listed by the name *Ci wujia* as a hypnotic sedative and an appetite stimulant (MHLW, 2012). Of late, an estrogenic activity of *E. senticosus* has been discovered. Lignins, specifically pinoresinol-glycosides, are transformed into enterolactone or enterodiol by human intestinal bacteria (Heinonen et al., 2001). Nowadays, *E. senticosus* is being utilized for the prevention or improvement of premenstrual syndrome (PMS), menstrual irregularity, and autonomic imbalance and indefinite complaint of menopausal females, in addition to its conventional application to improve adaptability to stress.

12.3.1.2 Oita Saffron (*Crocus sativus*)

Although *Crocus sativus* is the most expensive spice in the world and essential for making famous dishes like paella and bouillabaisse, it is also a traditional medicinal plant (Bagur, 2018). Major saffron-growing districts, such as Iran, Spain, and Kashmir, are famous and the saffron is cultivated

TABLE 12.2
Applications and Phytochemical Constituents of Native Medicinal and Aromatic Plants of Japan.

Common Name	Scientific Name	Part Used	Phytochemical Constituent	Application or Medicinal Action
Green tea	*Camellia sinensis*	Leaf	Epigallocatechin gallate (EGCG)	Anticancer, antiviral
Amur corktree	*Phellodendron amurense*	Bark	Berberine	Antibacterial, bitter stomachic
Bitter orange	*Citrus aurantium*	Peel	Synephrine, limonin	Sympathetic nervous system hyperactivity, blood pressure elevation, weight loss
Shell ginger	*Alpinia speciosa*	Leaf	Kawain, methysticin, yangonin	Sedative, euphoriant, perception improvement
Tachibana orange, shiikuwasha	*Citrus tachibana, Citrus depressa*	Peel	Nobiletin	Antidementia
Kudzu	*Puerariamontana* var. *lobata*	Root	Daidzein, genistein	Estrogenic
Common grape	*Vitis vinifera*	Leaf	Resveratrol	Anti-inflammatory, anticancer
Japanese geranium	*Geranium japonica thunbergii*	Aboveground part	Geraniin	Intestinal regulation, antiviral
Turmeric	*Curcuma longa*	Rhizome	Curcumin	Hepatic, cholagogue, anti-inflammatory, anticancer
Ginger	*Zingiber officinale*	Rhizome	Shogaol	Anti-inflammatory, analgesic, antirheumatic
Safflower	*Carthamus tinctorius*	Flower	Linoleic acid	Maintaining cell permeability and membrane fluidity
Camellia	*Camellia japonica*	Seed	Oleic acid	Lifestyle-diseases prevention, emollient
Wasabi	*Eutrema japonicum*	Aboveground part	Sinigrin	Isothiocyanate production
Citrus unshiu	*Citrus unshiu*	Fruit	Cryptoxanthin	Bone metabolism improvement, mucosa restoring and protection
Cat's whiskers	*Orthosiphon stamineus*	Leaf, stem	Potassium, sinensetin	Diuretic, antidementia
Mulberry	*Morus alba*	Leaf	Deoxynojirimycin (DNJ), iron, zinc, calcium	Suppression of sugar absorption, anti-anemic, immunopotentiation, dysgeusia, wound healing
Field Horsetail	*Equisetum arvense*	Whole plant	Silicon	Immunopotentiation, diuretic, collagen fortification
Summer squash	*Cucurbita pepo*	Seed	Cucurbitacin	Anti-inflammatory, anticancer, urethritis
Sesame	*Sesamum indicum*	Seed	Sesamin	Anti-inflammatory, liver protection

(Continued)

TABLE 12.2 (CONTINUED)
Applications and Phytochemical Constituents of Native Medicinal and Aromatic Plants of Japan.

Common Name	Scientific Name	Part Used	Phytochemical Constituent	Application or Medicinal Action
Hokkaido Siberian ginseng	*Eleutherococcus senticosus*	Stem	Pinoresinol	Estrogenic
Burdock	*Arctium lappa*	Root	Arctiin	Estrogenic
Japanese cherry	*Cerasus* × *yedoensis*	Leaf, flower	Coumarin	Antibacterial, anticoagulant, anti-edema, antispasmodic
Purple gromwell	*Lithospermum erythrorhizon*	Root	Shikonin	Anti-inflammatory, granulation
Alexandrian senna	*Senna alexandrina*	Foliole	Sennoside	Purgative
Japanese star swertia	*Swertia japonica*	Whole plant at flowering period	Secoiridoid	Promotes blood circulation
Sugar cane	*Saccharum officinarum*	Pith of the stem	Glycolic acid	Keratolysis
Rice	*Oryza sativa*	Bran	Ferulic acid	Anticancer, dementia prevention
Saffron	*Crocus sativus*	Pistil	Crocin	Promote blood circulation, brain function improvement
Dokudami, chameleon plant	*Houttuynia cordata*	Leaf	Quercitrin	Diuretic, laxative
Japanese mint	*Mentha arvensis*	Leaf	L-Menthol	Swelling and itching of the throat, flatulence, antidiarrhea
Silver birch	*Betula pendula*	Leaf	Quercetin	Anti-inflammatory, anti-allergic

outdoors. However, at the Taketa area of Oita prefecture, the saffron is kept in a hothouse before flowering and harvest (OBPD, 2022). This results in high-quality pistils. Saffron is known for the emmenagogue effect and removing blood stasis, and has been used for improvement of menopausal autonomic imbalance such as cold, hot flashes, and PMS in Japan. In recent years, it has come to light that certain constituents activate brain-derived neurotrophic factors (BDNF); subsequently, studies are being conducted on psychoneurotic symptoms, such as depression and dementia. Crocin, a water-soluble carotenoid pigment, and safranal, an essential oil ingredient, are involved in the functions of saffron. According to Dai et al. (2020), the effect of saffron in improving depression was better than a placebo in a meta-analysis of its functionality, an effectiveness which was equivalent to that of conventional antidepressants. No significant difference regarding side effects was found among saffron, placebo, and antidepressants. As a result, saffron has the potential to be an alternative antidepressant for treating mild and moderate depression disorders.

12.3.1.3 Kagoshima Mulberry (*Morus alba*)

Mulberry is the only plant on which the silkworm feeds, and it has supported silk production by sericulture. In 1996, a study of the functionality of mulberry leaf was started, and the result revealed that deoxynojirimycin (DNJ), which competitively inhibits binding between disaccharides and α-glucosidase (disaccharidase), suppresses sugar absorption (Junge et al., 1996). Thus, mulberry leaf tea has been utilized for prevention of lifestyle-related diseases, commencing with diabetes (Kijima et al., 2010). Unabsorbed glucidein in the small intestine is metabolized by intestinal bacteria in the large intestine, and short-chain fatty acids (SCFA), such as butyric and propionic acids, are produced (Ríos-Covián et al., 2016). Eventually, the intestinal pH decreases, and the intestinal environment is improved. Generally, polyphenol inhibits enzyme activity, and herbal tea inhibits disaccharidase activity. However, the activity of mulberry leaf is of unparalleled strength (Amin et al., 2020). Studies reveal that the effectiveness of mulberry leaf tea is equivalent to that of voglibose, which is a drug for treatment of diabetes (Miyahara et al., 2004). By the way, in "*KissaYōjōki*" ("Drinking Tea for Health"), written by Zen Master Eisai (1141–1215), mulberry leaf was recommended for diabetes disorders. The wisdom of people of olden times is a marvel.

12.3.1.4 Okinawa Shell ginger (*Alpinia zerumbet*)

Okinawa is located at the southern tip of Japan. Various medicinal plants, such as shell ginger, turmeric (*Curcuma longa*), cat's whiskers plant (*Orthosiphon aristatus*), and *Momordica charantia*, grow there. Shell ginger of the family Zingiberaceae, growing from East Asia to the tropics, is used for various purposes. Fiber is taken from the stem for clothing and wallpaper, and the leaf and seed are used in herbal teas. One of the traditional Okinawan foods, called *Moochie*, is a rice cake wrapped in shell ginger leaves, containing essential oils with antibacterial and antiseptic qualities. Recently, cosmetics using shell ginger extract have been developed because the leaves are rich in polyphenols. Okinawa Industrial Technology Center has explored the functionality of the flower, leaf, root, and fruit of shell ginger. As a result, α-amylase inhibitory activity in flowers, leaves, and roots and α-glucosidase inhibitory activity in flowers have been confirmed (Okinawa Prefecture, 2021). The results indicated that shell ginger is useful in preventing hyperglycemia and diabetes. An inhibitory activity of angiotensin-converting enzyme (ACE) was observed in leaves and roots. ACE is involved in regulating blood pressure, and hypertension can be improved by inhibition of ACE. Furthermore, inhibition of lipase activity was observed in the flowers, leaves, roots, and fruits. Lipase is a fat-degrading enzyme; therefore, dyslipidemia and obesity can be prevented by inhibiting this enzyme. People living in Okinawa are known for their longevity. Possibly, the daily consumption of herbal tea drinks by the Okinawans is a factor involved.

12.3.2 Native Aromatic Plants

12.3.2.1 Camphor (*Lindera umbellata*)

Japan was the largest producer of camphor in the world before World War II and the essential oil of *L. umbellata*, which is used as a soap fragrance, has been extracted in the Izu region and exported for 150 years. Recently, the physiological activities of domestically produced essential oils have attracted attention due to their contributions to health. For example, effects in preventing leukemia have been attributed to *L. umbellata* oil (Maeda and Katagata, 2011). In aromatherapy, medicinal constituents are absorbed through the skin and enter directly in the bloodstream. Therefore, appropriate essential oils should be selected or blended with due regard to their medical mode of action. Table 12.3 shows the functional components of essential oils and their therapeutic actions.

12.3.2.2 Hokkaido Japanese mint (*Mentha arvensis*)

Cultivation of Japanese mint began in 1817 in Okayama, then in Hiroshima and Yamagata, and eventually was started in Hokkaido in 1891. Before World War II, the essential oil of Japanese mint had occupied the world market. Around 1960, import of essential oil was liberalized, the domestic market share was taken away, and the export of peppermint (*Mentha piperita*) essential oil dropped sharply (Kitami Hakka Tsusho, 2022; Tree of Life Library, 2022). At the present time, most of the cultivation of Japanese mint and distillation of essential oil are confined to Hokkaido. Japanese mint has the higher menthol content, compared with the essential oil of peppermint (Sievers and

TABLE 12.3
Classification of Essential Oils (From Native Plant Species), Based on Their Medicinal Actions and Functional Components.

Groups	Subdivisions	Actions	Constituent of Essential Oil (Plant Name)
Hydrocarbon	Monoterpene	Forest bathing	Pinene (shiikuwasha, hinoki cypress, Sakhalin fir, kōyamaki, tachibana orange, shell ginger)
		Antibacterial, antiviral	Terpinene (tachibana orange, shiikuwasha, shell ginger)
		Digestive, promote blood circulation, anticancer	d-Limonene (yuzu, tachibana orange, shiikuwasha, Sakhalin fir, *Lindera umbellata,* kōyamaki, camphor)
		Fragrance	Myrcene (hinoki cypress), phellandrene (shell ginger)
		Local anesthesia, insect repellence	Camphene (Sakhalin fir, shell ginger)
	Sesquiterpene	Anti-inflammatory, sedative	Caryophyllene (kōyamaki, shell ginger)
		Stomachic, demulcent	Zingiberene (ginger)
		Forest bathing, stress relief, soporific	Thujopsene (hiba arborvitae)
Alcohol	Monoterpene alcohol	Stimulant	L-Menthol (Japanese mint, peppermint) α-terpineol (hinoki cypress, camphor, shell ginger), borneol (Shell ginger)
		Demulcent	Linalool (*Lindera umbellata,* tachibana orange, shell ginger), geraniol (*Lindera umbellata,* ginger)
		Antibacterial, anti-inflammatory, IgA induction	Hinokitiol, β-thujaplicin (hiba arborvitae, Taiwan cypress)
	Sesquiterpene alcohol	Fragrance	Cedrol (hiba arborvitae, kōyamaki)
Phenol, derivative	Phenylpropanoid	Antiseptic	Carvacrol (shell ginger)
	Phenol ether	Fragrance	Safrole (camphor)
Aldehyde	Monoterpene	Insect repellence	Citral (ginger)
Ketone	Monoterpene	Fragrance	Menthone (Japanese mint)
		Sedative after stimulant	Camphor (camphor, shell ginger)
Ester	Monoterpene	Fragrance	Bornyl acetate (Sakhalin fir) Geranyl acetate (ginger)
Oxide	Monoterpene	Demulcent	Cineole (*Lindera umbellata,* ginger, shell ginger)

Lowman, 1933). Menthol acts directly on smooth muscle, adjusts the influx of calcium ions into the cell, and causes antispasmodic effects (Amato et al., 2014). Therefore, the essential oil of Japanese mint is effective for digestive disorders, such as irritable bowel syndrome and colic, and is listed on the Japanese Pharmacopeia. In the West, the treatment of taking enteric-coated capsules of peppermint essential oil for irritable bowel syndrome is popular (Kline et al. 2001). In addition, the capsules are effective for gastrointestinal symptoms, such as heartburn, nausea, and diarrhea. In addition, essential oil of Japanese mint is used for overcoming drowsiness and improving concentration because of its function in brain activation. It also provides relief from pain, like a local anesthesia, hence it is used in lotions and ointments and massage for the treatment of headaches, muscle pains, and bruises.

12.3.2.3 Aomori hiba (*Thujopsis dolabrata*)

Aomori hiba is a tree of the Cupressaceae family that grows on the Shimokita Peninsula of Aomori prefecture and Iwate prefecture. It has been used as a material for constructing temples and Buddha statues because it has antibacterial and antiseptic properties. Consequently, the Golden Hall of Chusonji temple has been free from damage by mold and rotting fungi for more than 800 years. Hinokitiol (β-thujaplicin), present in essential oil of Aomori hiba has a strong antibacterial effect. Hinokitiol is also present in the essential oil of Taiwan Hinoki, but for some reason not in Kiso Hinoki. The essential oil of Aomori hiba exhibits an antibacterial action against *Staphylococcus epidermidis*, enterococci, and even methicillin-resistant *Staphylococcus aureus* (MRSA), which causes nosocomial infections. In recent years in animal experiments it was revealed that trans-nasal administration of hinokitiol can induce IgA (immunoglobulin A) (Kataoka et al., 2018). Vaccines for respiratory diseases, which are given in Japan as an intramuscular injection, can induce IgG (immunoglobulin G), but not IgA. Consequently, the vaccine cannot prevent the infection itself even if it prevents the exacerbation of the infectious disease. A vapor inhalation, using essential oil of Aomori hiba, may be useful for control of cold and influenza infections. In addition, a skin-whitening effect due to tyrosinase inhibition and a hair growth effect due to inhibition caused by binding to the androgen receptor of dihydrotestosterone by hinokitiol have been reported (Ishii et al., 1999; Takahashi et al., 2010).

12.3.2.4 Kiso Hinoki (*Chamaecyparis obtusa*)

Kiso Hinoki is also a cypress (member of the Cupressaceae family), growing in Taiwan and Japan. Japan is a forest-rich nation, with forests occupying 67% of the land. However, it is still backward as regards the use of forests. The Ministry of Agriculture, Forestry, and Fisheries started working on the diffusion of forest therapy and explored the physiological and psychological effects of the forest environment on humans. In consequence, hyperactivity of the parasympathetic nervous system decreased concentration of salivary cortisol, and activation of natural killer (NK) cells in people in natural Kiso Hinoki forests have been reported (Li et al., 2008). These effects are persistent: the activation of NK cells is maintained for more than 30 days after a forest bathing trip (Li, 2012). In behavioral pharmacological studies, an anti-anxiety effect was observed when mice inhaled the essential oil of Kiso Hinoki (Kasuya et al., 2015). In an examination of intracerebral transferability of the essential oil from Kiso Hinoki, α-pinene, which is one of the components of the essential oil of Kiso Hinoki, was detected in all regions of the brain: the cerebral cortex, hippocampus, hypothalamus, and medulla oblongata. As a result, the anti-anxiety effect was probably caused because α-pinene was uniformly transferred to the brain by inhalation. In Japan, Kiso Hinoki wood has been used in constructing bathtubs and cutting boards from olden times because Kiso Hinoki is long-lasting in water. The reason why the Hinoki bathtub is popular must be not only its durability, but also the belief that it can remove anxiety and stress.

12.3.2.5 Kochi Yuzu (*Citrus junos*)

The yuzu originated upstream of the Yangtze River in China and it is believed yuzu was introduced to Japan in the Nara or Asuka periods. In Japan, there is a custom of taking a yuzu bath at the winter solstice, a tradition which is said to have started in the Edo period. D-limonene, which is a principal component of essential oil of yuzu, promotes blood circulation and warms the core of the body. It is said that people who take a yuzu bath can fall fast asleep until the morning. The anti-anxiety effect of the essential oil of yuzu on mice was suggested on the basis of behavioral pharmacology (open-field and elevated plus maze) tests. On the other hand, no increase in spontaneous momentum was observed (Miyahara, 2011). In the elevated plus maze test, d-limonene showed the anti-anxiety effect, whereas l-limonene, an optical isomer of limonene, did not show it. In recent years, Japanese native plant essential oils, such as yuzu, Kiso Hinoki, Aomori Hiba, and *L. umbellata*, are getting popular among aromatherapists, perfumers, and chefs of foreign countries. In particular, the scent of the essential oil of yuzu is attracting great attention. Generally, essential oil of citrus is manufactured by squeezing the fruit. However, essential oil of yuzu is manufactured by distillation, because no furocoumarins, such as bergapten, which cause phototoxicity, remain in the essential oil after distillation, and, therefore, it can be used in cosmetics without any concern.

12.4 FUTURE CHALLENGES

MAPs are related to forest therapy, health, tourism, nursing care, elderly care, and mental health care, and challenges in MAPs research and business can be met by collaborative efforts. Challenges are likely to appear in health promotion, preventive medicine, and healing treatment. Already, the Japanese government has had a difficult medical expenses problem, and recommended preventive medicine (MHLW, 2017). Due to the high percentage of elderly people in the Japanese population, age-related diseases, such as dementia and lifestyle-related diseases, are on the increase. The spread of medicinal herbs and aromatherapy as integrative medicine is another challenge. Although the Japanese government recommends the use of medicinal herbs and aromatherapy in integrative medicine (MHLW, 2014), the actual use is still limited.

Usually, Japanese doctors and pharmacists have studied Kampo, which is part of the core curriculum in the medical education system, whereas other phytotherapies are not so well known (Murakami, 2012). Therefore, sharing the evidence-based data of cures from using medicinal herbs and aromatherapy obtained by non-medical professionals, such as herbalists, scientists, and therapists, with medical doctors is necessary.

Adding the skin-whitening and anti-aging properties of phytochemicals from MAPs to cosmetics, and the popularity of aromatic cosmetics for men may boost the growth of organic cosmetics considerably.

Recently, the industry of incorporating aroma into spatial fragrance, such as stage production and store branding, has been expanding, and it is increasing in adoption not only in hotels and commercial facilities but also at entrances, shared spaces, and relaxing space for employees of general companies.

Conservation of MAPs is another challenging area for future researchers. A project based on the idea of repeated use of a circular forestry was started in Hokkaido in 2000. Sakhalin fir (*Abies sachalinensis*) is a characteristic tree of Hokkaido and the essential oil distilled from its leaves and branches has an air-purifying effect (Inoue et al., 2018). Four kinds of tasks, namely planting, trimming, cutting, and effective use exist over about 10,000 ha and the cutting area will become the next planting area. The cutting–planting exercise is repeated every year somewhere in the 10,000 ha land area, and, eventually, the first planted trees become usable in about 60 years. The forest was certificated by the Forest Stewardship Council (FSC) in 2003, admitting the management of the land under environmental, social, and economical criteria (Figure 12.2). The forest resources are used for building materials, biomass, compost, charcoal, pyroligneous acid, smoking materials, ornaments, and cosmetic materials. The leaves and branches which are left after cutting and processing

FIGURE 12.2 Forest Stewardship Council (FSC) logo signifies that the product comes from responsible sources—environmentally appropriate, socially beneficial, and economically viable.

are used as materials having essential oil. This local project involves the whole community, giving rise to employment and leads society to become more active and nature loving.

Historically many medicinal botanical gardens can be seen, especially in the central region of Japan, because about 3,000 kinds of MAPs used to grow on Mt. Ibuki around 1700; today, about 280 of these species are left (Mt. Ibuki Driveway, 2022). At present, botanical gardens are striving to preserve the rare and endangered species in various parts of Japan, and genetic resources, such as populations of wild species, are being preserved at gene banks.

REFERENCES

AEAJ. 2019. *A Survey Report on Aromatics Market in Japan in 2018*. Aroma Environment Association of Japan. https://www.aromakankyo.or.jp/pdf/news/1238/FIX_2018%E5%B8%82%E5%A0%B4%E8%AA%BF%E6%9F%BB%E3%83%AA%E3%83%AA%E3%83%BC%E3%82%B9_v4.pdf.

Ahmad, S., Fazli, I.S., Jamal, A., Kamaluddin, Maaz, M., Iqbal, M., and Abdin, M.Z. 2006a. Cultivation of Babchi (*Psoralea corylifolia* L., Fabaceae) for the production of furocoumarin (psoralen) of therapeutic value. In: *Traditional Systems of Medicine*, edited by Abdin, M.Z., and Abrol, Y.P., 465–470. New Delhi: Narosa Publishing House.

Ahmad, S., Jamal, A., Fazli, I.S., Iqbal, M., Kamaluddin, and Abdin, M.Z. 2006b. Nutritional approach for improved production of Atrilal (*Ammi majus* L.). In: *Traditional Systems of Medicine*, edited by Abdin, M.Z., and Abrol, Y.P., 534–545. New Delhi: Narosa Publishing House.

Ahmad, A., Khan, I., Abrol, Y.P., and Iqbal, M. 2005. Role of sulphate transporter system in sulphur efficiency of mustard genotypes. *Plant Science* 169: 842–846.

Ali, G., Iqbal, M., and Srivastava, P.S. 2000. Experimental studies on Brahmi, the innocuous little herb. *Hamdard Medicus* 43: 41–45.

Ali, G., Mughal, M.H., Srivastava, P.S., and Iqbal, M. 1997b. Micropropagation of *Tribulus terrestris* L., an important medicinal plant. *Journal of Plant Biology* 40: 202–205.

Amato, A., Liotta, R., and Mule, F. 2014. Effects of menthol on circular smooth muscle of human colon: Analysis of the mechanism of action. *European Journal of Pharmacology* 740: 295–301. https://doi.org/10.1016/j.ejphar.2014.07.018.

Amin, A.R., Kassab, R.B., Abdel Monein, A.E., and Amin, H.K. 2020. Comparison among garlic, berberine, resveratrol, hibiscus sabdariffa, genus Zizyphus, hesperidin, red beetroot, Catha edulis, Portulaca oleracea, and mulberry leaves in the treatment of hypertension and type 2 DM: A comprehensive review. *Natural Product Communications* 15(4): 1–24.

Anis, M., and Iqbal, M. 1986. Antipyretic utility of some Indian plants in traditional medicine. *Fitoterapia* 57: 52–55.

Anis, M., Sharma, M.P., and Iqbal, M. 2000. Herbal ethnomedicine of the Gwalior-forest division in Madhya Pradesh, India. *Pharmaceutical Biology* 38(4): 241–253.

Atique, A., Iqbal, M., and Ghouse, A.K.M. 1985a. Use of *Annona squamosa* and *Piper nigrum* against diabetes. *Fitoterapia* 56: 190–192.

Atique, A., Iqbal, M., and Ghouse, A.K.M. 1985b. Ethnobotanical study of cluster fig (*Ficus racemosa*). *Fitoterapia* 56: 236–240.

Bagur, M.J., Alonso Salinas, G.L., Jimenez-Monreal, A.M., Chaouqi, S., Liorens, S., Martinez-Tome, M., and Alonso, G.L. 2018. Saffron: An old medicinal plant and a potential novel functional food. *Molecules* 23(1): 30. https://doi.org/10.3390/molecules23010030.

Bleakney, T.L. 2008. Deconstructiong an adaptogen: *Eleutherococcus senticosus*. *Holistic Nursing Practice* 22(4): 220–224. https://doi.org/10.1097/01.HNP.0000326005.65310.7c.

Cheng, Z., Zhang, Z., Han, Y., Wang, J., Wang, Y., Chen, X., Shao, Y. et al. 2020. A review on anti-cancer effect of green tea catechins. *Journal of Functional Food* 74: 104172.

Dai, L., Chen, L., and Wang, W. 2020. Safety and efficacy of saffron (*Crocus sativus* L.) for treating mild to moderate depression: A systematic review and meta-analysis. *The Journal of Nervous and Mental Disease* 208(4): 269–276.

ESCOP. 2009. *ESCOP Monographs: The Scientific Foundation for Herbal Medicinal Products*. 2nd ed. European Scientific Cooperative on Phytotherapy, Stuttgart: Georg Thieme Verlag.

Fujiwara, K., Imanishi, J., Watanabe, S., Ozawa, K., and Sakurada, K. 2011. Changes in attitudes of Japanese doctors toward complementary and alternative medicine – Comparison of surveys in 1999 and 2005 in Kyoto. *Evidence-Based Complementary and Alternative Medicine* 2011: Article ID 608921.

Funayama, S. 2015. *Science of Folk Medicine* [in Japanese]. Tokyo: SB Creative Corp.

Hayashi, S. 2017. Positioning of essential oils in herbal therapy and new findings in herbal therapy [in Japanese]. *Aromatopia* 142: 24–27.

Heinonen, S., Nurmi, T., Liukkonen, K., Poutanen, K., Wähälä, K., Deyama, T., Nishibe, S., and Adlercreutz, H. 2001. In vitro metabolism of plant lignans: New precursors of mammalian lignans enterolactone and enterodiol. *Journal of Agricultural and Food Chemistry* 49(7): 3178–3186.

Ikei, H., Song, C., Lee, J., and Miyazaki, Y. 2015. Comparison of the effect of olfactory stimulation by air-dried and high-temperature-dried wood chips of hinoki cypress (*Chamaecyparis obtuse*) on prefrontal cortex activity. *Journal of Wood Science* 61(5): 537–540. https://doi.org/10.1007/s10086-015-1495-6.

Ikei, H., Song, C., and Miyazaki, Y. 2016. Effects of olfactory stimulation by α-pinene on autonomic nervous activity. *Journal of Wood Science* 62(6): 568–572. https://doi.org/10.1007/s10086-016-1576-1.

Imanishi, J. 2017. Current status and future of herbal therapy [in Japanese]. *Aromatopia* 142: 2–5.

Inoue, M., Hayashi, S., and Craker, L.E. 2018. Production, cultivation, and conservation of domestic aromatic plants in Japan. In: *Recent Advances in Plant Research*, edited by Welch, B., and Wilkerson, M., 219–238. New York: Nova Science Publishers.

Iqbal, M., Parveen, R., Parveen, A., Parveen, B., and Aref, I.M. 2018. Establishing the botanical identity of plant drugs based on their active ingredients under diverse growth conditions. *Journal of Environmental Biology* 39(1): 123–136.

Ishii, K., Hirao, R., Yamashita, N., Nishimori, T., Yamamoto, Z., and Asanuma, Y. 1999. Androgen inhibitory effect of hair restorer compounding ingredients [in Japanese]. *Japanese Pharmacology and Therapeutics* 27(11): 1749–1755.

Junge, B., Matzke, M., and Stoltefuss, J. 1996. Chemistry and structure-activity relationships of glucosidase inhibitors. In: *Oral Antidiabetics*, edited by Kuhlmann, J., and Puls, W., 411–482. Berlin: Springer. https://doi.org/10.1007/978-3-662-09127-2_15.

Kasuya, H., Iida, S., Ono, K., Satou, T., and Koike, K. 2015. Intracerebral distribution of α-Pinene and the anxiolytic-like effect in mice following inhaled administration of essential oil from *Chamaecyparis obtuse*. *Natural Product Communications* 10(8): 1479–1482.

Kataoka, K., Kunitomo, E., Kobuchi, K., Uene, M., Doi, T., and Miyake, T. 2018. Effect of hinokitiol on salivary secretory-IgA antibody production system. *Journal of Dental Health* 68: 137–144.

Kijima, Y., Kimura, T., Nakagawa, K., Asai, A., Hasumi, K., Oikawa, S., and Miyazawa, T. 2010. Effects of mulberry leaf extract rich in 1-deoxynojirimycin on blood lipid profiles in humans. *Journal of Clinical Biochemistry and Nutrition* 47(2): 155–161.

Kitami Hakka Tsusho. 2022. *History of Mint in Japan and World* [in Japanese]. Kitami Hakka Tsusho. https://hakka.be/story/world/ (Accessed January 5, 2022).

Kline, R.M., Kline, J.J., Di Palma, J., and Barbero, G.J. 2001. Enteric-coated, pH-dependent peppermint oil capsules for the treatment of irritable bowel syndrome in children. *The Journal of Pediatrics* 138(1): 125–128.

Kotaro Pharmaceutical. 2022. *Houttuynia cordata - A Kotaro* [in Japanese]. Kotaro Pharmaceutical Co., Ltd. https://www.kotaro.co.jp/product/detail/g509.html (Accessed February 27, 2022).

Li, Q. 2012. *Forest Medicine*. New York: Nova Science Publishers.

Li, Q., Morimoto, K., Kobayashi, M., Inagaki, H., Katsumata, M., Hirata, Y., Hirata, K. et al. 2008. Visiting a forest, but not a city, increase human natural killer activity and expression of anti-cancer proteins. *International Journal of Immunopathology and Pharmacology* 21(1): 117–127.

Maeda, H., and Katagata, Y. 2011. A physiological function of essential oil in *Lindera umbellate* [in Japanese]. *Aroma Research* 12(3): 244–245.

MAFF. 2013. *Alliance among Medical, Welfare, Food, and Agriculture* [in Japanese]. Japan: Ministry of Agriculture, Forestry and Fisheries. www.maff.go.jp/j/keikaku/ifukushokunou.html (Accessed on February 23, 2022).

MHLW. 2012. *Crude Drugs*. Japan: Ministry of Health, Labour, and Welfare. https://jpdb.nihs.go.jp/kyokuhou/files/000203133.pdf (Accessed on February15, 2022).

MHLW. 2014. Evidence-based Japanese. *Integrative Medicine*. Japan: Ministry of Health, Labour and Welfare. https://www.ejim.ncgg.go.jp/en/index.html (Accessed February 23, 2020).

MHLW. 2017. *A System of Taxation of Self-Medication* [in Japanese]. Japan: Ministry of Health, Labour and Welfare. https://www.mhlw.go.jp/stf/seisakunitsuite/bunya/0000124853.html#h2_freel (Accessed February 25, 2022).

Miyahara, N. 2011. Study of chemical composition and emotional behavior of essential oil from Yuzu (*Citrus junos*) [in Japanese]. Master Dissertation, Toho University, Japan.

Miyahara, C., Miyazawa, M., Satoh, S., Sakai, A., and Mizusaki, S. 2004. Inhibitory effects of mulberry leaf extract on postprandial hyperglycemia in normal rats. *Journal of Nutritional Science and Vitaminology* 50(3): 161–164.

Mt. Ibuki Driveway. 2022. *A Treasure Trove of Medicinal Plants* [in Japanese]. Japan: Japan Automobile Road Corporation. www.ibukiyama-driveway.jp/enjoy/herb/index.html# (Accessed March 7, 2022).

Murakami, C. 2012. The present state of integrative medicine and problems in japan-strategies for promotion of integrative medicine. *Nihonbashi Gakkan University Bulletin No 11*: 85–94.

Nakamura, M., Sugita, M., Shibata, Y., and Ojima, T. 2017. Effect of b-cryptoxanthin erich Satsuma mandarin juice supplementation on pulse wave velocity: A randomized controlledtrial. *Journal of Nutrition and Intermediary Metabolism* 8: 8–13. https://www.sciencedirect.com/science/article/pii/S2352385916300226.

Nagasaki University. 2014. *Shimabara Medicinal Food and Nutrition Education Project* [in Japanese]. Nagasaki University. https://www.nagasaki-u.ac.jp/ciugc/shimabara.pdf (Accessed February 23, 2022).

Nishizawa, M. 2020. Increase of healthy life expectancy by studies on Indonesian and Japanese medicinal plants. *Journal of the Asia-Japan Research Institute of Ritsumeikan University* 2: 217–223.

NMJH. 2006. *Botanical Garden of Everyday Life*. National Museum of Japanese History, National Institute for Humanities. www.rekihaku.ac.jp/english/exhibitions/plant/index.html (Accessed January 7, 2022).

OBPD. 2022. *Saffron Oita Brand Promotion Division*. https://theoita.com/en/sanpinintro/3674/ (Accessed January 6, 2022).

Okinawa Prefecture. 2021. *Useful Plant Resources in Okinawa: No. 5 Gettou*. Okinawa Prefecture. www.pref.okinawa.jp/site/shoko/kogyo/shokuhinkagaku/rensai_okinawayuuyousyokubutu/rensai.html (Accessed December 11, 2021).

Parveen, A., Ahmad, M., Parveen, B., Parveen, R., and Iqbal, M. 2022. The traditional system of Unani medicine, its origin, evolution and Indianization: A critical appraisal. *Indian Journal of Traditional Knowledge* 21(3): 511–521.

Parveen, A., Parveen, R., Akhatar, A., Parveen, B., Siddiqui, K.M., and Iqbal, M. 2020a. Concepts and quality considerations in Unani system of medicine. *Journal of AOAC International* 103(3): 609–633. https://doi.org/10.5740/jaoacint.19-0284.

Parveen, B., Parveen, A., Parveen, R., Ahmad, S., Ahmad, M., and Iqbal, M. 2020b. Challenges and opportunities for traditional herbal medicine today, with special reference to its status in India. *Annals of Phytomedicine* 9(2): 97–112.

Perera, A., Ton, S.H., and Palanisamy, U.D. 2015. Perspectives on geraniin, a multifunctional natural bioactive compound. *Trends in Food Science and Technology* 44(2). https://doi.org/10.1016/j.tifs.2015.04.010.

Ríos-Covián, D., Ruas-Madiedo, P., Margolles, A., Gueimonde, M., de los Reyes-Gavilán, C.G., and Salazar, N. 2016. Intestinal short chain fatty acids and their link with diet and human health. *Frontiers in Microbiology* 7: 185. https://doi.org/10.3389/fmicb.2016.00185.

Saba, Iqbal, M., and Srivastava, P.S. 2000a. In vitro micropropagation of *Silybum marianum* L. from various explants, and sylibin content. *Journal of Plant Biochemistry and Biotechnology* 9: 81–87.

Saba, Pande, D., Iqbal, M., and Srivastava, P.S. 2000b. Effect of $ZnSO_4$ and $CuSO_4$ on regeneration and lepidine content in *Lepidium sativum* L. *Biologia Plantarum* 43(2): 253–256.

Sashida, Y., and Kihara, H. 2013. *Familiar Medicinal Plants* [in Japanese]. Tokyo: Heibonsha.

Sievers, A.F., and Lowman, M.S. 1933. Commercial possibilities of Japanese mint in the United States as a source of natural menthol. *Technical Bulletin* No. 378. Washington, DC: United States Department of Agriculture. https://naldc.nal.usda.gov/download/CAT86200372/PDF.

Takahashi, S., Kamiya, T., Saeki, K., Nezu, T., Takeuschi, S., Takasawa, R., Sunaga, S. et al. 2010. Structural insights into the hot spot amino acid residues of mushroom tyrosinase for the bindings of thujaplicins. *Bioorganic and Medicinal Chemistry* 18(22): 8112–8118.

TAY. 2022. *Yorii-machi Echinacea* [in Japanese]. The Tourism Association of Yorii. http://yorii-kanko.jp/echinacea.html (Accessed February 17, 2022).

TCDA. 2014. *Domestic Cultivation Project of Medicinal Animals and Plants* [in Japanese]. Tokyo Crude Drugs Association. https://www.tokyo-shoyaku.com/jigyo.php?mode=saibai (Accessed February 23, 2022).

Tree of Life Library. 2022. *Mints* [in Japanese]. Tree of Life Co., Ltd. www.treeoflife.co.jp/library/partner-farm/herbtravel/vol16.html (Accessed January 5, 2022).

Winston, D., and Maimes, S. 2011. *Adaptogens: Herbs for Strength, Stamina, and Stress Relief.* Translated by Kumagaya, C., and Hougan, N. Tokyo: Fragrance Journal Ltd.

Yoshitake, T. 2012. *Aromatic Plants* [in Japanese]. Tokyo: Hosei University Pr.

13 Ecological, Phytochemical and Pharmacological Profile of *Melissa officinalis* L., an Aromatic Medicinal Plant

S. Nizamudeen, Sana Nafees, and Huda Nafees

CONTENTS

13.1 INTRODUCTION

Melissa officinalis L. is an aromatic, perennial, medicinal plant, commonly known as lemon balm. In Asia, it is named as Arabian balm as this herb was brought to India from the Arabian sub-continent (Ainslie, 1826). The genus Melissa has about 3500 species, that grow extensively in European countries, the eastern Mediterranean region, and the western part of Asia (Dymock, 1893). *M.*

DOI: 10.1201/9781003226925-13

officinalis has been used for 2,000 years, as its description and uses were mentioned in Historica Planatrum (approx. 300 BC) and Materia Medica (approx. 50–80 BC) (Alina, 2016). This herb grows to a height of about 6 feet (Sawicka, 2020), its leaves and stems are greenish-black in color and bitter in taste. It gives a pleasant, lemon-scented aroma (Bhattacharjee, 2004) that attracts cats, which is why it is called "Billilotan" in the Unani system of medicine (Ghani, 1921). It is the one of the most abundant of the aromatic and medicinal plants, rich in essential oils, and used for culinary, medicinal, and ornamental purposes. Due to its abundance in the Mediterranean region, the aromatic oils from lemon balm are extracted in a large-scale manufacturing process as a profitable source of income (Bhattacharjee, 2005). *M. officinalis* is used medicinally in Unani, Ayurveda, Siddha, and Homeopathy systems of medicine (Abdel-Naime, 2020). *De-Materia Medica* and the classical text book of Unani medicine, "*Khazinul-Advia*," mentions this herb as a medicine to treat mental disorders and as an antidote for scorpion and spider bites (Ghani, 2010), whereas Galen did not describe this plant (Ibn Baitar,1197–1248 CE, Pub. 1985). Recent scientific studies have identified the individual bioactive metabolites that have been proven to have various pharmacological activities, like antimicrobial, antidepressant, and anticancer properties (Sharifi-Rad, 2021).

13.2 TAXONOMIC CLASSIFICATION

Melissa officinalis, a species of the family Lamiaceae in the group of angiosperms, is commonly known as lemon balm, gentle balm, citronellae, or balm mint. From the literature and the web resources, it is found that lemon balm is the most common name used. The complete taxonomical classification of *M. officinalis* given in Table 13.1.

13.3 VERNACULAR NAMES OF *MELISSA OFFICINALIS*

The drug is known by its common name in different countries and languages. A list of some common names in different countries is given in Table 13.2.

13.4 GEOGRAPHICAL DISTRIBUTION AND ECOLOGY

M. officinalis is a plant which is quite widely distributed, growing in the temperate climate region. Globally, this species is naturally distributed in Mediterranean countries, Europe, North America, North Africa, and across Asia, (Pouyanfar, 2018). In Asia, this species is naturally distributed across the Himalayas and in northern Indian states, i.e., in Sikkim, Garwal, Darjeeling, Kasi, Utttarkhand,

TABLE 13.1
Taxonomy Tree of *Melisa officinalis*

Kingdom	Plantae: Plantes, Planta, Vegetal, Plants
Sub-Kingdom	Viridiplantae—green plants
Infra kingdom	Streptophyta—land plants
Super division	Embryophyta
Division	Trachynophyta—vascular plants, trachenophytes
Subdivision	Spermatophytina—spermatophytes, seed plants, phanerogams
Class	Magnoliopsida
Super order	Asteranae
Order	Lamiales
Family	Lamiaceae—mints, menthes
Genus	*Melissa*
Species	*officinalis*

TABLE 13.2
The Different Common Names for *Melissa officinalis* in Many Countries

Language	Vernacular Name	Reference(s)
Unani	Malsoonan, malitana	Ibn Baitar, 1197–1248 CE, Pub.1985
Latin	Apiastrum, aarookhia	Dymock, 1893; Ibn Baitar, 1197–1248 CE, Pub.1985
Arabic	Baklat -el-Utrujya, buck litul faristum	Dymock, 1893; Ainslie, 1826; Khory, 1985
Persian	Badrunj buyeh	Ainslie, 1826; Khory, 1985; Harwi, 2008; Hussain, 1855
English	Melissa, common balm, lemon balm, balm mint	Khory, 1985
Hindi	Ram-Tulsi	Khory, 1985
Bombay	Badaranj	Khory, 1985
Tamil	Parsee cunjam koray	Ainslie, 1826
Dutch	Mekkasubza	Ainslie, 1826; Khory, 1985
French	Melisse	Ainslie, 1826
German	Melisse	Ainslie, 1826

and Kashmir at an altitude of 1,200 to 1,300 m and in some parts of southern India. However, *M. officinalis* is native to the eastern Mediterranean region, which is why it is commercially cultivated in Southwestern, central and Eastern European countries, for export to different industries associated with culinary, pharmacological, medial and fragrance purposes (Moradkhani, 2010).

13.5 PHARMACOGNOSTIC DESCRIPTION

13.5.1 Macroscopic Characteristics

The herb consists of stems, leaves, and fruit as shown in Figure 13.1; stem is hairy, quadrangular, and has a purplish hue (Dymock, 1893; Khory, 1985). The leaves are two inches long, petiolate, oblong, heart-shaped, and crenate at the base, with serrated margins (Khory, 1985; Bhattacharjee, 2004). The upper surface is characterized by scattered hairs, whereas the lower surface is characterized by glands on the nerve axils. The calyx is striated, hairy, and five-toothed (Khory, 1985;

FIGURE 13.1 Aerial parts of *Melissa officinalis*.

Dymock, 1893), each tooth with a strong nerve. The corolla is not as hairy as the calices, is rose colored or brownish purple, and rugous (Khory, 1985). Each fruit has four seeds, which are three-angled, with a white patch on either side of the hilum (Khory, 1985; Dymock, 1893) and attached to the bottom of the calices, erect, angular, and dark brown in color, similar in size to Malunga (Khory, 1985). Flowers are axillary, in clusters of six, on a short peduncle (Khory, 1985; Dymock, 1893), white or pale rose in color, tiny in size, and occur in summer (Bhattacharjee, 2004). The entire dried herb has a light brown color and a unique aromatic odor or fragrance, similar to that of banaphasha or lemon, taste is astringent and bitter, like that of nimbada (Khory, 1985).

13.5.2 Microscopic Characteristics

The upper part of the stem's transverse section has a quadrangular layout with four corners. It is made up of an outer epidermis, which is followed by a narrow cortex of sub-epidermal masses of collenchyma at the corners and parenchymatous cells toward the vascular tissue. A parenchymatous pericycle surrounds a discontinuous ring of vascular tissue, which is only present under each corner, as seen in the transverse section. In the center, there is also a wide, parenchymatous pith. The lower part of the stem's transverse section is quadrangular in shape, with non-glandular trichomes on the outer epidermis along with a few glandular hairs, followed by a narrow cortex that is differentiated into collenchyma and parenchyma. A continuous ring of parenchymatous pericycle with isolated groups of lignified fibers surrounds the vascular tissue. In the center of the transverse section, there is also a wide, parenchymatous pith (Abdel Naime, 2020).

13.5.3 Cultivation and Collection

Well-drained, deep, alluvial, fertile soil, and a sunny and warm location is ideal for cultivation of these plants. Propagation of these plants is effected by seeds or cuttings. Seeds are directly sown in well-prepared land or rooted cuttings are planted at a spacing of 50 × 50 cm.

13.5.4 Phytochemical Constituents

The primary constituents of lemon balm have been identified as hydroxycinnamic acid derivatives, specifically rosmarinic acid, caffeic acids, chlorogenic acid, and metrilic acid (Triantaphyllou et al., 2001; Ziakova et al., 2003), tannins (Felklova et al., 1969; Kucera et al., 1967), flavonoids, including quercitrin, rhamnocitrin (Emamghoreishi and Talebianpour, 2009), luteolin, luteolin 7-*O*-beta-D-glucopyranoside, apigenin 7-*O*-beta-D-glucopyranoside), and luteolin 3-O-beta-D-glucuronopyranoside (Khayyal, 2000; Patora, 2003), monoterpene glycosides, sesquiterpenes, including β-caryophyllene and germacrene (Mikus, 2000), triterpenes, including ursolic and oleanolic acids (Emamghoreishi and Talebianpour, 2009), and volatile oils which include citronellal, citral a (geranial), citral b (neral), methyl citronellate, ocimene, citronellol, geraniol, nerol, -caryophyllene, -caryophyllene oxide, linalool, and ethric oil (Mrlianová, 2002; Mulken, 1988). Tittel and co-workers determined the composition of the volatile oil, which accounted for 0.5–0.1% of the plant by weight, with citronellal, geranial, and neral accounting for 50–70 percent of this oil (Tittel et al., 1982).

The components of the plant's essential oil vary depending on climate, but citral (gerinal, neral) citronellal, and geraniol are the primary components. There are numerous factors that influence essential oil composition and yield, including light intensity, nutrients, temperature, cultural practice, genotype, plant part age, and harvesting time (Moradkhani, 2010).

There are numerous variations in the components of *M. officinalis*. Sixteen genotypes grown in the same area in Yugoslavia for two years yielded 0.046–0.246% oil. The constituents were the same for all oils, but the relative amounts varied greatly between genotypes, with citronellal accounting for 2.711–12.141%, linalool accounting for 1.501–6.380%, caryophyllene accounting for

1.210–19.073%, and geranyl acetate + citronellal accounting for 9.710–26.913%. A 5-year selection and improvement project for *M. officinalis* in Spain increased the essential oil concentration from 0.2–0.3% to more than 0.5% (Evans, 2009).

The essential oil content of the medicinal herb ranges from 0.02 to 0.30%, which is much lower when compared with the oil percentage of other herbs of the Lamiaceae family. As a result, the market price and production cost of *Melissa* essential oil are very high. According to Meftahizade et al. (2010), the principal constituents of the essential oil include citral (geranial and neral), citronellal, geraniol, beta-pinene, alpha-pinene, and beta-caryophyllene, accounting for 96% of the oil content. Carnat et al. (1998) investigated the chemical composition of lemon balm essential oil and discovered that citral (neral + geranial) accounted for 48% of the essential oil, citronellal accounted for 39.47%, and caryophyllene accounted for 2.37%. Sarer et al. (1991) discovered that the percentage content of the primary constituents include alpha-pinene (2.86%), beta-pinene (11.37%), linalool (2.74%), citronella (5.86%), borneol (0.62%), neral (12.22%), and geraniol (38.13%). Citral is commonly utilized in foods and cosmetics due to its strong lemon fragrance and flavor (Leung, 1996). Fresh plants of lemon balm contain total phenolic, L-ascorbic acid, and carotenoids (Sarer et al., 1991). Eugenyl glycoside has been extracted from the leaves of lemon balm. Lemon balm tea yielded 10 mg/l of essential oil (74% citral) and a high concentration of polyphenol components (Carnat, 1998). Steam distillates of lemon balm callus cultures yielded dehydroabietane and another diterpene hydrocarbon, with the relative quantities of the two chemicals fluctuating significantly throughout the cultivation cycle (Koch-Heitzmann, 1985).

Lemon balm extract included hydroxycinnamic acid derivatives and flavonoids, including luteolin-7-*O*-glucoside (0.0002%), caffeic acid, *m*-coumaric acid, eriodictyol-7-O-glucoside, naringin, hesperidin, rosmarinic acid, naringenin, and hesperetin (Dastmalchi, 2008).

13.6 PHARMACOLOGICAL ACTIVITIES

13.6.1 Cardiovascular Effects

M. officinalis is used in traditional medicine specifically to treat cardiovascular disease (Soltanpour, 2019). Many of the *in-vivo* studies have proven *M. officinalis* to be effective in treating different cardiac conditions. A study on reperfusion-induced, lethal, ventricular arrhythmias in rats showed that treatment with *M. officinalis* had mild protective effects in reducing arrhythmic conditions (Joukar et al., 2014). Aqueous extract of aerial parts of *M. officinalis* on myocardial injury induced by isoproterenol in the Wistar rat model, with or without cardiac insult, was examined. The results revealed that *M. officinalis* extract at a lower dose (50 mg/kg) improved the redox potential system, thereby reducing the insult to cardiac tissues, but the higher doses (200 mg/kg) of extract exaggerated the injury and increased the myocardial ischemia (showing substantial increases of cardiac troponin-I level (Joukar et al., 2015). In a $CaCl_2$.induced, arrhythmic rat model, an ethanolic extract of *M. officinalis* reduced the incidence of ventricular premature beats, ventricular fibrillation, and ventricular tachycardia (Akhondali et al., 2015). These anti-arrythmic and bradycardiogenic effects of *M. officinalis* might be due to β-adrenergic antagonist activity (Somova et al., 2003, 2004). In rat models, treatment with the aqueous extract of *M. officinalis* prolonged the QRS complex and S-T intervals in ECG; this might be due to high concentration of rosmarinic acid and phenols, which increases= the conducting potential of the heart through Na^+ and K^+ channels (Joukar and Asadipour, 2015). In an *in-vivo* study, rosmarinic acid from *M. officinalis* dramatically lowered the level of superoxide anions and the production of TBF-β1 and α smooth muscle actin (Karthik et al., 2012). Fibrosis is known to lead to the development of heart failure and cardiac arrhythmias in patients with heart disease. Rosmarinic acid is the major component in *M. officinalis* which can play a major role in these activities along with other activities like antoxidant and hypolipidemic activities. This explains the traditional use of *M. officinalis* in treatment of heart diseases, although further clinical studies are needed to gauge the dose- dependence of these properties. These *in-vivo* studies are in lower doses and administered in an animal model; more human clinical trials are needed to exert its effect on the human heart, as few studies have shown the extracts to induce cardiac rhythm, with still no significant clinical studies having been observed or proven to exert reduced heart rate in humans (Alijaniha et al., 2015; Gazola et al., 2004), despite the fact that a few studies have shown the extract to lessen the incidence of palpitation episodes in patients with benign palpitation (Alijaniha et al., 2015).

13.6.2 Antianxiety Effect

In an *in-vivo* study, the antidepressant effect of *M. officinalis* water extract was evaluated by its influence on the behavior and associated neurotransmitters of rats performing a forced swimming test (Lin et al., 2015). Another study, carried out by Feliu-Hemmelmann and co-workers, with aqueous extracts of *M. officinalis* and *Passiflora caeruela* was shown to reduce the plasma level of corticosterone, which is the significant mediator in reducing stress in mice (Feliu-Hemmemlmann et al., 2013). Many *in-vivo* studies on the brains of mice treated with *M. officinalis* have demonstrated its tranquilizing effects and a few recent studies have proven it to exert anxiolytic properties, rosmarinic acid is found to be the major factor in inhibiting the action of GABA (Awad et al., 2009). In another *in-vivo* study, the oral administration of *M. officinalis* hydroalcoholic and ethanolic extract was shown to cause anxiolytic effects which might be due to GABA-T Inhibition, which eventually increases the GABA levels in the brains of treated mice (Ibarra et al., 2010; Taiwo et al., 2012). The outcome of two clinical trials confirmed the evidence from the above-mentioned animal research (dosage range of 240 mg– 360 mg/kg bodyweight) for any clinical setting. In a randomized, double-blind, placebo-controlled crossover clinical trial, 18 healthy volunteers were subjected to laboratory-induced stress, confirmed by completing a 20-minute version of the Defined Intensity Stressor Simulation (DISS) battery, after subjects were treated with 300 mg and 600 mg doses of standard *M. officinalis* methanolic extract (separated by 7 days of extract application). The *M. officinalis* 600 mg treatment alleviated the negative mood effects of DISS with increased self-ratings of relaxation and decreased self-ratings of alertness (Kennedy, 2004). A 15-day prospective, open-label research was conducted to assess the efficacy of Cyracos (hydroalchololic extract of *M. officinalis*) on mild to moderate anxiety disorder in volunteers associated with sleep problems. The primary results showed that symptoms improved after using Cyracos, which reduced the stress-induced anxiety manifestations by 18% ($p<0.01$), anxiety-related symptoms by 15% ($p<0.01$), and sleeplessness by 42% ($p<0.01$). Individuals responded to treatment in as many as 95% of instances (19/20), with 70% (14/20) achieving full remission for anxiety, 85% (17/20) for sleeplessness, and 70% (14/20) for both (Cases, 2011). The evidence presented above illustrates the traditional use of *M. officinalis* in anxiety-related disorder to act as a sedative drug. Furthermore, evidence is required from randomized, controlled trials to understand the *M. officinalis* pharmacological action of anxiolytic effects and to assess anxiolytic properties of other active compounds.

13.6.3 Neuroprotective Effects

Many *in-vitro* and *in-vivo* studies in neuroprotection models in different countries support the use of *M. officinalis* in treating a number of CNS illnesses. *M. officinalis* methanol extract protects PC12 cells against H_2O_2 toxicity in MTT and LDH studies (the cell viability assays often used for determining cytotoxicity when exposed to cytotoxic substances). In addition, the aqueous and methanolic extracts significantly reduced intracellular reactive oxygen species (ROS) production, showing a significant neuroprotective impact (Lopez et al., 2009). In another study, the acidic fraction of the *M. officinalis* ethanolic extract, which comprises polyphenols, flavonoids, and terpenoids, was shown to protect PC12 cells from induced toxicity and oxidative stress, which confirms its antioxidant property (Sepand et al., 2013). It has been observed that the *M. officinalis* extract can shift [^{3}H] (–)-nicotine in human brain cell membranes containing acetylcholine receptors. The IC_{50} values for the [^{3}H]-nicotine-displacing activity of *M. officinalis* ethanol extract were less than 100 g/ml (Wake et al., 2000). Several investigations have shown that nicotine protects against A-induced toxicity and that pre-treatment with nicotine reduces A-induced stress and apoptosis in neurons (Khiara et al., 1997; Liu et al., 2004). As a result, the nicotine receptor-enhancing characteristics of *M. officinalis* extract can be linked to its protective action against Aβ-induced toxicity. Furthermore, *M. officinalis* aqueous extract protected primary neurons against apoptosis produced by 3,4-methylenedioxy-methamphetamine, with these activities probably due to free radical scavenging and monoamine oxidase (MOA) inhibition (Hassanzadeh et al., 2011). Pre and post treatment with rosmarinic acid in human primary neuronal cells have been found to retain neuronal integrity and prevent neurotoxicity caused by cigautoxin (Braidy et al., 2014). The high antioxidant and radical scavenging capabilities of hydroalcoholic extract of *M. officinalis*, as well as its principal phenolic compounds rosmarinic acid A, quercetin, gallic acid, caffeic acid, chlorogenic acid, and rutin, may account for some of the plant's neuroprotective action. *In-vitro* cytotoxicity experiments revealed that *M. officinalis* essential oil (EO) protects against hypoxia-induced neuronal death. The essential oil of *M. officinalis* has been shown to drastically reduce caspase-3 activity and TUNEL-positive cells. Moreover, there was a reduction in malondialdehyde and an increase in Trolox equivalent antioxidant capacity levels in the hippocampus of ischemic rats after 100 mg/kg EO therapy. Despite this, essential oil resulted in neurotoxicity in a primary neuronal culture system at a concentration of 0.1 mg/ml (Mahita et al., 2014) .These findings propose that *M. officinalis* essential oil has protective activity against a number of CNS diseases which are associated with ischemic brain injury and this protective mechanism might be due to inhibition of hypoxia-inducible factor-1 (HIF–I, a transcription factor that plays a major role in hypoxia-induced expression of genes), oxidative stress and apoptosis (Bayat et al., 2012). Numerous studies have shown that the oxygenated monoterpenes found in *M. officinalis* essential oil, such as citral isomers and citronellal, have modest antioxidant properties (Mimica-Dukic et al., 2004; Misharina et al., 2008; Ruberto et al., 2000). These findings support the traditional usage of *M. officinalis* in the treatment of neurological diseases such as dementia, epilepsy, stroke, and paralysis.

13.6.4 Effects on Memory and Improving Concentration

The cholinergic system's role in the memory-improving effect of *M. officinalis* extract was investigated. These findings indicate that *M. officinalis* can improve memory and that the extract's cholinergic properties may contribute to the memory-improving benefits seen in this study. *M. officinalis* extract has the potential to be therapeutic in relieving some of the memory impairment seen in Alzheimer's disease (Soodi et al., 2014).

A clinical experiment was conducted to see if a fixed combination of valerian and *Melissa* could improve concentration and reduce hyperactivity and impulsiveness. The percentage of children with strong/very strong symptoms of poor focus decreased from 75% to 14%, hyperactivity from 61% to 13%, and impulsiveness from 59% to 22%. Social behavior, sleep, and symptom burden were all assessed highly positively by parents. Mild temporary adverse medication responses were found in only two children (Gromball et al., 2014).

13.6.5 Antidepressant Effect

M. officinalis has been used as an energizing medicine for centuries, according to ancient medieval manuscripts. *M. officinalis* aqueous and methanol extracts might modestly inhibit monoamine oxidase (MAO)-A in *in-vitro* research, with the methanol extract being more powerful in inhibiting (MAO)-A (Lopez et al., 2009). Nonetheless, the predicted IC_{50} values for the methanol and aqueous extracts were 19.3 and 48.2 g/ml, respectively, doses that are unlikely to be reproducible in clinical settings even with highly bio-available preparations. In the Forced Swimming Test (FST), the *M. officinalis* ethanol extract also had antidepressant effects by increasing nor-ephinerine neurotransmission (Emamghoreishi and Talebianpour, 2009; Taiwo et al., 2012). In the FST, rosmarinic acid (RA) reduced the length of immobility in mice by an antidepressant mechanism other than MAO inhibition or monoamine transport (Takeda et al., 2002). Therefore, more studies need to be performed in order to understand the fundamental antidepressant mechanism of *M. officinalis* compounds and to evaluate the MOA-A inhibitory mechanism and the major polar compounds, other than RA, also have to be assessed for any synergistic effects. The above-mentioned animal studies support the traditional use of *M. officinalis* as an antidepressant medication (with doses ranging from 25 mg/kg to 300 mg/kg), although no IC_{50} or EC_{50} values were reported to allow better understanding and assessment of the translational value effects of the drug, despite the fact that such doses are theoretically applicable in humans, based on the principles of allometric scaling of experimental doses.

13.6.6 Anti-Inflammatory and Antinociceptive Effects

According to the available historical manuscripts, *M. officinalis* has been used in treating many inflammatory conditions like inflammation of the joints, to relive joint pain and asthma. Many of the pharmacological activities of this herb have been shown to have anti-inflammatory and antinociceptive properties, *viz*., an aqueous extract of *M. officinalis* applied to inflammagen induced paw edema in rats was shown to reduce edema significantly, with a decrease in nociceptive response (Bounihi, 2013). In another study, the ethanol extract of *M. officinalis* revealed dose-dependent antinociceptive effect in chemical nociception models in mice, by inhibiting the L-arginine–nitric oxide pathway and activating the cholinergic system. Furthermore, ethanol extract of *M. officinalis* displayed dose-dependent antinociceptive effects in chemical nociception rat models, by inhibiting the L-arginine–nitric oxide pathway and activating the cholinergic system, the RA in the extract was found to be responsible for antinociceptive actions. (Gugnski et al., 2009), whereas the IC_{50} values for the extract are quite high (about 200 mg/kg) as were those for the RA (2.6 mg/kg); these doses appeared to be clinically relevant. In a diabetic, hyperalgesic animal model, long-term oral administration of *M. officinalis* essential oil (EO) (0.01 to 0.04 mg/day) had considerable antinociceptive effect (Hasanein and Riahi, 2015). In addition, the EO significantly reduced and inhibited paw edema caused by carrageenan and experimental stress in rats. However, the tested doses of 200 mg/kg to 400 mg/kg seems to be too high to be used in human studies, in spite of the fact that no safety concerns were raised at doses as high as 2,000 mg/kg (Bounii et al., 2013). In an animal model of diabetic hyperalgesia, long-term oral administration of *M. officinalis* EO (0.01–0.04 mg/day) had a considerable antinociceptive effect (Hasanein and Riahi, 2015). The anti-inflammatory mechanism of EO might be due to the presence of the citral content, which has been shown to block TNF-α in RAW 264.7 cells stimulated by Lipopolysaccharide (LPS) and reduce IL-6 and IL-1 in lipopolysaccharide-stimulated peritoneal macrophages from the normal mice model (Bounihi et al., 2013).

13.6.7 Hypoglycemic Effects

Traditionally, *M. officinalis* has been used to treat diabetes, and many *in-vitro* and *in-vivo* studies have shown significant hypoglycemic effects. In a study conducted by Weidner et al., *Mellisa officinalis* ethanolic extract at a dose of 0.6 mg/l, when administered to human primary adipocytes, triggered selective expression of PPAR gene in the metabolically important target cells. In another study in insulin-resistant obese mice, *M. officinalis* ethanolic extract given at a dose of 200 mg/kg showed a hypoglycemic effect (Weidner et al., 2015). Furthermore, the essential oil from *M. officinalis* was shown to restore normal plasma glucose levels and lowered the bodyweight of diabetic rats, although further toxicological studies should confirm the safety of the essential oil for use in human studies (Parisa and Hassan, 2015). A high-fat diet was given to streptozotocin-induced diabetic rats, and the study revealed that ursolic acid and 30 natural penta-cyclic triterpenoid carboxylic acids found in *M. officinalis* were shown to have a hypoglycemic effect which might be associated with increased plasma levels and by enhanced insulin sensitivity (Jang et al., 2009). Another few components, like oleanolic acid, from *M. officinalis* was shown to improve insulin sensitivity to cellular function, thereby exhibiting the hypoglycemic effects via antioxidant enzymes, phase-II response gene expression, inhibition of NF-κB, suppression of the polyol pathway, AGE production, and hyperlipidemia (Castellano et al., 2013). Lipophilic triterpenoids in combination with essential oil from *M. officinalis* could be further investigated as a potential hypoglycemic agent.

13.6.8 Metabolic Effects

In the traditional system of medicine, *M. officinalis* is used in obesity and cardiovascular diseases, and recent scientific studies done by many researchers have shown the extract to have hypolipidemic effects. In a study conducted by Zarei et al. (2014), hypercholesterolemic mice were treated with the ethanolic extract of *M. officinalis* in comparison with the control group, which was treated with atorvastatin, and the extract was shown to reduce the level of liver enzymes (AST, ALT, ALP) (Zarei et al., 2014). In another *in-vivo* study, insulin-resistant obese mice were treated with ethanolic extract of *M. officinalis* and were shown to significantly reduce- the plasma triglycerides, LDL, VLDL levels (Weidner et al., 2014). In an *in-vivo* study by Changizi-Ashtiyani et al. (2013) on mice having hypercholesterolemia, treatment with an extract of either *M. officinalis* or barberry was shown to reduce serum cholesterol, LDL, and triglycerides. The study also revealed that the alcoholic extract of *M. officinalis* was shown to have more hypolipidemic activities than that of *Berberis vulgaris*, which might be probably due to increased thyroid hormone and antioxidant properties of the *M. officinalis* extract (Changizi-Ashtiyani et al., 2013). Application of the extract to mice was shown to reduce the plasma triglyceride levels which might be due to a possible effect of quercetin from the plant in inhibiting the lipid peroxidation (Jun et al., 2012). Hyperlipidemic rats treated with an aqueous extract of *M. officinalis* had lower total serum lipid levels, total cholesterol, hepatic enzyme activities and the extract was also shown to reduce lipid peroxidation and glutathione levels in hepatic tissues, thereby confirming the plant's hypolipidemic and hepatoprotective effects (Bolkent et al., 2005). In another study, administration of *M. officinalis* essential oil orally significantly reduced serum triglyceride (TGL) levels and cholesterol levels in a dose- and time-dependent manner in HepG-2 cells of mice. The hypo-triglyceridemic effects might be due to a decrease in the translocation of sterol regulatory element-binding enzyme-1c (SREBP-1c) and its target genes involved in the synthesis of fatty acids due to decreased P-300/CBP associated factor (PCAF) histone acetylates activity, which results in a decreased synthesis of hepatic fatty acids (Jun et al., 2012). Despite intriguing results, the dose of *M. officinalis* extract used in this and other, previous experiments was not enough to be implicated in clinical experimental studies; additional research is needed with more, lower doses to be used in humans.

13.6.9 Cytotoxic Effects

M. officinalis has been used since antiquity to treat many types of cancer (Javedi et al., 2015). *M. officinalis* and its main component, citral, has shown to promote apoptosis in GBM cell lines expressing MRP1.Citral enhanced ROS generation and inhibited MRP1 expression in the same research (Queiroz et al., 2014). In another study, citral was shown to inhibit the viability of 27 human tumor cell lines and a mouse melanoma cell line. Hydroalcoholic extracts of *M. officinalis* inhibited cell proliferation and promoted death in colon cancer cells through increased intracellular ROS production (Weidner et al., 2015), although the observed IC_{50} values for the cytotoxic action were greater than 100 g/ml, which is unlikely to be duplicated in clinical trials for therapeutic use in day-to-day clinical practice. Decoctions of *M. officinalis*, containing rosmarinic acid (RA) and lithospermic acid-A, inhibited the growth of various human tumor cell lines (Carocho et al., 2015). In a few studies, a 50% ethanol extract from *M. officinalis* was shown to exhibit cytotoxic effects on human colon cancer cell line, as assessed by cell viability assays (MTT and NR assays, commonly used to detect cytotoxicity after exposure to toxic compounds) (Encalada et al., 2011; Fotakis et al., 2006). The dichloromethane and n-hexane fractions of the plant extract inhibited K652 and Jurkat cells in a dose-dependent manner. The dichloromethane fraction effectively triggered apoptosis in a leukemia cell line by increasing the Bax / Bcl -2 ratio and up-regulating Fas and Bax mRNA expression, showing its ability to activate both extrinsic and intrinsic apoptotic pathways. On the other hand, the n-hexane fraction of the plant did not modify the expression of apoptosis- related genes, which suggests that the induction of apoptosis may not be the primary reason for suppression of cellular growth by this plant extract fraction (Ebrahimnezhad Darzi and Amirghofran, 2013). These findings suggest that the plant's apoptosis-inducing effects might be due to essential oil components and lipophilic components (found in dicholoromethane and n-hexane fractions) which can permeate into cellular membranes. In an *in-vivo* study, *M. officinalis* ethanol extract is shown to have anti-mutagenic and anti-genotoxic properties toward mice (De Carvalho et al., 2011). However, the effects were profound at a higher dose of 0.5 mg/kg of body weight and via parenteral administration, whereas this dose is not feasible for translation to humans; the extract was found to be more effective at reversing genotoxicity (assessed by the comet assay) than mutagenicity (assessed by the micronucleus assay) through the alkylating agent methyl methanesulfonate (MMS). As phenolics has been known to exhibit antioxidant activity, these could be the potential agents in the extract to achieve these properties (De Carvalho et al., 2011). Rosmarinic acid can reduce the chromosomal damage caused by doxorubicin and ethanol in micronucleus experiments in mice and V-79 cells (De Oliveira et al., 2012). Furthermore, gallic acid serves as an anticarcinogenic agent by modifying the function of DNA repair enzymes, influencing their expression (Abdel Wahed et al., 2007). As a result, the polyphenolic compounds, particularly rosmarinic acid, which are present at high concentrations in hydrophilic extracts of *M. officinalis*, are responsible for anticarcinogenic and antigenotoxic actions. Despite growing evidence for the cytotoxic properties of M. officinalis and its preparations and fractions, it should be noted that the researchers did not provide enough information on the cytotoxic activity of the extract tested on human cells, entailing even more analysis of the tested extracts' selective cytotoxicity.

13.6.10 Antioxidant Activity

Various studies have shown that oxidative stress plays a major role in etiology of diseases like neurological disease, cardiovascular disease, diabetes mellitus, and various type of cancers (Elnakish et al., 2013; Melo et al., 2011; Rains et al., 2011). The oxidative stress occurs when there is reduction in the activities of superoxide dismutase (SOD) and reduced glutathione (GSH) and this reduction results in accumulation of free radicals, particularly hydrogen peroxide and superoxide anion, generated in huge amounts, which may overcome endogenous protective antioxidants (like GSH) and antioxidant enzymes (such as SOD) (Slemmer et al., 2008). The antioxidant potential of different extracts of *M. officinalis* has been found in several *in-vitro* as well as *in-vivo* studies (Bayat et al., 2012) In an in-vitro study, the hydro-ethanol extract of *M. officinalis* showed antioxidant activity when assayed using iron(II) chelation, iron (III) reduction, 1,1 diphenyl 1-2 picrylhydrazyl (DPPH), ABTS, nitric oxide free radical scavenging, superoxide anion, and inhibition of β-carotene linoleic acid bleaching assays (Dastmalchi et al., 2008). Results showed that antioxidant property of a hydro-ethanol extract (90.43±1.55 μg/ml) of *M. offcinalis* was significantly more effective than caffeic acid and gallic acid (44.57±7.10 and 44.29± 1.92 %), respectively, and was statistically comparable with that of BHA (butylated hydroxy anisole, 96.08±1.58 %) and quercetin (98.46±0.89%) (Dastmalchi et al., 2008).

In another study, the high ROS-scavenging activity of an ethanolic extract of *M. officinalis* was found (IC_{50} = 7.58 μg/ml) in the DPPH assay. This activity may be attributed to the presence of monoterpene aldehydes and ketones (citronellal, menthone, citral, and isomenthone), as well as the mixture of mono- and sesquiterpene hydrocarbons (Mimica-Dukic et al., 2004). A single- arm clinical trial among radiology staff was conducted to evaluate the efficacy of *M. officinalis* infusion on the oxidative stress level. Results of the study showed there was a considerable improvement seen in the plasma levels of SOD, GSH, peroxidase, and catalase and also there was a reduction in the DNA damage, myeloperoxidase activity, and lipid peroxidation (Zeraatpishe et al., 2011). From the above data, it can be concluded that the phenolic compounds present in *M. officinalis* preparations exhibit antioxidant activity through free radical scavenging inhibition of lipid peroxidation and by increasing the levels of endogenous antioxidant enzymes. Therefore, the therapeutic potential of *M. officinalis* might be due, in part, to its antioxidant property in the prevention of oxidative stress diseases like cardiovascular diseases, neurodegenerative diseases, etc. (Bayat et al., 2012).

13.6.11 Antimicrobial Activity

An *in-vitro* antimicrobial study has been carried out using *M. officinalis* essential oil against Gram-negative bacteria. Results showed that EO exhibits notable antimicrobial activity against *Pseudomonas aeruginosa*, *Salmonella typhi*, *Escherichia coli*, *Salmonella enteritidis*, and *Shigella* strains, particularly on the multiply resistant strain of *Shigella sonei*. It was found the highest activity was observed on the multiply resistant strain of *S. snoei* and *E. coli* (Mimica-Dukic et al., 2004). In another study, it was also observed the essential oil also exhibited antibacterial activity against food-borne pathogens and spoilage bacteria with minimum inhibitory concentration (MIC) values ranging from 72.0 to 1000.3 μg/ml when compared with *Rosmarinus officinalis* MIC values ranging from 91.3 to 1113.3 μg/ml, but it is lower than that of the positive control which was ciprofloxacin at 2.5–62.2 μg/ml (Hussain et al., 2011). The antibacterial and antifungal activities of the essential oil of *M. officinalis* were exhibited by citral and citronellal present in it (Mimica Dukic et al., 2004). In another study, it was found that the petroleum ether, ethyl acetate, chloroform, or n-butanol extracts of *M. officinalis* exhibited moderate to strong antibacterial activity against *Sarcina lutea*, *Staphylococcus aureus*, and *Bacillus cereus* having zone inhibition diameters ranging from 10.7–19.3 mm, 10.0–16.3 mm and 8.0–14.0 mm, respectively, with the zone inhibition of the standard antibiotics, namely amoxicillin and penicillin, being 28.0–29.0 and 32.7–36.6, respectively, against the aforementioned bacteria. Of the three extracts, the results revealed that petroleum ether and ethyl acetate extracts exerted the greatest antibacterial activity (Canadanovic Brunet et al., 2008). Decoctions of *M. officinalis* contain rosmarinic acid and lithospermic acid-A are mostly active against *Pseudomonas aeruginosa*, *Salmonella typhimurium*, and *Penicillium funiculosum*, having MIC = 0.2, 0.2, 0.1 and MBC= 0.4, 0.4, 0.2, respectively. Results showed the antimicrobial potency of *M. officinalis* decoctions were comparable to or higher than those of standard streptomycin and ampicillin (Carocho et al., 2015). These findings support the traditional use of *M. officinalis* for the treatment of wounds, fevers, scrofula, etc. Additionally, the pharmacological properties from the *M. officinalis* preparations may serve as natural preservatives to be used in food and pharmaceutical products.

13.6.12 Antiviral Activity

The antiviral activity of *M. officinalis* leaves has been used since antiquity to treat patients bitten by rabid dogs. In 2008, in an *in-vitro* study, it was found that the hydroalcoholic and aqueous extracts of the plant exerted significant anti-HSV1 and anti-HSV2 activity (Mazzanti et al., 2008). Geuenich et al. (2008) concluded in their study that an increase in HSV1 virion density before the virus attached to the host cells, which might be considered to be probable antiviral activity. In another study, it was found that RA plays a major role in antiviral activity (Astani et al., 2014). In another study, the antiviral activity of the essential oil (EO) of *M. officinalis* was measured *in vitro* using the plaque reduction assay method. Results showed that the antiviral effect of the essential oil operated before adsorption of the virus but not after penetration into the host cell. This indicates that the EO has a direct antiviral activity on HSV (Schitzler et al., 2008). In 2004, a study on the effect of different concentrations of the essential oil from *M. officinalis* found that the EO inhibited the penetration of HSV2 and also found that this effect was inversely proportional to the concentration. Later, this study suggested that the citral and citronellal in the EO contributed to the antiviral activity because of protein synthesis inhibition in cells (Allahverdiyev et al., 2004). The two clinical trials were conducted on a topical cream, containing 1% of the dried extract of *M. officinalis*, and its effect against HSV infections, in which one trial was a multicenter study having 115 patients whereas the other was a double-blind, controlled study involving 116 patients. The results of both studies showed significant antiviral activity of the topical cream and also revealed that, for maximum efficacy the treatment should be initiated at an early stage (Wolbling and Leonhardt, 1994).

13.6.13 Antispasmodic Activity

In the traditional system of medicine, *M. officinalis* has been used as an antispasmodic agent, and also as a remedy for hiccups and gripe. In 2003, an *in-vivo* study was conducted on the rat ileum to evaluate the relaxant or antispasmodic effect of the essential oil (EO) of *M. officinalis* and its main component, citral. Results showed both the EO and citral could inhibit the response of ACh, KCL and 5-hydroxytryotamine (5-HT) in a concentration-dependent manner, indicating spasmolytic effect (Sadraei et al., 2003). In 2005, a clinical trial, which was a randomized, double-blind, placebo-controlled study, was conducted to evaluate the effectiveness of a polyherbal formulation containing *M. officinalis, Matricaria recutita, and Fieniculum vulgare* var. *dulce* was tested for the treatment of infantile colic. The results of the study revealed significant improvements in colic in breast-fed infants within one week of the initiation of the treatment (Savino et al., 2005). These studies suggest, the possible role of the plant in reducing intestinal contractions and also alleviating broncho-constrictive diseases such as chronic obstructive pulmonary diseases or asthma. A detailed study of these properties needs to be evaluated in future scientific studies.

13.6.14 Anti-Angiogenic Effects

Angiogenesis is defined as the process of formation of new blood vessels from pre-existing blood vessels; the deregulated and pathological angiogenesis is well established in the formation of cancer cells, the angiogenetic vessels supplying nutrients to the cancer and tumor cells, thereby facilitating the process of metastasis (Belal Al-Husein et al., 2012). The hydroalcoholic extract of *M. officinalis*, at a dose of 100 mg/kg/day, showed inhibition of laser- induced choroidal neovascularization (CNV) development in a dose-dependent manner (Lee et al., 2014). In another study, a polyherbal formulation containing *M. officinalis* has been evaluated for anti-angiogenic effect. Results showed that there was a reduction in the levels of mRNA of genes encoding angiogenic factors like VEGF A,B,C, and D, a reduction in fibroblast growth factor-2 (FGF 2), and also in MMPs (MMP 2 and MMP 9) in treated cells, whereas the mRNA levels of angiogenic inhibitors were increased in differentiated cells (Hong et al., 2011). Another study demonstrated that this plant also inhibited the growth of adipose tissue and thus reduced bodyweight gain, by inhibiting angiogenesis (Yoon and Kim, 2011).

13.6.15 Anti-Epileptic Activity

In 2012 Bhat et al. conducted a study of hydroalcoholic and aqueous extracts of *M. officinalis* on animal models of epilepsy, namely maximal electric shock seizure model and the pentylenetetrazole seizure method. Results showed that both the extracts, i.e., hydroalcoholic and the aqueous extracts of *M. officinalis* exhibited potent anti-epileptic activity (Bhat et al., 2012). In another study, pre-treatment with a hydroalcoholic extract of *M. officinalis* significantly affected the prevention of seizure symptoms in rats induced by pentylenetetrazole injection (Gorgich et al., 2012). Both the above *in-vivo* studies could be extrapolated to humans according to allometric scaling principles, although the confirmation of safety and efficacy requires accurate evaluation in clinical trials.

13.7 CONCLUSION AND FUTURE PERSPECTIVE

Melissa officinalis is one of the most widely used medicinal plants since antiquity to treat different diseases in various medical systems globally. This review article has aimed to describe the ecological distribution of the plant along with its traditional uses, pharmacology, phytochemistry, etc. Many *in-vitro* and *in-vivo* scientific studies have confirmed the plant's different biological activities, including cardiovascular, anxiolytic, neuroprotective, antibacterial, antiviral, anticancerous, and anti-inflammatory activities. These studies will establish a foundation on which to validate the use of *M. officinalis* in today's clinical scenario. Based on the different scientific studies, an oral dose of about 600 mg of the plant extract is sufficient and proven to be safe and effective in clinical human trials. For the topical usage of *M. officinalis* the formulations containing about 1% extract may be particularly effective in treating skin viral infections. The different pharmacological activities of *M. officinalis* are mainly attributed to its essential oils, especially the constituents rosmarinic acid, citral, and citronellal. However, further clinical studies need to be carried out to optimize dosage to treat different ailments, as many of the studies had identified larger dosages as being effective but which may not be suitable for its usage in humans, and these gaps have to be streamlined for appropriate pharmacological actions with advisable dosage form. In addition, the mechanism of action at the molecular level is not well established, where it has to be synergistically validated with intracellular biological actions and signaling of the compounds in targeting specific cells. Furthermore, safety and clinical efficacy from the plant's essential oil compounds have to be streamlined for treating cancers, neurological disorders, mood disorders, etc.

REFERENCES

Abdel Naime, W. A., Fahim, J. R., Fouad, M. A., & Kamel, M. S. (2020, October 1). Botanical studies on the stem and root of Melissa officinalis L. (lemon balm). *Journal of Advanced Biomedical and Pharmaceutical Sciences*, 3(4), 184–189.

Abdelwahed, A., Bouhlel, I., Skandrani, I., Valenti, K., Kadri, M., Guiraud, P., ... Chekir-Ghedira, L. (2007). Study of antimutagenic and antioxidant activities of gallic acid and 1, 2, 3, 4, 6-pentagalloylglucose from Pistacia lentiscus: Confirmation by microarray expression profiling. *Chemico-Biological Interactions*, 165(1), 1–13.

Ainslie, W. (1826). Materia indica vol-I.

Akhondali, Z., Dianat, M., Radan, M. (2015). Induced Arrhythmias in Rats. *Elec phys*, 7(1), 971.

Alijaniha, F., Naseri, M., Afsharypuor, S., Fallahi, F., Noorbala, A., Mosaddegh, M., ... Sadrai, S. (2015). Heart palpitation relief with Melissa officinalis leaf extract: Double blind, randomized, placebo controlled trial of efficacy and safety. *Journal of Ethnopharmacology*, 164, 378–384.

Alina, V. R., Muresan, C., Muste, S., Muresan, V., Petrut, G., & Muresan, A. E. (2016). A comprehensive review about therapeutic qualities of essential oil lemon balm (Melissa officinalis L.). *Hop and Medicinal Plants*, 1(2), 16–23.

Astani, A., Heidary Navid, M., & Schnitzler, P. (2014). Attachment and penetration of acyclovir-resistant herpes simplex virus are inhibited by Melissa officinalis extract. *Phytotherapy Research*, 28(10), 1547–1552.

Allahverdiyev, A., Duran, N., Ozguven, M. (2004). Antiviral activity of the volatile oils of Melissa officinalis L. against Herpes simplex virus type-2. *Phytomedicine*, 11, 657–661.

Awad, R., Muhammad, A., Durst, T., Trudeau, V. L., & Arnason, J. T. (2009). Bioassay-guided fractionation of lemon balm (Melissa officinalis L.) using an in vitro measure of GABA transaminase activity. *Phytotherapy Research: An International Journal Devoted to Pharmacological and Toxicological Evaluation of Natural Product Derivatives*, 23(8), 1075–1081.

Baitar, I. (1985). *Al-Jameul Mufridat al Adviawa Al-Aghzia*. (Urdu Translation), CCRUM, New Delhi, Vol. I, pp. 185, 187. (m.o).

Bayat, M., Azami Tameh, A., Hossein Ghahremani, M., Akbari, M., Mehr, S. E., Khanavi, M., & Hassanzadeh, G. (2012). Neuroprotective properties of Melissa officinalis after hypoxic-ischemic injury both in vitro and in vivo. *DARU: Journal of Pharmaceutical Sciences*, 20(1), 1–10.

Belal Al-Husein, M. S., Abdalla, M., Deremer, D. L., & Somanath, P. R. (2012). Antiangiogenic therapy for cancer: An update. *Pharmacotherapy*, 32(12), 1095–1111.

Bhat, J.U., Nizami, Q., Aslam, M., et al., 2012. Antiepileptic activity of the whole plant extract of Melissa officinalis in swiss albino mice. *International Journal of Pharmaceutical Sciences and Research*, 3, 886–889.

Bhattacharjee, S. K. (2004). *Handbook of Aromatic Plants.* Pointer Publishers, Jaipur, pp. 283–284.

Bhattacharjee, S. K. & De, L. C. (2005). *Medicinal Herbs and Flowers.* Aavishkar Publishers, Distributors Jaipur, India, pp.265.

Bounihi, A., Hajjaj, G., Alnamer, R., Cherrah, Y., & Zellou, A. (2013). In vivo potential anti-inflammatory activity of Melissa officinalis L. essential oil. *Advances in Pharmacological Sciences*, 101759. doi.org/10.1155/2013/101759.

Bolkent, S., Yanardag, R., Karabulut-Bulan, O. (2005). Protective role of Melissa officinalis L. extract on liver of hyperlipidemic rats: a morphological and biochemical study. *Journal of Ethnopharmacology*, 99, 391–398.

Braidy, N., Matin, A., Rossi, F., Chinain, M., Laurent, D., & Guillemin, G. J. (2014). Neuroprotective effects of rosmarinic acid on ciguatoxin in primary human neurons. *Neurotoxicity Research*, 25(2), 226–234.

Čanadanović-Brunet, J., Ćetković, G., Djilas, S., Tumbas, V., Bogdanović, G., Mandić, A., ... Čanadanović, V. (2008). Radical scavenging, antibacterial, and antiproliferative activities of Melissa officinalis L. extracts. *Journal of Medicinal Food*, 11(1), 133–143.

Carnat, A., Fraisse, D., & Lamaison, J. L. (1998). The aromatic and polyphenolic composition of lemon balm (Melissa officinalis L. subsp. officinalis) tea. *Pharmaceutics Acta Helvetiae*, 72, 301–305.

Carocho, M., Barros, L., Calhelha, R. C., Ćirić, A., Soković, M., Santos-Buelga, C., ... Ferreira, I. C. (2015). Melissa officinalis L. decoctions as functional beverages: A bioactive approach and chemical characterization. *Food and Function*, 6(7), 2240–2248.

Carvalho, N. C. D., Corrêa-Angeloni, M. J. F., Leffa, D. D., Moreira, J., Nicolau, V., Amaral, P. D. A., ... Andrade, V. M. D. (2011). Evaluation of the genotoxic and antigenotoxic potential of Melissa officinalis in mice. *Genetics and Molecular Biology*, 34(2), 290–297.

Cases, J., Ibarra, A., Feuillère, N., Roller, M., & Sukkar, S. G. (2011, January 1). Pilot trial of Melissa officinalis L. leaf extract in the treatment of volunteers suffering from mild-to-moderate anxiety disorders and sleep disturbances. *Mediterranean Journal of Nutrition and Metabolism*, 4(3), 211–218.

Castellano, J. M., Guinda, A., Delgado, T., Rada, M., & Cayuela, J. A. (2013). Biochemical basis of the antidiabetic activity of oleanolic acid and related pentacyclic triterpenes. *Diabetes*, 62(6), 1791–1799.

Changizi-Ashtiyani, S., Zarei, A., Taheri, S., Rezaei, A., Golshan, M., & Ghafarzadegan, R. (2013). A comparative study of hypolipidemic activities of the extracts of Melissa officinalis and Berberis vulgaris in rats. *Journal of Medicinal Plants*, 12(47), 38–47.

Dastmalchi, K., Dorman, H. D., Oinonen, P. P., Darwis, Y., Laakso, I., & Hiltunen, R. (2008). Chemical composition and in vitro antioxidative activity of a lemon balm (Melissa officinalis L.) extract. *LWT-Food Science and Technology*, 41(3), 391–400.

De Oliveira, N. C., Sarmento, M. S., Nunes, E. A., Porto, C. M., Rosa, D. P., Bona, S. R., ... Da Silva, J. (2012). Rosmarinic acid as a protective agent against genotoxicity of ethanol in mice. *Food and Chemical Toxicology*, 50(5), 1208–1214.

Dymock, W., Warden, C. J. H., & Hooper, D. (1893). *PharmacographiaIndica.* The Institute of Health and Tibbi Research, Hamdard National Foundation, Pakistan, Vol. III, p. 117.

Ebrahimnezhad Darzi, S., Amirghofran, Z. (2013). Dichloromethane fraction of Melissa officinalis induces apoptosis by activation of intrinsic and extrinsic pathways in human leukemia cell lines. *Immunopharmacology and Immunotoxicology*, 35, 313–320

Elnakish, M.T., Hassanain, H.H., Janssen, P.M. (2013). Emerging role of oxidative stress in metabolic syndrome and cardiovascular diseases: important role of Rac/NADPH oxidase. *Journal of Pathology*, 231, 290–300.

Emamghoreishi, M., & Talebianpour, M. S. (2009). Antidepressant effect of Melissa officinalis in the forced swimming test. *DARU: Journal of Pharmaceutical Sciences*, 17(1), 42–47.

Encalada, M. A., Hoyos, K. M., Rehecho, S., Berasategi, I., de Ciriano, M. G. Í., Ansorena, D., & Calvo, M. I. (2011). Anti-proliferative effect of Melissa officinalis on human colon cancer cell line. *Plant Foods for Human Nutrition*, 66(4), 328–334.

Evans, W. C. (2009). *Trease and Evans Pharmacognosy.* 16th ed., WB Saunders Elsevier Ltd, London, pp. 110, 491, 495.

Felklova, L., & Natherova, K. (1969). Duskova. *Ceskoslovenska Farmacie*, 18(9), 457–460.

Fotakis, G., & Timbrell, J. A. (2006). In vitro cytotoxicity assays: Comparison of LDH, neutral red, MTT and protein assay in hepatoma cell lines following exposure to cadmium chloride. *Toxicology Letters*, 160(2), 171–177.

Gazola, R., Machado, D., Ruggiero, C. (2004). Lippia alba, Melissa officinalis and Cymbopogon citratus: effects of the aqueous extracts on the isolated hearts of rats. *Pharmacology Research,* 50, 477–480.

Geuenich, S., Goffinet, C., Venzke, S., Nolkemper, S., Baumann, I., Plinkert, P., et al. (2008). Aqueous extracts from peppermint, sage and lemon balm leaves display potent anti-HIV-1 activity by increasing the virion density. *Retrovirology*, 5, 27.

Ghani, M. N. (1921). *Khazain ul Advia*. Sheikh Mohammad Bashir and Sons, Urdu, Lahore, Pakistan, p. 768.

Ghani, N. (2010). *Khazainul Advia*. CCRUM, New Delhi, Vol. II, p. 251.

Gorgich, E., Komeili, G., Zakeri, Z., & Ebrahimi, S. (2012). *Comparing Anticonvulsive Effect of Melissa officinalis Hydro-alcoholic Extract and Phenytoin in Rat. Journal of Health Scope.* 1(1):44–8

Gromball, J., Beschorner, F., Wantzen, C., Paulsen, U., & Burkart, M. (2014). Hyperactivity, concentration difficulties and impulsiveness improve during seven weeks' treatment with valerian root and lemon balm extracts in primary school children. *Phytomedicine*, 21(8), 1098–1103.

Guginski, G., Luiz, A. P., Silva, M. D., Massaro, M., Martins, D. F., Chaves, J., & Santos, A. R. (2009). Mechanisms involved in the antinociception caused by ethanolic extract obtained from the leaves of Melissa officinalis (lemon balm) in mice. *Pharmacology, Biochemistry, and Behavior*, 93(1), 10–16.

Harwi, Y. (2008). *Ainul Hayat*. Urdu Translation Ibn sina Academy Aligarh. pp. 158.

Hasanein, P., Riahi, H. (2015). Antinociceptive and antihyperglycemic effects of Melissa officinalis essential oil in an experimental model of diabetes. *Med. Princ. Pract.* 24, 47–52.

Hassanzadeh, G., Pasbakhsh, P., Akbari, M., Shokri, S., Ghahremani, M.,Amin, G., et al. (2011).Neuroprotective properties of melissa officinalis L. Extract against ecstasy-induced neurotoxicity. *Cell Journal*, 13(1), 25–30.

Hong, Y., Kim, M. Y., & Yoon, M. (2011). The anti-angiogenic herbal extracts Ob-X from Morus alba, Melissa officinalis, and Artemisia capillaris suppresses adipogenesis in 3T3-L1 adipocytes. *Pharmaceutical Biology*, 49(8), 775–783.

Hussain, M. M. (1855). *Tohfatul Momineen (Persian)*. Matbah Husaini Publicaion, New Delhi, pp. 43–44.

Hussain, A.I., Anwar, F., Nigam, P.S. (2011). Antibacterial activity of some Lamiaceae essential oils using resazurin as an indicator of cell growth. *LWT – Food Science and Technology,* 44, 1199–1206.

Ibarra, A., Feuillere, N., Roller, M., Lesburgere, E., & Beracochea, D. (2010). Effects of chronic administration of Melissa officinalis L. extract on anxiety-like reactivity and on circadian and exploratory activities in mice. *Phytomedicine*, 17(6), 397–403.

Jang, S. M., Yee, S. T., Choi, J., Choi, M. S., Do, G. M., Jeon, S. M., ... Lee, M. K. (2009). Ursolic acid enhances the cellular immune system and pancreatic β-cell function in streptozotocin-induced diabetic mice fed a high-fat diet. *International Immunopharmacology*, 9(1), 113–119.

Javadi, B., Iranshahy, M., & Emami, S. A. (2015). Anticancer plants in Islamic traditional medicine. *Complementary Therapies for the Body, Mind and Soul*, 111–144.

Joukar, S., Asadipour, H., Sheibani, M., Najafipour, H., Dabiri, S., (2016).The effects of Melissa officinalis (lemon balm) pretreatment on the resistance of the heart to myocardial injury. *Pharmaceutical Biology*, 54(6), 1005–1013, DOI: 10.3109/13880209.2015.1091845.

Joukar, S., Zarisfi, Z., Sepehri, G., & Bashiri, A. (2014). Efficacy of Melissa officinalis in suppressing ventricular arrhythmias following ischemia-reperfusion of the heart: a comparison with amiodarone. *Medical Principles and Practice*, 23(4), 340–345.

Jun, H. J., Lee, J. H., Jia, Y., Hoang, M. H., Byun, H., Kim, K. H., & Lee, S. J. (2012). Melissa officinalis essential oil reduces plasma triglycerides in human apolipoprotein E2 transgenic mice by inhibiting sterol regulatory element-binding protein-1c-dependent fatty acid synthesis. *Journal of Nutrition*, 142(3), 432–440.

Karthik, D., Arunkumar, E., Viswanathan, P. (2012). Rosmarinic acid treatment alleviates fibrotic changes in the myocardium induced in a rat model of insulin resistance. *Asian Pacific Journal of Tropical Diseases,* 2, S920–S926.

Khan, A. (2012).*Muheet-e-Azam*. CCRUM, New Delhi, Vol. I, pp. 528–530 (a.g).

Khayyal, M. T., El-Ghazaly, M. A., Kenawy, S. A., Seif-El-Nasr, M., Mahran, L. G., Kafafi, Y., & Okpanyi, S. N. (2000). Antiulcerogenic effect of some gastrointestinally acting plant extracts and their combination. *Arzneimittel-forschung*, 51(7), 545–553.

Khory, H. K., & Katrak, N. N. (1985). *Materia Medica of India and Therapeutics*. 3rd ed., Neeraj Publishing House, New Delhi, p. 484 (m.o).

Kihara, T., Shimohama, S., Sawada, H., Kimura, J., Kume, T., Kochiyama, H. ... Akaike, A. (1997). Nicotinic receptor stimulation protects neurons against β -amyloid toxicity. *Annals of Neurology*, 42(2), 159–163.

Koch-Heitzmann, I., & Czygan, F. (1985). Untersuchungen an Kalluskulturen von Melissa officinalis L.2[40c]:13–20.

Kucera, L. S., & Herrmann, E. C. (1967). Proceedings of the Society for Experimental Biology and Medicine. *Society for Experimental Biology and Medicine*, 124(3), 865–869.

Lee, E. K., Kim, Y. J., Kim, J. Y., Song, H. B., & Yu, H. G. (2014). Melissa officinalis extract inhibits laser-induced choroidal neovascularization in a rat model. *PLOS ONE*, 9(10), e110109.

Leung, A. Y., & Foster, S. (1996). *Encyclopedia of Common Natural Ingredients Used in Food, Drugs, and Cosmetics*. John Wiley & Sons, Inc, New York.

Lin, S. H., Chou, M. L., Chen, W. C., Lai, Y. S., Lu, K. H., Hao, C. W., & Sheen, L. Y. (2015). A medicinal herb, Melissa officinalis L. ameliorates depressive-like behavior of rats in the forced swimming test via regulating the serotonergic neurotransmitter. *Journal of Ethnopharmacology*, 175, 266–272.

Liu, Q., & Zhao, B. (2004). Nicotine attenuates β -amyloid peptide-induced neurotoxicity, free radical and calcium accumulation in hippocampal neuronal cultures. *British Journal of Pharmacology*, 141(4), 746–754.

Lopez, V., Martin, S., Gomez-Serranillos, M. P., Carretero, M. E., Jager, A. K., & Calvo, M. I. (2009). Neuroprotective and neurological properties of Melissa officinalis. *Neurochemical Research*, 34(11), 1955–1961.

Mahita, M., Abuhamdah, R., Howes, M.-J., et al., 2014. Identification of a novel GABAA receptor channel ligand derived from Melissa officinalis and Lavandula angustifolia essential oils. *European Journal of Medicinal Plants*, 4(7), 810.

Mazzanti, G., Battinelli, L., Pompeo, C., Serrilli, A. M., Rossi, R., Sauzullo, I., et al. (2008). Inhibitory activity of Melissa officinalis L. extract on Herpes simplex virus type 2 replication. *Natural Product Research*, 22(16), 1433–1440.

Meftahizade, H., Sargsyan, E., & Moradkhani, H. (2010). Investigation of antioxidant capacity of Melissa officinalis L. essential oils. *Journal of Medicinal Plants Research*, 4(14), 1391–1395.

Melo, A., Monteiro, L., Lima, R.M. (2011). Oxidative stress in neurodegenerative diseases: mechanisms and therapeutic perspectives. *Oxidative Medicine and Cellular Longevity*. 1–14.

Mikus, J., Harkenthal, M., Steverding, D., & Reichling, J. (2000). *Planta Medica*, 66(4), 366–368.

Mimica-Dukic, N., Bozin, B., Sokovic, M., & Simin, N. (2004). Antimicrobial and antioxidant activities of Melissa officinalis L. (Lamiaceae) essential oil. *Journal of Agriculture and Food Chemistry*, 52(9), 2485–2489.

Misharina, T. A., & Samusenko, A. L. (2008). Antioxidant properties of essential oils from lemon, grapefruit, coriander, clove, and their mixtures. *Applied Biochemistry and Microbiology*, 44(4), 438–442.

Moradkhani, H., Sargsyan, E., Bibak, H., Naseri, B., Sadat-Hosseini, M., Fayazi-Barjin, A., & Meftahizade, H. (2010). Melissa officinalis L., a valuable medicine plant: A review. *Journal of Medicinal Plants Research*, 4(25), 2753–2759.

Mrlianová, M., Tekel'ová, D., Felklová, M., Reinöhl, V., Tóth, J. (2002). The influence of the harvest cut height on the quality of the herbal drugs Melissae folium and Melissae herba. *Planta Medica*, 68(2), 178–180.

Mulkens, A., & Kapetanidis, I. (1988). Eugenylglucoside, a new natural phenylpropanoid heteroside from Melissa officinalis. *Journal of Natural Products*, 51(3), 496–498.

Parisa, H., & Hassan, R. (2015). *Antinociceptive and Antihyperglycemic Effects of Melissa officinalis Essential Oil in an Experimental Model of Diabetes. Medical principles and practice : international journal of the Kuwait University, Health Science Centre*, 24(1), 47–52. https://doi.org/10.1159/000368755

Patora, J., Majda, T., Góra, J., & Klimek, B. (2003). Variability in the content and composition of essential oil from lemon balm (Melissa officinalis L.) cultivated in Poland. *Acta Poloniae Pharmaceutica*, 60(5), 395–400.

Pouyanfar, E., Hadian, J., Akbarzade, M., Hatami, M., Kanani, M. R., & Ghorbanpour, M. (2018). Analysis of phytochemical and morphological variability in different wild-and agro-ecotypic populations of Melissa officinalis L. growing in northern habitats of Iran. *Industrial Crops and Products*, 112, 262–273.

Queiroz, R. M. D., Takiya, C. M., Guimarães, L. P. T. P., Rocha, G. D. G., Alviano, D. S., Blank, A. F., … Gattass, C. R. (2014). Apoptosis-inducing effects of Melissa officinalis L. essential oil in glioblastoma multiforme cells. *Cancer Investigation*, 32(6), 226–235.

Rains, J.L., Jain, S.K. (2011). Oxidative stress, insulin signaling, and diabetes. *Free Radical Biology and Medicine*, 50, 67–75.

Ruberto, G., & Baratta, M. T. (2000). Antioxidant activity of selected essential oil components in two lipid model systems. *Food Chemistry*, 69(2), 167–174.

Savino, F., Cresi, F., Castagno, E. (2005). A randomized double-blind placebo-controlled trial of a standardized extract of Matricariae recutita, Foeniculum vulgare and Melissa officinalis (ColiMil) in the treatment of breastfed colicky infants. *Phytotherapy Research,* 19, 335–340.

Sadraei, H., Ghannadi, A., & Malekshahi, K. (2003). Relaxant effect of essential oil of Melissa officinalis and citral on rat ileum contractions. *Fitoterapia*, 74(5), 445–452.

Sarer, E., & Kökdil, G. (1991). Constituents of the essential oil from Melissa officinalis. *Planta Medica*, 57(1), 89–90.

Sawicka, B., Pszczółkowski, P., Barbas, P., & Bienia, B. (2020). Botanical, geographical, herbicidal aspects of lemon balm (Melissa officinalis L.). *Acta Scientific Agriculture*, 4(5), 34–39.

Schnitzler, P., Schuhmacher, A., Astani, A., & Reichling, J. (2008). Melissa officinalis oil affects infectivity of enveloped herpesviruses. *Phytomedicine*, 15(9), 734–740.

Sepand, M. R., Soodi, M., Hajimehdipoor, H., Soleimani, M., & Sahraei, E. (2013). Comparison of neuroprotective effects of Melissa officinalis total extract and its acidic and non-acidic fractions against a β-induced toxicity. *Iranian Journal of Pharmaceutical Research: IJPR*, 12(2), 415.

Sharifi-Rad, M., Berkay Yılmaz, Y., Antika, G., Salehi, B., Tumer, T. B., Kulandaisamy Venil, C., ... Sharifi-Rad, J. (2021, January). Phytochemical constituents, biological activities, and health-promoting effects of the genus Origanum. *Phytotherapy Research*, 35(1), 95–121.

Slemmer, J. E., Shacka, J. J., Sweeney, M. I., & Weber, J. T. (2008). Antioxidants and free radical scavengers for the treatment of stroke, traumatic brain injury and aging. *Current Medicinal Chemistry*, 15(4), 404–414.

Soltanpour, A., Alijaniha, F., Naseri, M., Kazemnejad, A., & Heidari, M. R. (2019). Effects of Melissa officinalis on anxiety and sleep quality in patients undergoing coronary artery bypass surgery: A double-blind randomized placebo controlled trial. *European Journal of Integrative Medicine*, 28, 27–32.

Somova, L.O., Nadar, A., Rammanan, P. (2003). Cardiovascular, antihyperlipidemic and antioxidant effects of oleanolic and ursolic acids in experimental hypertension. *Phytomedicine,* 10, 115–121.

Somova, L.I., Shode, F.O., Mipando, M. (2004). Cardiotonic and antidysrhythmic effects of oleanolic and ursolic acids, methyl maslinate and uvaol. *Phytomedicine,* 11, 121–129.

Soodi, M., Naghdi, N., Hajimehdipoor, H., Choopani, S., & Sahraei, E. (2014). Memory-improving activity of Melissa officinalis extract in naïve and scopolamine-treated rats. *Research in Pharmaceutical Sciences*, 9(2), 107.

Taiwo, A. E., Leite, F. B., Lucena, G. M., Barros, M., Silveira, D., Silva, M. V., & Ferreira, V. M. (2012). Anxiolytic and antidepressant-like effects of Melissa officinalis (lemon balm) extract in rats: Influence of administration and gender. *Indian Journal of Pharmacology*, 44(2), 189.

Takeda, H., Tsuji, M., Inazu, M., Egashira, T., & Matsumiya, T. (2002). Rosmarinic acid and caffeic acid produce antidepressive-like effect in the forced swimming test in mice. *European Journal of Pharmacology*, 449(3), 261–267.

Tittel, G., Wagner, H., & Bos, R. (1982). Chemical composition of the essential oil from Melissa. *Planta Medica*, 46(2), 91–98.

Triantaphyllou, K., Blekas, G., & Boskou, D. (2001). Antioxidant properties of hydro extracts obtained from the species Lamiaceae. *International Journal of Food Sciences and Nutrition*, 52(4), 313–317.

Wake, G., Court, J., Pickering, A., Lewis, R., Wilkins, R., & Perry, E. (2000). CNS acetylcholine receptor activity in European medicinal plants traditionally used to improve failing memory. *Journal of Ethnopharmacology*, 69(2), 105–114.

Weidner, C., Wowro, S.J., Freiwald, A. (2014). Lemon balm extract causes potent antihyperglycemic and antihyperlipidemic effects in insulin-resistant obese mice. *Mol. Nutr. Food Res.* 58, 903–907.

Weidner, C., Rousseau, M., Plauth, A., Wowro, S. J., Fischer, C., Abdel-Aziz, H., & Sauer, S. (2015). Melissa officinalis extract induces apoptosis and inhibits proliferation in colon cancer cells through formation of reactive oxygen species. *Phytomedicine*, 22(2), 26270.

Wolbling, R.H., Leonhardt, K. (1994). Local therapy of herpes simplex with dried extract from Melissa officinalis. *Phytomedicine*, 1, 25–31.

Yoon, M., & Kim, M. Y. (2011). The anti-angiogenic herbal composition Ob-X from Morus alba, Melissa officinalis, and Artemisia capillaris regulates obesity in genetically obese ob/ob mice. *Pharmaceutical Biology*, 49(6), 614–619.

Zeraatpishe, A., Oryan, S., Bagheri, M. H., Pilevarian, A. A., Malekirad, A. A., Baeeri, M., & Abdollahi, M. (2011). Effects of Melissa officinalis L. on oxidative status and DNA damage in subjects exposed to long-term low-dose ionizing radiation. *Toxicology and Industrial Health*, 27(3), 205–212.

Ziakova, A., Brandsteterova, E., & Blahova, E. (2003). Matrix solid-phase dispersion for the liquid chromatographic determination of phenolic acids in Melissa officinalis. *Journal of Chromatography*, 983(1–2), 271–275.

14 Plants as Biomedicine for Skin Diseases in Pakistan

Khafsa Malik, Mushtaq Ahmad, Zabta Khan Shinwari, Muhammad Zafar, Neelam Rashid, and Shazia Sultana

CONTENTS

14.1 INTRODUCTION

Medicinal flora plays a significant role in the healthcare sector all over the globe. About 50,000 plant species are used as medicinal plants around the world (Schippmann et al., 2002). Currently, about 20% of the world's people use plants and plant products for medication purpose (Butler, 2004). WHO stated that, for basic health, 80% of people around the globe, especially the people of rural areas, depend on therapeutic plants for their medication (WHO, 2003). Plant-based phytochemicals are known to have anti-allergic, anti-inflammatory, antimicrobial, hepato-protective, anti-carcinogenic, and antioxidant actions (Malik et al., 2005). The method of preparation and administration of herbal products differ from region to region (Balemie et al., 2004; Saxena et al., 2014).

Many African countries, such as Tanzania, Morocco, Benin, Ghana and Sierra Leone, are the best examples for the market-oriented commercial cultivation of medicinal plants to meet the increasing demands for herbal medicine. However, more than one-third of 263 medicinal plant products are reportedly used in religious and social rituals. Many medicinal plants have also been ancient beliefs (Gurib-Fakim, 2006). In the Western nations, natural drugs are becoming preferred now over synthetic drugs for treatment of diseases. Significantly, dementia has been treated successfully by herbal medication.

At present, Indigenous medicine is also gaining popularity in the developed nations. Percentage of the population using herbal medicine is 42% in Canada, 38% in Belgium, 70% in Australia, and 75% in France (Cavero et al., 2013). WHO (2002) has specified the role of folk drugs by signifying their quality and efficacy through ethnomedicine.

In South Asia, people depend mainly on medicinal plants for managing their ailments through ethnomedicine or the local traditional systems of medicine (Atique et al., 1985; Anis and Iqbal, 1986, 1994; Pattanayak et al., 2003; Parveen et al., 2020a). In Asia, about 15,000 plant species are used in the healthcare system, 8,000 species are found in China and 7,000 in China. Around 70–85% of people of Nepal are dependent on therapeutic flora for their health (Manandhar, 2002).

In Pakistan, about 6,000 plant species are present, of which some 600 species are considered to be medicinally important (Hamayun, 2003), and most of them are used in curing numerous human diseases (Abbasi et al., 2010b). About 80% of the people living in rural areas of Pakistan depend on the Unani therapeutic system, which is based primarily on medicinal plants and their products (Khan et al., 2012; Parveen et al. 2020a, 2020b). In Pakistan there are nine major ecological zones, which are rich in plant diversity (Hamayun, 2003), and people residing in these areas use the local medicinal flora for treating most of their ailments (Ahmad et al., 2009).

DOI: 10.1201/9781003226925-14

Though much research has been conducted in Pakistan on the usage of medicinal plant species for curing various ailments, information pertaining to treatment of skin disorders by herbal medicine is scanty and dispersed. This chapter is an effort to fill this gap.

14.2 SKIN DISEASES—HERBAL TREATMENT

Diseases of skin affect individuals of different age groups and cause damage to human health in a number of ways (Grice et al., 2009). Skin disorders constitute 34% of all related ailments (Saikia et al., 2006). Skin infections are very common in rural people of world. Dermal diseases, like ringworm, skin disorder, skin allergy, eczema, wound, swelling, boils, dermatitis, leprosy and itching, arise because of microbial attacks and unfavorable environmental factors (Suresh et al., 1995).

In developed and less-developed countries, skin ailments present a main health concern and represent a high proportion of all ailments encountered (Abbasi et al., 2010a; Spiewak, 2000). Most of the patients in Pakistan face eczema, pigment disorders, psoriasis, fungal infections, and urticaria (Neamsuvan et al., 2015). The modern treatment of skin ailments depends on the type of skin disease to be treated. The diseases of skin mostly caused by bacterial and fungal attacks are cured by antibiotics like clotrimazole, tolnaftate, and gentamicin. However, current allopathic drug therapies are costly and are expensive to use. Thus, herbal medicine is used as an extremely necessary approach for curing these ailments by local vegetation of the area. Globally, people depend quite effectively on traditional herbal medicine in the local healthcare system (Sullivan and Shealy, 1997). WHO also documented the uses of therapeutic plants by the Indigenous people from numerous areas of the ecosphere.

Globally, different studies have reported on the practice of medicinal important plants in the treatment of dermal diseases (Egharevba and Ikhatua, 2008; Gonzalez-Aspajo et al., 2015; Njoroge and Bussmann, 2007; Omwenga et al., 2015; Süntar et al., 2012).

Currently in Pakistan, the Ministry of Health is supporting the use of medicinal plants in their healthcare system. Less work has been carried out on medicinal plant usage for skin diseases in Pakistan, with no specific review performed on skin ailments (Abbasi et al., 2010a; Mughal et al., 2013). In this regard, this chapter represents the first documentation of medicinal information of herbal drugs used in Pakistan for curing skin ailments.

The following is the list of plants that are used to cure skin ailments in Pakistan:

1. ***Adiantum venustum* D. Don**

   Family: Pteridaceae; **Synonym:** *Adiantum venustum* var. *breviserratum* Ching; **English name:** Pata, kakwa; **Flowering period:** April–June; **Altitude:** 300–3,600 m; **Life Form:** Herb; **Habitat:** Raised rocks and soils that are moist and have sufficient quantities of humus and dead leaves; **Phytochemicals:** Normethylpentacyclic, lanostane-type triterpenes, flavonoids, steroids, lipids, long-chain compounds, triterpenoids, phenylpropanoids, and alicyclic acids (Alam et al., 2000; Pan et al., 2011); **Toxicity:** Nontoxic, acute toxicity (Mannan et al., 2008; Tapwal et al., 2011); **Part Used:** Flowers, leaves; **Mode of Utilization:** Decoction, extract, **Disease Category:** Athlete's foot, skin scars.

   Recipes: (a) 15 g dried flowers are boiled with 300 ml water then the sieved mixture is used for athlete's foot for one week twice a day. **For children (5–15 years):** 100 ml infusion is used at evening and morning for three weeks for treatment of athlete's foot. **For adults (16–50 years):** 120 ml infusion material are taken three times a day for one week to treat athlete's foot. **For children (5–15 years):** 80 ml warm extract are taken by children each day for eight days for treating dermal scars. **For adults (16–50 years):** 100 ml hot extract are taken twice a day for nine days for patients who are suffering from skin scars (Figure 14.1).

2. ***Ajuga integrifolia* Buch.-Ham. ex D. Don**

   Family: Lamiaceae; **Synonym:** *Ajuga integrifolia* var. *canescens* (Benth.) Cufod. **English name:** Bugle weed; **Flowering period:** April–June; **Altitude:** 1,500 m; **Life**

FIGURE 14.1 *Adiantum venustum.*

Form: Herb; **Habitat:** Moist ground and grassy slopes; Phytochemicals: *Neo*-clerodane diterpenoids, ergosterol-5,8-endoperoxide, ajugarin-I, 8-*O*-acetylharpagide and several phytoecdysteroids (Cocquyt et al., 2011; Coll and Tandrón, 2008); **Toxicity:** Cytotoxicity and phytotoxicity (Rehman et al., 2015; **Part Used:** Flower, Leaf; **Mode of Utilization:** Powder, decoction; **Disease Category:** Acne, skin problems.

Recipes: (**a**) 20 g dried flowers of plant are ground into finely powdered form and one tablespoon is taken daily with water to treat acne. **For children (5–15 years):** One teaspoon powder is taken with 100 ml hot water for 13–15 days to treat acne. **For adults (16–50 years):** One tablespoon powder is given before mealtime along with 250 ml water for 26 days. (**b**) Leaves (25–30 g) of *Ajuga integrifolia* are mixed in 300 ml of water to make a decoction. **For children (5–15 years):** 100 ml decoction is taken for 10–15 days daily for skin diseases. **For adults (16–50 years):** 150 ml drink is given twice a day for 10–15 days to treat skin problem (Figure 14.2).

3. ***Anaphalis margaritacea* (L.) Benth. & Hook. f.**

Family: Asteraceae; **Synonym:** *Anaphalis margaritacea* subsp. *angustior* (Miq.) Kitam. **English name:** Pearly everlasting; **Flowering period:** July-September; **Altitude:** 1,000 m, **Life Form:** Herb, **Habitat:** Grows best in dry conditions, including open pastures, roadsides, and waste areas; **Phytochemicals:** Flavonoids, sitosterol-3-*O*-β-glucopyranoside,

FIGURE 14.2 *Ajuga integrifolia.*

scopoletin, and triterpenoids (Ahmed et al., 2004; Ren et al., 2009); **Toxicity:** Cytotoxicity (Murillo-Alvarez et al., 2001); **Part Used:** Flowers, leaves; **Mode of Utilization:** Decoction, extract; **Disease Category:** Skin rashes, skin infection.

Recipes: (a) 25 g dried flowers are boiled in 300 ml water and the filtrate ("infusion") is then used for one week two times daily to treat rashes. **For children (5–15 years):** 100 ml infusion is used in evening and morning to treat rashes for one week. **For adults (16–50 years):** 20 ml infusion is given three times a day for one week to adult patients suffering from rashes. **(b)** 15–20 g leaves are boiled in 400 ml for 10–20 minutes. Extract is strained and used hot in cases of skin infection. **For children (5–15 years):** 80 ml extract are taken by the child for five to seven days for treating skin infection. **For adults (16–50 years):** 100 ml warm extract are taken twice t a day for 5–6 days to patients suffering from dermal infection (Figure 14.3).

4. ***Arisaema jacquemontii* Blume**

Family: Araceae; **Synonym:** *Arisaema cornutum* Schott; **English name:** Jack-in-the-pulpit; **Flowering period:** June–July; **Altitude:** 2,400–4,000 m; **Life Form:** Herb; **Habitat:** In orchards, vineyards, roadside, ditch banks, and cropland; **Phytochemicals:** Tannins, flavonoids and phenolics (Baba and Malik, 2015); **Toxicity:** Phytotoxicity (Tanveer et al., 2014); **Part Used:** Fruits, roots; **Mode of Utilization:** Infusion, extracts; **Disease Category:** Skin disease, boils.

Recipes: (a) 30 g dried fruits are boiled with 350 ml water then filtered. This filtrate or infused drink is used to treat skin diseases for 10 days, once a day. **For children (5–15 years):** 60 ml infusion is used per day for treating skin disorders, for one week. **For adults (16–50 years):** 120 ml infusion given each day for 14 days to patients suffering from skin diseases. **(b)** 15–20 g roots are boiled in water (600 ml) for 5–10 minutes. The extract is filtered and used hot in the case of chronic boils. **For children (5–15 years):** 30 ml extract are taken each day for 3–4 days to treat chronic boils. **For adults (16–50 years):** 60 ml warm extract is given to adults once a day for 5–6 days for boils (Figure 14.4).

5. ***Argyrolobium roseum* (Cambess.) Jaub. & Spach**

Family: Fabaceae; **Synonym:** *Argyrolobium roseum* subsp. *elongatum* Ali; **English name:** Roseum; **Flowering period:** April–October; **Altitude:** 1,550 m; **Life Form:** Herb; **Habitat:** Present in dry places; **Phytochemicals:** Tannins, flavonoids, terpenes and alkaloids, (Khanum et al., 2013); **Toxicity:** Sub-acutely toxic (Ahmed et al., 2008); **Part Used:** Leaf, seed; **Mode of Utilization:** Infusions, paste, **Disease Category:** Skin diseases, wounds.

FIGURE 14.3 *Anaphalis margarita.*

FIGURE 14.4 *Arisaema jacquemontii.*

Recipes: **(a)** 10–15 g of seeds are infused in water for six days. Seeds are filtered and 3–4 seeds are given regularly for 21–22 days for treating skin diseases. **For children (5–15 years):** 3–4 seeds are given with 120 m milk once a day in the morning for two weeks to treat skin diseases. **For adults (16–50 years):** 4–5 seeds with 120 ml milk are taken by patients twice a day for 15–20 days. **(b)** 10–20 g fresh leaves are crushed for 4–5 minutes to make a paste which is applied in the case of bleeding or wounds. **For children (5–15 years):** 2–3 g paste is applied to the skin to treat wounds for 3–4 days. **For adults (16–50 years):** 2–3 g paste is used 3–4 times per day for treating wounds (Figure 14.5).

6. ***Brassica juncea* (L.) Czern.**

Family: Brassicaceae; **Synonym:** *Brassica juncea* var. *agrestis* Prain; **English name:** Mustard greens; **Flowering period:** March–May; **Altitude:** 1,000 m; **Life Form:** Herb; **Habitat:** Grows in normal habitat; **Phytochemicals:** 24-methylcholesta-5 and E-22-dien-3β-*o* (Matsumoto et al., 1983); **Toxicity:** Heavy metal contamination (Kumar et al., 2008); **Part Used:** Flower, leaf; **Mode of Utilization:** Decoctions, powder; **Disease Category:** Skin problems.

Recipes: **(a)** 15–20 g flowers are boiled then a decoction is used for skin diseases. **For children (5–15 years):** Three tablespoons of liquid decoction is taken by the child twice a day for two weeks for skin problems. **For adults (16–50 years):** 120 ml decoction is given three times a day for three weeks to patients suffering from dermal diseases. **(b)** 20–30 g

FIGURE 14.5 *Argyrolobium roseum.*

FIGURE 14.6 *Brassica juncea.*

of leaf powder is taken for skin problems for a week. **For children (5–15 years):** 10–15 g of leaf powder is added to 60 ml of water and taken once a day for four weeks for skin diseases. **For adults (16–50 years):** 4–5 seeds are chewed directly, or a powder of seeds is mixed in hot water and used for skin diseases (Figure 14.6).

7. ***Butea monosperma* (Lam.) Taub.**

Family: Fabaceae; **Synonym:** *Butea braamania* DC.; **English name:** Forest flame; **Flowering period:** March–April; **Altitude:** 1,500 m; **Life Form:** Tree; **Habitat:** Dry areas; **Phytochemicals:** Butrin, butein, coreopsin, steroids, isocoreopsin, flavonoids, aromatic compounds, sulfurein (glycoside), glucose, glycine, glucosides, leucocyanidin, chalcones, and aurones (Rana and Avijit, 2012); **Toxicity:** Chronic toxicity (Donga et al., 2011); **Part Used:** Leaf, seed; **Mode of Utilization:** Powder, Extracts; **Disease Category:** Insect bites, skin allergy. **Recipes: (a)** 10–20 g leaves are crushed for 8–10 minutes to make a paste, adding a little turmeric. This paste is used for insect bites. **For children (5–15 years):** 2–3 g of paste is used on the skin to treat an insect bite three times a day for 1–2 days. **For adults (16–50 years):** 2–3 grams paste is applied on skin twice a day for four days. **(b)** Two spoons of seed are immersed in 240 ml water and the extract is mixed with sugar, then drunk in the morning for 21 days for treating skin allergy. **For children (5–15 years):**120 ml of extract is taken orally before breakfast in the morning to treat skin allergies. **For adults (16–50 years):** 240 ml extract is given in the morning for 20–25 days (Figure 14.7).

FIGURE 14.7 *Butea monosperma.*

8. ***Carthamus oxyacantha* M. Bieb.**

    Family: Asteraceae; **Synonym:** *Carthamus oxyacantha* subsp. *noeana* Sostak.; **English name:** Wild safflower; **Flowering period:** April-August; **Altitude:** 500m; **Life Form:** Herb; **Habitat:** A noxious weed found among summer-season crops; **Phytochemicals**: Phenolics, terpenoids, ethyl acetate fractions, and alkaloidal skeleton (Raza et al., 2015); **Toxicity:** Hepatoprotective (Bukhsh et al., 2014); **Part Used:** Flower, leaf; **Mode of Utilization:** Powdered form, decoctions; **Disease Category:** Skin diseases.

    Recipes: **(a)** 20 g dried flowers are ground into powder form and one tablespoon is taken with water regularly to treat skin diseases. **For children (5–15 years):** One teaspoon powder is taken with 100 ml hot water each morning for 12 days to treat dermal ailments. **For adults (16–50 years):** One tablespoon powder is taken early morning with 200 ml warm water for 14–15 days regularly. **(b)** 25–30 g leaf are mixed in 300 ml water to make a decoction. **For children (5–15 years):** 80 ml decoction is taken daily for 12 days to cure skin disorders. **For adults (16–50 years):** 150ml) decoction is taken twice a day for 12 days for treating skin disorders (Figure 14.8).

9. ***Cichorium intybus* L.**

    Family: Asteraceae; **Synonym:** *Cichorium intybus* f. *alba* Farw.; **English name:** chicory; **Flowering period:** May to October; **Altitude:** 300–3600 m; **Life Form:** Herb; **Habitat:** Grassy meadow and arable land; **Phytochemicals:** cichorines, cyanidin 3-malonylglucoside,saccharides, flavonoids, essential oils, and methoxycoumarin; (Bridle et al., 1984; Street et al., 2013); **Toxicity:** Metal toxicity (Rozpądek et al., 2014); **Part Used:** Leaf; **Mode of Utilization:** Powder, paste, **Disease Category:** Skin burn, measles.

    Recipes:(a) 30 g dried flowers are ground and 1–2 tablespoons is taken by water regularly for treating candidiasis. **For children (5–15 years):** One teaspoon powder are taken with 100 ml hot water in the morning for twelve days for treating candidiasis. **For adults (16–50 years):** One tablespoon of powder given early in the morning along with 200 ml water for fifteen days regularly. **(b)** 10 g leaves are mixed in 300 ml of water to make a decoction. This liquid is strained and used twice a day for ten days for treating skin ailments. **For children (5–15 years):** 80 ml decoction is taken daily forl0 days for curing skin ailments. **For adults (16–50 years):** 150 ml decoction is taken twice a day forl0 days for treating skin infections (Figure 14.9).

10. ***Centaurea iberica* Trevir. ex Spreng.**

    Recipes: **(a)** 60 g dried leaves are ground to a fine powder and one tablespoon is used with 150 ml of water once a day in the morning for 15 days for treating skin wounds. **For**

FIGURE 14.8 *Carthamus oxyacantha.*

FIGURE 14.9 *Cichorium intybus.*

children (5–15 years): One teaspoon is given daily with 80 ml of water for 10 days for treating skin wounds. **For adults (16–50 years):** 180 ml is taken daily for 12–15 days for the treatment of skin wounds. **(b)** 25–30 g flowers are boiled with 400 ml water. This infusion is then filtered and used daily for 10–12 days to treat skin inflammations. **For children (5–15 years):** 80 ml drink is given daily for 7–8 days to treat skin inflammations. **For adults (16–50 years):** 200 ml of infused drink is taken daily for 12–14 days for treating skin inflammations (Figure 14.10).

11. ***Isodon rugosus (*Wall. ex Benth.)**

Family: Lamiaceae; **Synonym:** *Isodon plectranthoides* Schrad. ex Benth.; **English name:** Wrinkled-leaf isodon; **Flowering period:** March–October; **Altitude:** 1,700 m; **Life Form:** Shrub; **Habitat:** Scrub and savannah slopes at altitudes; **Phytochemicals:** Alkaloids, flavonoids, saponins, oils, glycosides, anthraquinones, terpenoids, sterols, and tannins (Zeb et al., 2014)**; Toxicity:** Cytotoxicity (Zeb et al., 2014); **Part Used:** Leaves; **Mode of Utilization:** Powder, paste; **Disease Category:** Skin burn, measles.

Recipes: (a) 30–40 g of leaves are crushed and a paste is made by adding 400 ml water. This paste is applied onto skin for skin burns. **For children (5–15 years):** Paste is applied for skin burn, for one week. **For adults (16–50 years):** Paste is applied to the patient's skin for two weeks. **(b)** Dried leaves (20–35 g) are cut with a knife, put into a crusher, and ground into a powder. 2–3 tablespoons of powder (8–12 g) is used with 115 ml water for about 8–12 days to treat measles. **For children (5–15 years):** 20 ml of extract is taken each

FIGURE 14.10 *Centaurea iberica.*

FIGURE 14.11 *Isodon rugosus.*

day for 5–6 days for treating measles. **For adults (16–50 years):** 50 ml of extract is taken by adults twice a day for 6–7 days for chronic measles (Figure 14.11).

12. ***Galium aparine* L.**

Family: Rubiaceae; **Synonym:** Not mentioned; **English name:** Cleavers; **Flowering period:** March–July, **Altitude**: 550 m; **Life Form:** Herb; **Habitat:** Coastal sand dune, open woodlands, rainforests, rocky sites, dry coastal vegetation, riparian areas, and wetlands **Phytochemicals**: Anthraquinone, polyphenols, carboxylic acids, aldehydes, sterols, nordamnacanthal, and methoxylated flavones (Mocan et al., 2016; Morimoto et al., 2002); **Toxicity:** Topical effect (Atmaca et al., 2016); **Part Used:** Leaves; **Mode of Utilization:** Powder, paste; **Disease Category:** Skin burns, measles.

Recipes: (a) 30 g dried flowers are boiled in 300 ml water then the filtrate (infusion) is used. **For children (5–15 years):** 100 ml infusion is used in evening and morning for treating boils for a two-week period. **For adults (16–50 years):** 100 ml) infusion is taken daily for one week by the patient suffering from skin diseases. **(b)** 15 g leaves are boiled in 400 ml water then simmered for 25 minutes. The extract is used warm for measles treatment. **For children (5–15 years):** 80 ml warm extracted materia is taken by children for 10 days for treating measles. **For adults (16–50 years):** 100 ml warm extract is taken twice daily for 9–10 days by patients suffering from measles (Figure 14.12).

13. ***Impatiens bicolor* Gilg. L.**

Family: Balsaminaceae; **Synonym:** *Impatiens amphorata* **English name:** Busy Lizzie; **Flowering period:** April to August; **Altitude**: 1500–1780 m; **Life Form:** Herb; **Habitat:** Grow in orchards, vineyards, roadsides, ditch banks, and cropland **Phytochemicals:** Phenolic compounds, flavonoids, triterpenes, carbohydrates, alkaloids, sterols, sugars, sterols, resins, tannins, lactones, proteins, and mucilage (Todd et al., 1995); **Toxicity:** Alkaloidic toxin (Todd et al., 1995); **Part Used:** Whole plants, leaves; **Mode of Utilization:** Powder, infusion; **Disease Category:** Skin infection.

Recipes: (a) 50 g dry complete plant is immersed in 200 ml of water to make an infusion and strained liquid is used twice a day for three weeks to treat skin disorders. **For children (5–15 years):** 100 ml of infused drink is used in the morning for 12 days for treating dermal disorders. **For adults (16–50 years):** 200 ml is taken early in the morning for 15–20 days to cure skin disorders. **(b)** 30–40 g leaves are boiled in 300 ml water for 13 min. This solution is filtered and used for 12 days to treat skin ailments. **For children**

FIGURE 14.12 *Galium aparine.*

FIGURE 14.13 *Impatiens bicolor.*

(5–15 years): 80 ml infused drink is given for 10 days at night to treat skin infections. **For adults (16–50 years):** 180ml) drink is given at night for 10 days to cure skin infections (Figure 14.13).

14. ***Coriandrum sativum L.***

Family: Apiaceae; **Synonym (s):** *Bifora loureiroi* Kostel; **English name:** Coriander; **Flowering period:** March to May; **Altitude:** 1000 m, **Life Form:** Herb; **Habitat:** Its usual habitat is that of fields, or gardens, regardless of the substrate; **Phytochemicals:** Flavonoids, isocoumarins 2-decenoic acid, volatile component, capric acid, and E-11-tetradecenoic acid (Asgarpanah and Kazemivash, 2012); **Toxicity:** Antifungal toxicity (Soares et al., 2012); **Part Used:** Seeds, leaves; **Mode of utilization:** Infusions, extracts; **D–isease Category:** Skin inflammation, dermatitis.

Recipes: (a) 30 g seeds are boiled in 300 ml of water then the decoction is strained. This infusion is used for its anti-inflammatory properties, twice a day for five weeks. **For children (5–15 years):** 100 ml infusion is used for gargling in the evening and morning for one week for treating skin inflammation. **For adults (16–50 years):** 200 ml infusion is given twice a day for one week to patients suffering inflammation for gargling. **(b)** 15–20 g leaves are boiled in 200 ml water for 10–20 minutes and the extract is filtered and used to treat dermatitis. **For children (5–15 years):** 80 ml warm extract is taken by the child twice

FIGURE 14.14 *Coriandrum sativum.*

a day for 5–7 days to treat dermatitis. **For adults (16–50 years):** 250 ml extract is given twice a day for 8–10 days for patients suffering from dermatitis (Figure 14.14).

15. ***Crotalaria medicaginea* Lam.**

Family: Fabaceae; **Synonym:** No synonyms recorded, **English name:** Rattlepods; **Flowering period:** March–August; **Altitude:** 100–1,400 m; **Life Form:** Herb; **Habitat:** Damp grassland, especially in floodplains, depression rivers, deciduous shrubland, roadsides, and fields; **Phytochemicals:** Flavonoids, phenolics, saponins, hexahydroxydiphenic (HHDP) acid and pyrrolizidine alkaloids (Fletcher et al., 2008), **Toxicity:** Toxicity in seeds (Williams and Molyneux, 1987); **Part Used:** Seeds, leaves; **Mode of utilization:** Powder, infusion; **Disease Category:** Skin diseases, wound healing.

Recipes: **(a)** 15–20 g dried seeds are ground for 10–15 minutes to make powder. Two tablespoons of the powdered form is used with water daily in morning to treat skin diseases. **For children (5–15 years):** One teaspoon powdered form is taken with 60 ml cold water in the morning for 10–15 days to treat skin diseases. **For adults (16–50 years):** One tablespoon powder is used with 250 ml water in the evening for 10–15 days regularly. **(b)** 20–30 g leaves are infused in 250 ml water for 1–2 days. Infusion is strained and used once a day at nighttime for seven days to induce wound healing. **For children (5–15 years):** 120 ml infusion is given at night for 5–7 days for treating wounds. **For adults (16–50 years):** 250 ml infusion is given at night for 710 days to treat wounds (Figure 14.15).

FIGURE 14.15 *Crotalaria medicaginea.*

REFERENCES

Abbasi, A.M., Khan, M., Ahmad, M., Zafar, M., Jahan, S., Sultana, S., 2010a. Ethnopharmacological application of medicinal plants to cure skin diseases and in folk cosmetics among the tribal communities of North-West Frontier Province, Pakistan. *Journal of Ethnopharmacology* 128(2), 322–335.

Abbasi, A.M., Khan, M.A., Ahmad, M., Qureshi, R., Arshad, M., Jahan, S., Zafar, M., Sultana, S., 2010b. Ethnobotanical study of wound healing herbs among the tribal communities in Northern Himalaya Ranges District Abbottabad, Pakistan. *Pakistan Journal of Botany* 6, 3747–3753.

Ahmed, A.A., Hussein, T.A., Mahmoud, A.A., Farag, M.A., Paré, P.W., Wojcińska, M., Karchesy, J., Mabry, T.J., 2004. Nor-ent-kaurane diterpenes and hydroxylactones from Antennaria geyeri and Anaphalis margaritacea. *Phytochemistry* 65(18), 2539–2543.

Ahmad, S., Islam, M., Bano, G., Aslam, S., Koukab, S., 2009. Seasonal variation in current season and dead Biomass of Chrysopogon aucheri (boiss) stapf. and Cymbopogon Jwarancusa (Jones) Schult. in Highland Balochistan, Pakistan. *Pakistan Journal of Botany* 41(2), 519–527.

Ahmed, Z., Bhagat, A., Gupta, O.P., Gupta, K.K., Ram, G., Qazi, G.N., 2008. Insulin secretagogue fraction of Argyrolobium roseum. *Diabetologia Croatica* 37, 1–12.

Alam, M.S., Chopra, N., Ali, M., Niwa, M., 2000. Normethyl pentacyclic and lanostane-type triterpenes from Adiantum venustum. *Phytochemistry* 54(2), 215–220.

Anis, M., Iqbal, M., 1986. Antipyretic utility of some Indian plants in traditional medicine. *Fitoterapia* 57, 52–55.

Anis, M., Iqbal, M., 1994. Medicinal plantlore of Aligarh, India. *International Journal of Pharmacognosy* 32, 59–64.

Atique, A., Iqbal, M., Ghouse, A.K.M., 1985. Use of *Annona squamosa* and *Piper nigrum* against diabetes. *Fitoterapia* 56, 190–192.

Baba, S.A., Malik, S.A., 2015. Determination of total phenolic and flavonoid content, antimicrobial and antioxidant activity of a root extract of Arisaema jacquemontii Blume. *Journal of Taibah University for Science* 9(4), 449–454.

Balemie, K., Kelbessa, E., Asfaw, Z., 2004. Indigenous medicinal plant utilization, management and threats in Fentalle area, Eastern Shewa, Ethiopia. *Ethiopian Journal of Biological Sciences* 3(1), 37–58.

Bridle, P., Loeffler, R.T., Timberlake, C.F., Self, R., 1984. Cyanidin 3-malonylglucoside in Cichorium intybus. *Phytochemistry* 23(12), 2968–2969.

Bukhsh, E., Malik, S.A., Ahmad, S.S., Erum, S., 2014. Hepatoprotective and hepatocurative properties of alcoholic extract of Carthamus Oxyacantha seeds. *African Journal of Plant Science* 8(1), 34–41.

Butler, M.S., 2004. The role of natural product chemistry in drug discovery. *Journal of Natural Products* 67(12), 2141–2153.

Cavero, R., Akerreta, S., Calvo, M., 2013. Medicinal plants used for dermatological affections in Navarra and their pharmacological validation. *Journal of Ethnopharmacology* 149(2), 533–542.

Cocquyt, K., Cos, P., Herdewijn, P., Maes, L., Van den Steen, P., Laekeman, G., 2011. Ajuga remota Benth.: From ethnopharmacology to phytomedical perspective in the treatment of malaria. *Phytomedicine* 18(14), 1229–1237.

Coll, J., Tandrón, Y.A., 2008. *neo*-Clerodane diterpenoids from *Ajuga*: Structural elucidation and biological activity. *Phytochemistry Reviews* 7(1), 25.

Donga, S., Shukla, V.J., Ravishankar, B., Ashok, B., Mishtry, I., 2011. Chronic toxicity study of Butea monosperma (Linn.) Kuntze seeds in albino rats. *Ayu* 32(1), 120.

Egharevba, R., Ikhatua, M., 2008. Ethno-medical uses of plants in the treatment of various skin diseases in Ovia North East, Edo State, Nigeria. *Research Journal of Agriculture and Biological Sciences* 4(1), 58–64.

Gonzalez-Aspajo, G., Belkhelfa, H., Haddioui-Hbabi, L., Bourdy, G., Deharo, E., 2015. Sacha Inchi Oil (Plukenetia volubilis L.), effect on adherence of Staphylococus aureus to human skin explant and keratinocytes in vitro. *Journal of Ethnopharmacology* 171, 330–334.

Grice, E.A., Kong, H.H., Conlan, S., Deming, C.B., Davis, J., Young, A.C., Bouffard, G.G., Blakesley, R.W., Murray, P.R., Green, E.D., 2009. Topographical and temporal diversity of the human skin microbiome. *Science* 324(5931), 1190–1192.

Gürbüz, İ., Yesilada, E., 2007. Evaluation of the anti-ulcerogenic effect of sesquiterpene lactones from Centaurea solstitialis L. ssp. solstitialis by using various in vivo and biochemical techniques. *Journal of Ethnopharmacology* 112(2), 284–291.

Gurib-Fakim, A., 2006. Medicinal plants: Traditions of yesterday and drugs of tomorrow. *Molecular Aspects of Medicine* 27(1), 1–93.

Hamayun, M., 2003. Ethnobotanical studies of some useful shrubs and trees of District Buner, NWFP, Pakistan. *Ethnobotanical Leaflets* 2003(1), 12.

Jabbar, A., Zaman, M.A., Iqbal, Z., Yaseen, M., Shamim, A., 2007. Anthelmintic activity of Chenopodium album (L.) and Caesalpinia crista (L.) against trichostrongylid nematodes of sheep. *Journal of Ethnopharmacology* 114(1), 86–91.

Khan, M., Khan, M.A., Mujtaba, G., Hussain, M., 2012. Ethnobotanical study about medicinal plants of Poonch valley Azad Kashmir. *Journal of Animal and Plant Sciences* 22, 493–500.

Khanum, R., Jahangir, M., Abbasi, M.A., Mazhar, F., Kausar, S., Riaz, T., Ajaib, M., 2013. Phytochemical screening and antioxidant evaluations of different fractions of Argyrolobium roseum. *Asian Journal of Chemistry* 25(13).

Kiefer, M., 2004. Review about Ginkgo biloba special extract EGb 761 (Ginkgo). *Current Pharmaceutical Design* 10(3), 261–264.

Kumar, K.V., Singh, N., Behl, H., Srivastava, S., 2008. Influence of plant growth promoting bacteria and its mutant on heavy metal toxicity in Brassica juncea grown in fly ash amended soil. *Chemosphere* 72(4), 678–683.

Malik, F., Hussain, S., Dil, A., Hannan, A., Gilani, A., 2005. Islamic Republic of Pakistan. Chapter 22. In: *WHO Global Atlas of Traditional, Complementary and Alternative Medicine (Map Volume).* WHO, Geneva, 275–283.

Manandhar, N.P., 2002. *Plants and People of Nepal.* Timber Press.

Mannan, M.M., Maridass, M., Victor, B., 2008. A review on the potential uses of ferns. *Ethnobotanical Leaflets* 2008(1), 33.

Matsumoto, T., Shimizu, N., Shigemoto, T., Itoh, T., Iida, T., Nishioka, A., 1983. Isolation of 22-dehydro-campesterol from the seeds of Brassica juncea. *Phytochemistry* 22(3), 789–790.

Mughal, S.B., Arshad, N., Shoaib, M., Irum, N., Hussnain, N., 2013. Ethnobotanical literature survey of plants used to cure skin diseases. *World Applied Sciences Journal* 27(4), 474–478.

Murillo-Alvarez, J.I., Encarnaci6n, D.R., Franzblau, S., 2001. Antimicrobial and cytotoxic activity of some medicinal plants from Baja California Sur (Mexico). *Pharmaceutical Biology* 39(6), 445–449.

Neamsuvan, O., Kama, A., Salaemae, A., Leesen, S., Waedueramae, N., 2015. A survey of herbal formulas for skin diseases from Thailand's three southern border provinces. *Journal of Herbal Medicine* 5(4), 190–198.

Njoroge, G.N., Bussmann, R.W., 2007. Ethnotherapeautic management of skin diseases among the Kikuyus of Central Kenya. *Journal of Ethnopharmacology* 111(2), 303–307.

Omwenga, E., Hensel, A., Shitandi, A., Goycoolea, F., 2015. Ethnobotanical survey of traditionally used medicinal plants for infections of skin, gastrointestinal tract, urinary tract and the oral cavity in Borabu sub-county, Nyamira County, Kenya. *Journal of Ethnopharmacology* 176, 508–514.

Pan, C., Chen, Y., Ma, X., Jiang, J., He, F., Zhang, Y., 2011. Phytochemical constituents and pharmacological activities of plants from the genus Adiantum: A review. *Tropical Journal of Pharmaceutical Research* 10(5), 681–692.

Parveen, A., Parveen, R., Akhatar, A., Parveen, B., Siddiqui, K.M., Iqbal, M., 2020a. Concepts and quality considerations in Unani system of medicine. *Journal of AOAC International* 103(3), 609–633.

Parveen, B., Parveen, A., Parveen, R., Ahmad, S., Ahmad, M., Iqbal, M., 2020b. Challenges and opportunities for traditional herbal medicine today, with special reference to its status in India. *Annals of Phytomedicine* 9(2), 97–112.

Pattanayak, S.K., Sills, E.O., Mehta, A.D., Kramer, R.A., 2003. Local uses of parks: Uncovering patterns of household production from forests of Siberut, Indonesia. *Conservation and Society* 1(2), 209.

Rana, F., Avijit, M., 2012. Review on Butea monosperma. *International Journal of Research in Pharmacy and Chemistry* 2, 1035–1039.

Raza, M.A., Mukhtar, F., Danish, M., 2015. Cuscuta reflexa and Carthamus Oxyacantha: Potent sources of alternative and complimentary drug. *Springerplus* 4(1), 76.

Rehman, N.U., Begum, N., Ali, L., Al-Harrasi, A., Abbas, G., Ahmad, S., Khan, A.L., Shinwari, Z.K., Hussain, J., 2015. Lipid peroxidation, antiglycation, cytotoxic, phytotoxic, antioxidant, antiplatelet and antimicrobial activities of Ajuga bracteosa against various pathogens. *Pakistan Journal of Botany* 47(3), 1195–1197.

Ren, Z.-Y., Wu, Q.-X., Shi, Y.-P., 2009. Flavonoids and triterpenoids from Anaphalis margaritacea. *Chemistry of Natural Compounds* 45(5), 728.

Rozpądek, P., Wężowicz, K., Stojakowska, A., Malarz, J., Surówka, E., Anielska, T., Ważny, R., Miszalski, Z., Turnau, K., 2014. Mycorrhizal fungi modulate phytochemical production and antioxidant activity of Cichorium intybus L. (Asteraceae) under metal toxicity. *Chemosphere* 112, 217–224.

Saikia, A.P., Ryakala, V.K., Sharma, P., Goswami, P., Bora, U., 2006. Ethnobotany of medicinal plants used by Assamese people for various skin ailments and cosmetics. *Journal of Ethnopharmacology* 106(2), 149–157.

Saxena, N., Yadav, V., Verma, R., 2014. Traditional knowledge of medicinal plants used to cure gastro intestinal problems in Jalaun district of Uttar Pradesh, India. *Journal of Medicinal Plants Studies* 2(4), 24–28.

Schippmann, U., Leaman, D.J., Cunningham, A., 2002. Impact of cultivation and gathering of medicinal plants on biodiversity: Global trends and issues. In: *Biodiversity and the Ecosystem Approach in Agriculture, Forestry and Fisheries*. FAO, 142–167.

Spiewak, R., 2000. Occupational skin diseases among farmers. In: *Occupational and Para-Occupational Diseases in Agriculture*. Institute of Agricultural Medicine, Lublin, 42–152.

Street, R.A., Sidana, J., Prinsloo, G., 2013. Cichorium intybus: Traditional uses, phytochemistry, pharmacology, and toxicology. *Evidence-Based Complementary and Alternative Medicine* 2013.

Sullivan, K., Shealy, C.N., 1997. *The Complete Family Guide to Natural Home Remedies*. Barnes & Noble Books.

Süntar, I., Akkol, E.K., Keles, H., Yesilada, E., Sarker, S.D., Baykal, T., 2012. Comparative evaluation of traditional prescriptions from Cichorium intybus L. for wound healing: Stepwise isolation of an active component by in vivo bioassay and its mode of activity. *Journal of Ethnopharmacology* 143(1), 299–309.

Suresh, B., Kalyanaraman, V., Dhanasekaran, S., Annadurai, K., Dhanaraj, S., Balasubramanian, S., 1995. Evaluation of Santolina oil in search of new drugs against candidiasis. *Indian Journal of Pharmacology* 27(3), 171.

Tanveer, M., Habib-Ur-Rehman, M.M., Choudhary, M., 2014. Immunomodulatory, antileishmanial and phytotoxicity of Arisaema jacquemontii Blume plant extracts. *Archives of Applied Science Research* 6(5), 12–17.

Tapwal, A., Garg, S., Gautam, N., Kumar, R., 2011. In vitro antifungal potency of plant extracts against five phytopathogens. *Brazilian Archives of Biology and Technology* 54(6), 1093–1098.

WHO, 2003. *The World Health Report 2003: Shaping the Future*. World Health Organization.

15 Chemical Characterization and Biological Analysis of the Essential Oil Derived from *Marlierea eugeniopsoides* on the Coast of Paraná, Brazil

Camila Confortin, Bruna Gonçalves Lopes, Ana Helena Loos Moritz, Wanderlei do Amaral, Luiz Everson da Silva, Michele Debiasi Alberton, and Deepak Chandran

CONTENTS

15.1 INTRODUCTION

Bioprospecting is a broad term that can be conceptualized in several ways depending on the study area. In general, bioprospecting is the search for organic compounds of use to mankind in microorganisms, plants, and animals (Spartaco, 2014). Each ecosystem and biome throughout the planet has unique characteristics, such as the presence of specific and biodiverse species.

Science is concerned with extracting important information from biodiversity, based on "associated traditional knowledge." This knowledge, which passes down over the generations, was built by the interaction between humans and both plants and animals with the environments around them, and has been a valuable guide in the use of genetic resources (Monteiro, 2014). Indeed, natural medicinal herbs were one of the first material resources derived from nature; medicinal herbs have been used since the dawn of humanity in therapies to alleviate and cure diseases,

Medicinal plants are a source of secondary metabolites, and the process of chemical characterization helps to identify groups of plant metabolites, with specific chemical constitution of interest.

DOI: 10.1201/9781003226925-15

It is well known that the investigation of plant-based natural products has led to the discovery of many new substances, applied as new therapeutic therapies, or developed as drugs to treat disease (Bernardes, 2017; Reginatto, 2017).

Many studies have shown that having access to biomes rich in diversity is an interesting resource for natural product researchers. Among the existing biomes in Brazil, the Atlantic Forest is one of the most important, being one of the richest in terms of biodiversity (Santos, 2010). Due to the considerable wealth of species of both fauna and flora, the levels of exploitation are increasing, currently leaving only a little of the original forest. For this reason, the Atlantic Forest biome has been included among the priority areas for environmental conservation as a biodiversity hotspot (Myers et al., 2000).

Added to this, the market for pharmaceutical products derived from plants is increasing as natural medicines are still an effective and culturally appropriate therapeutic option. The demand for the use of natural products is also growing, mainly due to problems for both human health and the environment attributed to numerous synthetic products (Calixto, 2019).

Assessment of the bioactivity of essential oils for their therapeutic potential is one alternative use for existing natural resources. Essential oils are a mixture of volatile chemical constituents, obtained from plants, where the structure of the compounds present in the oil is of immense importance and is consequently the object of intense research aiming at the identification of both chemical composition and biological activity (Valli et al., 2018).

15.2 BIOPROSPECTING AND THE USE OF NATURAL RESOURCES—SEARCH FOR BIOACTIVE SUBSTANCES

Since the conservation and preservation of biological diversity ensures the survival of the planet as a whole, a major strategic need for most technologically developed countries should be to protect this heritage. However, natural resources are also a source of a wide range of products of economic importance. One area gaining considerable prominence is medicines derived from natural sources, where much research has been conducted to incorporate phytomedicines into the pharmaceutical market. The cosmetics industry and the advanced biotechnology sector are also major users of products and services relying on biodiversity (Silva et al., 2017).

On a global scale, it is well known that the available natural resources have already reached their limits of use, due to indiscriminate exploitation. Thus, studies focusing on the potential that environmental resources can bring to society are increasingly being considered of utmost importance for understanding natural systems. The field of bioprospecting operates in an area concerned with sustainability and loss of biodiversity. The discovery, description, and use of different species of plants and microorganisms is an intrinsic part of human history. The term bioprospecting was formally coined in 1993 as "research and use of biodiversity for the discovery of commercially useful genetic resources and biochemical substances" (Laird, 2002).

Among the many purposes of bioprospecting are the search and development of new drugs for the benefit of health and human welfare, as well as considerable economic reward (Saccaro Junior, 2011).

Brazil, as the holder of immense levels of biodiversity, occupies a prominent role in the debate on the potentials of bioprospecting, holding in its territory some of the richest ecosystems in terms of the number and diversity of species. It is also considered a country of great socio-biodiversity, due to the traditional communities that use low-impact technologies and very rich communities of natural resources (Pereira, 2009).

The amount of research in the fields of chemical and biological activities of the native plants being carried out in the Atlantic Forest is growing, motivated by the immense natural diversity found there. These works can help in the preparation of conservation plans that encourage the sustainable and rational use of biodiversity (Cruz, 2017).

15.3 NATURAL PRODUCTS—MEDICINAL PLANTS, ESSENTIAL OILS, AND THE DEVELOPMENT OF PHARMACEUTICALS

Studies of natural resources allow the identification of new compounds and genetic resources. Given the importance of medicinal plants as a therapeutic and economic resource, it is important to establish management and/or cultivation protocols for these plant species that also maintain balanced ecosystems (Simões et al., 2003). In the past few decades, society has realized the economic importance of biodiversity and traditional knowledge to science and technology. Consequently, there has been intense proliferation of both natural resource-related information and traditional knowledge, emphasizing that such information can be useful in the agricultural field, as well as application of inherited community knowledge of biodiversity when using plants as medicines.

The use of natural products, especially in therapeutics, has been undertaken even in primitive civilizations, but it was only in the middle of the nineteenth century that studies were intensified to fully characterize and use the chemical compounds isolated from medicinal plants. Advances in science and technology have now allowed visualization of the immense chemical diversity that can be found in nature, as a source of important bioactive compounds (Liang et al., 2006).

Bioprospecting plays a significant role in the discovery of new molecules from natural resources. Bioprospecting is directly associated with the study of biodiversity, in order to search for new genetic and biochemical resources of economic and social value, and to develop new products, such as pharmaceuticals, cosmetics, and agricultural products, among others (Pereira, 2009). Medicinal plants are a significant component of biodiversity, as they have been used for a long time to cure a wide variety of diseases. They are also considered to be fundamental raw materials for the development of phytotherapies.

Medicinal plants have greatly contributed to the development of new therapeutic strategies through their secondary metabolites. These are known to act directly or indirectly on organism and thus inhibit or activate important molecular and cellular targets. Science seeks to make progress from the benefits that nature offers, but also aims to respect the culture of the people when using resources taken from nature (Firmo, 2011).

Ethnobotanical, chemical, and biological studies that identify uses and possible therapeutic forms of plant resources become of immense value, and can assist in the preparation of conservation plans, encouraging the sustainable and rational use of biodiversity, while contributing to the strengthening and growth of areas such as pharmacology. The number of research studies in the fields of chemical and biological activities of native plants carried out in parts of the Atlantic Forest is growing, motivated by the immense natural diversity found there (Cruz, 2017).

As a rapidly growing field, research into essential oils from native species has often found compounds in plants that have become excellent sources of new drugs and agricultural products. As a great source of secondary metabolites that have diverse functions, plants have become targets for such phytochemical research. These studies can be done from fresh or dried plant materials with various extractive processes and with the use of various separation and analytical tools for the characterization of metabolites. The investigation of natural resources has led to the development of several drugs, such as local anesthetics, anticholinergics, adrenergic blockers, antibiotics, and immunosuppressants (Simões et al., 2017). Essential oils are important in the pharmaceutical industry. Essential or volatile oils are complex, volatile, and lipophilic substances that are generally scented and liquid, and originate from plant metabolism, having the most striking characteristics of high volatility and an intense and usually pleasant aroma. They are typically composed of a diverse range of chemical constituents, classified according to their chemical structure (Simões et al., 2017).

Essential oils exhibit activity against a wide variety of microorganisms. The compounds and their percentages present in the oils vary according to the species considered, the conditions of collection and extraction, and the parts of the plant used. The main volatile compounds isolated from essential oils are terpenes and their oxygenated derivatives (Solórzano-Santos and Miranda-Novales, 2011).

Essential oils come from raw materials of plant origin, being a characteristic trait of certain taxonomic groups, being found in most plant structures, most often in leaves, flowers, and fruits (Henriques et al., 2009). The oils are obtained in several ways, but the most common way to obtain these end products is by steam distillation (hydrodistillation) method, although extraction using organic solvents, lipids, supercritical fluids, or microwaves can also be used (Henriques et al., 2009).

Different chemical compounds can be found, depending on the extraction method applied. Among the most common volatiles found in essential oils are monoterpenes, diterpenes, and sesquiterpenes, presenting considerable structural variability (Henriques et al., 2009).

Considering the extensive use of essential oils in industry, many native species have various uses, especially in the pharmaceutical industry. With complex and diverse chemical compositions, essential oils can give excellent results when used in research, and assaying for enzyme inhibitory, antibacterial, and antifungal activities, among many others.

The essential oil extracted from native species, may have in its composition several types of chemical compounds. An important factor related to the formation of these chemical compounds is the place where the species grows. Most native species used in industry have a strong interaction with the environment. Research efforts have provided information on the effect of environmental conditions on the secondary metabolism of plants in order to maximize the production of specific active constituents of the species. Increased production of compounds of interest is the result of advances in understanding the influence of external factors on the synthesis of secondary metabolites.

The secondary metabolites present in the essential oil come from the chemical interaction between the plant and the place it is located, as well as the season of the year, which can also influence the concentration of these metabolites, thus exerting influence on the secondary metabolism of the species and generating multiple compounds with distinct characteristics.

In the search for bioactive substances, capable of acting against various diseases on their own or in synergy with existing drugs, antimicrobial assays have become a common tool. With the increase in resistance of various microorganisms to synthetic antibiotic drugs on the market, research that looks for substances that come from nature has great relevance, because it can find substances that help in the treatment of diseases related to the most diverse microorganisms.

In the following section, studies conducted with the species *Marlierea eugeniopsoides* (Kausel & D. Legrand) D. Legrand, belonging to the Myrtaceae family, native to the Atlantic Forest located on the coast of the State of Paraná, Brazil, are highlighted.

15.4 THE MYRTACEAE FAMILY—ESSENTIAL OIL EXTRACTION, CHEMICAL CHARACTERIZATION, AND ASSOCIATED BIOLOGICAL ACTIVITIES.

The Myrtaceae is one of the most characteristic families of the Brazilian flora, having considerable potential for economic development in Brazil. This family includes many species of medicinal and ornamental plants, producers of wood and edible fruits, and species suitable for use in the manufacture of paper, charcoal, and wood. The family belongs to the order Myrtiflorae (Myrtales) and comprises about 150 genera with approximately 5,500 species, which are mainly distributed in tropical and subtropical regions of the world (Moresco, 2014).

The members of the myrtle family are woody, shrubby, or arboreal plants with entire leaves that have oil glands. The flowers are generally white but, in some cases, red. They produce baciform or capsular loculicidal fruits and characteristically have a smooth bark trunk, being renewed with each growing season (Moresco, 2014).

In Brazil, this family is one of the most important, comprising about 23 genera with approximately 130 species. Some Myrtaceae family species have been used to treat gastrointestinal problems or infections, and these activities may be focused on the astringent properties that the plants possess. Antimicrobial, antimutagenic, hypoglycemic, diuretic, antioxidant, and anti-inflammatory activities have also been reported from members of the myrtle family, where the parts most commonly used include the leaves and the fruits, the latter being edible in some species (Alcântara, 2010).

The genus *Marlierea* occurs from the central region to the south of the country (Messias, 2005). *Marlierea* is characterized by its arboreal or shrubby habit, the inflorescences in panicles (or fascicles), rarely in racemes, axillary and/or terminal, flower buds usually closed, and petals often absent, with the fruit being a berry (Rosário, 2004).

Among the various species of the genus *Marlierea*, the species *M. eugeniopsoides* was chosen for the essential oils research. The species, known locally as "Guamirim" or "Guamirim-branco," has its geographical distribution from São Paulo to Rio Grande do Sul, in the Atlantic Forest domain. It is a tree that grows to an average height of 5 m, with flowers and fruits that can be edible, and is recognizable by leaves which have a long acuminate apex, often oblong, with numerous secondary veins. In some specimens, the leaf glands are large and opaque, being visible to the naked eye on the abaxial side of the leaves (Lima, 2015; Sobral, 2015). This species is native to the Atlantic Forest and flowers between the months of November and January. Its fruits usually appear from March to August, which supply the needs of avifauna.

The essential oils are found in various plant organs related to the survival of the plant in its ecosystem. The secretory sacs are often found in leaves, roots, stems, and bark. The chemical constituents found in the plant kingdom are synthesized and degraded by numerous anabolic and catabolic reactions, which constitute plant metabolism (Steffens, 2010).

Plants produce an extremely large number of compounds, many of which have very complex chemical structures and often do not have a well-defined function in the producing organisms, and may not have a clear role in metabolism. These metabolites are called special or secondary metabolites that can have immense therapeutic relevance (Kreis et al., 2017).

These secondary metabolites differ from primary metabolites because they are not essential to the survival, growth, and development of the plant, having as functions the defense of the plant and attraction of pollinators, for example. All plants are capable of synthesizing secondary metabolites, but it is in wild plants that they are most commonly found, due to the fact that throughout their evolutionary cycle they have created adaptive strategies in order to compete with other organisms, thus ensuring their survival (Vizzotto et al., 2010).

The next section provides research results on the chemical composition and biological activity of the *M. eugeniopsoides* species in four different seasons of the year and thus establishes the effect that seasonality has on the essential oil.

15.5 COLLECTING THE PLANT MATERIAL, EXTRACTION OF THE ESSENTIAL OIL, AND ANALYSIS OF THE CHEMICAL COMPOSITION

To examine the influence of seasonality on essential oil production, collection of leaves was performed in the Saint-Hilaire/Lange National Park (under authorization from ICMBio number 49770-2) on the southern coast of the State of Paraná, over the four seasons of the year (Table 15.1). The material was collected and stored at the Laboratory of Chemical and Biological Pre-Analysis of the Federal University of Paraná, Coastal Sector, where the essential oil was extracted.

The essential oils were extracted from fresh leaves (100 g) by the hydrodistillation method, using a modified Clevenger apparatus connected to a 2-liter, round-bottomed flask. The extraction process was carried out for a period of 4 hours, keeping the mixture at boiling point. The obtained hydrosols were collected and centrifuged.

For chemical characterization, gas chromatography analysis was performed on a Shimadzu GCMS-TQ8040, Split 90 Chromatograph coupled to a mass spectrometer (Santos et al., 2014). After the chromatography procedure, the chromatograms were analyzed in the GC–MS Postrun Analysis program, from which the components present in the essential oil samples were identified. The spectral and retention index data were compared with the literature (Adams, 2017).

TABLE 15.1
Dates of Collection of *Marlierea eugeniopsoides* Materials and the Collection Period.

Season	Date	Period
Spring	24/11/2017	Morning
Winter	10/07/2017	Morning
Autumn	19/06/2018	Morning
Summer	10/03/2018	Morning

15.5.1 Results and Discussion

The chemical composition and essential oil yield from *M. eugeniopsoides* from the coast of Paraná are shown below (Table 15.2).

The highest essential oil yield (0.45%) was obtained in the winter season. Several factors can influence the yield and the chemical composition of the essential oils, including the season in which the collection of the material occurs. The timing of collection is often one of the most important factors, since the amount of oil and the nature of the active constituents are not constant throughout the year. In addition, the time of collection and the conditions in which the plant is located also affect the oil yield.

The compounds of interest in essential oils are volatile secondary metabolites, which are responsible for the therapeutic and biological effects. They represent the result of the chemical interface between the environment and the plant, where their synthesis is often affected by environmental conditions. Thus, there are temporal and spatial variations in the relative proportions of plant secondary metabolites, where, despite genetic control, the expression can undergo changes resulting from the interaction of biochemical, physiological, ecological, and evolutionary processes (Souza et al., 2011).

In total, eleven different compounds with concentrations above 1% were identified in the essential oil samples from leaves of *M. eugeniopsoides* collected over the four seasons. These compounds included a range of monoterpenes (45.45%), oxygenated sesquiterpenes (45.45%), and sesquiterpenes (9.09%), the latter being represented by only bicyclogermacrene. On the other hand, only compounds present in concentrations > 1% were considered. Therefore, analysis of the essential oil content from the leaves of *M. eugeniopsoides* showed that seven major compounds were identified in winter, seven in spring, ten in summer, and nine in autumn (Table 15.3).

TABLE 15.2
Essential Oil Yield of *Marlierea eugeniopsoides* Leaves Collected at Different Seasons.

Season	Yield (%)
Spring	0.29
Summer	0.17
Autumn	0.45
Winter	0.32

TABLE 15.3
Chemical Constituents Found in the Essential Oil of *Marlierea eugeniopsoides* at Concentrations > 1% Collected Across All Four Seasons of the Year.

Chemical constituents	Winter (%)	Summer (%)	Spring (%)	Autumn (%)
α-Pinene	42.0	42.75	54.85	32.61
ß-Pinene	22.20	25.24	32.44	28.60
α-Myrcene	3.15	1.27	1.96	10.46
α-Silvestrene	2.25	1.52	1.39	2.46
E-ß-Ocimene	4.96	5.9	2.08	6.37
Biciclogermacrene	8.33	11.06	5.96	10.66
α-Globulol	1.88	5.77	1.66	2.67
cis-Cadin-4-en-7-ol	0	3.42	0	1.18
Cubeban-11-ol	0	1.51	0	0
Rosifoliol	0	1.56	0	0
2-epi-a-cedren-3-one	0	0	0	4.99

In all seasons where the oil was collected and extracted, the major compound was α-pinene, followed by β-pinene. Bicyclogermacrene was also present in all seasons. In this context, we can visualize the importance of chemical phenotypic characteristic which is associated to the mixture of compounds. These findings are related with different pressures and adaptations to the abiotic and biotic factors acquired over the evolution of the species in their environment. Thus, in the autumn and summer seasons, there was a greater variety of compounds found in the essential oil samples than in the winter and spring seasons.

In winter, the compounds bicyclogermacrene, α-pinene, β-p and *E*-β-ocymene were the compounds occurring at highest concentrations, with α-pinene being the major compound, totaling almost half of the sample (42.0%), followed by its β-pinene isomer (22.20%). Finally, α-globulol appears as the minor of the compounds exceeding 1% in concentration (1.88%).

The sample harvested in spring had a qualitative essential oil composition similar to that of the winter one. However, the amounts detected of the individual compounds differed significantly. The main compound detected was α-pinene, which alone totaled more than half of the sample (54.85%), followed by β-pinene (32.44%). There was also a reduction in the concentrations of bicyclogermacrene and *E*-β-ocimene, detected to bw 5.62% and 2.08%, respectively.

Thus, sesquiterpenes and oxygenated sesquiterpenes were the main chemical species produced by the plant. Gonçalves (2015) found that the production of essential oils was often influenced by factors both internal and external to the plant, which influenced not only the type of compounds produced, but also the quantities in which they were found, resulting in significant chemical variability even within the same species. Other researchers contend (Sharifi-Rad et al., 2017) that the genetic constitution of different species influences metabolite production, and that plants can contain different essential oil contents, which vary from species to species, because each one of them has a specific life cycle. Both hypotheses are probably acting simultaneously.

Several studies on essential oils obtained from species of the Myrtaceae family have shown evidence of antibacterial, antifungal, bactericidal, and cytotoxic activities (Cerqueira, 2002). Thus, the understanding of the chemical composition of the essential oils of native species is of great importance for many branches of industry, because biological activity and chemical composition are often closely related (Henriques et al., 2009).

15.5.1.1 Enzymatic Inhibitory Activity of Essential Oils From the Species *Marlierea eugeniopsoides* Collected on the Coast of Paraná, Brazil

Many molecular targets, such as enzymes, have been studied when seeking to discover drugs and herbal medicines with inhibitory mechanisms toward target organisms. According to Monteiro (2018), "it is considered promising [when] the sample [is] capable of inhibiting at least 50% of the enzyme activity, in the concentration tested." Table 15.4 shows the results of the enzymatic inhibition of the essential oil samples used in the current study.

Based on inhibition of acetylcholinesterase (AChE) activity, all samples showed promising results, since, across each of the seasons, the essential oil samples collected from *M. eugeniopsoides* showed inhibition potential greater than 50%. Oils derived from plants harvested in the spring season had the lowest AChE inhibitory activity, whereas summer-season-derived oils had the highest inhibitory activity, accounting for 94.3% of the activity, closely followed by the autumn-derived essential oil, which showed a 93.56% activity (Table 15.4).

In the samples, the monoterpenes α- and β-pinene were detected as the main compounds present in essential oils in each of the seasons, yet the oil had a low content of sesquiterpenes. The α-pinene compound has also been described in previous studies as being an acetylcholinesterase inhibitor. Effects of *Salvia lavandulaefolia* (Spanish sage) essential oil and some of its terpenes had been examined *in vitro* in a human erythrocyte acetylcholinesterase model. That study showed that *S. lavandulaefolia* essential oil, α-pinene, 1,8-cineole, and camphor were considered to be reversible, non-competitive AChE inhibitors. These findings suggest that, if the inhibitory activity of the essential oil is mainly due to the main terpenoid inhibitory constituents identified, there is a great synergistic effect between the constituents (Alcântara et al., 2010; Perry et al., 2000).

The α-glucosidase enzyme inhibition by essential oils from *M. eugeniopsoides* showed significant variances among the seasons in which the essential oils were harvested. Unsatisfactory inhibition was observed in oils derived from summer and winter samples, which were well below the acceptable inhibition considered to be effective. In sharp contrast, however, the autumn- and spring-derived oils reached extremely satisfactory results with 100% and 96.61 efficacy in enzyme inhibition of α-glucosidase activity.

Spring-derived oil had the highest concentration of monoterpene compounds, consisting mainly of α-pinene (54.85%) and β-pinene (32.44%), which may have contributed to the better performance of this oil. In comparison, autumn-derived oils had a composition containing a variety of lower-concentration compounds, present in smaller amounts than in other seasons. For example, the sample of autumn-derived oils had a higher percentage of myrcene (10.46%) than did the samples from the other seasons analyzed.

TABLE 15.4
Acetylcholinesterase and α-Glucosidase Inhibitory Activity* of the Essential Oil from *Marlierea eugeniopsoides*.

	Acetylcholinesterase		α-Glucosidase	
Sample	**%**	**SD**	**%**	**SD**
Winter	89.0	± 1.3	15.0	± 6.7
Summer	94.3	± 0.2	8.1	± 0.8
Spring	78.5	± 1.3	96.6	± 0.2
Autumn	93.6	± 0.2	100.0	± 0.0
Positive control	92.8	± 8.9	64.8	± 0.1

* The percentage of inhibition refers to essential oils at a concentration of 1 mg/mlPositive control: Neostigmine (100 μg/mL) for acetylcholinesterase; Quercetin (1 mg/ml) for α-glucosidase. SD: Standard Deviation.

As mentioned above and confirmed by Gobbo-Neto and Lopes (2007), chemical composition of plants can be variable. Weather, seasons of the year, growing place, and other factors can change the chemical composition of the essential oil, and consequently, its biological activity. Therefore, the variety of compounds found in the four seasons studied could contribute to understanding the results found in the enzyme inhibition assays. For example, in the autumn sample, there was considerable quantitative distribution of the compounds. The autumn sample showed good enzymatic inhibition results, 100% for α-glucosidase and 93.56% for acetylcholinesterase activities. However, the summer sample exhibited 94.3% inhibition for acetylcholinesterase inhibition but only 8.08% α-glucosidase inhibition. The summer sample also had a varied number of compounds, which may have influenced the results of enzyme inhibition.

A study with extracts of *Hedychium gardnerianum* (Arruda, 2010) found that α-pinene was one of the main constituents of samples, and was shown to have a high ability to inhibit the AChE enzyme. A study by Souza (2011), working with extracts from native species of the Cerrado biome, including *Eugenia dysenterica* which belongs to the Myrtaceae family, evaluated their potential for enzymatic inhibition. *Eugenia dysenterica* oil produced 100% inhibition of the α-glucosidase enzyme, more than 80% inhibition of the amylase and tyrosinase activities, but less than 10% inhibition of AChE activity, the results for the latter enzyme being unsatisfactory.

15.6 CONCLUSION

Rational approaches for sustainable development must be considered to balance the increasing demand for natural resources to be used in a sensible way, both for the generation of new products and for a possible generation of income, against the desire to conserve a high level of natural biodiversity. Bioprospecting is the process of studying, researching, and analyzing resources that can potentially be deployed for these practical purposes.

Studies of essential oils within native plant species are important as a route to the discovery and development of new plant-derived drugs. The findings from the current study of the inhibitory effects of essential oil components, singly or in combination, on activity of target enzymes, could be of great value to the pharmaceutical industry, which is in a continual search for substances that help in the treatment of diseases.

The yields of essential oils obtained from *M. eugeniopsoides* was higher in the autumn season. Chemical composition of the essential oil differed markedly between seasons. Essential oils from *M. eugeniopsoides* contained two major compounds, the monoterpenes α-pinene and β-pinene. The highest percentages of these compounds was found in spring—54.85% for α-pinene, and 32.44% for β-pinene. In the autumn season, enzyme inhibition assays by essential oils showed a great potential for inhibition of α-lgucosidase (100%) and acetylcholinesterase (93.56%) enzymes. This study highlights the value of the biodiversity found in the Atlantic Forest biome of the coast of Paraná, which, along with bioprospecting studies conducted here, has shown that essential oils derived from the harvest of leaves of selected plants from these biomes have potential as phytotherapeutic agents.

REFERENCES

Adams, R.P. *Identification of essential oil components by gas chromatography/mass spectrometry*, 4.1 ed. Allured Pub. Corp., San Diego California, 2017.

Alcântara, J.M.; Yamaguchi, K.K.L.; Silva, J.R.A.; Veiga Junior, V.F. Composição química e atividade biológica dos óleos essenciais das folhas e caules de Rhodostemonodaphne parvifolia Madriñán (Lauraceae). *Acta Amazonica*, 40(3), 567–572, 2010.

Arruda, M.A.T.F.S. *Pesquisa de inibidores da acetilcolinesterase em Hedychiumgradnerianumdos açores*. 120 f. Dissertação (Mestrado em Ciências biomédicas)- Departamento de ciências tecnológicas e desenvolvimento, Universidade dos Açores, Ponta Delgada, 2010.

Bernardes, L.S.C.; Lang, K.L.; Petrovick, P.R.; Schenkel, E.P. Produtos naturais e o desenvolvimento de fármacos. In: Simões, C.M.O.; Schenkel, E.P.; Gosmann, G.; Mello, J.C.P.; Mentz, L.A.; Petrovick, P.R. (Orgs). *Farmacognosia: do produto natural ao medicamento.* Artmed, Porto Alegre, p. 486, 2017.

Calixto, J.B. The role of natural products in modern drug discovery. *Anais da Academia Brasileira de Ciências [online].* 91(Suppl 3), e20190105, 2019. [Accessed 1 May 2022]. Available from: https://doi.org/10.1590/0001-3765201920190105.

Cerqueira, M.D. *Estudo fitoquímico de Myrciarotundifolia (Berg.) Legrand (Myrtaceae).* Dissertação (Mestrado), Programa de Pós-graduação em química, Universidade Federal da Bahia, Salvador, 2002.

Cruz, A.L.D. *O uso de recursos vegetais pela comunidade do Perequê, Guarujá-SP.* 2017. 71 f. Dissertação (Mestrado) -- Universidade Santa Cecília, Programa de Pós-Graduação em Sustentabilidade de Ecossistemas Costeiros e e Marinhos, Santos, SP, 2017.

Firmo, W.C.A.; Menezes, V.J.M.; Passos, C.E.C.; Dias, C.N.; Alves, L.P.L.; Dias, I.C.L.; Neto, M.S.; Olea, R.S.G. Historical context, popular use and scientific conception on medicinal plants. *Cadernos de pesquisa, São Luís,* 18(especial), 90–95, 2011.

Gobbo-Neto, L.; Lopes, N.P. Plantas medicinais: Fatores de influência no conteúdo de metabólitos secundários. *Química Nova,* 30(2), 374–381, 2007.

Gonçalves, J.M. *Atividades biológicas e composição química dos óleos essenciais de achyroclinesatureoides (lam) dc. e ageratumconyzoides l encontradas no semiárido baiano.* 2015. 111 f. Tese (Doutorado) - Curso de Programa de Pós-graduação em Biotecnologia, Universidade Estadual de Feira de Santana, Feira de Santana, 2015.

Henriques, A.T.; Simões-Pires, C.A.; Apel, M.A. Óleos essenciais: importância e perspectivas terapêuticas. In: Yunes, R.A.; Cechimel Filho, V. (orgs.). *Química de produtos naturais, novos fármacos e a moderna farmacognosia,* 2nd ed. Universidade do Vale do Itajaí, Itajaí, pp. 219–256, 2009.

Kreis, W.; Munkert, J.; Pádua, R.M. *Biossíntese de metabólitos primários e secundários.* In: Simões, C.M.O.; Schenkel, E.P.; Gosmann, G.; Mello, J.C.P.; Mentz, L.A.; Petrovick, P.R. (Orgs). *Farmacognosia: do produto natural ao medicamento.* Artmed, Porto Alegre, p. 486, 2017.

Laird, S. *Biodiversity and traditional knowledge – Equitable partnerships in practice,* edited by Laird, S. Earthscan Publications Ltd., London and Sterling, 2002.

Liang, X.T.; Fang, W.S. *Medicinal chemistry of bioactive natural products.* Wiley Interscience, New York, 2006.

Lima, D.L.; Caddah, M.K.; Goldenberg, R. A família Myrtaceae na Ilha do Mel, Paranaguá, Estado do Paraná, Brasil. *Hoehnea,* 42(3), 497–519, 2015.

Messias, K.L.S. *Estudo fitoquímico e farmacológico das folhas e caules da Marlierea tomentosa camb.* 2005. 62f. Dissertação (Mestrado) em Ciências farmacêuticas, Universidade do Vale do Itajaí, Itajaí, 2005.

Monteiro, F.X. *O acesso e a utilização da biodiversidade brasileira e dos conhecimentos tradicionais Associados como instrumento de viabilização do desenvolvimento sustentável.* 2014. 108f. Dissertação (mestrado)- Pós-Graduação em Direito, Escola superior, dom Helder Câmara, Belo Horizonte, 2014.

Monteiro, J.S.M. *Valorização de plantas aromáticas catarinenses – estudo químico-biológico dos óleos essenciais de Piper mikanianum (KUNTH) STEUDEL e Drimysangustifolia MIERS.* 91f. Dissertação (Mestrado em Química) - Programa de Pós-Graduação em Química, Centro de Ciências Exatas e Naturais, Universidade Regional de Blumenau, Blumenau, 2018.

Moresco, H.H. *Estudo fitoquímico e biológico de quatro espécies da família myrtaceae.* 2014. 190 f. Tese (Doutorado) - Curso de Programa de Pós-Graduação em Química, Universidade Federal de Santa Catarina, Florianópolis, 2014.

Myers, N.; Mittermeier, R.A.; Mittermeier, C.G.; Fonseca, G.A.B.; Kent, J. Biodiversity hotspots for conservation priorities. *Nature,* 403(6772), 853–858, 2000.

Pereira, A.M. *Condicionantes institucionais para bioprospecção no Brasil.* Dissertação(mestrado)-Universidade Estadual de Campinas, Instituto de Economia, Campinas-SP, 2009.

Perry, N.S.L.; Houghtonp, J.; Theobald, A.; Jenner, P.; Perry, E.K. In-vitro inhibition of human erythrocyte acetylcholinesterase by Salvia lavandulaefolia essential oil and constituents terpenes. *Journal of Pharmacy End Pharmacology,* 52(7), 895–902, 2000.

Reginatto, F.H. Introdução à análise fitoquímica. In: Simões, C.M.O. et al. *Farmacognosia: do produto natural ao medicamento.* Artmed, Porto Alegre, p. 486, 2017.

Rosário, A.S. *O gênero marliereacambess. (myrtaceae) na amazônia brasileira.* 2004. 78 f. Dissertação (Mestrado) - Curso de Mestrado em Botânica, Universidade Federal Rural da Amazônia, Belém, 2004.

Saccaro Junior, N. L. *Desafios da bioprospecção no Brasil, Texto para Discussão.* No. 1569. Instituto de Pesquisa Econômica Aplicada (IPEA), Brasília, 2011.

Santos, B.C.B. *Fitoquímica e ensaios biológicos de óleos essenciais de Pimenta pseudocaryophyllus e Hedychium coronarium da Mata Atlântica do Estado de São Paulo.* Dissertação (mestrado), Universidade Estadual de Campinas, Instituto de Química, Campinas, 2010.

Santos, T.G.; Fukuda, K.; Kato, M.J.; Sartorato, A.; Duarte, M.C.T.; Ruiz, A.L.T.G.; Carvalho, J.E.; D.E.; Augusto, F.; Marques, F.A.; Maia, B.H.L.N.S. Characterization of the essential oils of two species of Piperaceae by one- and two-dimensional chromatographic techniques with quadrupole mass spectrometric detection. *Microchemical Journal*, 115, 113–120, 2014.

Sharifi-Rad, J.; Sureda, A.; Tenore, G.C.; Daglia, M.; Sharifi-Rad, M.; Valussi, M.; Tundis, R.; Sharifi-Rad, M.; Loizzo, M.R.; Ademiluyi, A.O.; Sharifi-Rad, R.; Ayatollahi, S.A.; Iriti, M. Biological activities of essential oils: From plant Chemoecology to traditional healing systems. *Molecules*, 22(1), 70, 2017. https://doi.org/10.3390/molecules22010070.

Silva, L.E.; Amaral, W.; Albuquerque, U.P. Uso sustentável da biodiversidade e conservação de recursos naturais. *Guaju*, Matinhos. 3(1), 2–10, 2017.

Simões, C.M.O.; Schenkel, E.P.; Gosmann, G.; Mello, J.C.P.; Mentz, L.A.; Petrovick, P.R. *Farmacognosia da planta ao medicamento*, 5ª edição. Editora da UFSC, Florianópolis, 2003.

Simões, C.M.O.; Schenkel, E.P.; Gosmann, G.; Mello, J.C.P.; Mentz, L.A.; Petrovick, P.R. (Orgs). *Farmacognosia: do produto natural ao medicamento.* Artmed, Porto Alegre, p. 486, 2017.

Solórzano-Santos, F.; Miranda-Novales, M.G. Essential oils from aromatic herbs as antimicrobial agents. *Current Opinion in Biotechnology*, 23, 1–6, 2011.

Souza, M.F.; Manganotti, S.A.; Souza, P.N.S.; Meira, M.R.; Matos, C.C.; Martins, E.R. Influência do horário de coleta, orientação geográfica e dossel na produção de óleo essencial de Cordiaverbenacea. *Biotemas, [s.l.]*, 24(1), 9–14, 21 fev. 2011. http://doi.org/10.5007/2175-7925.2011v24n1p9.

Souza, P.M. *Atividade de inibição enzimática por espécies vegetais do bioma cerrado.* 2011. 90 f., il. Dissertação (Mestrado em Ciências da Saúde)-Universidade de Brasília, Brasília, 2011.

Spartaco, A.F.; Silva, C.; Bigi, M.F. Bioprospecção e biotecnologia. *Parc. Estrat. Brasília-DF*, 19(38), 45–80, janvier–junio 2014.

Steffens, A.H. *Estudo da composição química dos óleos essenciais obtidos por destilação por arraste a vapor em escala laboratorial e industrial.* 2010. 68 f. Dissertação (Mestrado) - Curso de Programa de Pós-graduação em Engenharia e Tecnologia de Materiais, Pontifícia Universidade Católica do Rio Grande do Sul, Porto Alegre, 2010.

Valli, M.; Russo, H.M.; Bolzani, V. The potential contribution of the natural products from Brazilian biodiversity to bioeconomy. *Anais da Academia Brasileira de Ciências [online].* 90(Suppl 1), 763–778, 2018. [Accessed 1 May 2022]. Available from: https://doi.org/10.1590/0001-3765201820170653.

Vizzoto, M.; Krolow, A.C.R.; Weber, G.E.B. *Metabólitos secundários encontrados em plantas e sua importância.* Embrapa Clima Temperado, Pelotas, p. 16, 2010.

16 Production of Bioactive Secondary Metabolites in Medicinal and Aromatic Plants Cultivated in Turkiye

Esra Koç and Belgizar Karayiğit

CONTENTS

16.1 INTRODUCTION

Turkiye is a country rich in plant diversity due to the fact that it connects Asia and Europe, is at the juncture of three important phytogeographic regions, namely the Iran-Turan, the Europe-Siberia, and the Europe-Mediterranean regions, and has three different climates, namely Mediterranean, Oceanic, and Terrestrial (Graham et al., 2008; Öztürk and Ocak, 2020). Medicinal and aromatic plants are used in a variety of areas such as therapeutics (medicine), cosmetics (perfume), food (nutrients, spices, beverages, and teas), biofuel, textile, ornamental plants, apiculture, phytotherapy, agriculture, aromatic oil, furnishing, and dye production (Tulukçu and Sağdıç, 2011). Medicinal and aromatic plants in Turkiye are usually cultivated in the Marmara, Mediterranean, Aegean, Eastern Black Sea, and Southeastern Anatolia regions (Bayram et al., 2010). In Turkiye, there are more than 40 provinces cultivating medicinal and aromatic plants, with Antalya, Aydın, İzmir, Kocaeli, and Mersin being among the top provinces in terms of plant diversity.

In Turkiye, there are 347 medicinal and aromatic plant species which to date have been commercialized. Of these, 35 species are endemic and 104 are exported (Faydaoğlu and Sürücüoğlu, 2011), with 75% growing in the wild and 25% being cultivated (Yaldız and Çamlıca, 2018). The rich diversity of the flora has great economic potential in terms of medicinal and aromatic plants collected from nature and cultivation (Table 16.1). Plants like laurel, linden, mahaleb, rosemary, and sage are collected from the nature, whereas others, such as aniseed, cumin, fennel, mint, and poppy, are cultivated (Kırıcı, 2015). Not only plants in danger of extinction, but also plants that are highly commercialized are cultivated (Figure 16.1).

Turkiye is both an importer and an exporter of medicinal and aromatic plants. Turkiye exports medicinal and aromatic plants to nearly 100 countries, primarily Brazil, Canada, France, Germany, Italy, the USA and Poland. Turkiye ranks first in terms of thyme production worldwide (Öztürk et al., 2014). Although many provinces, such as Hatay, Manisa, Muğla and Uşak cultivate thyme, Denizli Province has the highest level of thyme cultivation (Tunca et al., 2017). In Turkiye, thyme ranks first

DOI: 10.1201/9781003226925-16

TABLE 16.1
Selected Medicinal and Aromatic Plants Producted in Turkiye (based on TUIK, 2019, 2020, 2021)

	2019		2020		2021	
Plant	**Cultivation Area (decare)**	**Production (tons)**	**Cultivation Area (decare)**	**Production (tons)**	**Cultivation Area (decare)**	**Production (tons)**
Thyme	157,074	17,965	184,711	23,866	199,573	21,174
Cumin	321,889	20,245	212,132	13,926	155,122	8,386
Black cumin	37,085	3,603	33,773	3,412	83,915	6,435
Anise	239,171	17,589	155,317	10,716	110,712	6,936
Heather	7,468	1,951	6,860	1,788	7,174	1,911
Fennel	33,859	4,655	22,204	4,365	13,285	2,507
Red pepper	119,409	240,656	119,869	256,735	126,580	284,694
Coriander	155	12	2455	188	2,612	253
Capers	25	3	32	3	121	18
Hop	3,307	1,800	3,308	1,908	3,185	1,861
Melissa	209	93	284	150	534	266
Nettle	0	0	1	0.1	5	0.5
Salvia	5,602	1,233	6,655	1,271	9,326	1,848
Lavender	11,903	1,462	22,188	3,499	35,810	6,108
Rose (for oil)	38,457	16,560	41,320	18,202	41,046	18,020
Poppy (capsul)	677,369	27,288	451,252	20,542	516,721	21,037
Parsley	–	92,954	–	97,760	–	108,604
Rocket	–	13,654	–	15,045	–	27,350
Purslane	–	4,931	–	5,702	–	12,232
Mint	–	16,011	–	23,471	–	26,438
Dill	–	8,740	–	8,267	–	10,432
Garlic (green)	–	23,351	–	28,552	–	46,454
Garlic (dry)	–	103,096	–	116,840	–	132,617

among the medicinal and aromatic plants exported (Kırıcı, 2015). Thyme cultivation is followed in decreasing order by aniseed, sage, cumin, fennel, laurel leaf, locust, mahaleb, mint, rosemary, and sumac (Bayraktar et al., 2017). Turkiye meets nearly all (90%) of the global laurel leaf demand (Özgüven et al., 2005), where the laurel plant grows widely at the low altitudes of the Aegean and Mediterranean regions and in areas representing the Mediterranean climate in the Black Sea region. Laurel leaf is mainly exported to Brazil, Poland, the USA, and Vietnam (Bayraktar et al., 2017). Revenue from thyme, laurel, saffron, and ginger leaf export alone is approximately 108 million US dollars (TUIK, 2019).

Another high-value medicinal and aromatic product exported by Turkiye is poppy (*Papaver somniferum* L.) (TMO, 2015), which has been cultivated for many years. In Turkiye, poppy is cultivated on an area of 451,252 acres (TUIK, 2020) and the United Nations recognizes Turkiye as a traditional poppy producer (Öztürk et al., 2014; Bulut and Arslan, 2016). In Turkiye, poppy is legally cultivated in a total of 13 provinces, including Afyonkarahisar, Amasya, Balıkesir, Burdur, Çorum, Denizli, Eskişehir, Isparta, Konya, Kütahya, Manisa, Tokat, and Uşak (Özgen et al., 2017).

FIGURE 16.1 Turkish provinces where medicinal and aromatic plants are grown in Turkiye. (**1:** *Cuminum cyminum*, **2:** *Nigella sativa*, **3:** *Papaver somniferum*, **4:** *Rosmanirus officinalis*, **5:** *Salvia officinalis*, **6:** *Melissa officinalis*, **7:** *Foeniculum vulgare*, **8:** *Origanum species*, *Satureja hortensis*, *Thymbra spicata*, *Coridothymus capitatus*, *Satureja spicigera*, *Thymus vulgaris*, **9:** *Lavandula officinalis*, **10:** *Mentha* sp., **11**: *Pimpinella anisum*, **12:** *Erica manipuliflora*, **13:** *Coriandrum sativum*, **14:** *Eruca vesicaria*, **15:** *Portulaca oleracea* (Kan et al., 2006; Beyzi and Gürbüz, 2010; Ipek and Gürbüz, 2010; Kaçar et al., 2010; Malayoğlu, 2010; Avcı, 2014; TOB, 2015; Akgören, 2011; Tripodi et al., 2017; Giray, 2018; Bozdemir, 2019; TUIK, 2020; Yıldırım et al., 2020).

16.1.1 Bioactive Compounds in Selected Medicinal and Aromatic Plants

Medicinal plants demonstrate therapeutic properties due to the biologically active compounds they synthesize. Alkoloids, flavonoids, phenolic compounds, saponins, steroids, and terpenes are all bioactive substances synthesized by medicinal plants. Secondary metabolites acquired from the organs of medicinal plants such as bulb, flower, leaf, rhizome, root, seed, and stem have a variety of properties including anticancer, antidiabetic (Shafay et al., 2021), anti-allergenic, antidepressant, antifungal, antihistaminic, antihypertensive, anti-inflammatory, antimalarial, antimicrobial, antioxidant, antispasmodic, antithyroid, and analgesic, and they can treat gastritis, ulcer, neurological disorders (Zachariah and Leela, 2018), and cardiovascular diseases (Aftab, 2019). Phytochemical volatile oils acquired from aromatic plants also exhibit a variety of properties, including antifungal, antimicrobial, antioxidant, and enzyme inhibition (Toker et al., 2015).

The volatile oils in thyme vary with the thyme species, containing mainly carvacrol and ortimol compounds with high antioxidant activities. *Origanum onites*, *Origanum vulgare*, *Origanum heracleoticum*, *Origanum minutifilorum*, *Origanum majorana*, *Origanum syriacum* oregano species contain carvacrol as the main component of the volatile oil (Dauqan et al., 2017; Gutierrez et al., 2017) (Table 16.2). The poppy plant is rich in a variety of secondary metabolites, particularly alkaloids such as codeine, morphine, noscapine, papaverine, and thebaine, which are used in medicine and pharmacy (Gürkan et al., 2003) (Table 16.2). Hence, poppy cultivation is subject to legislation due to the narcotic and addictive effects of morphine which is one of the alkaloids acquired from poppy capsules. In addition, bioactive compounds such as albumin, carsonic acid, fixed oil, laurenobiolid, lauric acid, menthol, menthon, mucilage, quercetrin, rosmanol, rosmarinic acid, sage carnosol, silymarin derived from flavonolignan, and taxifolin are contained in laurel. Bioactive

TABLE 16.2
Bioactive Compounds in Selected Medicinal and Aromatic Plants Cultivated in Turkiye

Plant	Bioactive Compounds	Medicinal Properties	Uses	References
Cuminum cyminum	Phenolics (flavonoids), Terpenes	Antidiabetic, Antioxidant, Anti-inflammatory, Cardioprotective, Dyspepsia, Digestive stimulant, Neuroprotective	Food, Pharmacy, Perfumery, Spice	Budak et al., 2020; Mnif and Aifa, 2015; Srinivasan, 2018;
Nigella sativa	Thymoquinone, Thymol, A-Hederin	Antidiabetic, Antihypertensive, Antioxidant, Antimicrobial, Anticarcinogenic, Anti-inflammatory, Analgesic, Gastroprotective, Hepatoprotective, Immuneostimulant, Neuroprotective, Pulmonary protective	Food, Spice	Srinivasan, 2018; Tabassum et al., 2018
Papaver somniferum	Alkaloids, e.g., Codeine, Morphine, Flavonoids, Phenylpropanoids Sanguinarine, Steroids	Antioxidant, Anticarcinogenic, Antimutagenic, Anti-inflammatory, Antimicrobial, Antiparasitic	Analgesic, Antitussive, Medicinal	Da Cheng et al., 2015
Rosmanirus officinalis	Carnosic acid, Phenolic acids, Flavonoids, Triterpenes	Anticancer, Antidiabetic, Antiobesity Antimicrobial, Antioxidant, Antiproliferative, Cardioprotective, Gastrointestinal disease, Genitory and urinary disorders Liver disease, Respiratory	Food flavoring, Culinary, Cosmetics, Medicinal, Perfumery	Borrás-Linares et al., 2014; Gezici et al., 2017; González-Vallinas et al., 2015; Moore et al., 2016; Sánchez-Camargo et al., 2016; Sedighi, 2015
Salvia officinalis	Flavonoids: Apigenin, Chlorogenic acid Quercetin, Rutin Terpenes and terpenoids: Carnasol, Cineole, Pinene, etc.	Anticancer, Antimutagenic, Antioxidant, Antiseptic, Cognitive and memory enhancing, Metabolic effects	Aromatic, Tea, Dye, Food, Landscape, Spice	Bağdat, 2006; Ghorbani, 2017; Hernandez-Saavedra et al., 2016

(Continued)

TABLE 16.2 (CONTINUED)
Bioactive Compounds in Selected Medicinal and Aromatic Plants Cultivated in Turkiye

Plant	Bioactive Compounds	Medicinal Properties	Uses	References
Melissa officinalis	Essential oils: Citronellal, Geraniol Sesquiterpenes: β-caryophyllenene, Caryophyllenene oxide Phenolics, Rosmaniric acid, Flavonoids	Antioxidant, Anticancer, Antidepressant, Anti-inflammatory, Antimicrobial, Anxiolytic, Antinociceptive Hypoglymic, Hypolipidemic, Spasmolytic	Food, Perfumery, Pharmacy, Spice, Tea	Hänsel et al., 2010; Kaçar et al., 2010; Miraj et al., 2016, Shakeri et al., 2016
Foeniculum vulgare	Essential oil: Estragol, Fenkon, *Trans*-anatol	Antitumor, Antigenotoxic, Antimicrobial, Antioxidant, Expectorant, Secretolytic	Drug, Food, Spice, Tea	Kan et al., 2006
Lavandula officinalis	Anthocyanins, Coumaric acid, Glycolic acid, Linalool, Linalyl acetate, Valeric acid	Anticonvulsive, Antioxidant, Antibacterial, Anti-inflammatory, Antidepressant, Local anesthetic, Sedative	Fragrances, Perfumery, Shampoos	Bozdemir, 2019; Giray, 2018; Hajhashemi et al., 2005
Mentha sp.	Lipophilic flavonoids, Phenolic compounds	Antioxidant, Antiulcer, Anti-inflammatory, Antidiabetogenic Cytoproductive, Chemopreventive, Heptoproductive	Gum, Medicine, Perfumery, Toothpaste	Mimica-Dukic and Bozin, 2008
Pimpinella anisum	Essential oil: *Trans*-anethol, Fatty acids: Petroselinic acid, Phenolic acids, Flavonol glycosides	Antioxidant, Antiseptic, Antimicrobial, Aperitif, Digestive, Antispasmotic, Expectorant, Anti-inflammatory, Diüretic agents	Spice, Toothpaste	Bettaieb Rebey et al., 2018; Shojaii and Fard, 2012
Erica manipuliflora	Anthocyanins, Coumarins, Flavonoids, Triterpenic compounds	Anticarcinogenic, Antiulcer, Antimicrobial Cyctotoxic	Balm, Tea	Şen et al., 2016
Eruca vesicaria	Glucosinolates, Phenolics: Polyglycosylated flavonols (Isorhamnetin, Kaempferol, and Quercetin)	Anticancer, Nutraceutical	Cosmetics, Deodorant, Food, Medicine	Bennett et al., 2006
Coriandrum sativum	Essential oils: Pinene, Terpinene Flavonoids, Polyphenol	Anti-inflammatory, Hypoglycemic, Hypocholesterolemic, etc.	Beverage, Food, Medicine, Perfumery, Spice	Beyzi and Gürbüz, 2010
Portulaca oleracea	Antocyanine, Ascorbic acid, α-Tokoferol asit, β-Carotene, Fenolic, Flavonoid,	Antioxidant, Anticancer, Anti-inflammatory, Antidiabetic, Analgesic, Antibacterial	Food, Medicinal plant	Karakaş, 2017

compounds in selected medicinal and aromatic plants grown in Turkiye, their medicinal properties, and their areas of use are presented in Table 16.2.

Among the main volatile oils exported from Turkiye are oils from rose and thyme, as well as lemon and orange from citrus fruits (Arslan et al., 2015). Rose oil is used in cosmetics and the perfume industry in many countries like Hong Kong, Japan, and Korea. As a result of the recent increase in interest in the production of volatile oils in Turkiye, volatile oils are also produced from aromatic plants such as aniseed, cumin, laurel, lemon leaf, mint, rosemary, and thyme in facilities established in the Antalya, Hatay, Manisa, Mersin, and Muğla provinces (Bayram et al., 2010).

16.1.2 Biotechnological Methods Used in the Production of Secondary Metabolites

Medicinal and aromatic plants have limited cultivated areas and most plant species used for this purpose are endemic. The excessive collection of these plants from nature by people and by pharmacy and cosmetics industries leads to the destruction and even extinction of endemic species. Plants are ripped from nature to acquire their secondary metabolites and the amount of secondary metabolites acquired is only 1% of the plant's dry weight. Due to such negativities, tissue and cell culture studies have become an important, developing technology to conserve native plant species.

Tissue culture techniques can remove the extinction threat from plants and enable plants to be harvested at almost any time and in large amounts. Micropropagation of plant cultures, using different plant tissue and cell cultures, is among one of the techniques used worldwide for producing pathogen-free or haploid plants (Avcı, 2014).

Although biologically active substances are synthesized by different organs of plants, such as bulb, shoot, flower, leaf, rhizome, root, seed, and stem, and can accumulate in these organs, not every secondary metabolite can be acquired using tissue culture methods. Developed tissue and cell culture studies offer different opportunities for the development of differentiated and organized cultures and undifferentiated and unorganized cultures for producing the appropriate secondary metabolite (Shafay et al., 2021).

Callus cultures and cell suspension cultures are undifferentiated *in-vitro* cultures. Although callus cultures make it possible to accumulate secondary metabolites, such as caffeine (Waller et al., 1983), capsaicin (Umamaheswai and Lalitha, 2007), catechin (Farzami, 2005), flavonoids and hyoscyamine (Maharik, 2009), and limonin (Barthe et al., 1987), cell suspension cultures make it possible to acquire other secondary metabolites, such as terpenoids (Misra et al., 2014), ajmalicine, berberine, ginseng, ubiquinone, paclitaxel, rosmarinic acid, serpentine, shikonin, and stilbene (Sara-Alsadat et al., 2015). Callus and cell suspension cultures have been successfully used as an alternative to whole-plant systems for producing commercially important bioactive compounds (Smetanska, 2008). For reasons such as the utilization of substrates which are used in secondary metabolite production in cultures that do not show a morphological differentiation tendency in other synthesis pathways, the lack of areas to storage secondary metabolites, and the inability to express the genes that are responsible for the biosynthesis of biologically active substances in undifferentiated tissues, fewer secondary metabolites can be produced from these cultures compared with differentiated cultures (Ruffoni et al., 2010; Isah et al., 2018).

Root cultures, shoot cultures, and somatic embryo cultures are differentiated and organized *in-vitro* tissue cultures. Many important secondary metabolites, such as anthraquinone, artemisinin (Wang et al., 2016), chlorogenic acid, cichoric acid (Fu et al., 2015), and reseprine are synthesized by the roots of plants in low amounts. In order to accumulate these secondary metabolites, it is necessary to remove the wild-grown plant from the soil and extract them from the roots, killing the plant. However, cultivation of these species under *in-vitro* conditions can assist in preventing the extinction of these plants while still harvesting these metabolites for sectors like the pharmaceutical, food, cosmetics, agriculture and medicine industries (Douglasa et al., 2014). Also, a variety of techniques can be applied to enhance the root development of plants, including altering the composition of the culture medium or adding an elicitor (Demirci et al., 2015).

Addition of plant growth regulators to the culture medium can enhance root development. Plant growth regulators increase the amount and activity of enzymes such as chlorophyllase, peroxidase, and phenylalanine ammonia-lyase, and cause an increase in the production of secondary metabolites (Abbasi et al., 2012). When plant growth regulators, such as benzyl amino purine and thidiazuron, were added to a nutrient medium prepared for cultivating *Hypericum perforatum* in an *in-vitro* medium, an increase in the synthesis of chlorogenic acid, flavonoids, hypericin, and phenolics in the roots was observed (Cui et al., 2013). Common plant growth regulators include benzyl amino pürine, gibberellic acid, indole-3-butyric acid, kinetin, 1- naphthalenic acid, and thidiazuron, which can increase root formation when added to the culture medium. Another intervention performed in the culture medium to increase root production is to add an elicitor to the *in-vitro* culture medium. Elicitors are any biotic or abiotic agents that can stimulate the plant defense system for the production of secondary metabolites. When abiotic elicitors such as UV, heavy metal supplement (Zhang et al., 2007), osmotic shock, or inorganic salts are added to the *in-vitro* culture medium, secondary metabolite accumulation will increase (Verpoorte 2013). For example, when elemental cerium is added to the tissue culture medium, the production of taxol (a precious metabolite in yew, *Taxus* spp.) will increase (Wu et al., 2001). Bacteria, fungi, and nematodes are all biotic elicitors. The amount of anthocyanin acquired by adding pectin to the culture medium prepared to cultivate *Vitis vinifera* was 2.5 greater than was produced in the medium in which the elicitor was not added. Different biosynthetic pathways can be induced in plants depending on the type of elicitor used, which makes elicitor applications one of the main strategies for increasing the production of secondary metabolites in *in-vitro* cell and tissue cultures (Cai et al., 2011; Srivastava and Srivastava, 2014). Adventitious roots are roots that develop from differentiated cells of an organ, such as a leaf, root, or stem. Due to their genetic and biosynthetic stability, adventitious roots are seen as an excellent source for the *in-vitro* production of secondary metabolites. Adventitious root cultures obtained by stimulation of any organ of the plant, such as leaf, stem, or root, can be used in the production of secondary metabolites from some plant species (Murthy et al., 201). The use of root cultures not only makes the production of whole plants unnecessary, but also increases the amount of secondary metabolites produced under *in-vitro* conditions relative to the amount of secondary metabolites produced by whole plants in nature (Srivastava and Srivastava, 2014).

One of the differentiated cell cultures, known as shoot (stem) culture, cultures the explants extracted from the shoot end meristem tissue in the medium. With shoot cultures, it is possible to acquire more secondary metabolites than those acquired in undifferentiated cultures (Erkoyuncu and Yorgancılar, 2015). Most bioactive compounds, such as bacoside A, carnasol, digitoxin, and hypericin, are produced in shoot cultures (Munish et al., 2015). Apical meristems of *Gentiana cruciata* were developed in shoot cultures to allow accumulation of secondary metabolites, such as amarogentin and iridoid glycosides (Hayta et al., 2011). The final class of differentiated cell cultures is somatic embryo culture (Fatima et al., 2015). Most phytochemical compounds, such as vinblastine and vincristine, are acquired from somatic embryo cultures. Cocoa oil is obtained from *Theobroma cacao* via somatic embryo cultures (Guillou et al., 2018).

Using plant cell and tissue cultures, it is possible to produce many economically important secondary metabolites, such as pharmaceutics and aromatic compounds. For those purposes, nearly 30,000 secondary metabolites such as alkaloids, flavonoids, lignin, organic acids, phenols, tannins, and terpenes are extracted from plants every year and new metabolites are added annually (Sood, 2020).

Bioactive compounds of plants are widely used in drug manufacturing. However, very small amounts of secondary metabolites are naturally obtained from plants and these amounts are not sufficient for drug manufacture (Georgiev, 2014). For example, alkaloids extracted from the *Catharantus roseus* plant are used in leukemia treatment. In order to acquire 1 g alkaloid required to meet the six-week needs of leukemia patients, 2 tons of plant leaves are required. This means production of medicinal plants must occur in large amounts which can be efficient on an agricultural scale. Micropropagation is an alternate mode of production used for reproducing plants via *in-vitro* cultures, generating solid, rapid, safe and homogeneous plants of a wide range of species. With this technique, it is also possible to produce secondary metabolites that are available in medicinal and aromatic plants (Khan et al., 2020). Under aseptic conditions, it is possible to micropropagate plants from explants derived from parts of plants with a totipotency property, such as callus, seed, or embryo (Abbasi et al., 2016). Plants developing under *in-vitro* conditions are transferred to *ex-vitro* conditions and allowed to adapt to natural conditions.

Another biotechnological method for producing secondary metabolites at higher amounts is a bioreactor. With plant tissue and cell cultures, secondary metabolites are produced in limited amounts and it is not possible to intervene in culture media because of contamination risk. Bioreactors can increase the amount of secondary metabolites produced, where the amount can reach thousands of liters. Bioreactors make it possible to increase commercial production from 0.75 ml to 75,000 l (Gerth et al., 2006; Gonçalves and Romano, 2018). This is because large bioreactors, which are tanks made of glass, plastic, or steel, allow for pH regulation, heat control, and addition of liquid culture without compromising the present aseptic conditions. In other words, bioreactors are a technological system with mechanical and electronic properties allowing for external physiochemical interventions. Bioreactor systems are a technology enabling the production of medicinal plants on an industrial scale and they are employed widely by the medicinal and cosmetics industry (Paek et al., 2005). For hairy root cultures; liquid-phase bioreactors, such as stirred tank reactors and bubble column reactors, and gas-phase bioreactors, such as nutrient mist reactors and liquid-gas-phase bioreactors are used. As hairy root cultures can develop even when the auxin hormone concentration decreases, they grow faster than other plant cultures. Also, with these cultures, it is possible to produce secondary metabolites similar to or in even greater amounts than in the natural plant. *Echinacea angustifolia* was cultivated in bioreactors and hairy root formation was stimulated via *Agrobacterium rhizogenes*, so that secondary metabolites such as caffeic acid and cichoric acid were obtained from the roots cultures produced (Cui et al., 2013). Plant cell suspensions and hairy root cultures are used as "natural bioactive factories" to produce plant-based compounds, and the use of a bioreactor system for these two methods is a noteworthy development for plant biotechnology and the commercial production of bioactive compounds (Sharma and Shahzad, 2013). The greatest production and the most successful commercial application of bioreactors was achieved in South Korea, where 45 tons (wet weight) of ginseng roots were produced. The largest plant cell culture facility in the world is in Germany where the facility commercially produces paclitaxel in 75,000 L capacity bioreactors (Yesil-Celiktas et al., 2010; Gonçalves and Romano, 2018). The greatest disadvantage of bioreactors is their high, initial setup costs as well as ongoing operation and control costs. However, they are popular with large companies due to the advantages of being fully controllable production environments and their capacity to make serial production in a very short time (Müller et al., 2008). As a result, companies in Japan and Germany produce secondary metabolites such as anthocyanins, arbitin, berberine, betacyanins, carthamin, digoxin, geraniol, ginseng, rosmarinic acid, and shikonin using plant cell and tissue cultures on a commercial scale (Ochoa-Villarreal et al., 2016).

Another biotechnological method used in the production of bioactive compounds is to add nanoparticles to the tissue culture medium to increase secondary metabolite production (Wang et al., 2016). Metal oxides, such as cadmium oxide, copper oxide, silicon dioxide, titanium oxide and zinc oxide, and metals, such as silver (silver nitrate) and gold, are now often used in the production of a variety of phytochemical compounds in *in-vitro* tissue cultures (Mohammed, 2015). When silver nanoparticles were applied to *Artemisia annua* hairy root cultures, the artemisinin concentration increased (Rivero-Montejo et al., 2021). When 40-nm cadmium oxide nanoparticles were added to barley (*Hordeum vulgare*) root culture, the amount of most secondary metabolites, including ferulic acid, linoleic acid, and palmitic acid increased (Večeřová et al., 2016). When silicon dioxide (5–15 nm) or titanium oxide (25 nm) nanoparticles were applied to the callus medium prepared to culture *Argania spinosa*, α-tocopherol production increased (Hegazi et al., 2020). A study conducted by Ahmad et al. (2020) reported an increase in the phenolic, antioxidant, and total flavonoid contents when ZnO (20–30 nm) or CuO (25–30 nm) nanoparticles were applied to the *Stevia rebaudiana* stem culture medium.

Genetic engineering studies and specific key transcription factors make secondary metabolite production possible in increased amounts by overexpressing appropriate biosynthetic genes *in-vitro* root cultures. The construct developed by placing the relevant genes within the boundaries of T-DNA via *Agrobacterium rhizogenes*, which is a plant pathogen and can infect the organs of plants such as leaf, root and stem and increase hairy rooting, is conveyed to the genome of the host plant. The *rolA*, *rolB*, *rolC*, and *rolD* oncogenes, which are known to be the moderators of plant growth and cell differentiation, and are carried on plasmids of *Agrobacterium rhizogenes*. Following the infection, these oncogenes are transferred into the host plant genome and facilitate the formation of roots with a constant growth potential (Bulgakov, 2008). *Rol* genes stimulate not only the hairy root formation but also the synthesis of different secondary metabolites. Signals formed by the *rolC* gene lead to the synthesis of phytoalexin-type secondary metabolites without intervening in the defense mechanism of the plant (Bulgakov et al., 2002, 2003). It is known that the *rolB* gene increases resveratrol production when introduced to *Vitis amurensis*. The *rolA* gene also stimulates nicotine production (Kiselev et al., 2007). Signals from the *rolD* gene stimulate proline production and encourage the development of roots. *Rol* genes not only have a collective effect but also antagonistic effects on each other. Signals formed by the *rolC* and *rolB* genes excessively increase the anthraquinone production in *Rubia cordifolia* (Bulgakov et al., 2002). When the *rolB* and *rolA* genes are introduced to *Ginseng* calli, they have an antagonistic effect on each other, so that ginsenoside production is not stimulated (Veena and Taylor, 2007). In hairy roots, which are induced as a result of infection of plants with *A. rhizogenes*, many important secondary metabolites, such as atropine, chlorogenic acid, cichoric acid, limonene, saponins and shikonin, are produced (Sharma and Shahzad, 2013).

CRISPR/Cas9 is a simple, accurate, and efficient gene editing technology currently paving the way for enhancing secondary metabolite pathways and in generating pharmacologically important metabolites (Dey, 2021). In order to increase the oleic acid content in *Camelina sativa* oil, fatty acid desaturase 2 genes were knocked out, using the CRISPR/Cas9 gene editing tool (Sainger et al., 2017). With a variety of technologies such as CRISPR/Cas9 and gene knockout (Jiang et al., 2017; Li et al., 2017) and targeted mutagenesis (Morineau et al., 2017), it is possible to produce a range of medically important metabolites such as alkaloids (Alagöz et al., 2016; Mercx et al., 2017), digosgenin (Feng et al., 2018), flavonoids (Mercx et al., 2017), phenolic acids (Li et al., 2017), and terpenes (Mercx et al., 2017) from medicinal plants such as *Camelina sativa* (Jiang et al., 2017; Morineau et al., 2017), *Nicotiana tabacum* (Mercx et al., 2017), *Papaver somniferum* (Alagöz et al., 2016), and *Salvia miltiorrhiza* (Li et al., 2017).

16.1.3 Production of Secondary Compounds from Aromatic and Medicinal Plants Cultivated in Turkiye by Biotechnological Methods

Studies have shown that a large number of secondary metabolites from medicinal plants have been successfully produced by tissue culture. Specific medicinal plants cultured in Turkiye via either tissue or cell culture studies have included *Achillea gypsicola*, *Capsicum annuum*, *Digitalis trojana*, *Daucus carota*, *Hypericum pruinatum*, *Hypericum perforatum*, *Hypericum adenotrichum*, *Isatis demiriziana*, *Physalis peruviana*, *Rosmarinus officinalis*, *Rubia tinctorum*, and *Vitis vinifera*. Both abiotic or biotic elicitors have been used to induce increased production of a number of secondary metabolites. For example, when melatonin was applied to the callus culture medium of *Rosmarinus officinalis*, the plant's rosmarinic acid and caffeic acid production doubled relative to the untreated control group. In addition, the melatonin-treated plants also started to produce volatile oils, such as linalool, methional, and styrene, which had not been present in the control group (Coskun et al., 2019). Likewise, after applying riboflavin at different concentrations and a UV-C elicitor at different times to *Daucus carota* callus culture, the anthocyanin content increased by 2.51 times at 4 mg riboflavin and by 2.17 times in response to a 15-minute UV-C application compared with the control group (Karataş et al., 2014). Another study which applied elicitors such as $CdSO_4$, fluorescent radiation, methyl jasmonate, or salicylic acid to the grapevine plant (*Vitis vinifera*) in a cell suspension culture medium found general increases in the total amount of secondary metabolites such as anthocyanins, phenolics, and *trans*-resveratrol compared with the control group (Cetin and Baydar, 2016). Methyl jasmonate applied to *Isatis demiriziana* cultured in *in-vitro* medium increased by 3.5 times the production of indirubin, which is commonly used in the treatment of diseases, such as chronic myelocytic leukemia and Alzheimer's disease, and also increased three fold the amount of triptantrin, which has many medicinal applications (Karakaş, 2019). Koç et al. (2020) found that putrescine and spermine applied as elicitors to a hot-pepper cell suspension culture at different concentrations increased capsaicin production, with spermine being more effective than putrescine. The most effective concentration for putrescine and spermine elicitors was 0.2 mg in both cases, and the most effective application time was 12 days. When *Physalis peruviana* was subjected to UV-B and short-term 45°C high-temperature stress *in vitro*, the amount of the plant's secondary metabolites, the withanolides, increased (Şahin et al., 2020).

Turkiye is an important center for *Hypericum* species. When the elicitor abscisic acid (ABA) was applied to *H. pruinatum* and *H. perforatum* cultures, the amount of pseudohypericin, which is used in drug manufacturing, increased (Önlü, 2019). In another study, *Hypericum adenotrichum* was cultured in medium to which chrome, mannan, pectin, polyethylene glycol (PEG), and sucrose were each applied as elicitors to increase the amount of secondary metabolites in the plant. The most effective elicitor for hypericin and pseudohypericin synthesis was pectin, whereas the most effective elicitor for the flavonoids hyperoside and isoquercitrin was chrome (Yamaner, 2011). In *Achillea gypsicola*, biotic (chitosan and yeast extract) and abiotic (cadmium chloride, light, methyl jasmonate, salicylic acid, silver nitrate, and sorbitol) elicitors were applied to cell suspension cultures to increase secondary metabolite accumulation. Light application increased the total phenolic and flavonol amount more than did any other elicitors. Although the yeast extract elicitor resulted in the greatest increases in concentrations of anthocyanins and camphor, chitosan, methyl jasmonate, and salicylic acid applications increased the concentrations of all secondary metabolites (Açıkgöz, 2017). Aras et al. (2018) found that salt stress applied to *in vitro* root cultures of *Rubia tinctorum* increased the total production of alizarin, anthraquinone, phenolic compounds, and purpurin. When both salicylic acid and temperature stress (45°C) were applied to *Digitalis trojana* callus cultures, good production of cardenolides was observed (Cingöz and Gurel, 2016). Islek et al. (2016) reported that an appropriate amount of thidiazuron applied to *Capsicum annuum* cell suspension culture medium led to an increase in capsaicin production. Immobilization of plant cells is another method used for enhancing secondary metabolite production in plant cell cultures. In *Rubia tinctorum*, the immobilization method increased alizarin production six fold and purpurin production 23 fold (Nartop, 2016). A similar study determined that application of $CdCl_2$ to the grape cell suspension culture medium increased the amount of phenolic compounds and tocopherols (Cetin et al., 2014). Islek et al. (2021) investigated the effects of immobilization and hydrostatic pressure applied to chili pepper cell suspension cultures on capsaicin production at different application times, and reported that the use of both immobilization and increased hydrostatic pressure led to notable increases in capsaicin amount. Callus was obtained from hypocotyl explants of pepper seedlings germinated *in vitro*, and then a cell suspension was prepared from the callus. Pressures of 50, 100, 200, 300, or 400 MPa were applied as elicitors to both free and immobilized cell suspensions, and the highest capsaicin amount was determined in the immobilized cell suspension cultures to which was applied 400 MPa.

Another common method used for increased biotechnological production of secondary metabolites is to modify the culture conditions to produce secondary metabolites from medicinal plants in cell or tissue cultures. Karatas (2021) determined the impacts of appropriate explant resources and auxin hormones (IAA, IBA, and NAA) on secondary metabolite contents of adventitious roots of *Rubia tinctorum in vitro*. He found that, after achieving adventitious root formation in the IBA medium, sub-culturing was useful to increase secondary metabolite production. *Digitalis* species are a medically important group because of the cardenolides (cardiac glycosides) they contain. Cardenolides have anticarcinogenic, acaricidal, and antibacterial properties. However, *Digitalis* species in Turkiye are endemic plants whose conservation status has been declared to be sensitive according to the Red Data Book of Turkish Plants (Ekim et al., 2000). An effective, fast, and highly productive *in vitro* reproduction protocol has been developed for *ex-situ* conservation of *Digitalis trojana*, which is endemic to Çanakkale and Balıkesir (Çördük and Aki, 2020). For this, adventitious shoots were obtained from leaf explants of plants germinated *in vitro* on Murashige & Skoog (MS) medium containing NA and BA. The greatest shoot production was reported in MS medium containing 3.0 mg/l BA + 0.1 mg/l NAA (Çördük and Aki, 2020). The researchers reported that this protocol proved to be a reliable method for this endangered, endemic species.

16.2 CONCLUSION

Although medicinal and aromatic plants have a wide area of usage, they are not often effectively cultivated commercially. More often than not, they are haphazardly collected from where they occur naturally for a variety of different purposes, which often leads to the destruction of natural ecosystems and the extinction of rare plants. In order to prevent such problems, it is necessary to protect medicinal and aromatic plants via advanced cultivation techniques and to train people not to collect these plants from nature whenever possible. Plant cell and tissue culture and genetic engineering techniques are now being intensively used for producing secondary metabolites from medicinal and aromatic plants. The amount of secondary metabolites produced using undifferentiated tissues like callus and cell suspension is less than the amount of secondary metabolites acquired using differentiated tissues like roots, shoots, and somatic embryos. When considering undifferentiated tissue cultures, cell suspension cultures grow more rapidly than callus cultures, so that the former are more desirable. Still, most metabolites are not produced in adequate amounts. The amount of secondary metabolites produced via root and shoot cultures is greater than the amount produced via somatic embryo cultures and may even be greater than the amount produced from plants in the nature. The greatest secondary metabolite production is observed in root cultures which have properties like genetic determination and rapid growth. It is necessary to choose plant tissue and cell cultures according to the properties of the target secondary metabolite to be produced and then to produce secondary metabolites using micropropagation or bioreactors on a commercial scale.

REFERENCES

Abbasi, B.H., Stiles, A.R., Saxena, P.K. and Liu, C.Z., 2012. Gibberellic acid increases secondary metabolite production in *Echinacea purpurea* hairy roots. *Application of Biochemistry and Biotechnology*, 168(7), 2057–2066.

Abbasi, B.H., Ali, H., Yücesan, B., Saeed, S., Rahman, K. and Khan, M.A., 2016. Evaluation of biochemical markers during somatic embryogenesis in *Silybum marianum* L. *3 Biotech*, 6(1), 1–8.

Açıkgöz, M.A., 2017. *Achillea gypsicola* türünde kallus kültürü ile sekonder Metabolit üretim potansiyelinin belirlenmesi [online]. Available from: http://earsiv.odu.edu.tr:8080/jspui/handle/11489/730 [Accessed 18 May 2022].

Aftab, T.A., 2019. Review of medicinal and aromatic plants and their secondary metabolites status under abiotic stress. *Journal of Medicinal Plants*, 7(3), 99–106.

Ahmad, M.A., Javed, R., Adeel, M., Rizwan, M., Ao, Q. and Yang, Y., 2020. Engineered ZnO and CuO nanoparticles ameliorate morphological and biochemical response in tissue culture regenerants of Candyleaf (*Stevia rebaudiana*). *Molecules*, 25(6), 1356.

Alagöz, Y., Gurkok, T., Zhang, B. and Unver, T., 2016. Manipulating the biosynthesis of bioactive compound alkaloids for next-generation metabolic engineering in opium poppy using CRISPR-Cas 9 genome editing technology. *Scientific Reports*, 6(1), 1–9.

Aras, Ö., Demirci, T. and Baydar, N.G., 2018. Effects of NaCl applications on root growth and secondary metabolite production in madder (*Rubia tinctorum* L.) root cultures. *International Journal of Secondary Metabolite*, 5(3), 210–216.

Arslan, N., Baydar, H., Kızıl, S., Karık, Ü., Şekeroğlu, N. and Gümüşçü, A., 2015. Tıbbi Aromatik Bitkiler Üretiminde Değişimler ve Yeni Arayışlar. *Ziraat Mühendisliği VIII. teknik kongresi*, 12 January 2015. Turkey.

Avcı, M., 2014. Türkiye'nin doğal egzotik ağaç ve çalıları. In: Ü. Akademik, ed. *Türkiye'nin Bitki Çeşitliliği ve Coğrafi Açıdan Değerlendirilmesi.* Ankara, Turkey: Orman Genel Müdürlüğü Yayınları, 28–53.

Bağdat, R.B., 2006. Tıbbi ve aromatik bitkilerin kullanım alanları, tıbbi adaçayı (*Salvia officinalis* L.) ve ülkemizde kekik adıyla bilinen türlerin yetiştirme teknikleri. *Tarla Bitkileri Merkez Araştırma Enstitüsü Dergisi*, 15(1–2), 19–28.

Barthe, G.A., Jourdan, P.S., McIntosh, C.A. and Mansell, R.L., 1987. Naringin and limonin production in callus culture and regenerated shoots from *Citrus* sp. *Journal of Plant Physiology*, 127(1–2), 55–65.

Bayraktar, Ö.V., Öztürk, G. and Arslan, D., 2017. Türkiye'de bazı tıbbi ve aromatik bitkilerin üretimi ve pazarlamasındaki gelişmelerin değerlendirilmesi. *Tarla Bitkileri Merkez Araştırma Enstitüsü Dergisi*, 26(2), 216–229.

Bayram, E., Kırıcı, S., Tansı, S., Yılmaz, G., Arabacı, O., Kızıl, S. and Telci, I., 2010. Tıbbi bitkilerin üretiminin arttırılması olanakları. *Ziraat Mühendisligi VII. Teknik Kongresi*, 11 January 2010. Ankara.

Bennett, R.N., Rosa, E.A., Mellon, F.A. and Kroon, P.A., 2006. Ontogenic profiling of glucosinolates, flavonoids, and other secondary metabolites in *Eruca sativa* (salad rocket), *Diplotaxis erucoides* (wall rocket), *Diplotaxis tenuifolia* (wild rocket), and *Bunias orientalis* (Turkish rocket). *Journal of Agricultural and Food Chemistry*, 54(11), 4005–4015.

Bettaieb Rebey, I., Bourgou, S., Aidi Wannes, W., Hamrouni Selami, I., Saidani Tounsi, M., Marzouk, B., Laure Fauconnier, M. and Ksouri, R., 2018. Comparative assessment of phytochemical profiles and antioxidant properties of Tunisian and Egyptian anise (*Pimpinella anisum* L.) seeds. *Plant Biosystems – An International Journal Dealing with all Aspects of Plant Biology*, 152(5), 971–978.

Beyzi, E. and Gürbüz, B., 2010. Ülkemizde kişniş *Coriandrum sativum* L. üretimi ve bitkinin genel özellikleri. *Ziraat Mühendisliği*, 354, 24–27.

Borrás-Linares, I., Stojanović, Z., Quirantes-Piné, R., Arráez-Román, D., Švarc-Gajić, J., Fernández-Gutiérrez, A. and Segura-Carreter, A., 2014. *Rosmarinus officinalis* leaves as a natural source of bioactive compounds. *International Journal of Molecular Sciences*, 15(11), 20585–20606.

Bozdemir, Ç., 2019. Türkiye'de yetişen kekik türleri, ekonomik önemi ve kullanım alanları. *Yüzüncü Yıl Üniversitesi Tarım Bilimleri Dergisi*, 29(3), 583–594.

Budak, I., Serim, A.T. and Asav, U., 2020. Ankara ili kimyon (*Cuminum cyminum* L.) tarlalarında bulunan yabancı otların tespiti. *Turkish Journal of Weed Science*, 23(2), 137–143.

Bulgakov, V.P., Tchernoded, G.K., Mischenko, N.P., Khodakovskaya, M.V., Glazunov, V.P., Radchenko, S.V., Zvereva, E.V., Fedoreyev, S.A. and Zhuravlev, Y.N., 2002. Effect of salicylic acid, methyl jasmonate, ethephon and cantharidin on anthraquinone production by *Rubia cordifolia* callus cultures transformed with the *rolB* and *rolC* genes. *Journal of Biotechnology*, 97(3), 213–221.

Bulgakov, V.P., Tchernoded, G.K., Mischenko, N.P., Shkryl, Y.N., Glazunov, V.P., Federoyev, S.A. and Zhuravlev, Z.N., 2003. Effects of Ca(2+) channel blockers and protein kinase/phosphatase inhibitors on growth and anthraquinone production in *Rubia cordifolia* callus cultures transformed by the *rolB* and *rolC* genes. *Planta*, 217(3), 349–355.

Bulgakov, V.P., 2008. Functions of rol genes in plant secondary metabolism. *Biotechnology Advances*, 26(4), 318–324.

Bulut, H.Y. and Arslan, N.T.D., 2016. Türkiye'de iklim faktörlerinin ve iklim değişikliğinin haşhaş (*Papaver somniferum* L.) bitkisinin verimi üzerine etkisi. *XII Ulusal Tarım Ekonomisi Kongresi*, 25 May 2016. Isparta.

Cai, Z., Riedel, H., Saw, N.M.M.T., Mewis, I., Reineke, K., Knorr, D. and Smetanska, I., 2011. Effects of elicitors and high hydrostatic pressure on secondary metabolism of *Vitis vinifera* suspension culture. *Process Biochemistry*, 46(7), 1411–1416.

Cetin, E.S., Babalik, Z., Hallac-Turk, F. and Gokturk-Baydar, N., 2014. The effects of cadmium chloride on secondary metabolite production in *Vitis vinifera* cv. cellsuspension cultures. *Biological Research*, 47(1), 1–6.

Cetin, E.S. and Baydar, N.G., 2016. Elicitor applications to cell suspension culture for production of phenolic compounds in grapevine. *Journal of Agricultural Sciences*, 22(1), 42–53.

Cingöz, G.S. and Gurel, E., 2016. Effects of salicylic acid on thermotolerance and cardenolide accumulation under high temperature stress in *Digitalis trojana* Ivanina. *Plant Physiology and Biochemistry*, 105, 145–149.

Çördük, N. and Aki, C., 2020. *In vitro* propagation of *Digitalis trojana* ıvanina an endemic medicinal plant of Turkey. In: S. Kumar, ed. *Endangered Plants*. London: IntechOpen, 3–12.

Coskun, Y., Duran, R.E. and Kilic, S., 2019. Striking effects of melatonin on secondary metabolites produced by callus culture of rosemary (*Rosmarinus officinalis* L.). *Plant Cell, Tissue and Organ Culture*, 138(1), 89–95.

Cui, H.J., Baque, M.A., Lee, E.J. and Paek, K.P., 2013. Scale-up of adventitious root cultures of *Echinacea angustifolia* in a pilot-scale bioreactor for the production of biomass and caffeic acid derivatives. *Plant Biotechnology Reports*, 7(3), 297–308.

Da Cheng, H., Xiao-Jie, G. and Pei, G.X., 2015. Phytochemical and biological research of Papaver pharmaceutical resources. *Medicinal Plants: Chemistry Biology and Omics*, 217–251.

Dauqan, E.M.A. and Abdullah, A., 2017. Medicinal and functional values of thyme (*Thymus vulgaris* L.) Herb. *Journal of Applied Biology and Biotechnology*, 5(2), 17–22.

Demirci, T., Özdamar, P. and Göktürk Baydar, N., 2015. Tıbbi ve aromatik bitkiler ile sebzelerde kök kaynaklı sekonder metabolitlerin üretiminin artırılmasına yönelik *in vitro* uygulamalar. *Türk Tarım-Gıda Bilim ve Teknoloji Dergisi*, 3(5), 261–270.

Dey, A., 2021. CRISPR/Cas genome editing to optimize pharmacologically active plant natural products. *Pharmacological Research*, 164, 105359.

Douglasa, M.H., Smallfield, B.M., Wallaceb, A.R. and Mc-Gimpseya, J.A., 2014. Saffron (*Crocus sativus* L.): The effect of mother corm size on progeny multiplication, flower and stigma production. *Scientia Horticulturae*, 166, 50–58.

Ekim, T., Koyuncu, M., Vural, M., Duman, H., Aytac, Z. and Adıguzel, N., 2000. Red data book of Turkish plants (Pteridophyta and Spermatophyta). In: S. Kumar, ed. *Endangered Plants*. London: IntechOpen, 246.

Erkoyuncu, M.T. and Yorgancılar, M., 2015. Bitki doku kültürü yöntemleri ile sekonder metabolitlerin üretimi. *Selçuk Tarım Bilimleri Dergisi*, 2(1), 66–76.

Farzami, M.S. and Ghorbant, M., 2005. Formation of catechin in callus cultures and micropropagation of *Rheum ribes* L. *Pakistan Journal of Biological Sciences*, 8(10), 1346–1350.

Fatima, S., Mujib, A. and Dipti, T., 2015. NaCl amendment improves vinblastine and vincristine synthesis in *Catharanthus roseus*: A case of stress signalling as evidenced by antioxidant enzymes activities. *Plant Cell, Tissue and Organ Culture*, 121(2), 445–458.

Faydaoğlu, E. and Sürücüoğlu, M.S., 2011. Geçmişten günümüze tıbbi ve aromatik bitkilerin kullanılması ve ekonomik önemi. *Kastamonu University Journal of Forestry Faculty*, 11(1), 52–67.

Feng, S., Song, W., Fu, R., Zhang, H., Xu, A. and Li, J., 2018. Application of the CRISPR/Cas9 system in *Dioscorea zingiberensis*. *Plant Cell, Tissue and Organ Culture*, 135(1), 133–141.

Fu, X., Yin, Z.P., Chen, J.G., Shangguan, X.C., Wang, X., Zhang, Q.F. and Peng, D.Y., 2015. Production of chlorogenic acid and its derivatives in hairy root cultures of *Stevia rebaudiana*. *Journal of Agricultural and Food Chemistry*, 63(1), 262–268.

Gerth, A., Schmidt, D. and Wilken, D., 2006. The production of plant secondary metabolites using bioreactors. *XXVII International Horticultural Congress-IHC2006: International Symposium on Plant Biotechnology*, August 2006.

Georgiev, M.I., 2014. Design of bioreactors for plant cell and organ cultures. In: K.Y. Paek, H.K. Murthy and J.J. Zhong, eds. *Production of Biomass and Bioactive Compounds Using Bioreactor Technology*. London: Springer, 3–16.

Gezici, S., Sekeroglu, N. and Kijjoa, A., 2017. *In vitro* anticancer activity and antioxidant properties of essential oils from *Populus alba* L. and *Rosmarinus officinalis* L. from South Eastern Anatolia of Turkey. *Indian Journal of Pharmaceutical Education and Research*, 51(3), 498–503.

Ghorbani, A. and Esmaeilizadeh, M., 2017. Pharmacological properties of *Salvia officinalis* and its components. *Journal of Traditional and Complementary Medicine*, 7(4), 433–440.

Giray, F.H., 2018. An analysis of world lavender oil markets and lessons for Turkey. *Journal of Essential Oil Bearing Plants*, 21(6), 1612–1623.

Gonçalves, S. and Romano, A., 2018. Production of plant secondary metabolites by using biotechnological tools. In: R. Vijayakumar and S.S.S. Raja, eds. *Secondary Metabolites: Sources and Application*. London: IntechOpen, 81–99.

González-Vallinas, M., Reglero, G. and de Molina, A.D., 2015. Rosemary (*Rosmarinus officinalis* L.) extract as a potential complementary agent in anticancer therapy. *Nutrition and Cancer*, 67(8), 1223–1231.

Graham, L.E., Graham, J.M. and Wilcox, L.W., 2008. *Bitki Biyolojisi*. Ankara: Palme yayıncılık, 351–357.

Guillou, C., Fillodeau, A., Brulard, E., Breton, D., Maraschin, S.D.F., Verdier, D., Simon, M. and Ducos, J.P., 2018. Indirect somatic embryogenesis of *Theobroma cacao* L. in liquid medium and improvement of embryo-to-plantlet conversion rate. *In Vitro Cellular and Developmental Biology – Plant*, 54(4), 377–391.

Gutierrez, G.E.P., Salas, M.A.P., Lopez, L.N., Mendoza, M.S.C., Olivo, G.V. and Heredia, J.B., 2017. Flavonoids and phenolic acids from oregano: Occurrence, biological activity and health benefits. *Plants*, 7(2), 2–23.

Gürkan, E., Öndersev, D.V., Ulusoylu, M., Göztaş, Z. and Dinçşahin, N., 2003. *Bitkisel tedavi. Marmara Üniversitesi Eczacılık Fakültesi Döner sermaye işletmesi*. İstanbul, Turkey: Teknik Eğitim Fakültesi Matbaa Birimi.

Hajhashemi, V., Ghannadi, A. and Sharif, B., 2005. Anti-inflammatory and analgesic properties of the leaf extracts and essential oil of *Lavandula angustifolia* Mill. *Journal of Ethnopharmacology*, 89(1), 67–71.

Hänsel, R., 2010. Saphonine. In: O. Sticher, ed. *Pharmakognosie-Phytopharmazie*. Berlin, Germany: Springer, 1007–1010.

Hayta, S., Akgun, I.H., Ganzera, M., Bedir, E. and Gurel, A., 2011. Shoot proliferation and HPLC-determination of iridoid glycosides in clones of *Gentiana cruciata* L. *Plant Cell, Tissue and Organ Culture*, 107(1), 175–180.

Hegazi, G.A., Ibrahim, W.M., Hendawy, M.H., Salem, H.M. and Ghareb, H.E., 2020. Improving α-tocopherol accumulation in *Argania Spinosa* suspension cultures by precursor and nanoparticles feeding. *Plant Archives*, 20, 2431–2437.

Hernandez-Saavedra, D., Perez-Ramirez, I.F., Ramos-Gomez, M., Mendoza-Diaz, S., Loarca-Pina, G. and Reynoso-Camacho, R., 2016. Phytochemical characterization and effect of *Calendula officinalis, Hypericum perforatum*, and *Salvia officinalis* infusions on obesity associated cardiovascular risk. *Medicinal Chemistry Research*, 25(1), 163–172.

Isah, T., Umar, S., Mujib, A., Sharma, M.P., Rajasekharan, P.E., Zafar, N. and Frukh, A., 2018. Secondary metabolism of pharmaceuticals in the plant *in vitro* cultures: Strategies, approaches, and limitations to achieving higher yield. *Plant Cell, Tissue and Organ Culture*, 132(2), 239–265.

Islek, C., Unal, B.T., Koc, E. and Kaya, D., 2016. Effect of thidiazuron used as an elicitor in the production of capsaicin on total protein and phenolic amounts, antioxidant enzyme activities of pepper plants. *International Journal of Environment, Agriculture and Biotecnology*, 1(3), 603–609.

Islek, C., Koc, E., Altuner, E.M. and Alpas, H., 2021. Using high hydrostatic pressure as an abiotic elicitor strategy for improving capsaicin production in free and immobilized cell suspension cultures of *Capsicum annuum* L. *High Pressure Research*, 42(2), 1–11.

Jiang, W.Z., Henry, I.M., Lynagh, P.G., Comai, L., Chaoon, E. and Weeks, D.P., 2017. Significant enhancement of fatty acid composition in seeds of the allohexaploid, *Camelina sativa*, using CRISPR/Cas9 gene editing. *Plant Biotechnology Journal*, 15(5), 648–657.

Kaçar, O., Göksu, E. and Azkan, N., 2010. Oğul otu (*Melissa officinalis* L.) yetiştiriciliğinde farklı bitki sıklıklarının bazı tarımsal özellikler üzerine etkisi. *Uludağ Üniversitesi Ziraat Fakültesi Dergisi*, 24(2), 59–71.

Kan, Y., Kartal, M., Aslan, S. and Yıldırım, N., 2006. Farklı koşullarda yetiştirilen rezene meyvelerinin uçucu yağ bileşenleri. *Ankara Eczacılık Fakültesi Dergisi*, 35(2), 95–101.

Karakaş, S., Çullu, M. and Dikilitaş, M., 2017. Comparison of two halophyte species (*Salsola soda* and *Portulaca oleracea*) for salt removal potential under different soil salinity conditions. *Turkish Journal of Agriculture and Forestry*, 41, 183–190.

Karakaş, Ö., 2019. Effects of methyl jasmonate and putrescine on tryptanthrin and ındirubin production in *in vitro* cultures of *Isatis demiriziana* mısırdalı. *International Journal of Secondary Metabolite*, 6(3), 241–250.

Karataş, İ., Elmastaş, M. and Karataş, R., 2014. Siyah havuç (*Daucus carota* ssp. sativus var. AtrorubensAlef) kallus kültüründe antosiyanin üretimine bazı uygulamaların etkisi. *Gaziosmanpaşa Bilimsel Araştırma Dergisi*, 9, 62–73.

Karatas, I., 2021. *Rubia tinctorum* L. (Kökboya) bitkisinin *in vitro* kök kültürlerinde hormon ve eksplant kaynaklarının sekonder Metabolit üretimi ve antioksidan aktivitelerine etkileri. *Kahramanmaraş Sütçü İmam Üniversitesi Tarım ve Doğa Dergisi*, 24(5), 939–947.

Khan, I., Khan, M.A., Shehzad, M.A., Ali, A., Mohammad, S., Ali, H., Alyemeni, M.N. and Ahmad, P., 2020. Micropropagation and production of health promoting lignans in *Linum usitatissimum*. *Plants*, 9(6), 728.

Kırıcı, S., 2015. Türkiye'de tıbbi ve aromatik bitkilerin genel durumu. *Türktop, Türkiye Tohumcular Birliği Dergisi*, 15, 4–11.

Kiselev, K.V., Dubrovina, A.S., Veselova, M.V., Bulgakov, V.P., Fedoreyev, S.A. and Zhuravlev, N.P., 2007. The *rolB* gene-induced overproduction of resveratrol in *Vitis amurensis* transformed cells. *Journal of Biotechnology*, 128(3), 681–692.

Koç, E., Işlek, C. and Karayiğit, B., 2020. Effects of spermine and putrescine polyamines on capsaicin accumulation in *Capsicum annuum* L. cell suspension cultures. *Acta Agriculturae Slovenica*, 115(2), 369–375.

Li, B., Cui, G., Shen, G., Zhan, Z., Huang, L., Chen, J.. and Qi, X., 2017. Targeted mutagenesis in the medicinal plant *Salvia miltiorrhiza*. *Scientific Reports*, 7(1), 1–9.

Maharik, N., Elgengaihi, S. and Taha, H., 2009. Anthocyanin production in callus cultures of *Crataegus sinaica* Bioss. *International Journal of Academic Research*, 1(1), 30–34.

Mercx, S., Smargiasso, N., Chaumont, F., De Pauv, E., Boutry, M. and Navarre, C., 2017. Inactivation of the β (1, 2)-xylosyltransferase and the α (1, 3)-fucosyltransferase genes in *Nicotiana tabacum* BY-2 cells by a multiplex CRISPR/Cas9 strategy results in glycoproteins without plant-specific glycans. *Frontiers in Plant Science*, 8, 403.

Mimica-Dukic, N. and Bozin, B., 2008. *Mentha* L. species (Lamiaceae) as promising sources of bioactive secondary metabolites. *Current Pharmaceutical Design*, 14(29), 3141–3150.

Miraj, S., Azizi, N. and Kiani, S., 2016. A review of chemical components and pharmacological effects of *Melissa officinalis* L. *Der Pharmacia Lettre*, 8(6), 229–237.

Misra, R.C., Maiti, P., Chanotiya, C.S., Shanker, K. and Ghosh, S., 2014. Methyl jasmonate-elicited transcriptional responses and pentacyclic triterpene biosynthesis in sweet basil. *Plant Physiology*, 164(2), 1028–1044.

Mnif, S. and Aifa, S., 2015. Cumin (*Cuminum cyminum* L.) from traditional uses to potential biomedical applications. *Chemistry and Biodiversity*, 12(5), 733–742.

Mohammed, A.E., 2015. Green synthesis, antimicrobial and cytotoxic effects of silver nanoparticles mediated by *Eucalyptus camaldulensis* leaf extract. *Asian Pacific Journal of Tropical Biomedicine*, 5(5), 382–386.

Morineau, C., Bellec, Y., Tellier, F., Gissot, L., Telemen, Z., Nogue, F. and Faure, J.D., 2017. Selective gene dosage by CRISPR-Cas9 genome editing in hexaploid *Camelina sativa*. *Plant Biotechnology Journal*, 15(6), 729–739.

Moore, J., Yousef, M. and Tsiani, E., 2016. Anticancer effects of rosemary (*Rosmarinus officinalis* L.) extract and rosemary extract polyphenols. *Nutrients*, 8(11), 731.

Munish, S., Ashok, A., Rajinder, G. and Sharada, M., 2015. Enhanced bacoside production in shoot cultures of *Bacopa monnieri* under the influence of abiotic elicitors. *Natural Product Research*, 29(8), 745–749.

Murthy, H.N., Dandin, V.S. and Paek, K.Y., 2016. Tools for biotechnological production of useful phytochemicals from adventitious root cultures. *Phytochemistry Reviews*, 15(1), 129–145.

Müller, J.L., Georgiev, M. and Bley, T., 2008. Metabolite and hormonal status of hairy root cultures of Devil's claw (*Harpagophytum procumbens*) in flasks and in a bubble column bioreactor. *Process Biochemistry*, 43(1), 15–23.

Nartop, P., 2016. Immobilization of *Rubia tinctorum* L. suspension cultures and biomass production. In: S.M. Jain, ed. *Protocols for In Vitro Cultures and Secondary Metabolite Analysis of Aromatic and Medicinal Plants: Second Edition*. New York: Springer, Humana press, 141–151.

Ochoa-Villarreal, M., Howat, S., Hong, S.M., Jang, M.O., Jin, Y.W., Lee, E.K. and Loake, G.J., 2016. Plant cell culture strategies for the production of natural products. *BMB Reports*, 49(3), 149–158.

Önlü, Ş., 2019. *Hypericumtürlerinde in vitro bitki rejenerasyonu ve sekonder metabolit üretimi* [online]. Available from: https://acikbilim.yok.gov.tr/handle/20.500.12812/44266 [Accessed 18 May 2022].

Özgen, Y., Arslan, N. and Bayraktar, N., 2017. Türkiye açısından önemli bitki haşhaşın önemi ve tarımı. *Ziraat Mühendisliği*, 364, 4–8.

Özgüven, M., Sekin, S., Gürbüz, B., Şekeroğlu, N., Ayanoğlu, F. and Ekrem, S., 2005. Tütün, tıbbi ve aromatik bitkiler üretimi ve ticareti. *Türkiye Ziraat Mühendisliği VI. Teknik Kongresi*, 3 January 2005. Ankara.

Öztürk, D. and Ocak, A., 2020. Bilecik florasına katkılar. *Journal of the Institute of Science and Tecnology*, 10(3), 1590–1596.

Öztürk, M., Temel, M. and Tınmaz, A.B., 2014. Türkiye'de kekik üretim ve pazarlaması. II. *Tıbbi ve Aromatik Bitkiler Sempozyumu*, 23 September 2014. Yalova.

Paek, K.Y., Chakrabarty, D. and Hahn, E.J., 2005. Application of bioreactor systems for large scale production of horticultural and medicinal plants. In: A.K. Hvoslef-Eide and V. Preil, eds. *Liquid Culture System for In Vitro Plant Propagation*. Dordrecht: Springer, 95–116.

Rivero-Montejo, S.D.J., Vargas-Hernandez, M. and Torres-Pacheco, I., 2021. Nanoparticles as novel elicitors to improve bioactive compounds in plants. *Agriculture*, 11(2), 134.

Ruffoni, B., Pistelli, L., Bertoli, A. and Pistelli, L., 2010. Plant cell cultures: Bioreactors for industrial production. In: M.S. Giardi, G. Rea and B. Berra, eds. *Bio-farms for Nutraceuticals: Functional Food and Safety Control by Biosensors*. Boston, MA: Springer, 203–221.

Sainger, M., Jaiwal, A., Sainger, P.A., Chaudhary, D., Jaiwal, R. and Jaiwal, P.K., 2017. Advances in genetic improvement of *Camelina sativa* for biofuel and industrial bio-products. *Renewable and Sustainable Energy Reviews*, 68, 623–637.

Sánchez-Camargo, A.P., Mendiola, J.A., Valdés, A., Castro-Puyana, M., García-Cañas, V., Cifuentes, A., Herrero, M. and Ibáñez, E., 2016. Supercritical antisolvent fractionation of rosemary extracts obtained by pressurized liquid extraction to enhance their antiproliferative activity. *The Journal of Supercritical Fluids*, 107, 581–589.

Sara–Alsadat, R., Ahmad, M. and Mokhtar, J.J., 2015. Paclitaxel production is enhanced in suspension cultured hazel (*Corylus avellana* L.) cells by using a combination of sugar, precursor, and elicitor. *Engineering in Life Sciences*, 15(2), 234–242.

Sedighi, R., Zhao, Y., Yerke, A. and Sang, S., 2015. Preventive and protective properties of rosemary (*Rosmarinus officinalis* L.) in obesity and diabetes mellitus of metabolic disorders: A brief review. *Current Opinion in Food Science*, 2, 58–70.

Shafay, S.E., El-Sheekh, M., Bases, E. and El-Shenody, R., 2021. Antioxidant, antidiabetic, anti-inflammatory and anticancer potential of some seaweed extracts. *Food Science and Technology*, 42, e20521.

Shakeri, A., Sahebkar, A. and Javadi, B., 2016. *Melissa officinalis* L.-A review of its traditional uses, phytochemistry and pharmacology. *Journal of Ethnopharmacology*, 188, 204–228.

Sharma, S. and Shahzad, A., 2013. Bioreactors: A rapid approach for secondary metabolite production. In: M. Shadid, A. Shahzad, A. Malik and A. Sahai, eds. *Recent Trends in Biotechnology Applications of Medicinal Plants*. Dordrecht: Springer, 25–49.

Shojaii, A. and Fard, M.A., 2012. Review of Pharmacological properties and chemical constituents of *Pimpinella anisum*. *International Scholarly Research Notices*, 2012, 510795.

Smetanska, I., 2008. Production of secondary metabolites using plant cell cultures. In: U. Stalh, U.E.B. Donalies and E. Nevoigt, eds. *Food Biotecnology*. Berlin: Springer, 187–228.

Srinivasan, K., 2018. Cumin (*Cuminum cyminum*) and black cumin (*Nigella sativa*) seeds: Traditional uses, chemical constituents, and nutraceutical effects. *Food Quality and Safety*, 2(1), 1–16.

Srivastava, S. and Srivastava, A.K., 2014. Effect of elicitors and precursors on azadirachtin production in hairy root culture of *Azadirachta indica*. *Applied Biochemistry and Biotechnology*, 172(4), 2286–2297.

Sood, H., 2020. Production of medicinal compounds from endangered and commercially important medicinal plants through cell and tissue culture technology for herbal industry. In: K. Sharma, K. Mishra, C. Danciu and K.K. Senepati, eds. *Bioactive Compounds in Nutraceutical and Functional Food for Good Human Health*. London: IntechOpen, 187–202.

Şahin, G., Telli, M., Ünlü, E.S. and Karakaş, F.P., 2020. Effects of moderate high temperature and UV-B on accumulation of withanolides and relative expression of the squalene synthase gene in *Physalis peruviana*. *Turkish Journal of Biology*, 44(5), 295–303.

Şen, B., Kessler, S., Gürdal, B., Kiemer, A. and Mat, A., 2016. The difference between the extracts of *Erica manipuliflora* in flowering and fruiting periods in terms of the cytotoxic effects. *İstanbul Üniversitesi Eczacılık Fakültesi Dergisi*, 46(2), 71–78.

Tabassum, H., Ahmad, A. and Ahmad, I.Z., 2018. *Nigella sativa* L. and its bioactive constituents as hepatoprotectant: A review. *Current Pharmaceutical Biotechnology*, 19(1), 43–67.

TMO, 2015. *Toprak mahsulleri ofisi. Haşhaş sektör raporu 2015* [online]. Available from: www.tmo.gov.tr [Accessed 18 May 2022].

Toker, R., Gölükcü, M. and Tokgöz, H., 2015. Tıbbi ve aromatik bitkilerin gıda sanayisinde kullanım alanları. *Türkiye Tohumcular Birliği Dergisi*, 4(15), 54–59.

TUIK, 2018. *Turkish Statistical Institute. Bitkisel üretim istatistikleri* [online]. Available from: www.tuik.gov.tr [Accessed 18 May 2022].

TUIK, 2019. *Turkish Statistical Institute. Bitkisel üretim istatistikleri* [online]. Available from: www.tuik.gov.tr [Accessed 18 May 2022].

TUIK, 2020. *Turkish Statistical Institute. Bitkisel üretim istatistikleri* [online]. Available from: www.tuik.gov.tr [Accessed 18 May 2022].

Tulukçu, E. and Sağdıç, O., 2011. Konya'da aktarlarda satılan tıbbi bitkiler ve kullanılan kısımları. *Erciyes Universitesi Fen Bilimleri Dergisi*, 27(4), 304–308.

Tunca, H. and Yeşilyurt, M.E., 2017. *Türkiye ve dünya'da kekik, DTB Raporu, Denizli* [online]. Available from: www.researchgate.com [Accessed 18 May 2022].

Umamaheswai, A. and Lalitha, V., 2007. *In vitro* effect of various growth hormones in *Capsicum annuum* L. on the callus induction and production of Capsiacin. *Journal of Plant Sciences*, 2(5), 545–551.

Večeřová, K., Večeřa, Z., Dočekal, B., Oravec, M., Pompeiano, A., Triska, J. and Urban, O., 2016. Changes of primary and secondary metabolites in barley plants exposed to CDO nanoparticles. *Environmental Pollution*, 218, 207–218.

Veena, V. and Taylor, G.C., 2007. *Agrobacterium rhizogenes*: Recent developments and promising applications. *In Vitro Cellular and Developmental Biology – Plant*, 43(5), 383–403.

Verpoorte, R., 2013. Secondary metabolites. In: R. Verpoorte, A.W. Alfermann, eds. *Metabolic engineering of plant secondary metabolism*. Netherlands: Springer Science, Business Media

Waller, G.R., MacVean, C.D. and Suzuki, T., 1983. High production of caffeine and related enzyme activities in callus cultures of *Coffea arabica* L. *Plant Cell Reports*, 2(3), 109–112.

Wang, P., Lombi, E., Zhao, F.J. and Kopittke, P.M., 2016. Nanotechnology: A new opportunity in plant sciences. *Trends in Plant Science*, 21(8), 699–712.

Wang, S., Towler, M.J. and Weathers, P.J., 2016. Root regulation of artemisinin production in *Artemisia annua*: Trichome and metabolite evidence. *Planta*, 244(5), 999–1010.

Wu, J., Wang, C. and Mei, X., 2001. Stimulation of Taxol production and excretion in *Taxus* spp cell cultures by rare earth chemical lanthanum. *Journal of Biotechnology*, 85(1), 67–73.

Yaldız, G. and Çamlıca, M., 2018. Türkiye'de tıbbi ve aromatik bitkilerin üretimi ve ticareti. *Bahçe*, 47, 224–229.

Yamaner, Ö., 2011. *Hypericum adenotrichum* Spach'un doku kültürü teknikleri ile çoğaltılması ve *in vitro* koşullarda sekonder metabolit değişiminin araştırılması [online]. Available from: http://adudspace.adu.edu.tr:8080/jspui/handle/11607/958 [Accessed 18 May 2022].

Yesil-Celiktas, O., Gurel, A. and Vardar-Sukan, F., 2010. Large scale cultivation of plant cell and tissue culture in bioreactors. *Transworld Research Network*, 37(667), 2.

Yıldırım, M.U., Sarihan, E.O. and Khawar, K.M., 2020. Ethnomedicinal and traditional usage of Saffron (*Crocus sativus* L.) in Turkey. Saffron: Academic Press, 21–31.

Zachariah, T.J. and Leela, N.K., 2018. Spices: Secondary metabolites and medicinal properties. In: A.B. Sharangi, ed. *Indian Spices: The Legacy. Production and Processing of India's Treasured Export*. Cham: Springer, 277–316.

Zhang, C.H., Fevereiro, P.S., He, G. and Chen, Z., 2007. Enhanced paclitaxel productivity and release capacity of *Taxus chinensis* cell suspension cultures adapted to chitosan. *Plant Science*, 172(1), 158–163.

17 Use of Medicinal and Aromatic Plant Metabolites in Cosmetics and Skin Diseases

Fatih Uckaya and Meryem Uckaya

CONTENTS

DOI: 10.1201/9781003226925-17

17.1 INTRODUCTION

Medicinal and aromatic plants (MAPs) have been used globally as therapeutics since time immemorial. Fellows (1991) defines a medicinal plant as a plant that contains a substance or substances that beneficially modulate the physiology of sick mammals and can be used by people for this purpose. According to Farnsworth and Soejarto (1991), any plant that purports to have medicinal properties, i.e., containing ingredients which are either health-related or have been proven to be useful or are used as drugs according to Western standards, is a medicinal plant. Plainly speaking, medicinal and aromatic plants are those that carry secondary metabolites such as alkaloids, terpenoids, or phenolic substances with pharmacological activity in organs such as leaves, stems, bark, flowers, fruits, seeds, roots, rhizomes, corms, and tubers (Baydar, 2013).

Complementary medicine today is heavily dependent on certain substances produced by plants. However, the fact that some currently clinically beneficial chemicals are derived from plants that were not previously classified as having medicinal properties changes the parameters of what a medicinal plant is. Based on the folklore prevalent in various tribal belts in the world about the medicinal utility of different wild-growing plant species, phytochemical analysis of those species in scientific laboratories is verifying the folk claims in most cases. The result is a consistent and massive increase in the number of medicinally acknowledged plants (Iqbal et al., 2018).

As far as is known, plants have been indisputably the most important source of medicine for all of mankind since the beginning of evolutionary history (Atique et al., 1985; Beigh et al., 2002), and this observation remains valid for the majority of the world's population. In 1985, WHO estimated that 3,200 million people around the world relied primarily on traditional medicines for their basic health needs (Farnsworth et al., 1985).

Similarly, aromatic plants are defined as plants that produce volatile compounds and thus are the sources of large amounts of essential oils. An essential oil is a concentrated, hydrophobic liquid that can evaporate easily and is the essence of herbal fragrances. Aromatic plants are primarily used in cosmetics, perfume production, flavorings in the food industry, and aromatherapy in medicinal products. Commonly known aromatic herbs include chamomile, lavender, peppermint, and rosemary.

Medicinal and aromatic plants have been known to be used as therapeutic, preventive, or complementary foods in many diseases since ancient times. Also, these plants are used as sweeteners in food. In addition, certain MAPs, with their essential oil content, are also included as raw materials in such fields as the cosmetic industry or paint industry.

The term "metabolite" generally refers to small molecules. Metabolites have various functions in plants such as fuel, structure, signals, stimulatory and inhibitory effects on enzymes, their own catalytic activities, and defense against and various interactions with other organisms. Metabolites found in plants are divided into primary metabolites and secondary metabolites. Primary metabolites are biological macromolecules composed of carbohydrates, proteins, fats and nucleic acids. They are used by the plant for essential energy needs, for metabolism, formation of body structure, and repair activities. Although secondary metabolites are very limited in plants, they play protective roles in plants and are important for their pharmacological effects.

17.1.1 Secondary Metabolites

Plants produce a variety of organic compounds, most of which are not directly involved in the processes of plant growth and development. These substances, called secondary metabolites, are distributed differently among plants, often within limited taxonomic groups. The increasing commercial importance of secondary metabolites has led to a great interest in the production of bioactive plant metabolites using tissue culture technology.

Tissue culture technology in plants can be routinely established under sterile conditions for either propagation or extraction of secondary metabolites from leaves, stems, roots, and meristems. It is known that *in vitro* production of secondary metabolites can be made in plant cell suspension cultures obtained from commercial and medicinal plants.

Unlike primary metabolites, the absence of secondary metabolites does not result in instant death. Rather, it causes long-term deterioration, or perhaps no significant change, in the survival, fertility, or aesthetics of the plant. Secondary metabolites are generally confined to a narrow cluster of species within a phylogenetic group. They also play important roles in plant defense against herbivores and in interspecific defense.

Secondary metabolites are biosynthetically derived from primary metabolites. Their distribution among plants is limited. They differ in quality and quantity for a particular plant species growing in different places. They are synthesized by specific cell types at different plant developmental stages.

Medicinal plants are rich in secondary metabolites, including alkaloids, glycosides, amines, steroids, flavonoids, and related metabolites, which are widely used in the pharmaceutical industry. Today, their use continues without decline, and, according to the World Health Organization (WHO), about 80% of the planet's population is still dependent on herbal medicines for health care, especially in developing countries (Colin, 2006; Gurib-Fakim, 2006). Aromatic plants have been used in the Middle East since 5000 BC for both their preservative and medicinal properties as well as for enhancing the aroma and flavor of foods (Chang, 2000; Li, 2006; Piccaglia et al., 1993).

In order to better understand the possible medicinal values of plants, it is necessary to know their chemical composition. The main life functions in which metabolites play a role are cell division and growth, respiration, storage, and reproduction. Primary metabolites include sugars, amino acids, Krebs cycle intermediates, proteins, nucleic acids, and small molecules such as polysaccharides. Consequently, primary metabolites are similarly present in all living cells (Seigler, 1995).

Secondary plant metabolites are divided into several classes based on their chemical structure. These are classified as follows:

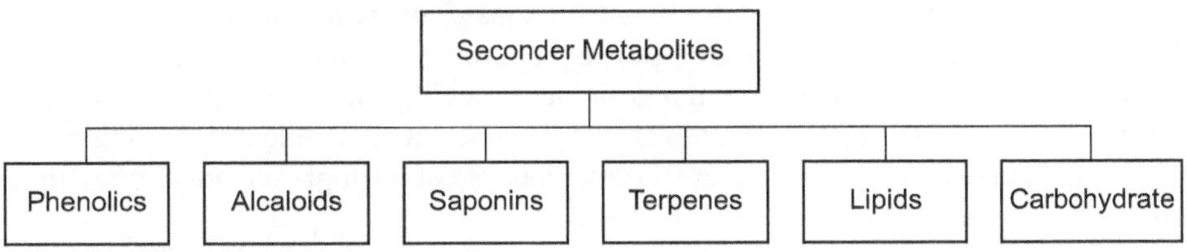

Phenolics

Phenolic compounds constitute the largest group of secondary metabolites found in plants. Their common feature is that they have at least one phenol group or aromatic ring (Figure 17.1). In addition to including simple structures with a single aromatic ring, there are also very complex polymeric molecular phenolic structures. Phenolic substances determine the color, taste, and aroma of many herbs, foods, and beverages.

Some phenolic compounds are very important pharmacologically. For example, quercetin is important due to its anti-inflammatory activity, silybin for its antihepatotoxic properties, genistein and daidzein for their phytestrogenic activity, and naringenin for its insecticidal properties (Goławska et al., 2014). Most of the phenolic molecules, of which the most effective are flavonoids,

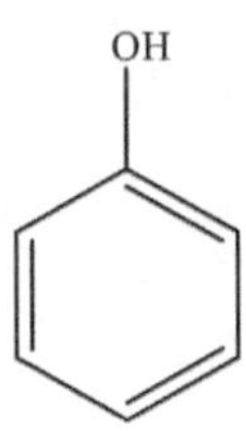

FIGURE 17.1 Phenol.

also have strong free radical scavenging activity. Phenolics can be classified based on their structure as follows:

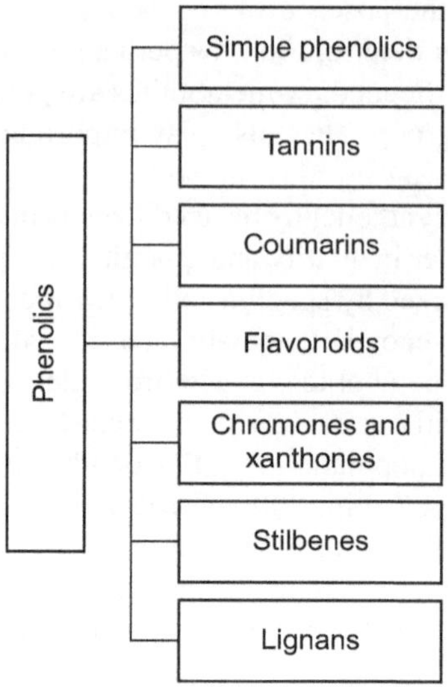

Phenolic acids are simple phenolic molecules found in all plants. Free phenol molecules are found in very few numbers in nature. Gallic acid, a prominent example, is particularly common (Figure 17.2). The gallic acid molecule has inhibitory effects against bacteria, viruses, fungi, inflammation, tumors, and mutagenic factors.

The pharmacological activities of many medicinal plants depend on their simple phenolic content. The antimicrobial and diuretic activities of *Arctostaphylos uva-ursi* are due to the phenolic molecules it contains (Eric, 2002). Red peppers contain capsaicinoids and, accordingly, show circulatory stimulant, rubefacient, and analgesic activities (Spiller et al., 2008). In addition, the anthelmintic activity of *Dryopteris filix-mas* can be given as examples of the pharmacological effects of simple phenolics (Hoffmann, 2003). An example of a simple phenolic, quassin, is shown in Figure 17.3.

Tannins are polyphenolic compounds with protein precipitation properties. They have been used since ancient times to transform raw animal hides into leather. However, this is not the case for many tannin-like substances due to their structure and biosynthetic origin (Hagerman and Butler, 1981). Tannins are divided into two main groups: hydrolyzable tannins and dense tannins. Hydrolyzable tannins consist of phenolic acid molecules, such as gallic, which are linked by ester bonds to a central glucose molecule. The two main types of hydrolyzable tannins are gallotanins and ellagitanins, which are composed of gallic acid and benzyl acid units, respectively. Geraniin, one of the ellagitannins, was isolated from *Geranium* species and its structure has been elucidated (Catarino et al.,

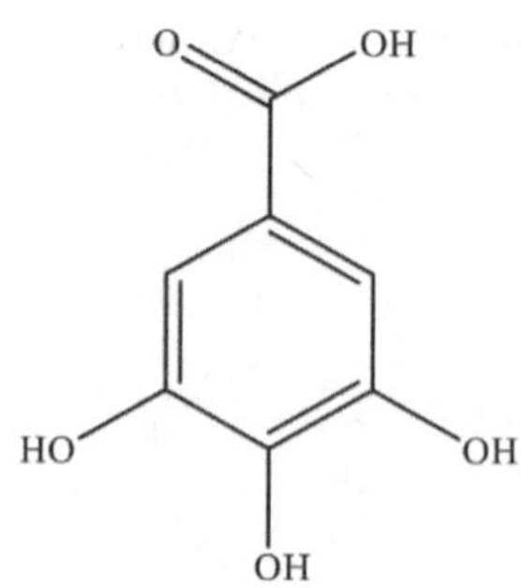

FIGURE 17.2 Gallic acid.

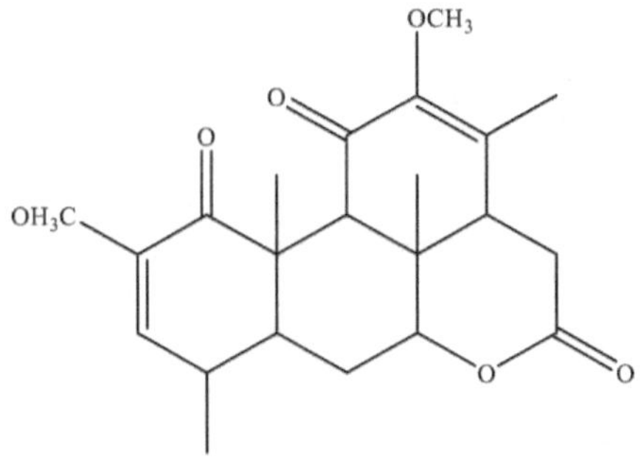

Quassin

FIGURE 17.3 Examples of simple phenolics.

2017). Condensed tannins, on the other hand, are compounds that resemble oligomeric flavonoid molecules (Puneet et al., 2013).

Coumarin molecules are derivatives of benzo-α-pyrone, which are the lactone of *O*-hydroxycinnamic acid. Approximately 1,000 natural coumarin molecules have been isolated from plants found in nature. The coumarin molecule is found in about 150 species belonging to more than thirty plant families. Alfalfa or melilot are examples of plants which are particularly rich in coumarins (Hoffmann, 2003). Aesculetin and umbelliferone, which are common examples of coumarin, are found in plants in free form or in the form of glycosides (Evans, 2009). It has been reported that coumarins have activities that inhibit inflammation, coagulation, cancer, and Alzheimer's disease (Xu et al., 2015). Some examples of coumarin are shown in Figure 17.4.

Flavonoids make up the largest group of phenols. More than 2,000 flavonoid compounds have been identified, of which about 500 are in the free form (Evans, 2009). Flavonoids are molecules that usually contain a benzyl ring at the 2, 3, or 4 positions in their structure. Flavonoids can be divided into classes such as anthocyanins, flavones, and flavonols according to the oxidation of the central C ring. Their similarity with flavones is that they are yellow in color. It is one of the largest phenolic groups commonly found in nature. However, they occur in young tissues and cell sap of higher plants. The biological activities of flavonoid-containing plant species such as *Glycyrrhiza glabra* and *Ginkgo biloba* have been investigated and reported to be used in pharmacologically effective drugs (Montanher et al., 2007). Some plant species containing flavonoids are included in the British Pharmacopoeia. The most prominent examples of these are *Betula pendula, Calendula officinalis, Leonurus cardiaca* and *Passiflora edulis.* This group is mostly known for its anti-inflammatory, anti-allergenic, antithrombotic, and vasoprotective properties, inhibition of tumor growth and protective properties of gastric mucosa in terms of biological activity (Serafini et al., 2010). Two examples of flavonoids are shown in Figure 17.5.

Chromone and xanthone compounds are structural derivatives of benzo-γ-pyrone and have no pharmaceutically significant effects. For example, cloves contain eugenin and mustard seeds contain khellin (Evans, 2009). Furanochromones have been identified as the active components of *Ammi visnaga* fruits and are complex molecules. *Polygala nyikensis* is used to treat skin problems in some mountainous areas. Accordingly, it has recently been shown that the root of the plant exhibits antifungal activity due to the presence of xanthones (Susana et al., 2011).

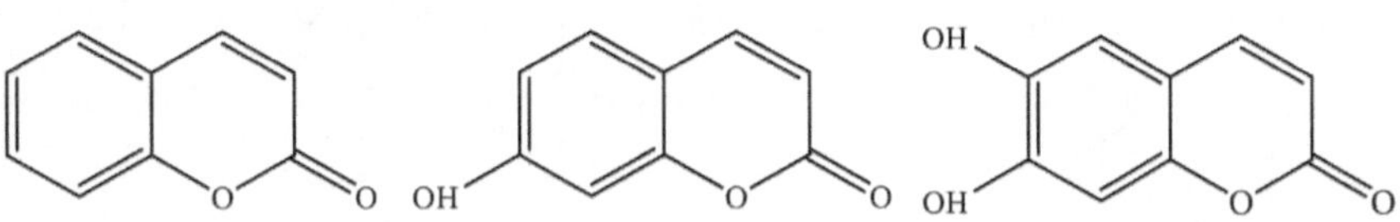

Coumarin Umbelliferone Esculetin

FIGURE 17.4 Examples of coumarins.

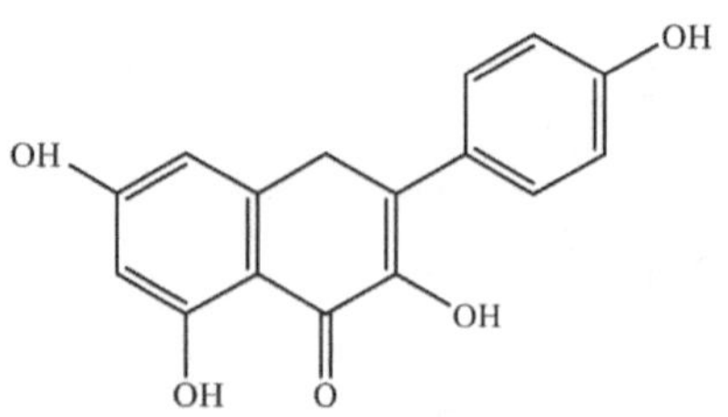

Kaempferol

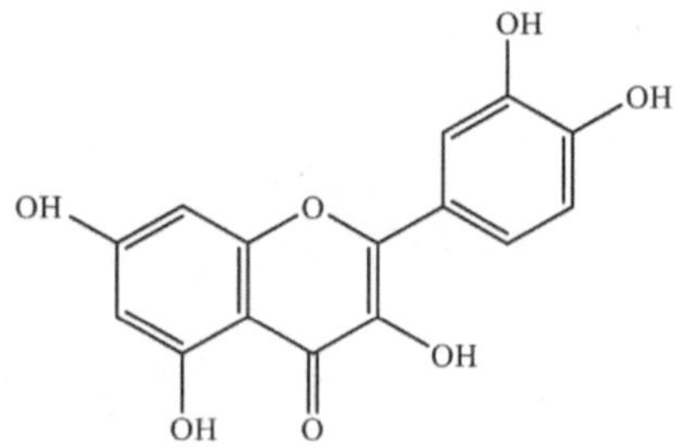

Quercetin

FIGURE 17.5 Examples of flavonoids.

Stilbenes are a group of secondary metabolites commonly found in plants as heartwood components. Stilbene has been reported in *Pinus* species (Seigler, 1995; Gehm et al., 1997).

Lignans are dimeric compounds formed mainly by combining two molecules of a phenylpropene derivative. They are common in members of the Asteraceae, Pinaceae, and Rutaceae families. The most prominent plant sources are *Achillea lingulata* and *Fagara heitzii*. Lignans generally show antimicrobial, antifungal, and cytotoxic activity (Mbaze et al., 2009; Saeidnia et al., 2011; Sharma et al., 2008).

Alkaloids are organic compounds containing at least one nitrogen atom in their heterocyclic ring. Except that it is known that all alkaloids contain nitrogen, it has not been possible to make a clear definition about them. The main types of alkaloids are acridones, aromatics, carbolines, ephedras, ergots, imidazoles, indoles, bisindoles, indolizidines, manzamines, oxindoles, quinolines, quinozolines, phenyl isoquinolines, phenylethylamines, purolininepyridines, loindoles, pyridines, and simple tetrahydroisoquinolines (Tadeusz, 2015).

Alkaloids

Alkaloids have a wide variety of pharmacological activities, including analgesia, local anesthesia, cardiac stimulation, respiratory stimulation, and vasoconstriction, muscle relaxation, antineoplastic, hypertensive, and hypotensive properties. They show toxicity, cytotoxicity, and many pharmacological activities in vertebrates (Seigler, 1995; Hoffmann, 2003).

Nicotine is abundant in tobacco plants and other *Nicotiana* species. The plant has calming properties (Figure 17.6). It is also extremely toxic and causes respiratory paralysis when used in high doses (Benowitz, 2009).

Caffeine is found in botanically unrelated species such as coffee, tea, mate, guarana, and colas. It is a molecule associated with chlorogenic acid found in raw coffee beans (Figure 17.6). Caffeine and other compounds that make up the most intense part of coffee flavor are released by roasting. The caffeine molecule exerts diuretic and stimulating effects on respiratory, cardiovascular, and central nervous systems (Benowitz, 1990).

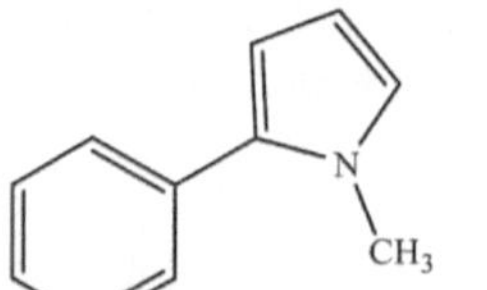

Nicotine

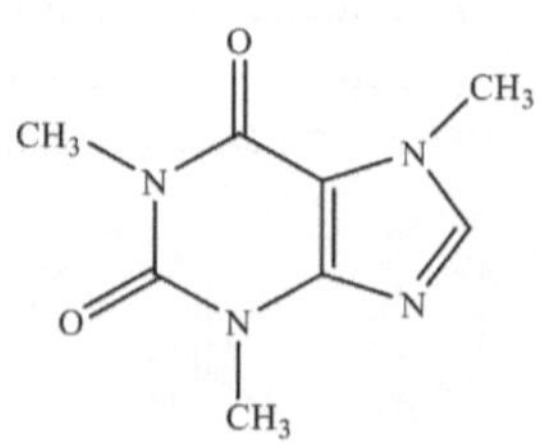

Caffeine

FIGURE 17.6 Examples of alkaloids.

Saponins

Saponins are compounds that have a steroid or triterpenoid attached to a carbohydrate unit and a polycyclic aglycone moiety. These carbohydrate units consist of a monosaccharide or oligosaccharide chain. These molecules allow foam formation in aqueous solutions. Saponins are found in plants from about 90 different families (Assa et al., 1973).

Saponins have many pharmacological effects. Some of these are antitumor, sedative, expectorant, and analgesic activities. Glycyrrhizin molecule is obtained from *Glycyrrhizae radix*. It is used in the treatment of chronic hepatitis and cirrhosis. The molecule also has an expectorant and antitussive effect. Some saponins, such as those from *Bupleurum falcatum*, have anti-inflammatory properties. Another saponin, aescin from horse chestnut, shows similar properties. It has also been reported that the aescin molecule has an anti-edematous effect (Güçlü-Üstündağ and Mazza, 2007).

Terpenes

Terpene molecules make up the largest group of secondary compounds found in plants. They are all derived from 5-carbon isoprene units that are chemically combined in different ways (Hoffmann, 2003). Classification of terpenes is made according to the number of isoprene units in their structures.

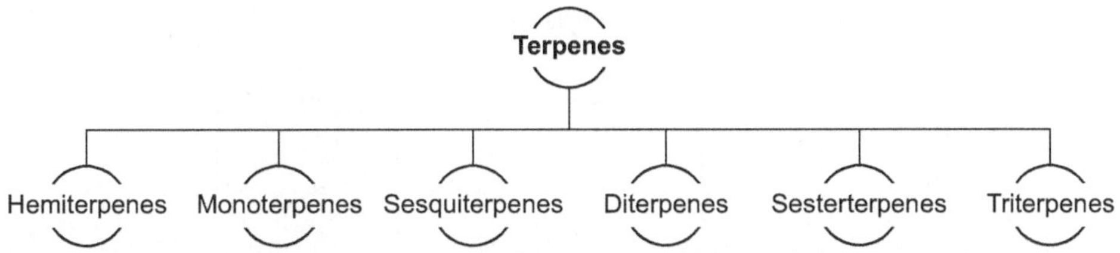

Hemiterpenes consist of units of terpene molecules containing a single isoprene unit. Angelic acid and isovaleric acid are called hemiterpenoids because these two single-isoprene unit molecules also contain oxygen (Seigler, 1995).

Monoterpenes are molecules that consist of two isoprene units and have the molecular formula $C_{10}H_{16}$ (Figure 17.7). They are among the most important components of essential oils found in plants. Geraniol, a monoterpene, is widely found in nature. Monoterpenes also include molecular groups such as unsaturated hydrocarbons, alcohols, alcohol esters, aldehydes, and ketones. Monoterpenes, like camphor and menthol compounds, have anti-irritant and antipruritic effects (Yi et al., 2004).

Sesquiterpenes have the molecular formula $C_{15}H_{24}$ and consist of three isoprene units (Figure 17.8). There are more than 200 structural sesquiterpene species based on their biogenetic structure. Some sesquiterpene lactone molecules show antibacterial, antifungal, and antiprotozoan activities. The helenalin molecule found in *Arnica montana* flowers also has anticardiotonic effects (Hikino, 1985; Seigler, 1995).

Diterpenes are molecules that consist of four isoprene units and have the chemical formula $C_{20}H_{32}$ (Figure 17.9). These are divided into two classes: acyclic and macrocyclic compounds. Diterpenes can have six-membered ring structures or fused five- or seven-membered ring structures. They exist in the molecule as substituents such as esters or epoxides (Hoffmann, 2003). Diterpenoids are found

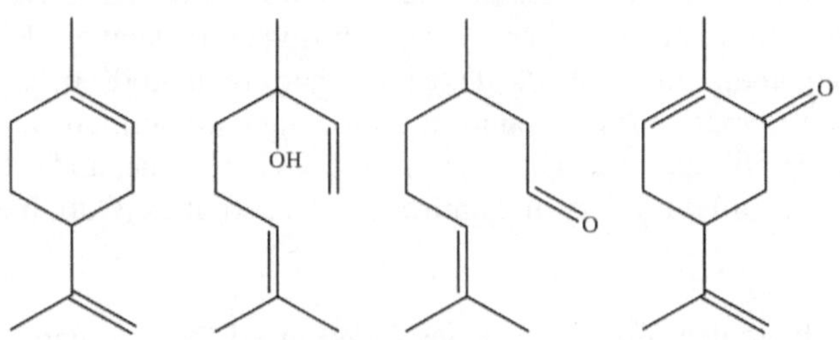

Limoene Linalool Citronellal Carvone

FIGURE 17.7 Examples of monoterpenes.

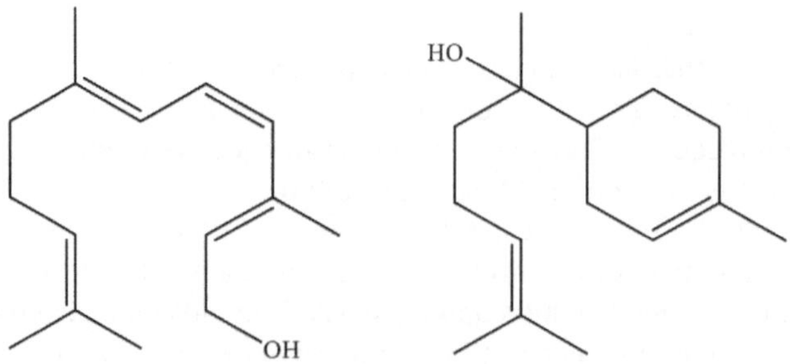

Farnesol Farnesol

FIGURE 17.8 Examples of sesquiterpenes.

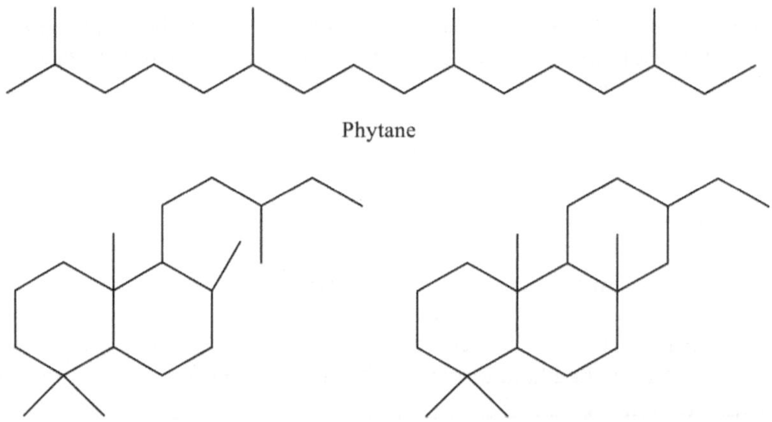

Phytane

Labdane Abiatane

FIGURE 17.9 Examples of diterpenes.

as active molecules in some plants. Vitamin A is a diterpenoid and is called carotene. *Teucrium chamaedrys* belongs to the Labiatae family and produces neoclerodan-type diterpenes, which are used medicinally as diaphoretic and antirheumatic (Bağcı et al., 2010). Like all terpene groups, diterpenes have shown some pharmacological properties such as analgesic, antibacterial, antifungal, and anti-inflammatory activities (Hoffmann, 2003).

Sesterterpenes have 25 carbon atoms and 5 isoprene units. A well-known example of sesterterpenoids is geranyl farnesol. This molecule is isolated from the seed oils of *Camellia sasanqua* of the Theaceae family (Akihisa et al., 1999). The geranyl farnesol molecule has been reported to exert cytotoxic activity in rat leukemic M1 cells (Ishukura et al., 1984).

Triterpenes have the molecular formula $C_{30}H_{48}$ and consist of six isoprene units (Figure 17.10). Squalene is a typical linear triterpene. It is the main component of shark liver oil. It is formed as a result of the reduction reaction of two farnesyl pyrophosphate molecules. Triterpenes make up an important part of lipids found in all plants. There are more than 4,000 triterpenoids isolated from plants in nature. These compounds are found in both plants and animals as precursors of steroid molecules (Seigler, 1995). β-boswellic acid, an ursane-type triterpene, and α-boswellic acid, an oil-type triterpene, have anti-inflammatory and antirheumatic activities (Culioli et al., 2003).

Lipids

Lipids are one of the biological macromolecules found in all living organisms. First of all, these molecules are divided into saturated and unsaturated fats. Structurally, lipids can be classified as triglycerides and membrane lipids. Lipids, as the main component of all biological membranes, form

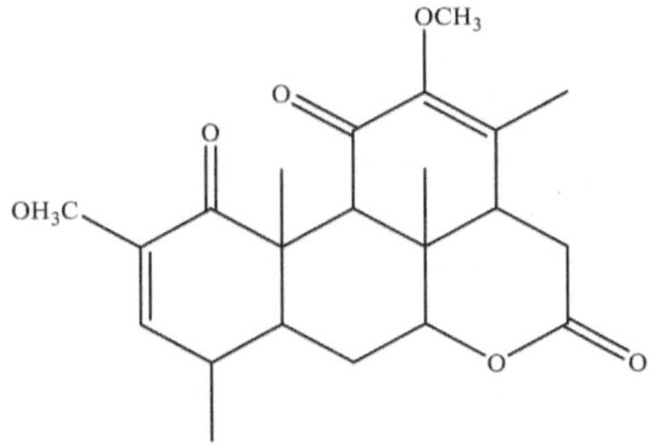

Quassin

FIGURE 17.10 Examples of triterpenes.

the cell structure and help protect the cell against external factors. They are also used as energy storage for various biological activities and fuel for cellular activities (Subramaniam et al., 2011). Although lipids are examples of the primary metabolites of plants, recent studies have also reported that these molecules also have pharmacological effects as secondary metabolites (Fahy et al., 2009).

Carbohydrates

Carbohydrates are found in all living organisms on the planet. Carbohydrates are the starting point of all phytochemicals and all animal biochemicals, as well as being the first product of photosynthesis in plants. Cellulose is a homopolysaccharide formed by combining glucose molecules. They are the structures that make the plants hard and strong. It is also the single, most abundant organic substance on the earth. Although carbohydrates are primary metabolites, they act as secondary metabolites via glycosidation linkages (Asif et al., 2011).

17.1.2 Functions of Secondary Metabolites

Many secondary metabolites have signal functions that affect the activities of other cells, control their metabolic activities, and coordinate the development of the whole plant. (Mansfield, 2000). Plants use secondary metabolites, such as essential oils, flavonoids, or tetraterpenes, to attract insects for pollination. Similarly, they use secondary compounds to attract other animals for the propagation of seeds. In this case, secondary metabolites act as signaling compounds. In addition, today's drugs and food supplements containing terpenoid, alkaloid, or flavonoid molecules are used to treat or prevent various diseases (Raskin et al., 2000). In particular, some of these compounds are effective in preventing certain types of cancer (Reddy et al., 2003; Watson et al., 2001).

It is estimated that 14–28% of higher plant species are used medicinally and 74% of pharmacologically active plant-derived components were discovered after ethnomedicinal uses of plants have been followed (Ncube et al., 2008). Secondary metabolites are metabolic intermediates or products of general metabolism that are found as products of differentiation in a limited number of taxonomic groups; they are not essential for the growth and survival of the producer organism. Secondary metabolites can be found in plants in different ratios. They provide pharmacological efficacy according to their rates of presence. In research, their molecules are isolated, and their biological activities and clinical effects are studied. Due to the low amount of these substances in plants and the losses occurring during isolation, they are synthesized by chemical methods or produced by tissue culture methods. In Table 17.1, some plants, their active ingredients, and the tissue culture types which have been used, are presented. The advantage of the chemical method is that it is cheap and practical, but there are also side effects of the production of synthetic molecules. The tissue culture method is an expensive method. However, it seems more advantageous since natural molecules and their appropriate isomers are obtained (Wang et al., 2015).

TABLE 17.1
Secondary Metabolites from Plant Cell, Tissue, and Organ Cultures

Plant Name	Active Ingredient	Culture type (Reference)
Adhato davasica	Vasine	Shoot culture (Shalaka et al., 2009)
Agastache rugosa	Rosmarinic acid	Hairy root (Lee et al., 2007)
Ammi majus	Umbelliferone	Shootlet (Krolicka et al., 2006)
	Triterpenoid	Suspension (Staniszewska et al., 2003)
Angelica gigas	Deoursin	Hairy root (Xu et al., 2008)
Arachis hypogaea	Resveratol	Hairy root (Kim et al., 2008)
Artemisia annua	Artemisinin	Callus (Baldi and Dixit, 2008)
Aspidosperma ramiflorum	Ramiflorin	Callus (Olivira et al., 2001)
Azadirachta indica	Azadirachtin	Suspension (Sujanya et al., 2008)
Brucea javanica	Cathin	Suspension (Wagiah et al., 2008)
Bupleurum falcatum	Saikosaponins	Root (Kusakari et al., 2000)
Camellia chinensis	Flavones	Callus (Nikolaeva et al., 2009)
Capsicum annum	Capsaicin	Callus (Umamaheswai, A. and Lalitha, 2007)
Cassia acutifolia	Anthraquinones	Suspension (Nazif et al., 2000)
Cassia senna	Anthraquinones	Hairy root (Shrivastava et al., 2006)
Catharanthus roseus	Indole alkaloids	Suspension (Zhao et al., 2001)
	Vincristine	Suspension (Lee-Parsons and Rogce, 2006)
	Catharathine	Suspension (Ramani and Jayabaskaran, 2008)
Cayratia trifoliata	Stilbenes	Suspension (Roat and Ramawat, 2009)
Centella asiatica	Asiaticoside	Hairy root (Kim et al., 2007), Callus (Kiong et al., 2005)
Drosera rotundifolia	7-Methyljuglone	Shoot culture (Hohtola et al., 2005)
Eleutherococcus senticosus	Eleuthrosides	Suspension (Shohael et al., 2005)
Eriobotrya japonica	Triterpenes	Callus (Shohael et al., 2005)
Fabiana imbricata	Rutin	Callus and cell suspenson (Schmeda-Hirschmann et al., 2004)
Fagopyrum esculentum	Rutin	Hairy root (Lee et al., 2008)
Fritillaria unibracteata	Alkaloids	Multiple shoot (Gao et al., 2004)
Gentiana macrophylla	Glucosides	Hairy root (Tiwari et al., 2007)
Gentianella austriaca	Xanthone	Multiple shoot (Vinterhalter et al., 2008)
Glycyrrhiza glabra	Glycyrrhizin	Hairy root (Mehrotra et al., 2008)
Gymnema sylvestre	Gymnemic acid	Callus (Gopi and Vatsala, 2006)
Hemidesmus indicus	Lupeol, Rutin	Shoot culture (Misra et al., 2005)
Hypericum perforatum	Hypericin	Multiple shoot (Kornfeld et al., 2007)
Mentha arvensis	Terpenoids	Shoot (Phatak and Heble, 2002)
Momordica charantia	Flavonoids	Callus (Agarwal and Kamal, 2007)

There are more than 3,000 plants used for their essential oils, and approximately 300 of them are used commercially for aroma and fragrance purposes (Van de Braak and Leijten, 1999). Some biological activities and various uses of medicinal and aromatic plants depend on these oils (Table 17.2).

17.2 SECONDARY METABOLITES IN COSMETICS

The desire to look beautiful is a human need and is as old as the origin of humanity. Cosmetics are substances specially prepared to add beauty to people in general and increase their attractiveness.

TABLE 17.2
Activities and Uses of Medicinal and Aromatic Plants

Main activities	Uses
Antioxidant activity	Cosmetic industry, perfumes, skin products
Antimicrobial activity	Pharmaceutical industry, medicines
Cytotoxic activity	Food industry, flavours, fragrances, preservatives

Prehistoric people used tree leaves and animal skins to protect themselves from the harmful effects of nature. At that time, common vegetation and minerals were used for cosmetic purposes (Jain and Namita, 1979). The use of traditional systems has stood the test of time. Only a few new drugs and modern herbal cosmetics have emerged in these systems. There are many medicinal plants that promote hair growth, keeping it soft and silky, and not allowing hair to turn gray at an early stage. Over the centuries, experiences continued to accumulate, and now cosmetology has become a well-defined science in its own right. The study of primitive cultures has also been an indication that cosmetic beautification forms are applied in all areas of the business world (Upadhyay, 1985).

In Ayurvedic and Unani systems of medicine, herbs have been an integral part of the health system since ancient times (Iqbal et al., 2018; Parveen et al., 2020a, 2020b). In addition to health care, medicinal plants beautify the body with various cosmetics and colors. The synthetic chemical compounds could produce no suitable match for the safety and efficacy of natural products (Chunekar and Pandey, 1969).

According to the Consumer Products Regulations 1989, a cosmetic product means any substance or preparation intended to be applied to any external surface of the human body (i.e., the epidermis, hair system, nails, lips and external genitalia) or teeth. Except for therapeutic purposes, the product is primarily used to clean its target, scent it, protect it, keep it in good condition, change its appearance, or combat body odor or perspiration (Dweck, 1996).

New bioactive compounds can be obtained from algae found in the sea and plants grown in the soil. Popular ingredients include Chinese herbs, vitamins, minerals, antioxidants, enzymes, hormones, and many natural substances. Today, many new products containing natural oils and herbs have been introduced to the market and the numbers are likely to increase in the future. Before the methods for synthesizing substances with similar properties were discovered, plants were the main source and basis of all cosmetics (Dweck, 1996).

The use of plant extracts in cosmetic formulations has been increasing in recent years, as animal-derived extracts are less preferred. This is partly due to reports of bovine spongiform encephalitis. The negative effects of some products of animal origin need to be eliminated or replaced. If synthetic chemicals are to be preferred, their use is limited due to their side effects. Natural molecules obtained from plant extracts open new horizons for further studies. The plant extracts, extraction method, herb-solvent ratio, and content of active substances are not sufficiently defined for the field of cosmetics. However, for cosmetic products color, fragrance, transparency, or the stability of active ingredients over time are often a limiting factor. Plant extracts differ in their content from purified and isolated therapeutic agents in several respects. First, they have a lower concentration of the central compound than the pure chemical synthetic molecules. Second, the medicinal plants often contain additional active molecules that are both chemically and therapeutically effective (Aburjai and Natsheh, 2003).

In cosmetics, herbal extracts are used as (a) total extracts and (b) selective extracts. Total extracts essentially apply essences based on historical traditions. On the other hand, selective extracts are mostly used because of the effectiveness of their specific activity. For instance, licorice is preferred for irritated skin, ginkgo as a free radical scavenger, blueberries for skin lightening, walnuts for skin tanning, and wheat germ to stimulate cell proliferation. Wherever possible, the selection of plant materials is based on their ethnobotanical use. Quoting the data published by raw material suppliers is not appropriate. These data are often very difficult to prove before they are used in official texts (Dweck, 1997a).

17.2.1 Some Secondary Metabolites Used for Cosmetic Purposes

Acacia concinna DC. (Mimosaceae):

This saponin has been isolated from seeds of the plant. It is a mixture of acacinin-A and acacinin-B. Acacinin-A consists of acacic acid and a carbohydrate mixture containing glucose, xylose, and fructose. Akasic acid, lactone, and sapogenin-B were isolated from the husk membrane of seeds. The seed pods, known as shikakai, are used as detergents and as a fish poison. A decoction of the capsules is a laxative and relieves bile. The seeds are roasted and eaten. The seed pods' dry powder is used to stimulate hair growth and prevents hair cracks, hair loss, and dandruff, and maintains the original hair color (Sharma et al., 2003) (Figure 17.11).

Callicarpa macrophylla Vahl (Verbenaceae)

A new tetracyclic diterpene, kalliterpenone, and its acetate have been isolated from *C. macrophylla* and their stereostructures determined. Other compounds, such as luteolin, apigenin, luteolin-7-*O*-glucuronide, and apigenin-7-*O*-glucuronide, have also been isolated from the dried fruits. The roots give an aromatic oil used for stomach ailments and the leaves are heated and applied to rheumatic joints. The dried fruit powder is mixed with creams to treat acne, pimples, blemishes, and other allergic skin spots (Sharma et al., 2003) (Figure 17.12).

Curcuma amada Roxb. (Zingiberaceae)

Campesterol, stigmasterol, β-sitosterol, cholesterol, and fatty acids are purified from its rhizomes. The fatty acids contain saturated, straight-chain monoenoic and dienoic acids. In the Ayurvedic system of medicine, dry rhizomes of the plant are recommended for stomach problems. These rhizomes are also used for bruises and sprains. Their powder is used to treat acne and blemishes on the face (Sharma et al., 2003) (Figure 17.13).

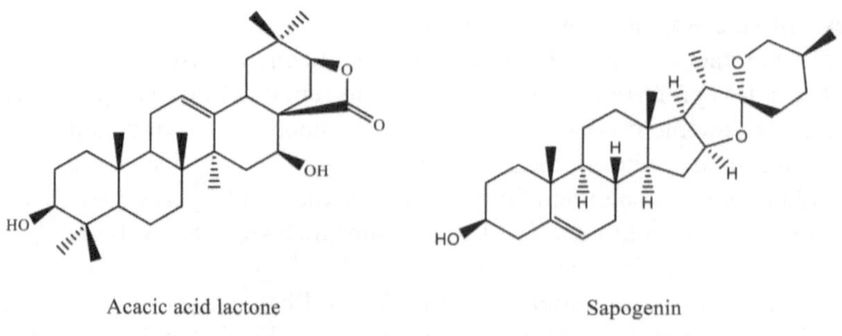

FIGURE 17.11 Structure of active compounds of *Acacia concinna*.

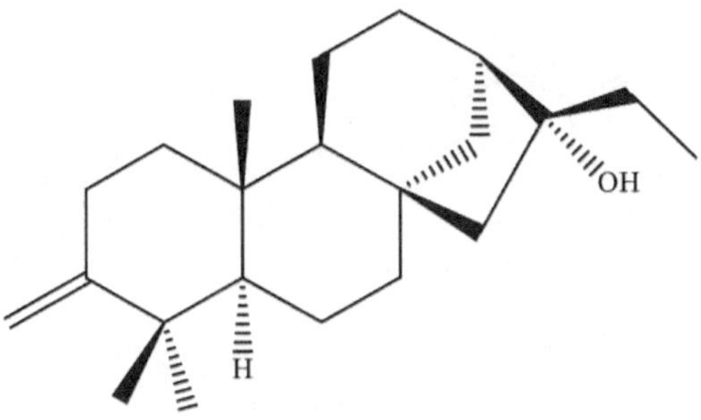

Calliterpenone

FIGURE 17.12 Structure of active compounds of *Callicarpa macrophylla*.

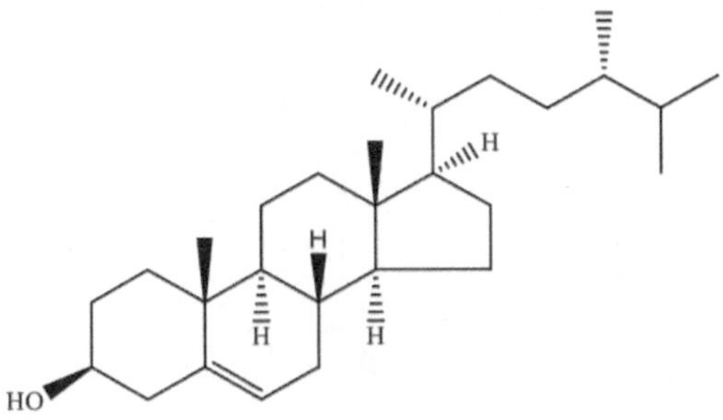

Campesterol

FIGURE 17.13 Structure of active compounds of *Curcuma amada*.

Curcuma longa Linn. (Zingiberaceae)

The main components of the dry rhizome of the plant are D-α-phellandrene, D-sabinene, cineol, borneol, and zingiberene. In the Indian systems of medicine, turmeric is used as an abdominal pain reliever, tonic, and blood purifier. Fresh rhizome juice is used as an antiparasitic in many skin diseases. The rhizomes contain curcumin, an essential oil and crystalline coloring agent used as a carminative, stomachache reliever, and tonic. Dried rhizome powder has been used for cosmetic purposes since the ancient times. Turmeric paste with a little lime is a popular treatment for sprains and bruises. The rhizome is used to enhance the color of the skin (Sharma et al., 2003) (Figure 17.14).

Emblica officinalis Gaertn. (Euphorbiaceae)

Phyllembic acid is one of the main components of the fruits. Trigalloylglucose, terchebin, corilagin, and ellagic acid are among the molecules isolated from fruits, which are used as an astringent, cooling, diuretic, and laxative. The fruit is a rich source of vitamin C. The oil obtained from its seeds is especially recommended for hair care. Dried fruit powder is used in hair dyes and shampoos, and as a tonic for hair roots (Sharma et al., 2003) (Figure 17.15).

Rubia cordifolia Linn. (Rubiaceae)

The plant contains a coloring agent that is a mixture of purpurin and munjistin. Some parts of the plant contain alizarin and its glycosides. Cristofin molecule, a new anthraquinone derivative, is also obtained from this plant and used for the dyeing of coarse, cotton fabrics. The plant roots are tonic, antidysenteric, antiseptic, and decongestant. The boiled leaves and stems work as a vermifuge. In addition, plant roots are used for coloring medicinal oils. For the treatment of dark spots on the face, the powder of dried roots is mixed with honey and applied to the skin (Sharma et al., 2003) (Figure 17.16).

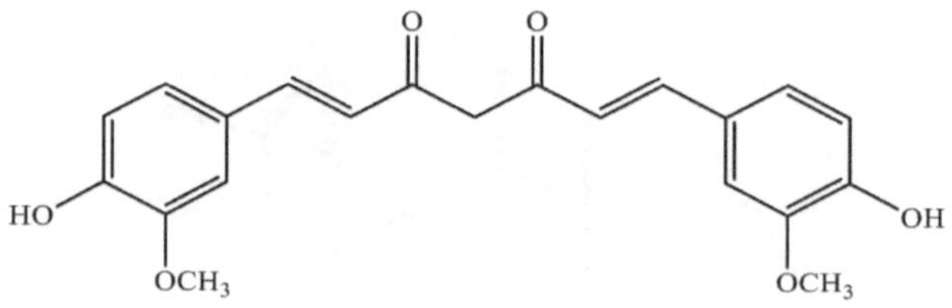

Curcumin

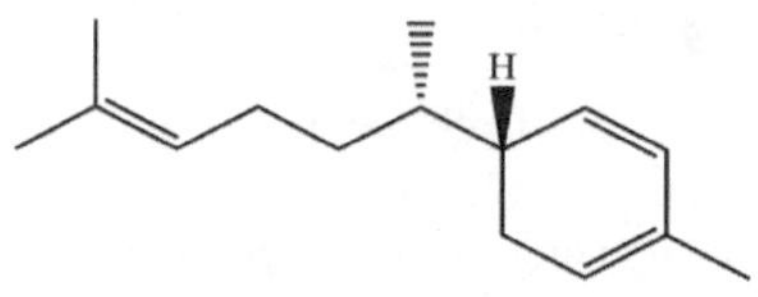

Zingiberene

FIGURE 17.14 Structure of active compounds of *Curcuma longa*.

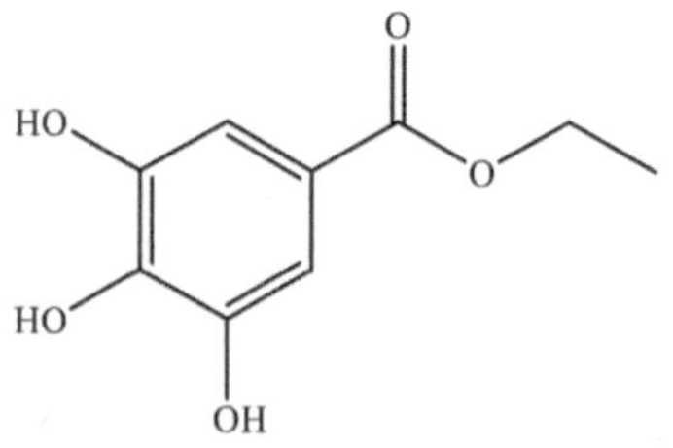

Phyllembic acid

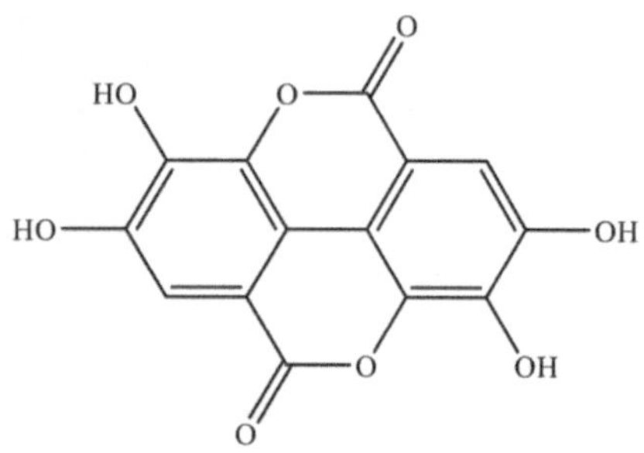

Ellagic acid

FIGURE 17.15 Structure of active compounds of *Emblica officinalis*.

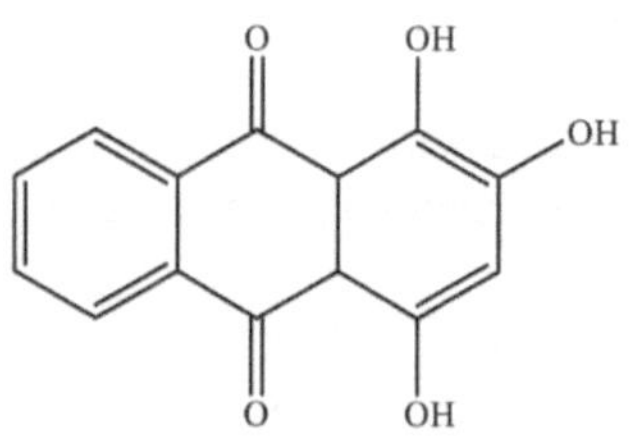

Purpurin

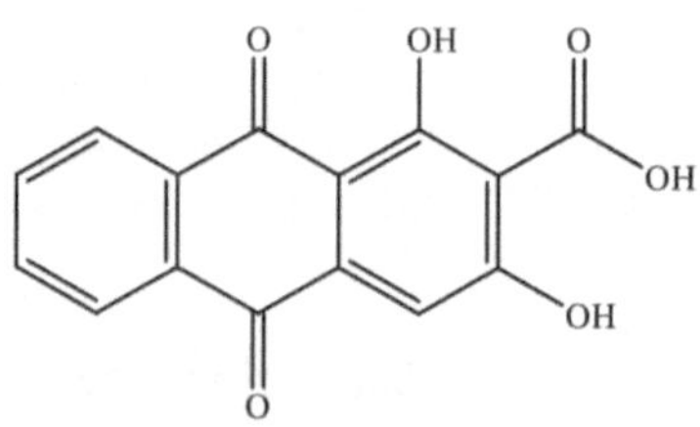

Munjistin

FIGURE 17.16 Structure of active compounds of *Rubia cordifolia*.

17.3 SKIN DISEASES AND SECONDARY METABOLITES

The skin, the largest organ of the body, creates a strong protective barrier against external influences. Some important functions of the skin include regulating the body's temperature and fluid balance, and protecting the body against harmful microbes, chemicals, and sun rays. The skin consists of three layers, the epidermis, dermis, and hypodermis. In the dermis layer, there is connective tissue containing elastin and collagen fibers in particular, as well as blood vessels and special nerve endings. The epidermis is the outermost layer of the skin. The stratum corneum is the outermost part of the epidermis that interacts directly with the environment. This part is not only a collection of dead cells, but also a complex organ that maintains the homeostatic balance of the skin. Disruption of the

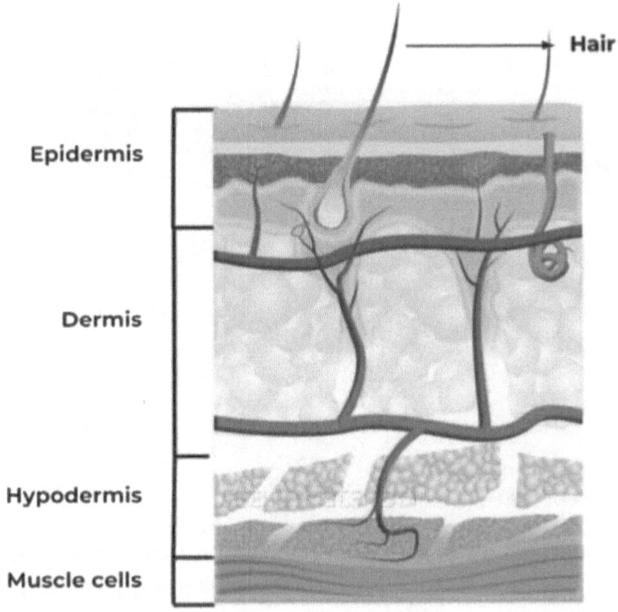

FIGURE 17.17 Structure of skin.

homeostatic balance occurs during the aging process. Accordingly, rapid changes in the external environment directly affect the epidermis and dermis (Tranggono, 2000) (Figure 17.17).

Active compounds of natural origin are added to cosmetic formulations to protect the skin against exogenous or endogenous harmful agents, as well as to rebalance the dermal homeostasis lipids that change with dermatosis and aging. These formulations are characterized by a lipid composition close to human sebum. Therefore, partial treatment of the outermost layers of the epidermis with a large amount of phospholipids is achieved by forming a thin monomolecular layer that supplies the skin with natural moisture. Compounds in the chemical composition of phospholipids provide protection against oxidation, harmful sun rays, inflammation, and radical molecules (Orthoefer, 2002; Passi, 2002).

The mucilage, consisting of polysaccharides and complex sugars, relieves dryness and forms a membrane covering the nerves. Fatty acid- and triglyceride-rich seed oils reduce the transepidermal water loss. They are also used as emollients to keep the skin moist and during the formation of skin and hair. Plants with anti-inflammatory activity generally contain high concentrations of flavonoid molecules, which are very rich in tannins, and are used to tighten and tone the skin. It is beneficial to use medicinal herbs containing antimicrobial and antifungal biocides for the healing of infected skin (Aburjai et al., 2003) (Figure 17.18).

17.3.1 Acne vulgaris and Secondary Metabolites

Acne vulgaris is a multifactorial disease caused by the chronic, inflammatory effect of pilosebaceous follicles (Liu and Huang, 2013; Song et al., 2011). Acne is one of the most common skin diseases in the world, affecting all age groups (Zouboulis et al., 2005). It is very common in teenagers during adolescence. This is because of their androgen production during this period (Wu et al., 2007). The incidence of acne is higher in adult women than in men. It has been reported that the incidence rate of the disease is approximately 12% (Perkins et al., 2012). Approximately 50 million people in the United States are exposed to this disease (Zaenglein et al., 2016). It has been reported that the total cost of acne treatment is approximately 3 billion dollars per year (Bhate and Williams, 2013). However, acne does not threaten life seriously, but it causes a decrease in

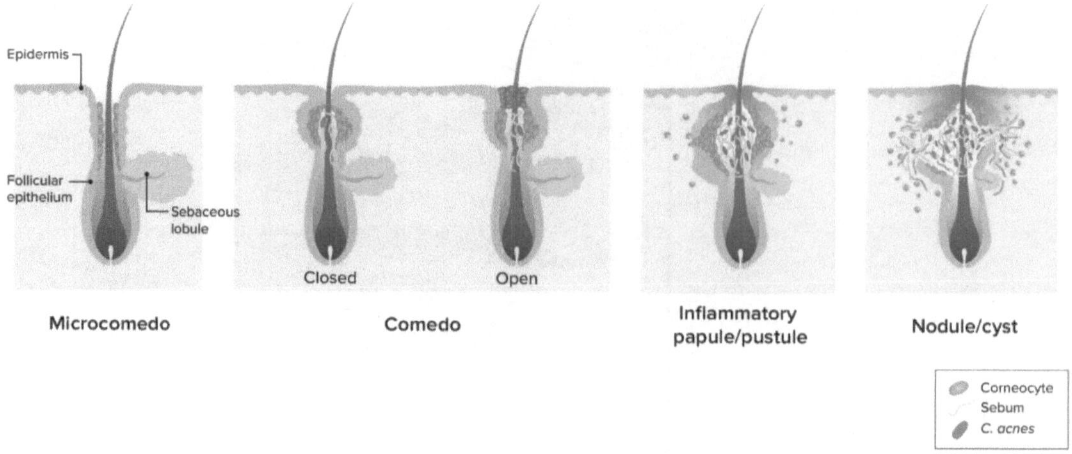

FIGURE 17.18 Formation of acne vulgaris.

self-confidence, increased anxiety disorder, and alienation from society in patients (Rapp et al., 2006). The main pathogenic factors that play a role in acne formation are the proliferation and colonization of *Propionibacterium acnes* bacteria, abnormal follicular hyperkeratinization, increased sebum production, and inflammatory mechanisms (Zaenglein et al., 2016). In addition, diet and family genetic structure significantly affect the development of acne lesions (Kapoor and Saraf, 2011). The classification of acne vulgaris is made according to the factors of comedones, papules, pustules, cysts, nodules, and scar formation (Tahir et al., 2016). Due to the negative impact of acne on quality of life, drug complications, and drug therapy problems such as inadequate response, it seems necessary to investigate new pharmacological agents for the treatment of this disorder. In this context, medicinal plants constitute a valuable resource for the discovery of new drugs that can be used to treat acne vulgaris (Azimi et al., 2012; Bahramsoltani et al., 2014; Hosein Farzaei et al., 2016; Williams et al., 2012).

Alpha-mangostin

Alpha-mangostin is a natural xantonoid molecule extracted from various parts of the mangosteen tree (Jung et al., 2006). It has been shown in the literature that film-forming solutions containing an α-mangostin-rich extract inhibit the growth of *P. acnes* (Asautjarit et al., 2014). Alpha-mangostine prevented activation of the TAK1-NF-κB pathway and reduced inflammation (Zou et al., 2019).

Ampelopsin

Ampelopsin molecule is a type of favonoid that regulates levels of plasma lipid and blood sugar. The molecule has pharmacological activities, such as antitumor, neuroprotective, and hepatoprotective activity (Kou and Chen, 2012). Ampelopsin suppressed ROS-mediated activation of the PI3K/Akt/NF-kB signaling pathways (Qi et al., 2012).

Artonin E

Artonin E is a prenylated flavonoid molecule extracted from *Artocarpus elasticus*. The molecule has biological properties, such as antimalarial, antifungal, antibacterial, anti-estrogenic and anticancer activities (Etti et al., 2017). Artonin E showed antimicrobial activity comparable to clindamycin (Nayak et al., 2017). It also inhibited arachidonate 5-lipoxygenase activity (Reddy et al., 1991).

Berberine

Berberine is an isoquinoline alkaloid. The molecule was extracted from *Berberis vulgaris* L. It increased bacterial death by macrophages by increasing adenosine monophosphate-activated protein kinase signaling (Li et al., 2017). It inhibited proinflammatory IL-6 and TNF-α protein secretion (Spatuzza et al., 2014) and modulated the Sirt1/NF-κB signaling pathway in macrophages (Zhang et al., 2017). The molecule showed an inhibitory effect against oxidative stress (Sadeghnia et al., 2017; Sun et al., 2017). It suppressed lipogenesis of sebaceous glands and also reduced sebum production (Seki and Morohashi, 1993).

Curcumin

Curcumin is a polyphenolic compound. The molecule has extensive pharmacological effects. It has been extracted from the rhizomes of *Curcuma longa* L., which has antiproliferative, anticarcinogenic, and antibacterial activities (Farzaei et al., 2018; Haghi et al., 2017). Curcumin-loaded particles accumulated significantly in the skin and inhibited the growth of *P. acnes*. In addition, curcumin-loaded myristic acid microemulsions showed an inhibitory effect against *Staphylococcus epidermidis* (Liu and Huang, 2012). Curcumin has been shown to have antioxidant effects by activating the NF-κB pathway (Dai et al., 2018) and inhibiting the secretion of IL-1β, IL-6, and TNF-α (Zhang and Li, 2018). In addition, curcumin suppressed infection by regulating the NF-kB signaling pathway (Yu et al., 2018).

Ellagic Acid

The ellagic acid molecule is a benzopyran-derived polyphenol. The molecule was isolated from *Punica granatum* L. (Hosein Farzaei et al., 2016). The application of tetracycline and ellagic acid by mixing the molecules in certain proportions made *P. acnes* more sensitive to antibiotics and the human immune system (Sivasankar et al., 2016). Ellagic acid decreased the expression of COX-2, iNOS, NF-κB, TNF-α, IL-1ß and IL-6 (Umesalma and Sudhandıran, 2010).

Fustin

The fustin molecule is a type of flavonoid obtained from the plant *Intsia palembanica* Miq. The molecule inhibited *P. acnes* lipase activity more potently than tetracycline (Batubara et al., 2014). It also inhibited IL-6 mRNA expression (Moon et al., 2015).

Lupeol

The lupeol molecule is a pentacyclic lupane-type triterpene. The molecule has lipid-lowering, anticancer, and antimicrobial activities. Lupeol has been described in numerous plant species, including *Solanum melongena* L. (Liu et al., 2013). Lupeol pacified the *P. acnes* infection by decreasing IL-12 and TNF-α and increasing IL-10 (Zhu et al., 2016). It reduces ROS levels and lipid peroxidation with its antioxidant power (Prasad et al., 2008).

Myricetin

The myricetin molecule is a common flavonoid and is at the forefront with its nutraceutical value (Semwal et al., 2016). It suppresses NF-κB activation in the MAPK signaling pathway (Hou et al., 2018). In addition, it shows an anti-inflammatory effect in human keratinocytes through inhibition of production pathways stimulated by TNF-α (Da Lee and Lee, 2016). It has also shown antioxidant activity by suppressing the induction of STAT1 activation (Cho et al., 2016), upregulation of MAPK signaling pathways (Kang et al., 2010), and scavenging ROS (Chobot and Hadacek, 2011).

Resveratrol

Resveratrol molecule is a phytoalexin polyphenol naturally produced by different plants. It shows inhibitory effect against free radicals, oxidation, and inflammation (Farzaei et al., 2018). Resveratrol has shown strong antibacterial activity against *P. acnes* (Farzaei et al., 2018). For the molecule, higher concentrations (200 mg/l) show a bactericidal effect, whereas lower resveratrol concentrations (50 and 100 mg/l) show a bacteriostatic effect (Docherty et al., 2007).

Terchebulin

The terchebulin molecule is an ellagitannin with a terchebulic acid moiety. The molecule is an acne-healing agent with inhibitory activity against lipase enzyme and oxidation. It exhibits a potent bacterial inhibition activity against *P. acnes* (Muddathir et al., 2013).

Ursolic acid

The ursolic acid molecule is a pentacyclic triterpenoid obtained from various plants. It has many biological functions, including antimicrobial, anti-wrinkle, and anticancer activities (Sultana, 2011). Ursolic acid also shows anti-inflammatory activity. It does this by suppressing IL-1ß and IL-6 (Zhang et al., 2018) and inhibiting STAT3 and NF-KB activities (Peng et al., 2016). It also shows antioxidant activity through MAPK modulation (Ma et al., 2014).

17.3.2 Atopic Dermatitis and Secondary Metabolites

Atopic dermatitis is a skin disease that causes itchy and chronic inflammation (Miodovnik et al., 2012). Although its pathogenesis has not been fully elucidated, various studies have reported that its pathogenesis is related to factors such as allergies, genetic factors, environmental factors, and immunological functions (Albanesi et al., 2007). Eczema is characterized by some typical symptoms such as itching, skin redness, increased skin thickness, and skin dryness caused by inflammatory responses (Wilmer and Luster, 1995). Infiltration of inflammatory cells, including macrophages, mast cells, and eosinophils, mediates eczema formation (Howell et al., 2008).

Keratinocytes form the densest group of epidermal cells. These cells play a critical role in the pathogenesis of inflammatory skin diseases, including atopic dermatitis (Chang et al., 2003). Keratinocytes produce different cytokines and chemokines in response to signals sent by TNF-α and interferon-gamma (Xia et al., 1995). Different cytokines and chemokines are considered among the important regulators of eczema pathogenesis. In addition, stimulation of keratinocytes by TNF-α and IFN-γ can lead to activation of different signaling pathways, such as NF-kB, mitogen-activated protein kinases, and signal transducer and activator of transcription (STAT) 1 pathways (Kumar and Pandey, 2013).

The natural product isolated from plant sources is responsible for the realization of different pharmacological activities. It has been reported in the literature that many flavonoids have anti-oxidative, hepatoprotective, antibacterial, antiviral, anticancer, anti-inflammatory, and anti-photoaging activities (Kwon et al., 2012; Mathieson and Hehre, 1994) (Table 17.3).

Some medicinal herbs reportedly cause liver damage. The use of these herbs, together with drugs such as methotrexate, will be harmful. Examples of these are echinacea, chaparral, german, and ragwort, etc. (Borins, 1998). For example, herbs containing linolenic acid, such as evening primrose oil, which is used to treat dermatitis, psoriasis, and xerosis, lower the seizure threshold. Therefore, it may be necessary to increase anticonvulsant doses (Ferguson et al., 1997; Choi and Kim, 2014).

17.4 CONCLUSION

For the prevention and treatment of acute, chronic and mild disease types, herbal medicines, extracts of plant materials, or plant secondary metabolites are widely used. Similarly, numerous *in vitro* and *in vivo* studies have been conducted for the treatment of mild or severe skin inflammatory disorders, including chronic wounds, psoriasis, allergic reactions, and atopic dermatitis, proving the pharmacological significance and therapeutic feasibility of these plant products. Some of the most dermatologically important drug interactions include the immunostimulating effects of echinacea, astragalus, licorice root, alfalfa sprouts, and vitamin E. They may also reduce the effectiveness of zinc, corticosteroids, and immunosuppressants (Miller, 1998).

TABLE 17.3
Some of *In Vivo* Studies for the Pharmacological and Therapeutic Effects of Secondary Metabolites for the Treatment of Atopic Dermatitis

Secondary Metabolites	Route of Administration	Structure of active compounds	Ref.
Hirsutenone	Topical	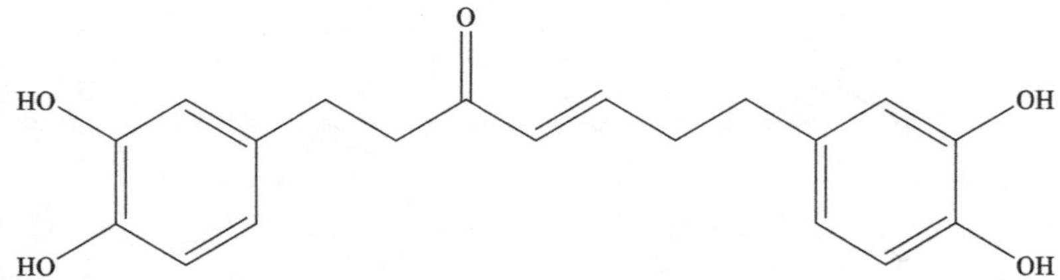	(Joo et al., 2009)
Aloperine	Topical	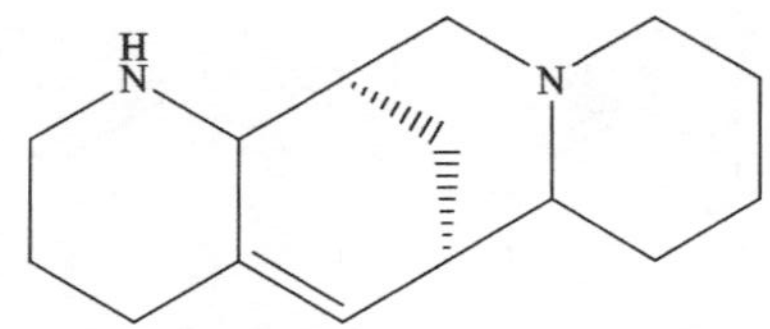	(Yuan et al., 2011)
Baicalein	Topical	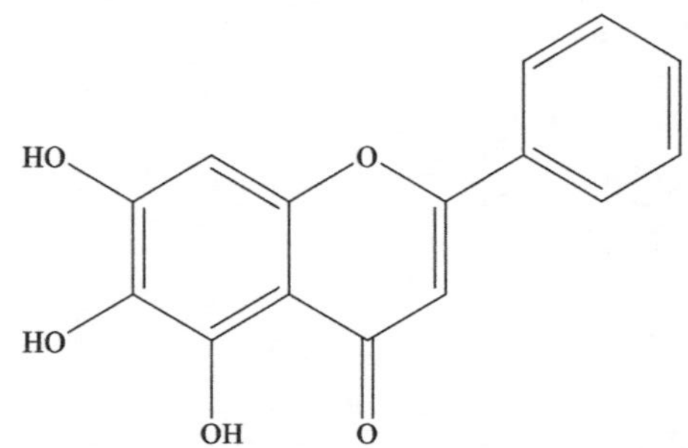	(Yun et al., 2010)

(*Continued*)

TABLE 17.3 (CONTINUED)

Some of *In Vivo* Studies for the Pharmacological and Therapeutic Effects of Secondary Metabolites for the Treatment of Atopic Dermatitis

Secondary Metabolites	Route of Administration	Structure of active compounds	Ref.
Conjugated linoleic acid	Oral	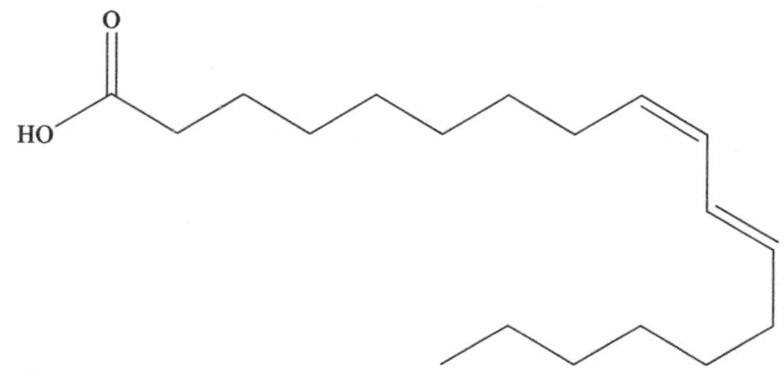	(Noli et al., 2007)
Epigallo-catechin-3-gallate	Topical	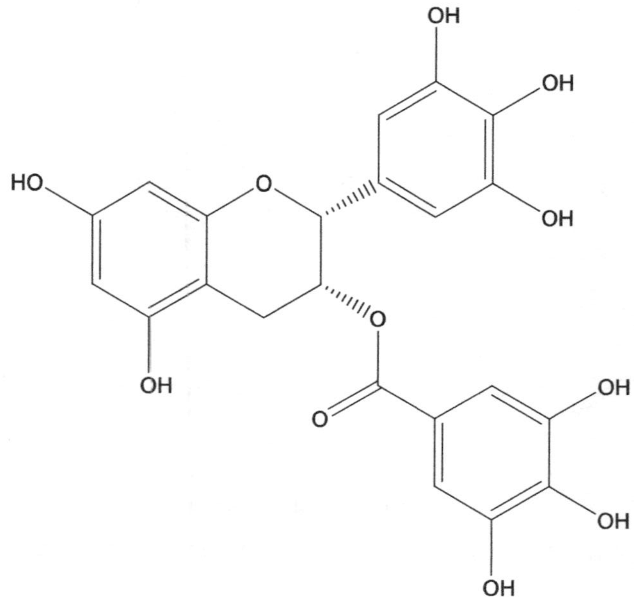	(Noh et al., 2008)

(Continued)

TABLE 17.3 (CONTINUED)
Some of *In Vivo* Studies for the Pharmacological and Therapeutic Effects of Secondary Metabolites for the Treatment of Atopic Dermatitis

Secondary Metabolites	Route of Administration	Structure of active compounds	Ref.
Fisetin	Topical	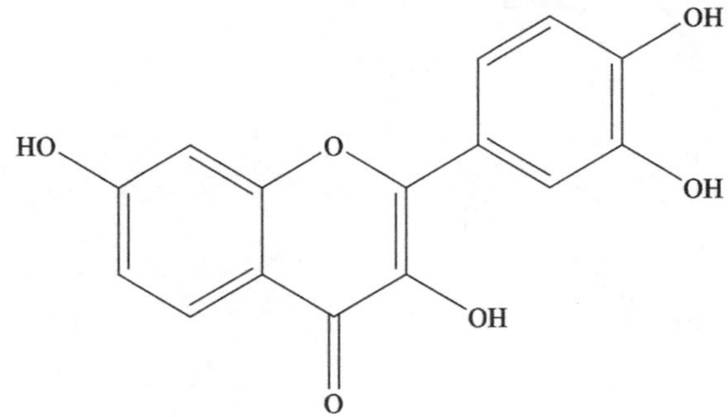	(Kim et al., 2014)
Galangin	Topical	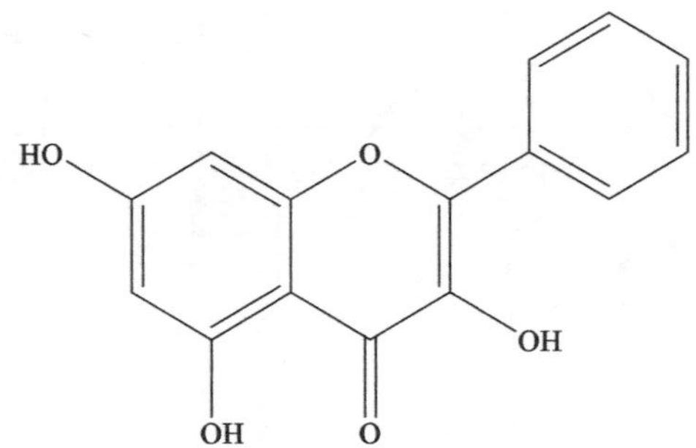	(Choi and Kim, 2014)

(*Continued*)

TABLE 17.3 (CONTINUED)
Some of *In Vivo* Studies for the Pharmacological and Therapeutic Effects of Secondary Metabolites for the Treatment of Atopic Dermatitis

Secondary Metabolites	Route of Administration	Structure of active compounds	Ref.
Oleanolic acid	Oral	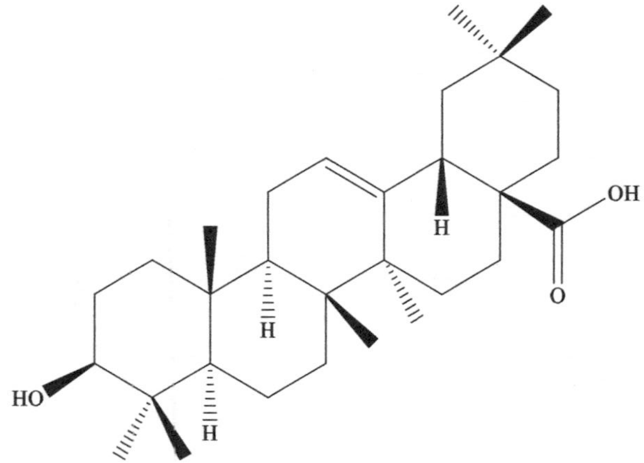	(Choi et al., 2013)
Quercetin-3-O-(200-gallat e)-a-l-rhamnopyranoside	Topical	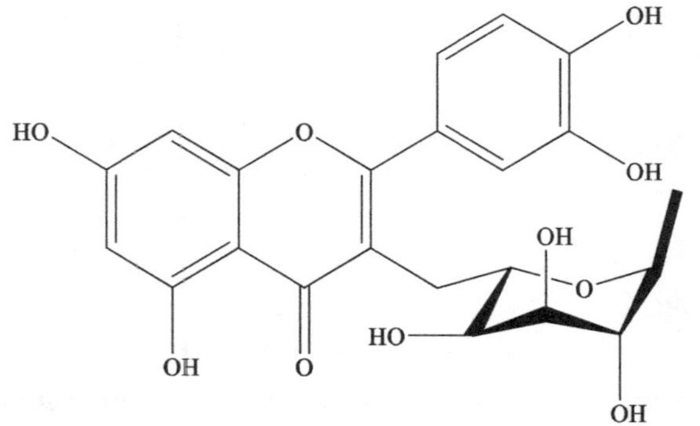	(Park et al., 2015)

(*Continued*)

TABLE 17.3 (CONTINUED)
Some of *In Vivo* Studies for the Pharmacological and Therapeutic Effects of Secondary Metabolites for the Treatment of Atopic Dermatitis

Secondary Metabolites	Route of Administration	Structure of active compounds	Ref.
Resveratrol	Topical	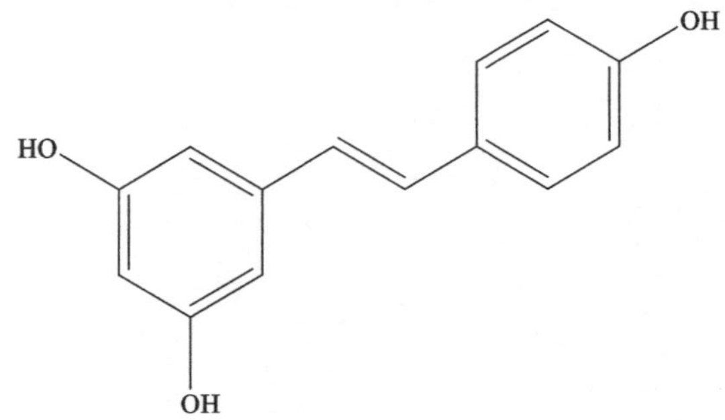	(Karuppagounder et al., 2014)
Rosmarinic acid	Intraperitoneal injection	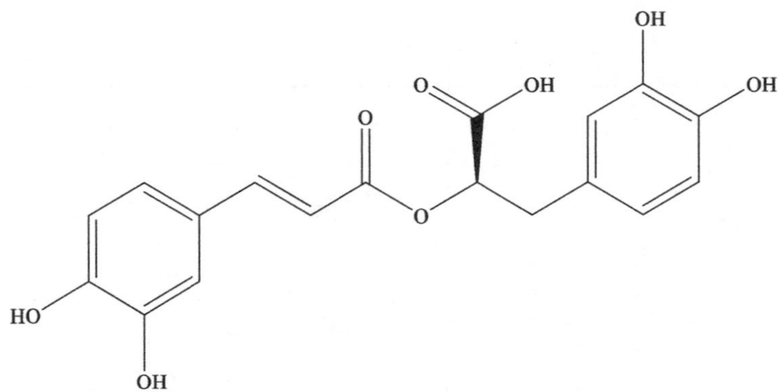	(Jang et al., 2011)

(Continued)

TABLE 17.3 (CONTINUED)
Some of *In Vivo* Studies for the Pharmacological and Therapeutic Effects of Secondary Metabolites for the Treatment of Atopic Dermatitis

Secondary Metabolites	Route of Administration	Structure of active compounds	Ref.
Silymarin	Topical	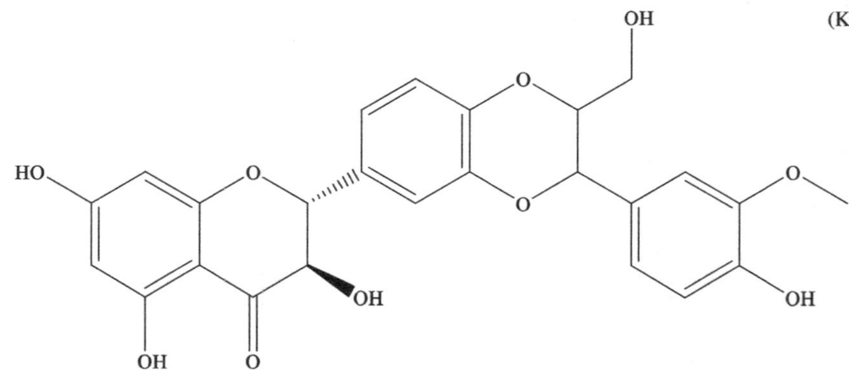	(Kang et al., 2008)

Some non-pharmacological approaches are considered beneficial in the treatment of mild to moderate skin diseases. However, pharmacological treatments are needed in the treatment of chronic skin diseases. It has been found that pharmacological agents show great efficacy for skin treatment. Their therapeutic applicability is limited by their use and few off-target side effects associated with patient non-adherence. Alternatively, the oral, intravenous, or topical use of natural herbal remedies, herbal extracts, creams, ointments, solutions, and capsules has gained considerable acceptance for skin treatment over the years. Literature analysis indicates that the application of natural herbal formulations as alternative and complementary medicines yields positive results. Also, numerous studies have demonstrated that the use of natural remedies in the treatment of atopic diseases is safe, cost-effective, systemically non-toxic, patient-compatible, and effective, as compared with a conventional therapeutic regimen. Isolating the secondary metabolite from plants or using the extracts or essential oils of plants containing the secondary metabolites with high pharmacological effects will make these formulations more effective. Obtaining the secondary metabolites from plants and using them in formulations in line with the studies conducted indicates that the choice in the future will go in this direction.

REFERENCES

Aburjai, T., Natsheh, F.M. "Plants used in cosmetics." *Phytother. Res.*, 17(9), 987–1000, 2003.

Agarwal, M., Kamal, R. "Studies on flavonoid production using in vitro cultures of *Momordica charantia* L." *Ind. J. Biotechnol.*, 6, 277–279, 2007.

Akihisa, T., Koike, K., Kimura, Y., Sashida, N., Matsumoto, T., Ukiya, M., Nikaido, T. "Acyclic and incompletely cyclized triterpene alcohols in the seed oils of Theaceae and Gramineae." *Lipids*, 34(11), 1151–1175, 1999.

Albanesi, C., De, P.O., Girolomoni, G. "Resident skin cells in psoriasis: A special look at the pathogenetic-functions of keratinocytes." *Clin. Dermatol.*, 25(6), 581–588, 2007.

Asasutjarit, R., Larpmahawong, P., Fuongfuchat, A., Sareedenchai, V., Veeranondha, S. "Physicochemical properties and anti-Propionibacterium acnes activity of film-forming solutions containing alpha-mangostin-rich extract."*AAPS Pharm. Sci. Tech.*, 15(2), 306–316, 2014.

Asif, H.M., Akram, M., Saeed, T.M., Khan, I., Akhtar, N., Ur Rehman, R., Ali Shah, S.M., Ahmed, K., Shaheen, G. "Review paper carbohydrates." *Int. Res. J. Biochem. Bioinforma.*, 1(1), 1–5, 2011.

Assa, Y., Shany, S., Gestetner, B., Tencer, Y., Birk, Y., Bondi, A. "Interaction of alfalfa saponins with components of the erythrocyte membrane in hemolysis." *Biochim. Biophys. Acta*, 307(1), 83–91, 1973.

Atique, A., Iqbal, M., Ghouse, A.K.M. "Use of *Annona squamosa* and *Piper nigrum* against diabetes." *Fitoterapia*, 56, 190–192, 1985.

Azimi, H., Fallah-Tafti, M., Khakshur, A.A., Abdollahi, M. "A review of phytotherapy of acne vulgaris: Perspective of newpharmacological treatments." *Fitoterapia*, 83(8), 1306–1317, 2012.

Bagci, E., Yazgın, A., Hayta, S., Cakılcıoglu, U. "Composition of the essential oil of *Teucrium chamaedrys* L. (Lamiaceae) from Turkey." *J. Med. Plant Res.*, 4(23), 2587–2589, 2010.

Bahramsoltani, R., Farzaei, M.H., Rahimi, R. "Medicinal plants and their natural components as future drugs for the treatment of burn wounds: An integrative review." *Arch. Dermatol. Res.*, 306(7), 601–617, 2014.

Baldi, A., Dixit, V.K. "Enhanced artemisinin production by cell cultures of *Artemisia annua*." *Curr. Trends Biotechnol. Pharmacol.*, 2, 341–348, 2008.

Batubara, I., Kuspradini, H., Muddathir, A.M., Mitsunaga, T. "*Intsia palembanica* wood extracts and its isolated compounds as *Propionibacterium acnes* lipase inhibitor." *J. Wood Sci.*, 60(2), 169–174, 2014.

Baydar, H. *Tıbbive Aromatik Bitkiler Bilimive Teknolojisi.* Süleyman Demirel Üniversitesi, Ziraat Fakültesi. Isparta, Yayın No: 51, 2013.

Beigh, S.Y., Nawchoo, I.A., Iqbal, M. "Herbal drugs in India: Past and present uses." *J. Trop. Med. Plants*, 3, 197–204, 2002.

Benowitz, N.L. "Clinical pharmacology of caffeine." *Annu. Rev. Med.*, 41(1), 277–288, 1990.

Benowitz, N.L. "Pharmacology of nicotine: Addiction, smoking-induced disease, and therapeutics." *Annu. Rev. Pharmacol. Toxicol.*, 49, 57–71, 2009.

Bhate, K., Williams, H. "Epidemiology of acne vulgaris." *Br. J. Dermatol.*, 168(3), 474–485, 2013.

Borins, M. "The dangers of using herbs: What your patients need toknow." *Postgrad. Med.*, 104(1), 91–100, 1998.

Catarino, M.D., Silva, A.M.S., Cruz, M.T., Cardoso, S.M. "Antioxidant and anti-inflammatory activities of *Geranium robertianum* L. decoctions." *Food Funct.*, 8(9), 3355–3365, 2017.

Chang, F., Steelman, L.S., Shelton, J.G., Lee, J.T., Nayolanic, P.M., Blalock, W.L., Franklin, R., McCubrey, J.A. "Regulation of cell cycle progression and apoptosis by the Ras/Raf/MEK/ERK pathway." *Int. J. Oncol.*, 22(3), 469–480, 2003.

Chang, J. "Medicinal herbs: Drugs or dietary supplements?" *Biochem. Pharmacol.*, 59(3), 211–219, 2000.

Cho, B.O., Yin, H.H., Park, S.H., Byun, E.B., Ha, H.Y., Jang, S.I. "Anti-inflammatory activity of myricetin from Diospyros lotus through suppression of NF-kB and STAT1 activation and Nrf2-mediated HO-1 induction in lipopolysaccharide-stimulated RAW264.7 macrophages." *Biosci. Biotechnol. Biochem.*, 80(8), 1520–1530, 2016.

Chobot, V., Hadacek, F. "Exploration of pro-oxidant and antioxidant activities of the flavonoid myricetin." *Redox Rep.*, 16(6), 242–247, 2011.

Choi, J.K., Kim, S.H. "Inhibitory effect of galangin on atopic dermatitis-like skin lesions." *Food Chem. Toxicol.*, 68, 135–141, 2014.

Choi, J.K., Oh, H.M., Lee, S., Park, J.-W., Khang, D., Lee, S.W., Lee, W.S., Rho, M.-C., Kim, S.-H. "Oleanolic acid acetate inhibitsatopic dermatitis and allergic contact dermatitis in a murine model." *Toxicol. Appl. Pharmacol.*, 269(1), 72–80, 2013.

Chunekar, K.C., Pandey, G.S. "Bhavprakash nighantu commentary." *Choukhamba Prakashan Varanasi*, 49–56, 1969.

Collin, H. Herbs, spices and cardiovascular disease. In: Peter, K.V., ed. *Handbook of Herbs and Spices*, Vol. 3. Woodhead Publishing Limited, Cambridge, 126–137, 2006.

Culioli, G., Mathe, C., Archier, P., Vieillescazes, C. "A lupane triterpene from frankincense (*Boswellia sp.*, Burseraceae)." *Phytochemistry*, 62(4), 537–41, 2003.

Da Lee, H., Lee, C.S. "Flavonoid myricetin inhibits TNF-α-stimulated production of inflammatory mediators by suppressing the Akt, mTOR and NF-kappaB pathways in human keratinocytes." *Eur. J.Pharmacol.*, 784, 164–172, 2016.

Dai, J., Gu, L., Su, Y., Wang, Q., Zhao, Y., Chen, X., Deng, H., Li, W., Wang, G., Li, K. "Inhibition of curcumin on influenza A virus infection and influenzal pneumonia via oxidativestress, TLR2/4, p38/JNK MAPK and NF-kappaB pathways." *Int. Immunopharmacol.*, 54, 177–187, 2018.

Docherty, J.J., McEwen, H.A., Sweet, T.J., Bailey, E., Booth, T.D. "Resveratrol inhibition of *Propionibacterium acnes*." *J. Antimicrob. Chemother.*, 59(6), 1182–1184, 2007.

Dweck, A.C. "African plants." *Cosmet. Toiletries*, 112, 41–51, 1997a.

Dweck, A.C. "Botanicals – Research of actives." *Cosmet. Toiletries*, 111, 45–57, 1996.

Eric, Y. "Botanical medicines for the urinary tract." *World J. Urol.*, 20(5), 285–293, 2002.

Etti, I.C., Abdullah, R., Kadir, A., Hashim, N.M., Yeap, S.K., Imam, M.U., Ramli, F., Malami, I., Lam, K.L., Etti, U., Waziri, P., Rahman, M. "The molecular mechanism of the anticancer effect of Artonin E in MDA-MB 231 triplenegative breast cancer cells." *PLOS ONE*, 12(8), e0182357, 2017.

Evans, W.C. *Pharmacognosy*. 16th ed. Edinburgh, London, New York, Philadelphia, St Louis, Sydney, Toronto, 2009.

Fahy, E., Subramaniam, S., Murphy, R.C., Nishijima, M., Raetz, C.R., Shimizu, T., Spener, F., van Meer, G., Wakelam, M.J., Dennis, E.A. "Update of the LIPID MAPS comprehensive classification system for lipids."*J. Lipid Res.*, 50, 9–14, 2009.

Farzaei, M.H., Bahramsoltani, R., Rahimi, R. "Phytochemicals as adjunctive with conventional anticancer therapies." *Curr. Pharm. Des.*, 22(27), 4201–4218, 2016.

Farzaei, M.H., Rahimi, R., Nikfar, S., Abdollahi, M. "Effect ofresveratrol on cognitive and memory performance and mood: A meta-analysis of 225 patients." *Pharmacol. Res.*, 128, 338–344, 2018.

Farzaei, M.H., Zobeiri, M., Parvizi, F., El-Senduny, F.F., Marmouzi, I., Coy-Barrera, E., Naseri, R., Nabavi, S.M., Rahimi, R., Abdollahi, M. "Curcumin in liver diseases: A systematic review of the cellular mechanisms of oxidative stress and clinical perspective." *Nutrients*, 10, 7, 2018.

Fellows, L.E. *Pharmaceuticals from Traditional Medicinal Plants and Others: Future Prospects.* A paper presented at the symposium "New Drugs from Natural Sources" sposored by I.B.C. Technical Services Ltd., London, June 13th–14th 1991, Royal Botanic Garden, Kew.

Ferguson, J.E., Chalmers, R.J.G., Rowlands, D.J. "Reversible dilated cardiomyopathy following treatment of atopic eczema with Chinese herbal medicine." *Br. J. Dermatol.*, 136(4), 592–593, 1997.

Fransworth, N.R., Akerele, O., Bingel, A.S. "Medicinal plants in therapy." *Bull. Word Health Organiz.*, 63, 965–981, 1985.

Fransworth, N.R., Soejarto, D.D. "Global importance of medicinal plants." In: Akerele, O., Heywood, V., Synge, H., eds. *Conservation of Medicinal Plants.* Cambridge University Press, Cambridge, 25–51, 1991.

Gao, S.L., Zhu, D.N., Cai, Z.H., Jiang, Y., Xu, D.R. "Organ culture of a precious Chinese medicinal plant – *Fritillaria unibracteata*." *Plant Cell Tiss. Org. Cult*, 59, 197–201, 2004.

Gehm, B.D., McAndrews, J.M., Chien, P., Jameson, J.L. "Resveratrol, a polyphenolic compound found in grapes and wine, is an agonist for the estrogen receptor." *Proc. Natl Acad. Sci. U. S. A.*, 94(25), 14138–14143, 1997.

Goławska, S., Sprawka, I., Łukasik, I., Goławski, A. "Are naringenin and quercetin useful chemicals in pest-management strategies?" *J. Pest Sci. (2004)*, 87(1), 173–180, 2014.

Gopi, C., Vatsala, T.M. "In vitro studies on effects of plant growth regulators on callus and suspension culture biomass yield from *Gymnema sylvestre* R. Br." *Afr. J. Biotechnol.*, 5, 1215–1219, 2006.

Güçlü-Üstündağ, Ö., Mazza, G. "Saponins: Properties, applications and processing."*Crit. Rev. Food Sci. Nutr.*, 47(3), 231–258, 2007.

Gurib-Fakim, A. "Medicinal plants: Traditions of yesterday and drugs of tomorrow." *Mol. Aspects Med.*, 27(1), 1–93, 2006.

Hagerman, A.E., Butler, L.G. "The specificity of proanthocyanidin-protein interactions." *J. Biol. Chem.*, 10(256), 4494–4497, 1981.

Haghi, A., Azimi, H., Rahimi, R. "A comprehensive review on pharmacotherapeutics of three phytochemicals, curcumin,quercetin, and allicin, in the treatment of gastric cancer."*J. Gastrointest. Cancer*, 48(4), 314–320, 2017.

Hikino, H. "Recent research on oriental medicinal plants." In: Wanger, H., Hiniko, H., Farnsworth, N.R., eds. *Economic and Medicinal Plant Research*, Vol. 1. Academic Press, London, 53–85, 1985.

Hoffmann, D. *Medical Herbalism: The Science and Practice of Herbal Medicine Healing Arts Press One Park Street*. Healing Arts Press, Rochester, Vermont, 2003.

Hohtola, A., Jalonen, J., Tolnen, A., Jaakola, L., Kamarainen, T., Pakonen, M., Karppinen, K., Laine, K., Neubauer, P., Myllykoshi, L., Gyorgy, Z., Rautio, A., Peltonen, O. "Natural product formation by plants, enhancement, analysis, processing and testing." In: Jalkanen, A., Nygren, P., eds. *Sustainable Use Renewable Natural Resources – From Principles to Practices*. University of Helsinki Publication, Helsinki, 34–69, 2005.

Hou, W., Hu, S., Su, Z., Wang, Q., Meng, G., Guo, T., Zhang, J., Gao, P. "Myricetin attenuates LPS-induced inflammation in RAW 264.7 macrophages and mouse models." *Future Med. Chem.*, 10(19), 2253–2264, 2018.

Howell, M.D., Fairchild, H.R., Kim, B.E., Bin, L., Boguniewicz, M., Redzic, J.S., Hansen, K.C., Leung, D.Y. "Th2 cytokines act on S100/A11 to downregulate keratinocyte differentiation." *J. Investig. Dermatol.*, 128(9), 2248–2258, 2008.

Iqbal, M., Parveen, R., Parveen, A., Parveen, B., Aref, I.M. "Establishing the botanical identity of plant drugs based on their active ingredients under diverse growth conditions." *J. Environ. Biol.*, 39(1), 123–136, 2018.

Ishikura, H., Mochizuki, T., Izumi, Y., Usui, T., Sawada, H., Uchino, H. "Differentiation of mouse leukemic M1 cells induced by polyprenoids." *Leuk. Res.*, 8(5), 843–852, 1984.

Jain, S.K., Namita, D. "Some eth nobotanical notes from north eastern India." *Econ. Bot.*, 33(1), 52, 1979.

Jang, A.H., Kim, T.H., Kim, G.D., Kim, J.E., Kim, H.J., Kim, S.S., Jin, Y.-H., Park, Y.S., Park, C.-S. "Rosmarinic acid attenuates 2,4-dinitrofluorobenzene-induced atopic dermatitis in NC/Nga mice." *Int. Immunopharmacol.*, 11(9), 1271–1277, 2011.

Joo, S.S., Kim, S.G., Choi, S.E., Kim, Y.-B., Park, H.Y., Seo, S.J., Choi, Y.W., Lee, M.W., Lee, D.I. "Suppression of T cell activation by hirsutenone, isolated from the bark of *Alnus japonica*, and its therapeutic advantages for atopic dermatitis." *Eur. J. Pharmacol.*, 614(1–3), 98–105, 2009.

Jung, H.-A., Su, B.-N., Keller, W.J., Mehta, R.G., Kinghorn, A.D. "Antioxidant xanthones from the pericarp of *Garcinia mangostana* (Mangosteen)." *J. Agric. Food Chem.*, 54(6), 2077–2082, 2006.

Kang, J.S., Yoon, W.K., Han, M.H., Lee, H., Lee, C.W., Lee, K.H., Han, S.-B., Lee, K., Yang, K.-H., Park, S.-K., Kim, H.M. "Inhibition of atopic dermatitis by topical application of silymarin in NC/Nga mice." *Int. Immunopharmacol.*, 8(10), 1475–1480, 2008.

Kang, K.A., Wang, Z.H., Zhang, R., Piao, M.J., Kim, K.C., Kang, S.S., Kim, Y.W., Lee, J., Park, D., Hyun, J.W. "Myricetin protects cells against oxidative stress-induced apoptosis via regulation of PI3K/Akt and MAPK signaling pathways." *Int. J. Mol. Sci.*, 11(11), 4348–4360, 2010.

Kapoor, S., Saraf, S. "Topical herbal therapies an alternative and complementary choice to combat acne." *Res. J. Med. Plants*, 5(6), 650–659, 2011.

Karuppagounder, V., Arumugam, S., Thandavarayan, R.A., Pitchaimani, V., Sreedhar, R., Afrin, R., Harima, M., Suzuki, H., Nomoto, M., Miyashita, S., Suzuki, K., Watanabe, K. "Resveratrol attenuates HMGB1 signaling and inflammation in house dust mite-induced atopic dermatitis in mice." *Int. Immunopharmacol.*, 23(2), 617–623, 2014.

Kim, G.-D., Lee, S.E., Park, Y.S., Shin, D.-H., Park, G.G., Park, C.-S. "Immunosuppressive effects of fisetin against dinitrofluorobenzene-induced atopic dermatitis-like symptoms in NC/Nga mice." *Food Chem. Toxicol.*, 66, 341–349, 2014.

Kim, J.S., Lee, S.Y., Park, S.U. "Resveratol production in hairy root culture of peanut, *Arachys hypogaea* L. transformed with different *Agrobacterium rhizogenes* strains." *Afr. J.Biotechnol.*, 7, 3788–3790, 2008.

Kim, O.T., Bang, K.H., Shin, Y.S., Lee, M.J., Jang, S.J., Hyun, D.Y., Kim, Y.C., Senong, N.S., Cha, S.W., Hwang, B. "Enhanced production of asiaticoside from hairy root cultures of *Centella asitica* (L.) Urban elicited by methyl jasmonate." *Plant Cell Rep.*, 26(11), 1914–1949, 2007.

Kiong, A.L., Mahmood, M., Fodzillan, N.M., Daud, S.K. "Effects of precursor supplementation on the production of triterpenes by *Centella asiatica* callus culture." *Pak. J. Biol. Sci.*, 8(8), 1160–1169, 2005.

Kornfeld, A., Kaufman, P.B., Lu, C.R., Gibson, D.M., Bolling, S.F., Warber, S.L., Chang, S.C., Kirakosyan, A. "The production of hypericins in two selected *Hypericum perforatum* shoot cultures is related to differences in black gland culture." *Plant Physiol. Biochem.*, 45, 2432, 2007.

Kou, X., Chen, N. "Pharmacological potential of ampelopsinin Rattan tea." *Food Sci. Hum. Wellness*, 1(1), 14–18, 2012.

Krolicka, A., Kartanowicz, R., Wosinskia, S., Zpitter, A., Kaminski, M., Lojkowska, E. "Induction of secondary metabolite production in transformed callus of *Ammi majus* L. grown after electromagnetic treatment of the culture medium." *Enz. Microb. Technol.*, 39, 1386–1389, 2006.

Kumar, S., Pandey, A.K. "Chemistry and biological activities of flavonoids: An overview." *Sci. World J.*, 2013, 162750, 2013.

Kusakari, K., Yokoyama, M., Inomata, S. "Enhanced production of saikosaponins by root culture of *Bupleurum falcatum* L. using two step control of sugar concentration." *Plant Cell Rep.*, 19(11), 1115–1120, 2000.

Kwon, D.-J., Bae, Y.-S., Ju, S.M., Goh, A.R., Youn, G.S., Choi, S.Y., Park, J. "Casuarinin suppresses TARC/CCL17 and MDC/CCL22 production via blockade ofNF-κB and STAT1 activation in HaCaT cells." *Biochem. Biophys. Res. Commun.*, 417(4), 1254–1259, 2012.

Lee, S.Y., Cho, S.J., Park, M.H., Kim, Y.K., Choi, J.I., Park, S.U. "Growth and rutin production in hairy root culture of buck weed (*Fagopyruumesculentum*)." *Prep. Biochem. Biotechnol.*, 37, 239–246, 2007.

Lee, S.Y., Xu, H., Kim, Y.K., Park, S.U. "Rosmarinic acid production in hairy root cultures of *Agastache rugosa* Kuntze." *World J. Microbiol.Biotechnol.*, 20, 969–972, 2007.

Lee-Parsons, C.W.T., Rogce, A.J. "Precursor limitations in methyljasmonate-induced *Catharanthus roseus* cell cultures." *Plant Cell Rep.*, 25(6), 607–612, 2006.

Li, C.-G., Yan, L., Jing, Y.-Y., Xu, L.-H., Liang, Y.-D., Wei, H.-X., Hu, B., Pan, H., Zha, Q.-B., Ouyang, D.-Y., He, X.-H. "Berberine augments ATP induced inflammasome activation in macrophages by enhancing AMPK signaling." *Oncotarget*, 8(1), 95–109, 2017.

Li, T.S.C. "The range of medicinal herbs and spices." In: Peter, K.V., ed. *Handbook of Herbs and Spices*, Vol. 3. Woodhead Publishing Limited, Cambridge, 113–125, 2006.

Liu, C.H., Huang, H.Y. "Antimicrobial activity of curcumin-loaded myristic acid microemulsions against *Staphylococcus epidermidis*." *Chem. Pharm. Bull.*, 60(9), 1118–1124, 2012.

Liu, C.H., Huang, H.Y. "In vitro anti-propionibacterium activity by curcumin containing vesicle system."*Chem. Pharm. Bull.*, 61(4), 419–425, 2013.

Liu, F., He, Y., Liang, Y.,Wen, L.,Zhu, Y., Wu, Y., Zhao, L., Li, Y., Mao, X., Liu, H. "PI3-kinase inhibition synergistically promoted the anti-tumor effect of lupeol in Hepatocellular carcinoma."*Cancer Cell Int.*, 13(1), 108, 2013.

Ma, J.Q., Ding, J., Zhang, L., Liu, C.M. "Ursolic acid protects mouse liver against CCl4-induced oxidative stress and inflammation by the MAPK/NF-kappaB pathway." *Environ. Toxicol. Pharmacol.*, 37(3), 975–983, 2014.

Mansfield, J.W. "Antimicrobial compounds and resistances. The role of phytoalexins and phytoanticipins." In: Slusarenko, A.J., Fraser, R.S.S., Vanloon, L.C., Fraser, R.S., eds. *Mechanism of Resisyance to Plant Diseases*. Springerverlag, New York, 325–363, 2000.

Mathieson, A.C., Hehre, E.J. "A comparison of the marine algae from the Goleta Slough and adjacent open coast of Goleta/Santa Barbara, California with those in the Southern Gulf of Maine." *Biodivers. Herit. Libr.*, 96(887), 207–258, 1994.

Mbaze, L.M., Lado, J.A., Wansi, J.D., Shiao, T.C., Chiozem, D.D., Mesaik, M.A., Choudhary, M.I., Lacaille-Dubois, M.A., Wandji, J., Roy, R., Sewald, N. "Oxidative burst inhibitory and cytotoxic amides and lignans from the stem bark of *Fagaraheitzii* (Rutaceae)." *Phytochemistry*, 70(11–12), 1442–1447, 2009.

Mehrotra, S., Kukreja, A.K., Khanuja, S.P.S., Mishra, B.N. "Genetic transformation studies and scale up of hairy root culture of *Glycyrrhiza glabra* in bioreactor." *Elec. J. Biotechnol.*, 11, 717–728, 2008.

Miller, L.G. "Herbal medicinals: Selected clinical considerations focusing on known or potential drug-herb interactions." *Arch. Intern. Med.*, 158(20), 2200–2208, 1998.

Miodovnik, M., Koren, R., Ziv, E., Ravid, A. "The inflammatory response of keratinocytes and its modulationby vitamin D: The role of MAPK signaling pathways." *J. Cell. Physiol.*, 227(5), 2175–2183, 2012.

Misra, N., Misra, P., Datta, S.K., Mehrotra, S. "In vitro biosynthesis of antioxidants from *Hemidesmus indicus* R. Br. cultures In vitro." *Dev. Biol. Plant.*, 41, 285–290, 2005.

Montanher, A.B., Zucolotto, S.M., Schenkel, E.P., Fröde, T.S. "Evidence of anti-inflammatory effects of *Passiflora edulis* in an inflammation model." *J. Ethnopharmacol.*, 109(2), 281–288, 2007.

Moon, J.E., Shin, J.-H., Kwon, O., Kim, J.Y., Jomf, J. "A standardized extract of *Rhus verniciflua* stokes protects Wistar rats against lipopolysaccharide-induced acute inflammation." *J. Med. Food*, 18(11), 1223–1230, 2015.

Muddathir, A.M., Yamauchi, K., Mitsunaga, T. "Anti-acne activity of tannin-related compounds isolated from *Terminalia laxiflora*." *J. Wood Sci.*, 59(5), 426–431, 2013.

Nayak, M., Nagarajan, A., Majeed, M., Nagabhushanam, K., Choudhury, A.K. "In vitro anti-acne activity of phytoactives from the stem bark of *Artocarpus hirsutus* Lam. and characterisation of pyranocycloartobiloxanthone A as a mixture of two anomers." *Nat. Prod. Res.*, 1–5, 2017.

Nazif, N.M., Rady, M.R., Seif, M.M. "Stimulation of anthraquinone production in suspension cultures of *Cassia acutifolia* by salt stress." *Fitoterapia*, 71(1), 34–40, 2000.

Ncube, N.S., Afolayan, A.J., Okoh, A.I. "Assessment techniques of antimicrobial properties of natural compounds of plant origin: Current methods and future trends." *Afr. J. Biotechnol.*, 7(12), 1797–1806, 2008.

Nikolaeva, T.N., Zagoskina, N.V., Zaprometov, M.N. "Production of phenolic compounds in callus cultures of tea plant under the effect of 2,4-D and NAA." *Russ. J. Plant Physiol.*, 56(1), 45–49, 2009.

Noh, S.U., Cho, E.A., Kim, H.O., Park, Y.M. "Epigallocatechin-3-gallate improves Dermatophagoides pteronissinus extract-induced atopic dermatitis-like skin lesions in NC/Nga mice by suppressing macrophage migration inhibitory factor." *Int. Immunopharmacol.*, 8(9), 1172–1182, 2008.

Noli, C., Carta, G., Cordeddu, L., Melis, M.P., Murru, E., Banni, S. "Conjugated linoleic acid and black currant seed oil in the treatment of canine atopic dermatitis: A preliminary report." *Vet. J.*, 173(2), 413–421, 2007.

Olivira, A.J.B., Koike, L., Reis, F.A.M., Shepherd, S.L.K. "Callus culture of *Aspidosperma ramiflorum* Muell.-Arg.: Growth and alkaloid production."*Acta Sci.*, 23, 609–612, 2001.

Orthoefer, F.T. "High phospholipid-containing dermatological compositions." *US Pat. Appl. Publ.*, 9, US2002012648, 2002.

Park, E.J., Kim, J.Y., Jeong, M.S., Park, K.Y., Park, K.H., Lee, M.W., Joo, S.S., Seo, S.J. "Effect of topical application ofquercetin-3-O-(2″-gallate)-α-l-rhamnopyranoside on atopic dermatitisin NC/Nga mice." *J. Dermatol. Sci.*, 77(3), 166–172, 2015.

Parveen, A., Parveen, R., Akhatar, A., Parveen, B., Siddiqui, K.M., Iqbal, M. "Concepts and quality considerations in Unani system of medicine." *J. AOAC Int.*, 103(3), 609–633, 2020a.

Parveen, B., Parveen, A., Parveen, R., Ahmad, S., Ahmad, M., Iqbal, M. "Challenges and opportunities for traditional herbal medicine today, with special reference to its status in India." *Ann. Phytomedicine*, 9(2), 97–112, 2020b.

Passi, S. "Cosmetic formulations comprising lipids, antioxidants, anti-inflammatory, and anti-free-radical agents." *PCT Int. Appl*, 16, WO0226207, 2002.

Peng, J., Ren, X., Lan, T., Chen, Y., Shao, Z., Yang, C. "Renoprotective effects of ursolic acid on ischemia/reperfusion induced acute kidney injury through oxidative stress, inflammation and the inhibition of STAT3 and NFkappaB activities." *Mol. Med. Rep.*, 14(4), 3397–3402, 2016.

Perkins, A.C., Maglione, J., Hillebrand, G.G., Miyamoto, K., Kimball, A.B. "Acne vulgaris in women: Prevalence across the lifespan." *J. Womens Health*, 21(2), 223–230, 2012.

Phatak, S.V., Heble, M.R. *Organogenesis and Terpenoid from Plant Tissue Culture*. Claredon Press: Oxford, 1–21, 2002.

Piccaglia, R., Marotti, M., Giovanelli, E., Deans, S.G., Eaglesham, E. "Antibacterial and antioxidant properties of Mediterranean aromatic plants." *Ind. Crops Prod.*, 2(1), 47–50, 1993.

Prasad, S., Kalra, N., Singh, M., Shukla, Y. "Protective effects of lupeol and mango extract against androgen induced oxidative stress in Swiss albino mice." *Asian J. Androl.*, 10(2), 313–318, 2008.

Puneet, G., Aditi, S., Vinayak, K., Anant, G.N., Samir, D., Nikhil, M. "*Camellia sinensis* (tea): Implications and role in preventing dental decay." *Pharmacogn. Rev.*, 7(14), 152–156, 2013.

Qi, S., Xin, Y., Guo, Y., Diao, Y., Kou, X., Luo, L., Yin, Z. "Ampelopsin reduces endotoxic inflammation via repressing ROS-mediated activation of PI3K/Akt/NF-κB signaling pathways." *Int. Immunopharmacol.*, 12(1), 278–287, 2012.

Ramani, S., Jayabaskaran, C. "Enhanced catharathine and vindoline production in suspension cultures of *Catharanthus roseus* by ultraviolet-B light." *J. Mol. Signal.*, 3, 9–14, 2008.

Rapp, S.R., Feldman, S.R., Graham, G., Fleischer, A.B., Brenes, G., Dailey, M. "The acne quality of life index (Acne-QOLI)." *Am. J. Clin. Dermatol.*, 7(3), 185–192, 2006.

Raskin, I., Ribnicky, D.M., Komarnytsky, S., Ilic, N., Poulev, A., Borisjuk, N., Brinker, A., Moreno, D.A., Yakoby, R.N. "Plant and human health in the twenty-first century." *Trends Biotechnol.*, 20, 522–531, 2000.

Reddy, G.R., Ueda, N., Hada, T., Sackeyfio, A.C., Yamamoto, S., Hano, Y., Aida, M., Nomura, T. "A prenylflavone, artonin E, as arachidonate 5-lipoxygenase inhibitor." *Biochem. Pharmacol.*, 41(1), 115–118, 1991.

Reddy, L., Odhav, B., Bhoola, K.D. "Natural product for cancer prevention: Global perspective." *Pharmacol. Ther.*, 99(1), 1–13, 2003.

Roat, C., Ramawat, K.G. "Elicitor-induced accumulation of stilbenes in cell suspension cultures of *Cayratia trifoliata* (L.) Domin." *Plant Biotechnol. Rep.*, 3(2), 135–138, 2009.

Sadeghnia, H.R., Kolangikhah, M., Asadpour, E., Forouzanfar, F., Hosseinzadeh, H. "Berberine protects against glutamate induced oxidative stress and apoptosis in PC12 and N2a cells." *Iran. J. Basic Med. Sci.*, 20(5), 594–603, 2017.

Saeidnia, S., Gohari, A.R., Mokhber-Dezfuli, N., Kiuchi, F. "A review on phytochemistry and medicinal properties of the genus *Achillea*." *Daru*, 19(3), 173–186, 2011.

Schmeda-Hirschmann, G., Jordan, M., Gertn, A., Wilken, D., Hormazabal, E., Tapia, A.A. Secondary metabolite content in Fabiana imbricate plants and in vitro cultures." *Z. Naturforsch.*, 5, 48–54, 2004.

Seigler, D.S. *Plant Secondary Metabolism.* Springer Science Business Media, New York, 1995.

Seki, T., Morohashi, M. "Effect of some alkaloids, flavonoids and triterpenoids, contents of Japanese–Chinese traditional herbal medicines, on the lipogenesis of sebaceous glands." *Skin Pharmacol. Physiol.*, 6(1), 56–60, 1993.

Semwal, D.K., Semwal, R.B., Combrinck, S., Viljoen, A. "Myricetin: A dietary molecule with diverse biological activities." *Nutrients*, 8(2), 90, 2016.

Serafini, M., Peluso, I., Raguzzini, A. "Flavonoids as anti-inflammatory agents." *Proc. Nutr. Soc.*, 69(3), 273–278, 2010.

Shalaka, D.K., Sandhya, P. "Micropropagation and organogenesis in *Adhatoda vasica* for the estimation of vasine." *Pharmacog. Magaz*, 5, 539–363, 2009.

Sharma, L., Agarwal, G., Kumar, A. "Medicinal plants for skin and hair care." *Indian J. Tradit. Knowl.*, 2(1), 62–68, 2003.

Sharma, P.R., Shanmugavel, M., Saxena, A.K., Qazi, G.N. "Induction of apoptosis by a synergistic lignan composition from *Cedrus deodara* in human cancer cells." *Phytother. Res.*, 22(12), 1587–1594, 2008.

Shohael, A.M., Chakrabarty, D., Yu, K.W., Hahn, E.J., Paek, K.Y. "Application of bioreactor system for largescale production of *Eleutherococcus sessiliflorus* somatic embryos in an air-lift bioreactor and production of eleutherosides." *J. Biotechnol.*, 120(2), 228–236, 2005.

Shrivastava, N., Patel, T., Srivastava, A. "Biosynthetic potential of in vitro grown callus cells of *Cassia senna* L. var. senna." *Curr. Sci.*, 90, 1472–1473, 2006.

Sivasankar, C., Maruthupandiyan, S., Balamurugan, K., James, P.B., Krishnan, V., Pandian, S.K. "A combination of ellagic acid and tetracycline inhibits biofilm formation and the associated virulence of *Propionibacterium acnes* in vitro and in vivo." *Biofouling*, 32(4), 397–410, 2016.

Song, Y.-H., Cai, H., Gu, N., Qian, C.-F., Cao, S.-P., Zhao, Z.-M. "Icariin attenuates cardiac remodelling through down-regulating myocardial apoptosis and matrix metalloproteinase activity in rats with congestive heart failure." *J. Pharm. Pharmacol.*, 63(4), 541–549, 2011.

Spatuzza, C., Postiglione, L., Covelli, B., Ricciardone, M., Benvenuti, C., Mondola, P., Belfiore, A. "Effects of berberine and red yeast on proinflammatory cytokines IL-6 and TNF-alpha in peripheral blood mononuclear cells (PBMCs) of human subjects." *Front. Pharmacol.*, 5, 230, 2014.

Spiller, F., Alves, M.K., Vieira, S.M., Carvalho, T.A., Leite, C.E., Lunardelli, A., Poloni, J.A., Cunha, F.Q., de Oliveira, J.R. "Anti-inflammatory effects of red pepper (*Capsicum baccatum*) on carrageenan- and antigen-induced inflammation." *J. Pharm. Pharmacol.*, 60(4), 473–478, 2008.

Staniszewska, I., Krolicka, A., Mali, E., Ojkowska, E., Szafranek, J. "Elicitation of secondary metabolites in vitro cultures of *Ammi majus* L." *Enz. Microbiol. Technol*, 33, 565–568, 2003.

Subramaniam, S., Fahy, E., Gupta, S., Sud, M., Byrnes, R.W., Cotter, D., Dinasarapu, A.R., Maurya, M.R. "Bioinformatics and systems biology of the lipidome." *Chem. Rev.*, 111(10), 6452–6490, 2011.

Sujanya, S., Poornasri, D.B., Sai, I. "In vitro suspension cultures of *Azadirachta indica*." *J. Bio Sci.*, 33, 113–120, 2008.

Sultana, N. "Clinically useful anticancer, antitumor, and antiwrinkle agent, ursolic acid and related derivatives as medicinallyimportant natural product." *J. Enzyme Inhib. Med. Chem.*, 26(5), 616–642, 2011.

Sun, Y., Yuan, X., Zhang, F., Han, Y., Chang, X., Xu, X., Li, Y., Gao, X. "Berberine ameliorates fatty acid-induced oxidative stress in human hepatoma cells." *Sci. Rep.*, 7(1), 11340, 2017.

Susana, J., Beatriz, G.M., Fabiana, C.M., Maria, A.D., Moacir, G.P. "Antifungal activity of five species of Polygala." *Braz. J. Microbiol.*, 42(3), 1065–1075, 2011.

Tadeusz, A. *Alkaloids: Chemistry, Biology, Ecology, and Applications*. 2nd ed. Elsevier, Amsterdam, Netherlands, 2015.

Tahir, I., Khan, M.R., Shah, N.A., Aftab, M. "Evaluation of phytochemicals, antioxidant activity and amelioration of pulmonary fibrosis with *Phyllanthus emblica* leaves." *BMC Complement. Altern. Med.*, 16(1), 406, 2016.

Taylor, E.J., Yu, Y., Champer, J., Kim, J. "Resveratrol demonstrates antimicrobial effects against *Propionibacterium acnes* in vitro." *Dermatol. Ther. (Heidelb)*, 4(2), 249–257, 2014.

Tiwari, K.K., Trivedi, M., Guang, Z.C., Guo, G.Q., Zheng, G.C. "Genetic transformation of *Gentiana macrophylla* with *Agrobacterium rhizogenes*: Growth and production of secoiridoid glucoside gentiopicroside in transformed hairy root cultures." *Plant Cell Rep.*, 26(2), 199–210, 2007.

Tranggono, R.I.S. "The influence of SPA products treatment for skin-care." *J. Appl.Cosmetol.*, 18, 177–184, 2000.

Umamaheswai, A., Lalitha, V. "In vitro effect of various growth hormones in *Capsicum annum* L. on the callus induction and production of capsaicin." *J. Plant Sci.*, 2(5), 545–551, 2007.

Umesalma, S., Sudhandiran, G. "Differential inhibitory effects of the polyphenol ellagic acid on inflammatory mediators NF-κB, iNOS, COX-2, TNF-α, and IL-6 in 1, 2-dimethylhydrazine-induced rat colon carcinogenesis." *Basic Clin. Pharmacol. Toxicol.*, 107(2), 650–655, 2010.

Upadhyay, V.P. *Plants as Cosmetics*. International Seminar on Medicinal Plants. Mungpoo. Govt. of West Bengal, February 1985.

Van de Braak, S.A.A.J., Leijten, G.C.J.J. *Essential Oils and Oleoresins: A Survey in The Netherlands and Other Major Markets in the European Union*. CBI, Centre for the Promotion of Imports from Developing Countries, Rotterdam, Netherlands, 1999.

Vinterhalter, B., Jankovic, T., Sovikin, L., Nikolic, R., Vinterhalter, D. "Propagation and xanthone content of *Gentianella austiaca* shoot cultures." *Plant Cell Tiss. Org. Cult*, 94, 329–335, 2008.

Wagiah, M.E., Alam, G., Wiryowidagdo, S., Attia, K. "Imporved production of the indole alkaloid cathin-6-one fromcell suspension cultures of *Brucea javanica* (L.) Merr." *Ind. J. Sci. Technol.*, 1, 1–6, 2008.

Wang, J., Qian, J., Yaol, L., Lu, Y. "Enhanced production of flavonoids by methyl jasmonate elicitation in cell suspension culture of *Hypericum perforatum*." *Bioresour. Bioprocess.*, 2(1), 5, 2015.

Watson, A.A., Fleet, G.W.J., Asano, N., Molyneux, R.J., Nash, R.J. "Polyhydroxy lated alkaloid–natural occurrence and therapeutic applications." *Phytochemistry*, 56(3), 265–295, 2001.

Williams, H.C., Dellavalle, R.P., Garner, S. "Acne vulgaris." *Lancet*, 379(9813), 361–372, 2012.

Wilmer, J.L., Luster, M.I. "Chemical induction of interleukin-8, a pro-inflammatory chemokine, in human epidermal keratinocyte cultures and its relation to cytogenetic toxicity." *Cell Biol. Toxicol.*, 11(1), 37–50, 1995.

Wu, T.-Q., Mei, S.-Q., Zhang, J.-X., Gong, L.-F., Wu, F.-J., Wu, W.-H., Li, J., Lin, M., Diao, J.-X. "Prevalence and risk factors of facial acne vulgaris among Chinese adolescents." *Int. J. Adolesc. Med. Health*, 19(4), 407–412, 2007.

Xia, Z., Dickens, M., Raingeaud, J., Davis, R.J., Greenberg, M.E. "Opposing effects of ERK and JNK-p38 MAP kinases on apoptosis." *Science*, 270(5240), 1326–1331, 1995.

Xu, H., Kim, Y.K., Suh, S.Y., Udin, M.R., Lee, S.Y., Park, S.U. "Deoursin production from hariy root culture of *Angelica gigas*." *J. Korea Soc. Appl. Biol. Chem*, 51, 349–351, 2008.

Xu, L., Wu, Y., Zhao, X., Zhang, W. "The study on biological and pharmacological activity of coumarins." In: *Asia-Pacific Energy Equipment Engineering Research Conference*, Atlantis Press, Zhuhai, China, 135–138, 2015.

Yi, Z., Wang, Z., Li, H., Liu, M. "Inhibitory effect of tellimagrandin I on chemically induced differentiation of human leukemia K562 cells." *Toxicol. Lett.*, 147(2), 109–119, 2004.

Yu, S., Wang, M., Guo, X., Qin, R. "Curcumin attenuates inflammation in a severe acute pancreatitis animal model by regulatingTRAF1/ASK1 signaling." *Med. Sci. Monit.*, 24, 2280–2286, 2018.

Yuan, X.Y., Ma, H.M., Li, R.Z., Wang, R.-Y., Liu, W., Guo, J.-Y. "Topical application of aloperine improves 2,4-dinitrofluorobenzene-induced atopic dermatitis-like skin lesions in NC/Nga mice." *Eur. J. Pharmacol.*, 658(2–3), 263–269, 2011.

Yun, M.Y., Yang, J.H., Kim, D.K., Cheong, K.-J., Song, H.-H., Kim, D.-H., Cheong, K.-J., Kim, Y.-I., Shin, S.-C. "Therapeutic effects of baicalein on atopic dermatitis-like skin lesions of NC/Nga mice induced by Dermatophagoides pteronyssinus." *Int. Immunopharmacol.*, 10(9), 1142–1148, 2010.

Zaenglein, A.L., Pathy, A.L., Schlosser, B.J., Alikhan, A., Baldwin, H.E., Berson, D.S., Bowe, W.P., Graber, E.M., Harper, J.C., Kang, S., Keri, J.E., Leyden, J.J., Reynolds, R.V., Silverberg, N.B., Stein Gold, L.F., Tollefson, M.M., Weiss, J.S., Dolan, N.C., Sagan, A.A., Stern, M., Boyer, K.M., Bhushan, R. "Guidelines of care for the management of acne vulgaris." *J. Am. Acad. Dermatol.*, 74(5), 945–973.e33, 2016.

Zhang, H., Shan, Y., Wu, Y., Xu, C., Yu, X., Zhao, J., Yan, J., Shang, W. "Berberine suppresses LPS-induced inflammation through modulating Sirtl/NF-kappaB signaling pathway in RAW264.7 cells." *Int. Immunopharmacol.*, 52, 93–100, 2017.

Zhang, Z., Li, K. "Curcumin attenuates high glucose-induced inflammatory injury through the reactive oxygen species-phosphoinositide3-kinase/protein kinase B-nuclear factor-kappaB signaling pathway in rat thoracic aorta endothelial cells." *J. Diabetes Investig.*, 9(4), 731–740, 2018.

Zhang, Z., Zhang, H., Chen, R., Wang, Z. "Oral supplementation with ursolic acid ameliorates sepsis-induced acute kidney injury in a mouse model by inhibiting oxidative stress and inflammatory responses." *Mol. Med. Rep.*, 17(5), 7142–7148, 2018.

Zhao, J., Zhu, W., Hu, Q. "Enhanced catharanthine production in *Catharanthus roseus* cell cultures by combined elicitor treatment in shake flasks and bioreactors." *Enz. Microb. Technol.*, 28(7–8), 673–681, 2001.

Zhu, Y., Li, X., Chen, J., Chen, T., Shi, Z., Lei, M., Zhang, Y., Bai, P., Li, Y., Fei, X. "Thepentacyclic triterpene lupeol switches M1 macrophages to M2 and ameliorates experimental inflammatory bowel disease." *Int. Immunopharmacol.*, 30, 74–84, 2016.

Zou, W., Yin, P., Shi, Y., Jin, Y., Gao, Q., Li, J., Liu, F. "A novel biological role of α-Mangostin via TAK1-NF-κB pathway against inflammatory." *Inflammation*, 42(1), 103–112, 2019.

Zouboulis, C.C., Eady, A., Philpott, M., Goldsmith, L.A., Orfanos, C., Cunliffe, W.C., Rosenfield, R. "What is the pathogenesis of acne?"*Exp. Dermatol.*, 14(2), 143, 2005.

18 Nutraceutical Perspectives of the Coastal Sand Dune Wild Legume *Sesbania bispinosa*

Suvarna Jeppu Shreelalitha and Kandikere Ramaiah Sridhar

CONTENTS

18.1 INTRODUCTION

Legumes exhibit agriculturally valuable traits for exploitation as cash crops, cover crops, green manures, intercrops, and mulches. Several regionally cultivated ethnic wild legumes are beneficial in terms of nutrition as well as traditional medicine. Seeds of wild legumes usually consist of an adequate quantity of protein, energy, fat, and bioactive compounds, thus serving as a precious source of food and fodder. As animal-based foodstuffs are becoming expensive, legumes occupy the foremost place in the food and agricultural sector. However, a variety of antinutritional components present in legume seeds prevents their direct utilization as a source of food or feed without suitable processing. Different strategies can be employed to reduce or eliminate the antinutritional principles to below threshold level. There is ample scope to utilize unprocessed food sources through diet management for safe consumption by exploiting antinutritional compounds as bioactive principles to combat several human ailments.

DOI: 10.1201/9781003226925-18

The Indian subcontinent has been segregated into ten biogeographic zones, which are further divided into twenty-five provinces with more than 400 biomes (Singh and Chaturvedi, 2017). The coastline of India is one of the major biogeographic zones and serves as the cradle of floral, faunal, and microbial diversity. About 30 wild legumes are popularly known in India as nutritional sources amongst the tribals (Arora et al., 1980; Gunjatkar and Vartak, 1982; Viswanathan et al., 1999, 2001; Janardhanan et al. 2003; Vadivel et al. 2011). The sand dunes of the coasts of Southwest India harbor many wild legumes, which are agriculturally, nutritionally, and medicinally versatile (Rao and Meher-Homji, 1985; Arun et al., 1999; Rao and Sherieff, 2002; Bhagya and Sridhar, 2009).

Sesbania (Family, Fabaceae; Subfamily, Papilionoideae; Tribe, Robinieae) consists of nearly 60 species (annuals, perennials, herbs, shrubs, and trees) distributed in tropical and subtropical regions (Evans, 1990; Veasey et al., 1999). They have economically viable applications in agroforestry, cover crop, green manure, fodder, nutrition, pharmaceuticals and industries (Anderson, 1989; Ndoye et al., 1990; Shahjalaland Topps, 2000; Hossain and Becker, 2001; Palaniappan and Siddeswaran, 2001; Mandal et al., 2003; Vadivel and Biesalski, 2010; Pollard et al., 2011; Sridhar et al., 2016; Shreelalitha and Sridhar, 2016, 2018, 2019; Chanda et al., 2017; Sarkar et al., 2017; Sarwar et al., 2017; Anita and Sridhar, 2021). The agricultural, nutritional, and pharmaceutical importance of *Sesbania* spp. have been documented (Sridhar et al., 2016; Bhat and Karim, 2009; Anita and Sridhar, 2020, 2021). In southern India, the mature *Sesbania* seeds are consumed on cooking by tribal sects like Khari and Ghonda (Siddhuraju et al., 1995). *Sesbania bispinosa* is commonly called spiny Sesbania and is prevalent in saline-prone coastal sand regions of the southwest coast of India, which is tolerant to the adverse climatic geographic and soil conditions (e.g., high temperature, alkaline pH, saline soils, and marshy regions) (Prasad, 1993; Ipor and Oyen, 1997). The traits of *Sesbania* to withstand extreme conditions will be of immense significance in sustainable agriculture and rehabilitation of saline-prone or marine-influenced agricultural terrain. The Plant Genetic Resources Conservation Unit (PGRCU) in the University of Georgia maintained *S. bispinosa* as one of the important industrially valuable germplasm resources (Morris, 1999). It has also been considered one of the 27 economically feasible legumes (Singh et al., 2007). There seem to be only meager reports on the nutraceutical qualities of *S. bispinosa* grown in saline-prone habitats. Thus, the present study attempts to assess the nutritional and bioactive potential of uncooked and cooked dry and ripened seeds of *S. bispinosa* established on the coastal sand dunes of Southwest India.

18.2 NUTRITIONAL EVALUATION

18.2.1 Seed Source and Process

Ripened and dry pods of *Sesbania bispinosa* (Jacq.) W. Wight Pers. were harvested from five locations along the hind dunes of Someshwara (12°47′N, 74°52′E), Southwest coast of India during the post-monsoon season (Figure 18.1a–d). The seeds were detached from the pods (Figure 18.1e, f), sun dried until the moisture content was below 10%, followed by assessment of dimensions and dry mass. Seed lots from each sampling location were divided into two groups: the first group has been considered as the control (uncooked), whereas the second group was pressure cooked (6.5 l, Deluxe stainless steel; *TTK* Prestige™, Prestige Ltd., India) after soaking in freshwater (1:3 v/v) and sun drying. After milling the dried seeds, the flour was preserved in air-tight containers under refrigeration.

18.2.2 Proximal and Mineral Analysis

The moisture content of the seed flours was determined gravimetrically after oven drying (100°C; Scientronic SBIM-25; New Delhi, India) until it reached a constant weight. The crude protein (N′6.25) content of the flours was assessed by the micro-Kjeldahl method by Humphries (1956). The total lipid content of flours was evaluated gravimetrically on extraction with petroleum ether (AOAC, 1995).

FIGURE 18.1 Stands of *Sesbania bispinosa* growing on the coastal sand dunes of the Southwest coast of India (a and b), ripened pods (c), dry pods (d), ripened seeds (e), and dry seeds (f).

The contents of crude fiber and ash were determined gravimetrically (AOAC, 1995). The carbohydrate content and calorific value were calculated (Müller and Tobin, 1980; Ekanayake et al., 1999). Mineral composition of the seed flours was assessed using an atomic absorption spectrophotometer (GBC 902; Melbourne, Australia) after digestion in an acid mixture (nitric acid, sulfuric acid, and perchloric acid) (AOAC, 1995). Assessment of total phosphorus by the vanado-molybdophosphoric acid method (KH_2PO_4 as standard) was achieved by absorbance at 420 nm (AOAC, 1995).

18.2.3 Protein Fractions and Amino Acid Analysis

To determine albumin, globulin, prolamin and glutelin contents was based on the method described by Gheyasuddin et al. (1970) and the nitrogen content of each protein fraction was assessed by the micro-Kjeldahl method (Humphries, 1956). The amino acids of flours of cooked seeds were evaluated based on Hofmann et al. (1997, 2003). Hydrolysis of flours was carried out using hydrochloric acid followed by removal of acid content in a rotary evaporator (Büchi Laboratoriumstechnik AG RE121; Flawil, Switzerland) fitted with a diaphragm vacuum pump (MC2C; Vacuubrand GmbH, 97877 Wertheim, Germany). Derivatization of amino acids was carried out by esterification with trifluoroacetylation (Brand et al., 1994) with the addition of an internal standard (*trans*-4-(aminomethyl)-cyclohexanecarboxylic acid) (Aldrich; purity, 97%). The GC-C-IRMS/MS (gas chromatography-combustion-isotope ratio mass spectrometry) measurements were performed for qualitative as well as quantitative assays of amino acids using the gas chromatograph (Hewlett-Packard 58590 II) linked through a split with a combustion interface to the IRMS system (GC-C-II to MAT 252, Finnigan MAT; Germany) for determination of nitrogen via a transfer line with a mass spectrometer (GCQ, Finnigan MAT; Germany).

18.2.4 Protein Digestibility

In vitro protein digestibility (IVPD) was established based on Akeson and Stahmann (1964) and the protein content in the digest was assessed by the micro-Kjeldahl method (Humphries, 1956). The essential amino acid (EAA) score, as well as the protein digestibility corrected amino acid score (PDCAAS) of the EAA requirement for human adults (FAO-WHO, 1991), were calculated. The protein efficiency ratio (PER) was determined based on the composition of amino acids in the flours (Alsmeyer et al., 1974).

18.3 EVALUATION OF BIOACTIVE COMPONENTS

The content of total phenolics in the flours was estimated by extraction with methanol (Rosset et al., 1982). The orthodihydric phenolic content in flours was assessed according to Mahadevan and Sridhar (1985). The tannin content was evaluated by the vanillin-HCl method (Burns 1971). The L-canavanine content of flours was assessed by the procedure of D'Cunha and Sridhar (2010). The vitamin C content in flours was assayed according to the protocol of Roe (1954). The total antioxidant activity of flours was determined following the procedure of Prieto et al. (1999), whereas the reducing power assay was performed by the method described by Oyaizu (1986). The trypsin inhibition activity of flours was determined by the enzymatic assay (Kakade et al., 1974), whereas the hemagglutinin activity was carried out by the protocol of Occenã et al. (2007).

18.4 DATA ANALYSIS

The seed characteristics, proximate composition, minerals, protein fractions, protein digestibility, and bioactive components of uncooked and cooked seed flours were compared by *t*-test using Statistica version 8.0 (2008). Only cooked and uncooked flour (from either dry or ripened seeds) were compared by *t*-test for amino acid composition.

18.5 SEED FEATURES AND NUTRITIONAL QUALITIES

18.5.1 Seed Features

Stands of *Sesbania bispinosa* are common in the coastal hind dunes of the Someshwara region (Figure 18.1a, b). Being an annual legume, it grows in the rainy season (June to September), produces inflorescences during the post-monsoon season (October to December), followed by the formation of pods and seeds. The pods transform from green to brown on ripening to black on drying (Figure 18.1c, d). The bullet-shaped ripened seeds are partially green, reddish-brown, whereas the dry seeds are black (Figure 18.1e, f). Ripened and dry seeds show a distinct spherical hilum region. The length (*l*), breadth (*b*), length/breadth (*l/b*) ratio and dry mass of ripened seeds were each significantly higher than for dry seeds (Table 18.1).

TABLE 18.1
Distinctive Features of Ripened as Well as Dry Seeds of *Sesbania bispinosa* (n=25; mean±SD)

	Ripened seeds	Dry seeds
Length, *l*(cm)	0.39±0.03**	0.28±0.02
Breadth, *b* (cm)	0.21±0.03**	0.16±0.01
Thickness, *t* (cm)	0.14±0.02	0.14±0.005
l/b ratio[a]	1.88±0.22*	1.78±0.10
Dry weight/seed(g)	0.008±0.001**	0.007±0.001

Asterisks indicate significant differences between ripened and dry seeds (*t*-test: * $p<0.05$; ** $p<0.001$)

[a] Length-Breadth ratio

TABLE 18.2
Proximal Composition of Uncooked as Well as Cooked Seed Flours of *Sesbania bispinosa* on a Dry Weight Basis (n=5; mean±SD)

	Ripened seeds		Dry seeds	
	Uncooked	**Cooked**	**Uncooked**	**Cooked**
Moisture (%)	6.06±0.52**	5.29±0.29	9.84±0.38***	7.02±0.24
Crude protein (g/100 g)	26.44±0.73**	23.82±0.39	31.35±0.73***	26.97±1.14
Total lipid (g/100 g)	3.13±0.15	4.26±0.12***	2.33±0.11	2.86±0.06**
Crude fiber (g/100 g)	9.36±0.97	10.11±1.27*	13.84±0.17	16.15±0.25***
Ash (g/100 g)	3.22±0.20	3.14±0.64	2.89±0.19	2.78±0.15
Carbohydrates (g/100 g)	57.81±1.66	58.62±1.39	49.48±0.70	51.56±1.29
Calorific value (kJ/100 g)	1518±21	1522±19	1439±5**	1413±3

Asterisks indicate significant differences between cooked and uncooked flour from either ripened or dry seeds (*t*-test: $^*p<0.05$, $^{**}p<0.01$, $^{***}p<0.001$)

18.5.2 Proximal and Mineral Composition

The moisture content of uncooked seeds significantly decreased in response to cooking (Table 18.2). The crude protein content was greater in dry seeds than in ripened seeds. Cooking resulted in a significant increase in total lipid, crude protein, and crude fiber contents. There was no significant change in ash and carbohydrate contents following cooking. Cooking did not alter the calorific value of ripened seeds, while it significantly reduced the calorific value of dry seeds.

Minerals content of ripened and dry seeds significantly decreased on cooking except for calcium, phosphorus, and magnesium in ripened seeds (Table 18.3). In both types of seed, potassium, magnesium, copper, zinc, and manganese contents were higher than or comparable to NRC-NAS (1989) standards. The iron content of ripened seeds was higher than the NRC-NAS (1989) pattern. The selenium content was also higher than the range suggested by Pennington and Young (1990). In both seed types, the Na/K ratio was <1, whereas the Ca/P ratio was >1.

18.5.3 Protein Fractions and Amino Acids

Albumin and glutelin contents showed a significant increase in cooked seeds relative to dry seeds, whereas the effects were opposite in globulin content and there was no significant change in prolamin content (Table 18.4). Cooking increased the albumin/glutelin ratio in both seed types.

Among the amino acids, tryptophan was not detectable in either sample of cooked seeds (Table 18.5). Glutamic acid was the most common amino acid, followed by aspartic acid and arginine. Amino acid contents were significantly higher in cooked ripened seeds than in dry seeds. However, the ratio of EAA/TAA among the seeds was comparable.

Among the EAA, glycine, tyrosine, histidine, and arginine contents were similar to or higher in soybean (Bau et al., 1994) and wheat (USDA, 1999). Similarly, the aspartic acid, serine, threonine, alanine, glycine, valine, tyrosine, lysine, histidine, and arginine contents were comparable or higher than that of wheat. The isoleucine and histidine contents of both cooked seeds were higher than in

TABLE 18.3
Mineral Composition of Uncooked as Well as Cooked, Ripened Seed Flours of *Sesbania bispinosa* on a Dry Weight Basis (mg/100 g) (n=5; mean±SD).

	Ripened seeds		Dry seeds		NRC-NAS Pattern[a, b]
	Uncooked	Cooked	Uncooked	Cooked	
Sodium	104.18±6.60*	82.33±16.73	109.07±7.17**	76.52±4.24	120-200[a]
Potassium	652.01±41.28**	476.89±96.89	642.68±42.26**	485.67±26.89	500-700[a]
Calcium	287.12±18.18	265.29±53.90	286.59±18.85*	227.51±12.60	600[a]
Phosphorus	238.48±15.10	232.01±47.14	232.05±15.26*	197.33±10.93	500[a]
Magnesium	131.38±8.32	128.73±26.16	137.73±9.06*	119.14±6.60	60[a]
Iron	18.19±1.15*	10.45±2.12	8.12±0.53*	6.57±0.36	10[a]
Copper	1.92±0.12**	1.40±0.28	1.47±0.10*	1.15±0.06	0.6-0.7[a]
Zinc	7.34±0.46*	5.90±1.20	5.86±0.39*	5.03±0.28	5.0[a]
Manganese	6.71±0.42*	5.58±1.13	5.97±0.39*	4.88±0.27	0.3-1.0[a]
Selenium	0.75±0.003***	0.64±0.002	2.29±0.004***	1.51±0.004	0.05-0.2[b]
Na/K ratio	0.16	0.17	0.17	0.16	
Ca/P ratio	1.20	1.14	1.24	1.15	

Figures across the rows showing asterisks indicate a significant difference between the corresponding cooked and uncooked seeds (*t*-test: *p<0.05, **p<0.01, ***p<0.001)

[a] NRC-NAS (1989) recommended pattern for human infants

[b] Recommended pattern by Pennington and Young (1990)

TABLE 18.4
Fractions of Protein of Uncooked and Cooked Seed Flours of *Sesbania bispinosa* (g/100 g) (n=5; mean±SD).

	Ripened seeds		Dry seeds	
	Uncooked	Cooked	Uncooked	Cooked
Albumin (A)	8.24±0.66	9.98±0.70*	8.51±0.43	9.45±0.54*
Globulin (G)	10.18±1.11**	5.48±0.32	11.11±0.65***	5.19±0.67
Prolamin	1.61±0.37	0.95±0.39	2.67±0.58	2.79±1.27
Glutelin	2.30±0.36	4.28±0.59**	5.11±0.83	6.14±0.62*
A/G ratio[a]	0.82	1.83	0.77	1.85

Figures across the rows showing asterisks indicate a significant difference between the corresponding cooked and uncooked seeds (*t*-test: *p<0.05, **p<0.01, ***p<0.001)

[a] Albumin-Globulin ratio

uncooked seeds, whereas the valine content of cooked ripened seeds was comparable to the FAO-WHO (1991) standard.

The EAA score of isoleucine and histidine in cooked ripened seeds and histidine in cooked dry seeds exceeded 100 (Table 18.6). The PDCAAS of histidine was highest, followed by isoleucine in cooked ripened and dry seeds. However, the PER in both seeds was <2. The IVPD of both seeds

TABLE 18.5
Amino Acid Composition of Cooked, Ripened as Well as Dry Seed Flours of *Sesbania bispinosa* in Comparison With Other Sources of Food (g/100 g protein; n=5; mean±SD).

	Ripened seeds	Dry seeds	Soybean[a]	Wheat[b]	FAO-WHO[c]
Essential Amino Acids					
Threonine	2.75±0.04*	2.46±0.09	3.76	2.2–3	3.4
Valine	3.37±0.05*	3.07±0.11	4.59	3.7–4.5	3.5
Cystine	0.63±0.01*	0.58±0.02	1.70	1.6–2.6	2.5[d]
Methionine	0.77±0.12*	0.71±0.03	1.22	0.9–1.5	
Isoleucine	2.86±0.05*	2.59±0.10	4.62	3.4–4.1	2.8
Leucine	5.04±0.08*	4.54±0.17	7.72	6.5–7.2	6.6
Tyrosine	2.25±0.04*	2.02±0.08	1.24	1.8–3.2	6.3[e]
Phenylalanine	3.50±0.06*	3.18±0.12	4.84	4.5–4.9	
Tryptophan	ND	ND	3.39	0.7–1	1.1
Lysine	4.64±0.07*	4.09±0.15	6.08	1.8–2.4	5.8
Histidine	2.44±0.04*	2.15±0.08	2.50	1.9–2.6	1.9
Non-Essential Amino Acids					
Glutamic acid	15.71±0.25*	13.53±0.50	16.90	35.5–36.9	
Aspartic acid	7.43±0.12*	6.67±0.25	11.30	3.7–4.2	
Serine	3.91±0.06*	3.50±0.13	5.67	3.7–4.8	
Proline	3.32±0.05*	2.96±0.11	4.86	11.4–11.7	
Alanine	3.62±0.06*	3.30±0.12	4.23	2.8–3	
Glycine	5.28±0.08*	4.76±0.18	4.01	3.2–3.5	
Arginine	7.18±0.12*	6.28±0.23	7.13	3.1–3.8	
EAA/TAA ratio[f]	0.43	0.44	–	–	–

Figures across the rows showing asterisks indicate a significant difference between the corresponding ripened and dry seeds (t-test: $*p<0.01$)

[a] Bau et al. (1994)
[b] USDA (1999)
[c] FAO-WHO (1991)
[d] Cystine+Methionine
[e] Tyrosine+Phenylalanine
[f] Essential amino acid/Total amino acid
ND, Not detectable

significantly increased on cooking, and the value was higher in cooked ripened than in dry seeds (Table 18.7).

18.6 BIOACTIVE COMPONENTS

Among the bioactive components, all except for the hemagglutinin activity of uncooked seeds significantly decreased on cooking (Table 18.8). Total phenolic content was high in uncooked ripened seeds, whereas L-canavanine and vitamin C contents of dry seeds were higher than in ripened seeds. Orthodihydric phenols and tannins were low in quantity. The trypsin inhibition activity was higher in dry than in ripened seeds, and this activity showed a significant decrease in cooking. The total antioxidant activity was higher in dry than in ripened seeds, and it decreased significantly in cooked seeds. The reducing power of dry seeds was higher than that in ripened seeds, which was

TABLE 18.6
Score of Essential Amino Acids (EAA), Protein Digestibility Corrected to Amino Acid score (PDCAAS) and Protein Efficiency Ratio (PER) of Seed Flours of *Sesbania bispinosa*.

	Ripened seeds	Dry seeds
EAA Score		
Threonine	80.90	72.26
Valine	96.18	87.71
Cystine + Methionine	56.29	51.49
Isoleucine	102.29	92.53
Leucine	76.40	68.85
Tyrosine + Phenylalanine	91.30	82.48
Lysine	79.95	70.49
Histidine	128.48	113.06
PDCAAS		
Threonine	55.36	43.95
Valine	65.82	53.35
Cystine + Methionine	38.42	31.32
Isoleucine	70.0	56.29
Leucine	52.28	41.88
Tyrosine + Phenylalanine	62.48	50.17
Lysine	54.71	42.88
Histidine	87.92	68.77
PER		
PER_1	1.5	1.3
PER_2	1.6	1.4
PER_3	0.9	0.6
Average PER	1.3	1.1

ND, Not detectable

TABLE 18.7
The *In-Vitro* Protein Digestibility (IVPD) (%) of Uncooked as Well as Cooked Seed Flours of *Sesbania bispinosa* (n=5; mean±SD).

Ripened seeds		Dry seeds	
Uncooked	Cooked	Uncooked	Cooked
55.75±1.31	68.43±5.88*	52.38±4.12	60.83±1.44*

Figures across the rows showing asterisks indicate a significant difference between the corresponding cooked and uncooked seeds (*t*-test: *$p<0.01$)

higher in uncooked compared with cooked seeds, which gradually increased with an increase in the amount of flour (Figure 18.2). There was no change in hemagglutinin activity against the A^+ blood group between uncooked and cooked seeds, whereas, for the B^+ blood group, it decreased by up to 50%. Although there was no change in hemagglutinin activity against O^+ blood group in ripened seeds, it was completely eliminated in dry seeds following cooking.

TABLE 18.8
Bioactive Components of Uncooked as Well as Cooked Seed Flours of *Sesbania bispinosa* (n=5, mean±SD).

	Ripened seeds		Dry seeds	
	Uncooked	Cooked	Uncooked	Cooked
Total phenolics (%)	2.07±0.09*	1.23±0.02	1.89±0.03**	1.34±0.05
Orthodihydric phenols (%)	0.07±0.003*	0.05±0.01	0.05±0.002*	0.03±0.004
Tannins (%)	0.07±0.003*	0.04±0.003	0.08±0.01*	0.03±0.003
Canavanine (%)	0.11±0.001*	0.09±0.003	0.12±0.001*	0.07±0.005
Vitamin C (mg/100 g)	1.39±0.01*	1.26±0.01	4.30±0.01**	4.17±0.01
Total antioxidant activity (μM-eq vitamin C/g)	245.08±1.64*	218.23±0.63	277.23±1.64**	252.44±0.38
Trypsin inhibition activity (TIu/mg)	3.41±0.07**	1.77±0.12	6.70±0.11**	2.73±0.09
Hemagglutinin activity[a] (Hu/g)				
A+	20	20	20	20
B+	40	20	40	20
O+	20	20	20	0

Figures across the rows showing asterisks indicate a significant difference between the corresponding cooked and uncooked seeds (*t*-test: *p<0.01, **p<0.001)
[a] Hemagglutinin activities are mean of three independent observations
ND, Not detectable

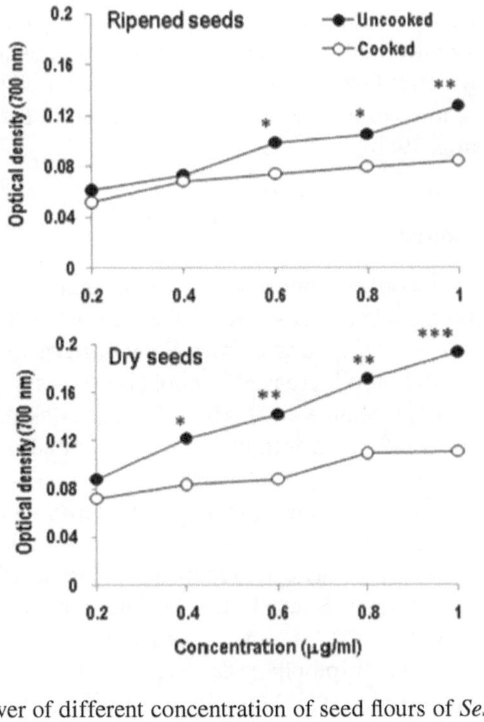

FIGURE 18.2 Reducing power of different concentration of seed flours of *Sesbania bispinosa* (n=5, mean; *t*-test: *p<0.05, **p<0.01, ***p<0.001).

18.7 DISCUSSION

18.7.1 Proximal Qualities

The crude protein content of seed flours of *S. bispinosa* is higher than in many edible crop legumes (e.g. chickpea, green gram, pigeon pea, and winged bean) (Jambunathan and Singh, 1980; Nwokolo, 1987; Baudoin and Maquet, 1999; Viswanathan et al., 2001). However, it is lower than in *Sesbania rostrata* and *Sesbania sesban* (Hossain and Becker, 2001; Siddhuraju et al., 2002; Pugalenthi et al., 2004). Although the total lipid and crude fiber contents of ripened and dry seeds of *S. bispinosa* significantly increased on cooking, their quantities were still lower than *S. rostrata* and *S. sesban* (Hossain and Becker, 2001; Siddhuraju et al., 1995, 2002; Pugalenthi et al., 2004).The carbohydrate content in *S. bispinosa* seeds is comparable with other *Sesbania* spp. (*S. rostrata* and *S. sesban*) (Hossain and Becker 2001; Siddhuraju et al., 1995, 2002; Pugalenthi et al., 2004). *Sesbania* seeds are a potential source of the food-grade polysaccharide galactomannan, which has several food, health, and industrial applications (Vietmeyer, 1986; Pollard et al., 2011). The calorific value of *S. bispinosa* is lower than for other *Sesbania* spp. (*S. rostrata* and *S. sesban*) (Hossain and Becker, 2001; Siddhuraju et al., 1995, 2002; Pugalenthi et al., 2004). Such differences in proximal qualities of *S. bispinosa* seeds from other *Sesbania* spp. may be attributed to its adaptation to a saline-prone habitat.

18.7.2 Mineral Composition

Legumes seeds are known for being rich in potassium, magnesium, iron, zinc, and calcium (Salunkhe et al., 1985). Although there is a significant loss of minerals following cooking of seeds of *S. bispinosa*, the contents of many minerals (potassium, iron, magnesium, copper, zinc, and manganese) are similar to or higher than the NRC-NAS (1989) standards. The selenium content of *S. bispinosa* is higher than the range recommended by Pennington and Young (1990). It is also known that iron, zinc, manganese, and selenium serve as antioxidants as well as strengthening the immune system (Talwar et al., 1989; Combs and Gray, 1998). According to Chaturvedi et al. (2004), magnesium, zinc, and selenium can counter cardiomyopathy, muscle degeneration, immunological dysfunction, and congenital malformations. The uncooked and cooked seeds of *S. bispinosa* possess a low Na/K ratio (<1), which is favorable to combating high blood pressure (Yusuf et al., 2007). The high Ca/P ratio (>1) is known to prevent the loss of calcium in the urine and helps to restore it in the bones (Shills and Young, 1988).

18.7.3 Protein Bioavailability

Among fractions of protein in uncooked ripened or dry seeds, globulin (G) was higher than albumin (A), while it was the opposite in cooked seeds. The highest globulin content in uncooked dry seeds agrees with data from a *S. bispinosa* landrace from Southeastern India (Siddhuraju et al., 1995; Pugalenthi et al., 2004). The A/G ratio increased in cooked seeds compared with uncooked seeds, which denotes decreased globulin content in seeds following cooking. Such loss of globulins in cooked seeds results in lower toxicity from antinutritional compounds as they are usually associated with the globulin fraction.

In cooked ripened and dry seeds, glutamic acid was the amino acid with the highest content, as also reported in *S. rostrata* as well as *S. sesban* (Hossain and Becker, 2001). The glutamic acid content was also highest followed by aspartic acid in uncooked seeds (*S. bispinosa*) in Southeastern India (Siddhuraju et al., 1995; Pugalenthi et al., 2004). The cooked ripened and dry seeds were devoid of tryptophan corroborating the studies carried out on *S. bispinosa* of Southeast India (Pugalenthi et al., 2004). However, tryptophan was present in *S. bispinosa* of Southeastern and Northeastern India (Siddhuraju et al., 1995; Hossain and Becker, 2001). Except for sulfur amino acids, other EAAs were in adequate quantities in *S. bispinosa* of Southeastern India (Siddhuraju et al., 1995). The quantity of tyrosine and arginine in *S. bispinosa* is higher than in soybean (Bau et al. 1994), whereas the contents of aspartic acid, threonine, alanine, glycine, tyrosine, lysine, histidine, and arginine are similar to or higher than those in wheat (USDA 1999). Contents of isoleucine, as well as histidine, are also similar to or higher than the FAO-WHO (1991) recommended patterns for human adults.

The EAA ratio was higher in cooked, ripened seeds compared with dry seeds based on FAO-WHO (1991) standard. The score for histidine exceeded 100 in both cooked seeds, while it occurred for isoleucine only in cooked ripened seeds. The PDCAAS pattern also followed the EAA score. The average PER of cooked seeds was below 1.5, which indicates poor quality of the protein (Friedman, 1996). However, PER_1 and PER_2 in cooked, ripened seeds were 1.5 and 1.6, respectively, which denotes moderate protein quality. These findings are also supported by significantly increased IVPD of cooked seeds, which is a sign of substantial loss of antinutritional compounds. As seen in the present study, cooking also significantly increased the IVPD of another *S. bispinosa* of Southeast India (Sidduraju et al., 1995). The differences in protein bioavailability in *S. bispinosa* seeds in our study may differ from those of other landraces or *Sesbania* spp. due to their adaptability to saline-prone sand dunes.

18.7.4 Fatty Acid Composition

Cooking the ripened seeds of *S. bispinosa* resulted in increased linoleic, α-linolenic, and eicosatrienoic acids (Shreelalilitha and Sridhar, 2016). Interestingly, the docosahexaenoic acid content was higher in uncooked compared with cooked, ripened seeds. The cooked, dry seeds possess γ-linolenic as well as docosadienoic acids. The quantity of α-linolenic and docosahexaenoic acids were significantly increased on cooking the dry seeds. Increased fatty acid ratios Total Unsaturated Fatty Acids/Total Saturated Fatty Acids and Total Polyunsaturated Fatty Acids/Total Monounsaturated Fatty Acids (TUFA/TSFA and TPUFA/TMUFA) were detected in cooked, ripened seeds, whereas the decreased ratio of ω-fatty acids (ω-6/ω-3) in cooked, dry seeds fulfill the FAO/WHO (1998) recommended standards as well as the ratio suggested by Simopoulos (2002).

18.7.5 Bioactive Components

On cooking, the bioactive compounds of seeds significantly decreased in ripened and dry seeds. The quantity of total phenolics in ripened and dry seeds was lower than in *S. rostrata* and *S. sesban*, as well as another landrace of *S. bispinosa* from Southeastern India (Siddhuraju et al., 1995, 2002; Hossain and Becker, 2001; Pugalenthi et al., 2004). The canavanine content of *S. bispinosa* is lower than in *S. rostrata* and *S. bispinosa* of Southeast India (Siddhuraju et al., 2002), which is similar to the seeds of *Vicia articulata* and *Vicia ervilia*. The range of L-canavanine contents in *S. bispinosa* is nutritionally secure and also beneficial in treating pancreatic cancer as well as combating colon cancer (Thomas et al., 1986; Enneking and Wink, 2000; Bence et al., 2003). Higher vitamin C content in uncooked and cooked, dry seeds reflected the higher total antioxidant activity as well as the greater reducing power. The trypsin inhibitory activity of cooked, ripened and dry seeds is lower than in the *S. bispinosa* landrace of Southeast India (Siddhuraju et al., 1995). The hemagglutinin activity in both seeds did not change on cooking except for a 50% decrease in the B^+ group hemagglutinin, although cooking dry seeds lost their hemagglutinin activity against the O^+ group, indicating the necessity to modify the seed processing method to eliminate the lectins. It would be worthwhile comparing the bioactive components of seeds of *S. bispinosa* landraces from saline-prone habitats with other landraces growing in non-saline regions.

18.8 CONCLUSION

Sesbania has several established applications in agriculture, food, health, and industry. The ripened as well as the dry seeds of *Sesbania bispinosa* are good sources of protein, fiber, energy, many essential minerals, several essential amino acids, and improved *in vitro* protein digestibility, with essential fatty acids. The presence of many bioactive components resulted in a significant increase in total antioxidant activity as well as reducing power. The ripened and dry seeds of *S. bispinosa* met several standards of FAO-WHO (1991, 1998). In addition to serving as a potential, ethnically valued, nutritional sources, the seeds of *S. bispinosa* fulfill the pharmaceutical values needed to combat human ailments. Future investigations need to follow different processing strategies (e.g., germination, cooking, and fermentation) and diet management to extend the nutraceutical value of *Sesbania* seeds for human nutrition and health security.

ACKNOWLEDGMENTS

The first author was awarded a Junior Research Fellowship under the scheme Research Fellowship in Sciences for Meritorious Students by the University Grants Commission, New Delhi. The second author is grateful for the award of a UGC-BSR Faculty Fellowship by the University Grants Commission, New Delhi, and to Mangalore University for the award of the adjunct professorship.

REFERENCES

Akeson, W.R. & M.A. Stahmann. 1964. A pepsin pancreatin digest index of protein quality. *Journal of Nutrition* 83(3): 257–261.

Alsmeyer, R.H., A.E. Cunningham & M.L. Happich. 1974. Equations predict PER from amino acid analysis. *Food Technology* 28: 34–38.

Anderson, D.M.W. 1989. *Sesbania* species as sources of gum exudates and seed galactomannan gums. In: Macklin, B. & D.O. Evans (Eds.), *Perennial Sesbania species in Agroforestry Systems*. NFTA Special Publication # 90-01, Waimanalo, USA, pp. 99–194.

Anita, D.D. & K.R. Sridhar. 2020. Nutraceutical potential of ripened beans of mangrove wild legume *Sesbania speciosa*. In: Patra, J.K., R.R. Mishra & H. Thatoi (Eds.), *Biotechnological Utilization of Mangrove Resources*. Elsevier Inc., Academic Press, Oakville, Ontario, Canada, pp. 243–259.

Anita, D.D. & K.R. Sridhar. 2021. Nutritional and bioactive components of ethnically valued seeds of mangrove wild legume *Sesbania speciosa*. In: Mohan, V.R., P.S. Tresina & A. Doss (Eds.), *The Phytochemical and Pharmacological Aspects of Ethnomedicinal Plants*. Apple Academic Press, Oakville, Ontario, Canada, pp. 389–405.

AOAC. 1995. *Official Methods of Analysis*, 16th edition. Association of Official Analytical Chemists, Washington, DC.

Arora, R.K., K.P.S. Chandel, B.S. Joshi & K.C. Pant. 1980. Rice bean: Tribal pulse of eastern India. *Economic Botany* 34(3): 260–263.

Arun, A.B., K.R. Beena, N.S. Raviraja & K.R. Sridhar. 1999. Coastal sand dunes - A neglected ecosystem. *Current Science* 77: 19–21.

Bau, H.M., C. Vallaume, C.-F. Lin, J. Evrard, B. Quéméner, J. Nicolas & L. Méjean. 1994. Effect of solid state fermentation using *Rhizophus oligosporus* sp. T-3 on elimination of antinutritional substances and modification of biochemical constituents of defatted rape seed meal. *Journal of the Science of Food and Agriculture* 65(3): 315–322.

Baudoin, J.P. & A. Maquet. 1999. Improvement of protein and amino acid content in seeds of food legumes – A case study in *Phaseolus*. *Biotechnologie, Agronomie, Socéité et Environment* 3: 220–224.

Bence, A.K., V.R. Adams & P.A. Crooks. 2003. L-canavanine as a radiosensitization agent for human pancreatic cancer cells. *Molecular and Cellular Biochemistry* 244(1–2): 37–43.

Bhagya, B. & K.R. Sridhar. 2009. Ethnobiology of coastal sand dune legumes of southwest India. *Indian Journal of Traditional Knowledge* 9: 611–620.

Bhat, R. & A.A. Karim. 2009. Exploring the nutritional potential of wild and underutilized legumes. *Comparative Reviews in Food Science and Food Safety* 8(4): 305–331.

Brand, W.A., A.R. Tegtmeyer & A. Hilkert. 1994. Compound-specific isotope analysis, extending towards 15N/14N and 13C/12C. *Organic Geochemistry* 21(6–7): 585–594.

Burns, R. 1971. Methods for estimation of tannins in grain sorghum. *Agronomy Journal* 63(3): 511–512.

Chanda, S.C., A.K.M.A. Prodhan & A.K.M.G. Sarwar. 2017. Screening *Sesbania* accessions based on early biomass yield. *Journal of the Bangladesh Agricultural University* 15: 188–192.

Chaturvedi, V.C., R. Shrivastava & R.K. Upreti. 2004. Viral infections and trace elements: A complex interaction. *Current Science* 87: 1536–1554.

Combs, G.F. & W.P. Gray. 1998. Chemopreventive agents: Selenium. *Pharmacology and Therapeutics* 79(3): 179–192.

D'Cunha, M. & K.R. Sridhar. 2010. L-canavanine and L-arginine in two wild legumes of the genus *Canavalia. IIOAB Journal* 1: 29–33.

Ekanayake, S., E.R. Jansz & B.M. Nair. 1999. Proximate composition, mineral and amino acid content of mature *Canavalia gladiata* seeds. *Food Chemistry* 66(1): 115–119.

Enneking, D. & M. Wink. 2000. Towards the elimination of anti-nutritional factors in grain legumes. In: Knight, R. (Ed.), *Current Plant Science and Biotechnology in Agriculture*, Vol. 34. Kluwer Academic Publishers, Dordrecht, pp. 375–384.

Evans, D.O. 1990. What is *Sesbania*? Botany, taxonomy, plant geography and natural history of the perennial members of the genus. In: Macklin, B. & D.O. Evans (Eds.), *Perennial Sesbania species in Agroforestry Systems*. Nitrogen Fixing Tree Association, Waimanalo, pp. 5–16.

FAO-WHO. 1991. *Protein Quality Evaluation: Reports of a Joint FAO-WHO Expert Consultation, Food and Nutrition Paper # 51.* Food and Agricultural Organization of the United Nations, FAO, Rome.

FAO-WHO. 1998. *Preparation and Use of Food-Based Dietary Guidelines*. Report of Joint FAO-WHO Consultation, Technical Report Series # 880, FAO, Geneva.

Friedman, M. 1996. Nutritional value of proteins from different food sources - A review. *Journal of Agricultural and Food Chemistry* 44(1): 6–29.

Gheyasuddin, S., C.M. Cater & K.F. Mattil. 1970. Effect of several variables on the extractability of sunflower seed proteins. *Journal of Food Science* 35(4): 453–456.

Gunjatkar, N. & V.D. Vartak. 1982. Enumeration of wild legumes from Pune District, Maharashtra. *Journal of Economic and Taxonomic Botany* 3: 1–9.

Hofmann, D., M. Gehre & K. Jung. 2003. Sample preparation techniques for the determination of natural 15N/14N variations in amino acids by gas chromatography-combustion-isotope ratio mass spectrometry (GC-C-IRMS). *Isotopes in Environmental and Health Studies* 39(3): 233–244.

Hofmann, D., K. Jung, J. Bender, M. Gehre & G. Schüürmann. 1997. Using natural isotope variationsof nitrogen in plants as an early indicator of air pollution stress. *Journal of Mass Spectrometry* 32(8): 855–863.

Hossain, M.A. & K. Becker. 2001. Nutritive value and antinutritional factors in different *Sesbania* seeds and their morphological fractions. *Food Chemistry* 73(4): 421–431.

Humphries, E.C. 1956. Mineral composition and ash analysis. In: Peach, K. & M.V. Tracey (Eds.), *Modern Methods of Plant Analysis*, Vol. 1. Springer, Berlin, pp. 468–502.

Ipor, I.B. & L.P.A. Oyen. 1997. *Sesbania* Adanson. In: Hanum, F. & L.J.G. Van der Maesen (Eds.), *Plant Resources of South-East Asia*. Backhuys Publishers, Leiden, the Netherlands, pp. 236–240.

Jambunathan, R. & U. Singh. 1980. Studies on Desi and Kabuli chickpea (*Cicerarietinum*) cultivars. 1. Chemical composition. In: *Proceedings of the International Workshop on Chickpea Improvement*. International Crop Research Institute of Semi-Arid Tropics, Hyderabad, pp. 61–66.

Janardhanan, K., V. Vadivel & M. Pugalenthi. 2003. Biodiversity in Indian underexploited/tribal pulses. In: Jaiwal, P.K. & R.P. Singh (Eds.), *Improvement Strategies for Leguminosae Biotechnology*. Kluwer Academic Publishers, Dordrecht, the Netherlands, pp. 353–405.

Kakade, M.L., J.J. Rackis, J.E. McGhee & G. Puski. 1974. Determination of trypsin inhibitor activity of soy products, a collaborative analysis of an improved procedure. *Cereal Chemistry* 51: 376–382.

Mahadevan, A. & R. Sridhar. 1985. *Methods in Physiological Plant Pathology*, 3rd edition. Sivakami Publications, Madras, India.

Mandal, U.K., G. Singh, U.S. Victor & K.L. Sharma. 2003. Green manuring: Its effect on soil properties and crop growth under rice-wheat cropping system. *European Journal of Agronomy* 19(2): 225–237.

Morris, J.B. 1999. Legume genetic resources with novel value added industrial and pharmaceutical use. In: Janick, J. (Ed.), *Perspectives on New Crops and New Uses*. ASHS Press, Alexandria, pp. 196–201.

Müller, H.G. & G. Tobin. 1980. *Nutrition and Food Processing*. Croom Helm Ltd., London.

Ndoye, I., K. Tomekpe, B. Dreyfus & Y.R. Dommergues. 1990. *Sesbania* and *Rhizobium* symbiosis: Nodulation and nitrogen fixation. In: Macklin, B. & D.O. Evans (Eds.), *Perennial Sesbania species in Agroforestry Systems*. Nitrogen Fixing Tree Association, Wamanalo, pp. 31–38.

NRC-NAS. 1989. *Recommended Dietary Allowances*. National Academy Press, Washington, DC.

Nwokolo, E. 1987. Nutritional evaluation of pigeon pea meal. *Plant Foods for Human Nutrition* 37(4): 283–290.

Occenã, I.V., E.R.E. Majica & F.E. Merca. 2007. Isolation of partial characterization of a lectin from the seeds of *Artocarpus camansi* Blanco. *Asian Journal of Plant Sciences* 6(5): 757–764.

Oyaizu, M. 1986. Studies on products of browning reactions: Antioxidative activities of products of browning reaction prepared from glucosamine. *Japanese Journal of Nutrition* 44: 307–315.

Palaniappan, S.P. & K. Siddeswaran. 2001. Regional overview on green manure in rice-based cropping systems. In: Gowda, C.L.L., A. Ramakrishna, O.P. Rupela & S.P. Wani (Eds.), *Legumes in Rice Based Cropping Systems in Tropical Asia: Constraints and Opportunities*. International Crops Research Institute for Semi-Arid Tropics, Patancheru, Hyderabad, pp. 126–135.

Pennington, J.A.T. & B. Young. 1990. Iron, zinc, copper, manganese, selenium and iodine in foods from the United States total diet study. *Journal of Food Composition and Analysis* 3(2): 1661–1684.

Pollard, M.A., P. Fischer & E.J. Windhab. 2011. Characterization of galactomannans derived from legume endosperms of genus *Sesbania* (Faboideae). *Carbohydrate Polymers* 84(1): 550–559.

Prasad, M.N.V. 1993. Bioresource potential of *Sesbania bispinosa* (Jacq.) W.F. Wight. *Bioresource Technology* 44(3): 251–254.

Prieto, P., M. Pineda & M. Aguilar. 1999. Spectrophotometric quantitation of antioxidant capacity through the formation of a phosphomolybdenum complex: Specific application to the determination of vitamin E. *Analytical Biochemistry* 269(2): 337–341.

Pugalenthi, M., V. Vadivel, P. Gurumoorthi & K. Janardhanan. 2004. Comparative nutritional evaluation of little known legumes, *Tamarindus indica*, *Erythrina indica* and *Sesbania bispinosa*. *Tropical and Subtropical Agroecosystems* 4: 107–123.

Rao, T.A. & V.M. Meher-Homji. 1985. Strand plant communities of the Indian sub-continent. *Proceedings of the Indian Academy of Science* 94(2–3): 505–523.

Rao, T.A. & A.N. Sherieff. 2002. *Coastal Ecosystem of the Karnataka State, India II - Beaches*. Karnataka Association for the Advancement of Science, Bangalore.

Roe, J.H. 1954. Chemical determination of ascorbic, dehydroascorbic and diketogluconic acids. In: Glick, D. (Ed.), *Methods of Biochemical Analysis*, Vol. 1. InterScience Publishers, New York, pp. 115–139.

Rosset, J., F. Bärlocher & J.J. Oertli. 1982. Decomposition of conifer needles and deciduous leaves in two Black Forest and two Swiss Jura streams. *Internationale Reveue der Gesamten Hydrobiologie* 67: 695–711.

Salunkhe, D.K., S.S. Kadam & J.K. Chavan. 1985. Chemical composition. In: Salunkhe, D.K., S.S. Kadam & J.K. Chavan (Eds.), *Postharvest Biotechnology of Food Legumes*. CRC Press Inc., Boca Raton, FL, pp. 33–35.

Sarkar, M., S. Sutradhar, A.K.M.G. Sarwar, M.N. Uddin, S.C. Chanda & M.S. Jahan. 2017. Variation of chemical characteristics and pulpability of dhaincha (*Sesbania bispinosa*) on location. *Journal of Bioresources and Bioproducts* 2: 24–29.

Sarwar, A.K.M.G., S.M.Z. Hossain & S.C. Chanda. 2017. Effect of Dhaincha accessions on soil health and grain yield of rice. *Journal of Bioscience and Agriculture Research* 13(2): 1140–1145.

Shahjalal, M. & J.H. Topps. 2000. Feeding *Sesbania* leaves as a sole feed on growth and nutrient utilization in goats. *Asian-Australian Journal of Animal Sciences* 13(4): 487–489.

Shills, M.E.G. & V.R. Young. 1988. Modern nutrition in health and disease. In: Neiman, D.C., D.E. Buthepodorth & C.N. Nieman (Eds.), *Nutrition*. WmC Brown Publishers, Dubugue, IA, pp. 276–282.

Shreelalitha, S.J. & K.R. Sridhar. 2016. Composition of fatty acids of the lipids extracted from seeds of *Sesbania bispinosa* grown on the Indian coastal sand dunes. *Current Nutrition and Food Science* 12(1): 50–55.

Shreelalitha, S.J. & K.R. Sridhar. 2018. Physical and cooking properties of seeds of two wild legume landraces of *Sesbania*. *International Journal of Agricultural Technology* 14: 363–376.

Shreelalitha, S.J. & K.R. Sridhar. 2019. Functional attributes of seeds of two coastal germplasms of *Sesbania*. In: Öztürk, M. & K.R. Hakeem (Eds.), *Plants and Human Health, Volume 2, Phytochemistry and Molecular Aspects*. Springer Nature, Singapore, pp. 159–172.

Siddhuraju, P., O. Osoniyi, H.P.S. Makkar & K. Becker. 2002. Effect of soaking and ionising radiation on various antinutritional factors of seeds from different species of an unconventional legume, *Sesbania* and a common legume, green gram (*Vigna radiata*). *Food Chemistry* 79(3): 273–281.

Siddhuraju, P., K. Vijayakumari & K. Janardhanan. 1995. Studies on the underexploited legumes, *Indigofera linifolia* and *Sesbania bispinosa*: Nutrient composition and antinutritional factors. *International Journal of Food Sciences and Nutrition* 46(3): 195–203.

Simopoulos, A.P. 2002. The importance of the ratio of omega-6/omega-3 essential fatty acids. *Biomedicine and Pharmacotherapy* 56(8): 365–379.

Singh, J.S. & R.K. Chaturvedi. 2017. Diversity of ecosystem types in India: A review. *Proceedings of the Indian National Science Academy* 83: 569–594.

Singh, R.J., G.H. Chung & R.L. Nelson. 2007. Landmark research in legumes. *Genome* 50(6): 525–537.

Sridhar, K.R., D.D. Anita & S.D. Ghate. 2016. Fatty acid composition of mangrove wild legume seeds (*Sesbania speciosa*) in Southwestern India. *Recent Patents in Food and Nutrition Agriculture* 8(2): 124–131.

Statistica, Version # 8. 2008. StatSoft Inc., Tulsa, OK.

Talwar, G.P., L.M. Srivastava & K.D. Mudgil. 1989. *Textbook of Biochemistry and Human Biology*, 2nd edition. Prentice Hall of India Private Ltd, New Delhi.

Thomas, F.A., G.A. Rosenthal, D.V. Gold & K. Dickey. 1986. Growth inhibition of a rat colon tumor by L-canavanine. *Cancer Research* 46(6): 2898–2903.

USDA. 1999. Nutrient data base for standard reference release # 13, food group 20: Cereal grains and pasta. *Agriculture Handbook* # 8–20. Agricultural Research Service, US Department of Agriculture, USA.

Vadivel, V. & H.K. Biesalski. 2010. HPLC analysis of bioactive compounds in ten different wild type underutilized legume grains. *IIOAB Journal* 1: 17–24.

Vadivel, V., W. Stuetz, V. Scherbaum & H.K. Biesalski. 2011. Total free phenolic content and health relevant functionality of Indian wild legume grains: Effect of indigenous processing methods. *Journal of Food Composition and Analysis* 24(7): 935–943.

Veasey, E.A., E.A. Schammass, R. Vencovsky, P.S. Martins & G. Bande. 1999. Morphological and agronomic characterization and estimates of genetic parameters of *Sesbania* Scop. (Leguminosae) accessions. *Genetics and Molecular Biology* 22(1): 81–93.

Vietmeyer, N.D. 1986. Lesser-known plants of potential use in agriculture and forestry. *Science* 232(4756): 1379–1384.

Viswanathan, M.B., D. Thangadurai & N. Ramesh. 2001. Biochemical evaluation of *Neonotoniawightii* (Wight and Arn) Lackey (Fabaceae). *Food Chemistry* 75(3): 275–279.

Viswanathan, M.B., D. Thangadurai, K. Tamilvendan & N. Ramesh. 1999. Chemical analysis and nutritional assessment of *Teramnus labialis* (L.) Spreng. (Fabaceae). *Plant Foods for Human Nutrition* 54(4): 345–352.

Yusuf, A.A., B.M. Mofio & A.B. Ahmed. 2007. Proximate and mineral composition of *Tamarindus indica* Linn 1753 seeds. *Scientific World Journal* 2: 1–4.

19 General Plant Geography Formation of the Aegean Region and the Use of Aegean Herbs in Local Cuisine and for Medicinal Purposes

Feyza Candan

CONTENTS

19.1 INTRODUCTION

Plants are a very important part of the environment we live in and they have always attracted people's attention due to their benefits to humans and because they cover a large proportion of the earth. Therefore, studies on botany, which is one of the branches of biology, carried out in line with these interests, date back to ancient times.

Botanical studies began with simple but everyday classifications of plants according to some properties, such as edible/non-edible, poisonous/non-poisonous, medicinal/non-medicinal. Later, research endeavors related to expectations and purposes in this direction were developed in parallel with scientific progress and technical abilities.

Floral studies, which aim to list the plants in a region, have provided information about some characteristics of the plants on which detailed studies will be made, such as where and in which season they grow. In recent years, research on each country's flora around the world continues to progress by including more detailed information.

Many scientific studies have been carried out on plants in Türkiye. Most of the first plant studies about Türkiye, which consisted of lists of flora stemming from foreign researchers' botanical trips covering Türkiye as a whole, included general information. In 17001–1702, the botanist Pitton

DOI: 10.1201/9781003226925-19

de Tournefort published details of the plants he collected from North and Northeast Anatolia as a floral list. This was followed by Sibthorp, an English botanist, who collected plants from İzmir and Uludağ at the end of the eighteenth century. Later, various researchers contributed to these studies by collecting plants from different regions: Aucher Eloy, who traveled through Anatolia in the 1830s, mostly from the border regions; the Australian, Kotschy, from the Taurus region in 1836; Genevan botanist Edmont Boissier, the author of "Flora Orientalis," from Izmir and its surroundings; and, in 1845–1851, Heldreich from around Izmir.

Later, the Flora of Türkiye was revealed first in the work entitled "Flora of Turkey and East Aegean Islands," prepared under the editorship of Davis (1965–1985). The discoveries that took place in the 28 years after the publication of the tenth volume (Davis, 1988), which was published in addition to this work, were introduced with the volume prepared by Turkish botanists and added to this series as the eleventh volume (Güner et al., 2000), and many supplements and new volumes have been presented (Güner, 2012, 2014, 2018). However, about 12,000 plant taxa, including endemic and relict taxa, which are in the rich floristic structure of Türkiye, are and will be the subject of various research studies for many purposes.

19.2 GENERAL PLANT GEOGRAPHY AND FORMATION OF THE AEGEAN REGION

In the Aegean Region, the Mediterranean climate expands into the interior regions by the effect of the valleys. The Inner West Anatolian section is under the influence of the continental climate. The vegetation of the Aegean Region falls under the title of the Mediterranean Plant Geography region. The dominant trees in the natural vegetation of the Aegean Region are red or Turkish pine (*Pinus brutia*). Red pines are Mediterranean-climate trees, which are drought tolerant and demand light. These trees are found up to 1000 m on the south-facing slopes of the Bozdağlar, Aydın Mountains, and Menteşe Region mountains; and, starting from the coastal zone, they expand into the valleys where the Mediterranean climate is introduced. The lower layer of the red pine forests is rich in maquis species. In the Aegean Region, the slopes of the mountains are covered with red pines up to the Inner West Anatolia. In Inner West Anatolia, the climate and conditions are not suitable for the productive cultivation of red pine.

The plant formations of the Aegean Region were examined in six categories: humid, semi-humid, and dry forests, maquis plants, garigue plants, and Alpine plants.

The dominant elements of humid forests are plants such as *Fagus orientalis*, *Populus tremula*, *Acer* sp., *Carpinus betulus*, *Taxus baccata*, and *Castanea sativa*.

The main elements of semi-humid forests are *Pinus sylvestris* above 1500 m, and, in other places, plants such as *Castanea sativa*, *Salix alba*, *Crataegus monogyna*, *Prunus spinosa*, *Rubus* spp, *Fagus orientalis*, *Tilia tomentosa*, *Carpinus betulus*, *Populus tremula*, *Sorbus* spp, *Cornus* sp, and *Cercis* sp.

Dry forests appear on the southern slopes of the mountains. The dominant elements of these forests are *Pinus nigra*, *Pinus brutia*, *Quercus cerris*, *Quercus infectoria* and *Quercus libani*, *Juniperus excelsa*, and *Juniperus foetidissima*.

The maquis plants are usually found in the lower parts of the forests on the southern slopes of the mountains and expand inland along the stream valleys. The places where *Pinus brutia* forests were destroyed are covered with maquis. The upper limit of maquis communities can be up to 800 m in the south and 500 m in the north. These plants are generally *Arbutus unedo*, *Arbutus andrachne*, *Pistacia terebinthus*, *Spartium junceum*, *Erica arborea*, *Cercis siliquastrum*, *Styrax officinalis*, *Laurus nobilis*, *Juniperus oxycedrus*, *Cistus creticus*, *Quercus coccifera*, and *Quercus ilex*.

Phrygian plants emerge as dwarf bush communities where maquis are destroyed. They are also called garigue. These plant communities are generally: *Sarcopoteriumspinosum*, *Asphodelus aestivus*, *Cistus* sp., and members of the Fabaceae family.

Alpine plants include plants above 1700 m. These are generally *Astragalus* spp., *Acantholimon* spp., *Verbascum* spp., and *Festuca* spp.

19.2.1 Plant Geography of Kütahya Plain and Its Surroundings

Kütahya Plain and its surroundings are an area where the climates in the Mediterranean, Black Sea, and Central Anatolia regions are all intertwined. As a result, the vegetation of the area bears the characteristics of all three regions. The vegetation of this region can be examined in two groups, namely semi-humid and dry forests, considering the climatic and edaphic conditions. The semi-humid forest area is the wettest part of the region. The amount of precipitation is around 800–1,000 mm; at the summit, this value exceeds 1,000 mm. Dry forests occur at 1,500–1,600 m of the mountains. Dry forests are common throughout the area, except for parts of the mountains higher than 1,500–1,600 m (Dönmez, 2014).

19.2.2 Plant Geography of the Area Between Gediz and Büyük Menderes

The land encompassing the vicinities of İzmir, the Urla Peninsula, the Küçükmenderes plain, Aydın, Söke and Milet regions include the Gediz River from the north, the Büyükmenderes River from the south, and the Alaşehir Gully from the east. This area is divided into four parts, namely semi-humid forests, dry forests, maquis, and garigue areas.

The distribution areas of semi-humid forests are mainly the northern faces of the mountainous areas. The semi-humid forests, which are scattered on the northern faces and stretching as fragmented strips, generally do not reach the summits. Dry forests are the dominant plant community of the mountainous areas in the region. These forests, which are poor in terms of species diversity at high altitudes, deprived of forest plants and sparse in appearance, attract attention with the diversity of maquis elements at low altitudes. It is common in the coastal part where the forest has been destroyed and in the low levels of the mountainous areas in the maquis study area.

Toward the east, due to the milder Mediterranean climate and the decrease in sea influence, the typical Mediterranean species disappear, and maquis also loses its dominance at this altitude. The garigue formation occurs where the maquis has been heavily destroyed. However, *Tilia argentea*, an auxin element that emerged in chestnut assemblages in the field, is noteworthy for being a relict species from a more humid period in this field (Dönmez, 2014).

19.2.3 Plant Geography of the Area between Büyük Menderes River and Upper Dalaman Stream

The vegetation of this area is considered to be semi-humid dry forests, maquis, and alpine plants. Semi-humid forests were able to grow in much narrower areas than the dry forests in the valleys of Babadağ and Akdağ masses open to the north. In addition, it is located in the form of islets in the black pine forests, which are the dominant formations of these masses. The dominant tree of semi-humid forests is *Castanea sativa*.

Dry forests, which start at 500–600 m in mountainous masses, above the maquis level, under the dominance of coniferous trees, expand by the summits, except in the valleys where the semi-humid forest parts are located. Dominant elements are *P. brutia* at lower levels and *Pinus nigra* at higher levels.

The places where the maquis formation covers the destruction areas of the red pine forests is common in the area of the southern skirts of the mountainous masses and the valleys where the sea effect is observed. It is common at 500–600 m, although some elements may reach elevations up to 1,000 m. Alpine plants in the region spread after 2,000 meters, which is the upper limit of the forest (Dönmez, 2014).

19.2.4 Plant Geography of the Upper Büyük Menderes Basin

The plant formations of this region were studied as humid-semi-humid-dry forests, bush communities, and alpine plants.

Humid forests spread on the north faces of east–west-oriented Murat Mountain (2,312 m), which is the highest elevation in the study area, whereas semi-humid forests spread on the south faces of this mountain. The plant species in the valleys of Murat Mountain open to the north are similar to the humid forests of the Black Sea region. The humid forest belt, which starts at 1,300–1,400 m on the northern faces of the mountain, descends to 1,000 m along the valleys. Other tree species that spread in these forests, the dominant element of which is *Fagus orientalis*, are hygrophilous plants such as *Populus tremula*, *Acer hyrcanum*, *Carpinus betulus*, *Taxus baccata*, and *Castanea sativa*.

The following are the dominant elements of the dry forests, which have a wide distribution on the plateau and in the hilly areas on the southern sides of the mountainous masses: *Pinus brutia* mostly in the lower levels, *Pinus nigra* above 1400–1500 m, and *Juniperus excelsa* and *Juniperus foetidissima* at its peaks.

Shrub formation in the field occurs mostly where dry forests have been destroyed. The dominant species are mostly *Quercus coccifera* and *Juniperus oxycedrus*. However, in some parts, such as in the Büyük Menderes valley, the shrubs are in great diversity. Examples of common species, such as maquis found here, are *Cercis siliquastrum*, *Pistacia terebinthus*, *Olea oleaster*, and *Arbutus andrachne* (Dönmez, 2014).

19.2.5 Plant Geography of the Area Between Büyük Menderes and Gökova

The plant formation of this land can be studied in only two groups, namely dry forests and maquis, since the land is located in a section without high elevations.

In dry forests, *Pinus brutia* and *P. pinea* are seen at low elevations, and *P. nigra* is seen at higher elevations. Dry forests are mostly species-poor, devoid of undergrowth, and sparsely structured, and they are partially diversified, with most xeric oak communities (*Quercus pubescens*, *Quercus infectoria*, *Quercus cerris*) occurring in narrow areas. Red pine forests start from 250–300 m in undamaged places and spread up to 1,100–1,200 meters on the northern and southern faces of masses such as Beşparmak Mountains, Ilbira Mountain, Gökbel Mountain, and Marçal Mountain.

P. pinea starts above the red pine level and goes up to 700–900 m. The existence of assemblages located in the pine assemblages is noteworthy. *P. nigra* forests dominate the area from 1,100 m above the masses. Larches descend from peaks to 1,050 m on the south faces. The lower layer of larch forests is much poorer than red pines. The species seen in this section are only the most xeric species of maquis, such as *Quercus coccifera* and *Juniperus oxycedrus*.

The maquis formation in the field not only forms the bush layer of the red pine forests, but also densely covers the areas where these forests have been destroyed. The species in this community, where the real maquis is present with almost all its elements, are shrubs that make up the lower layer of the aforementioned red pine forests (Dönmez, 2014).

19.2.6 Plant Geography of the Upper Gediz Basin

The vegetation of this land was distinguished as humid-semi-humid-dry forests, shrub-maquis formation, and alpine plants. The distribution areas of humid and semi-humid forests are very limited in the region. These forests are on the northern and northeastern faces of Murat Mountain (2,309 m), which constitutes the highest point of the basin, at levels higher than 1,600 m.

The dominant element of the humid and semi-humid forests is *Fagus orientalis*. Also, *Populus tremula*, *Carpinus betulus*, and *Acer platanoides* trees are seen in the beech forests.

The part that is distinguished as a semi-humid forest area in this area is the places from 1,700 m of Murat Mountain to the upper limit of the forest.

Dry forests are the most widespread plant community in the field. The distribution areas of dry forests are the parts of the mountainous masses other than the humid and semi-humid forest areas and the plateau areas at the foot of the masses.

Shrub communities consist of mostly hygrophilous species (*Corylus avellana, Sorbus torminalis, Vaccinium arctostaphylos, Viburnum lantana, Prunus divericata, Rhus cotinus, Lonicera caucasica, Cornus mas, Cornus sanguinea, Fraxinus ornus, Fraxinus angustifolia*) in the lower layer of humid-semi-humid forest and their destroyed areas, and maquis elements spreading in the lower layer of dry forests and destroyed areas.

Alpine plants in the field are represented by various grasses from the forest border to the peaks of Murat Mountain and Şaphane Mountain, both of which have an altitude of more than 2,000 m (Günal, 2003).

19.2.7 Plant Geography of Honaz and its Surroundings

The plant communities of this region were studied as dry forests, shrub-maquis communities, and Alpine formations. Honaz Mountain (2,528 m), located in the inner western Aegean part, draws attention with its special structure in terms of vegetation. Hygrophilous species, mostly Black Sea species, exist in the valleys on the northern sides of this mass, where the dominant plant formation is dry forests.

The dominant elements of dry forests are *Pinus brutia* in the lower elevations and *Pinus nigra* in the upper elevations. The red pine forests starting from the skirt of the mass increase up to 1,200 m in the north and 1,400 m in the south, except for the destroyed areas. *P. nigra* forests occupy the largest part of the dry forest areas. These forests, which dominate the field above the red pine level, stretch up to the alpine level.

Scattered among *P. nigra* trees are oaks (*Quercus infectoria, Q. pubescens, Q. cerris, Quercus frainetto*) in the valleys on the north sides, ash (*Fraxinus ornus, F. angustifolia*), elm (*Ulmus glabra, Ulmus minor*), maple (*Acer hyrcanum* subsp. *keckianum*), linden (*Tilia rubra* subsp. *caucasica*), chestnut (*Castanea sativa*), and *Sorbus torminalis*.

The bush formation is represented by maquis species (*Q. coccifera, Cercis siliquastrum, Pistacia terebinthus, Phillyrea latifolla, Juniperus oxycedrus, Cistus salviifolius, Arbutus andrachne*) on the northern skirts of Honaz Mountain and by two maquis members (*Q. coccifera, J. oxycedrus*) on the southern skirts.

Alpine formation starts from 2,000 m above Mount Honaz and stretches up to the summits. However, in the areas where the forest has been destroyed, there are parts where it starts from 1,700 m below its natural limit.

Another interesting feature of Mount Honaz in terms of plant geography is the plant community that occurs only in the valleys on the northern sides of this mass, in which Mediterranean species (*Pistacia terebinthus, P. terebinthus* subsp. *pallesiana, Phillyrea lotifolia, Cistus salviifolius, Quercus. coccifera, Cercis siliquastrum, Juniperus oxycedrus, Arbutus andrachne*) and the hygrophilous species (*Mespilus germanica, Cornus mas, C. sanguinea, Corylus avellana, Sorbus torminalis, Populus tremula*) of the Black Sea spread together.

The Euro-Siberian elements and the hygrophilous species (*Acer hyrcanum* subsp. *keckianum, Tilia rubra* subsp. *caucasica, Mespilus germanica, Cornus mas, Cornus sanguinea, Corylus avellana, Populus tremula*) unique to the Black Sea, which continue to exist in the valleys on the northern skirts of the Honaz mass, must be the relicts of plants growing on Mount Honaz under more humid conditions than today (Büyükoğlan, 2010).

19.3 USE OF AEGEAN HERBS IN LOCAL CUISINE AND FOR MEDICINAL PURPOSES

Today, plants are used in many areas for different purposes. Edible herbs, which are nutritional and energy sources, have an important place in our lives both in terms of ensuring daily nutritional diversity and their contribution to a healthy diet (Ceylan et al., 2019). As a result of the data obtained from studies investigating the positive impacts of vegetables and plants on human health today, it is seen that there have been some changes in the eating habits of people, and the conclusion that annual vegetable consumption per capita in the world increases each year also confirms these changes (Kök et al., 2020).

Türkiye is in a very advantageous geographical location in terms of natural resources, and, as one of the benefits of this location, its lands contain many wild plants and herb species. Some of these herbs are recognized by most people, whereas others are used only in a small region. Wild plants and herbs add diversity to our lives in many ways with their medicinal properties, the aromas they add to the meals, and the diversification of the table in the absence of vegetables (Kök et al., 2020). Anatolian people started to use wild plants as medicine and food in ancient times, and in addition to benefiting from wild-grown plants, they also made efforts to produce various species (Karaca et al., 2015).

Plant diversity is an invaluable treasure for Türkiye. Wild plants constitute a remarkable part of the plant richness in Anatolia. The fact that herbs and herb dishes occupy an important place in Türkiye and especially in Aegean cuisine is one of the results of this diversity (Karaca et al., 2015).

The word 'herb' is generally used for the meaning of volunteer plants that are not grown intentionally by people. Self-growing wild plants have nutrients that contribute to the human diet and have protective and improving effects on human health when recognized, collected, and consumed appropriately. In old times, as it was unlikely to find fresh vegetables and fruits and to access treatment when required, especially during winter months in our inner regions, local people found the solution was to gather herbs that were green as the weather gets warmer to remedy the deficiencies of many vitamins, mainly vitamin C, and to benefit from their medicinal properties (Kök et al., 2020).

Countless herbs grow in various forms all over the world. These herbs are used differently in each country according to their unique cuisine. In our country, the dishes made especially from wild plants are traditionally and commonly called herb dishes. Most of the plants used for preparing herb dishes are herbaceous; however, young and tender shoots of plants in the form of semi-bush, bush, or tree (leafy and leafless) are also used. As a result, the concept of herb dishes is a general term attributed to the dishes made using herbaceous parts of plants (Tuzlacı, 2011). There are a great number of different herbs grown in the Aegean Region of Türkiye, and the dishes made with them cause the Aegean cuisine to be called "green cuisine" (Kök et al., 2020). Some medicinal plants, which are generally grown in western Anatolia and used by local people, trading in small markets established in cities and districts, and evaluated as medicinal herbs in the literature, are given in alphabetical order (Table 19.1).

Turkish cuisine is defined as preparing, cooking, and preserving food and beverages that are consumed for essential nutrition by people living in Türkiye, including the tools and equipment, and techniques necessary for these processes, and all practices and beliefs that develop around the kitchen (Solmaz and Altıner, 2018). Turkish cuisine constitutes one of the most reputable culinary cultures of the world that is formed by the combination of different food cultures, differing from region to region, and spreading over a wide geography. Turkish cuisine has both influenced and been influenced by Balkan and Middle Eastern cuisines. Each region in Türkiye has its own food culture, such as Black Sea cuisine, Eastern cuisine, and Aegean cuisine. In the formation of local cuisines, factors, such as historical process, geographical location, and national or ethnic origin, play an important role. Reflecting the habits and lifestyles of different communities, the cuisine paves the way for the emergence of service characters with products suitable for geographical differences. For this reason, our country, which includes many different cultures, has a deep-rooted and rich history containing many differences in terms of culinary culture (Kök et al., 2020).

19.4 CONCLUSION

Consumption of green leafy herbs causes positive effects on human health as they contain vitamins, minerals, and elements that are important for human health. However, the use of high amounts of

TABLE 19.1
Plant Species Belonging to the Aegean Region of Türkiye and Method Used

Species	Family	English Name	Sporing/Flowering Time	Parts Used	Method Used	Purposes in Literature
Asplenium ceterach	Aspleniaceae	Miltwaste	April–May	Herbs	Infusion tea	Diuretic, hemorrhoids, kidneysand and stone, cold and flu treatmnet, cough suppressant, stomach diseases, wound treatment, bronchitis, prostate treatment
Equisetum arvense	Equisetaceae	Horsetail	March–April	Herbs	Infusiontea	Hemostatic, gout and rheumatism, hemorrhoids, liver diseases, sedative, wound treatment
Althaea officinalis	Malvaceae	Marshmallow	June–August	Flowers	Infusion tea	Diarrhea, cough suppressant, sedative, diuretic, laxative, gum treatment
Arbutus unedo	Ericaceae	Strawberry tree	October–April	Fruits	Raw	Antiseptic, hemostatic, diuretic, asthma, stomach diseases, sugardropper in the blood
Asparagus acutifolius	Asparagaceae	Wild asparagus	August–September	Roots	Boiled root water can be drunk	Allergy, cold and flu treatment
Asphodelus aestivus	Xanthorrhoeaceae	Common asphodel	March–June	Rhizomes	Squashed rhizomes as a cream, directly eaten	Antispasmodic, diuretic, hemorrhoids, rheumatism, eczema, abscess treatment, anti-inflammatory, antioxidant
Calendula arvensis	Asteraceae	Field marigold	January–June	Flowers	Dried flowers as a cream	Skin diseases, anti-inflammatory, antispasmodic, hemostatic, analgesic, laxative, vesicular stomatitis treatment, gallbladder treatment
Carduus nutans subsp. *nutans*	Asteraceae	MuskThistle	June–July	Branches with flowers	Five minutes decoction	Blood purifier, antipyretic, arteriosclerosis treatment
Centaurium erythraea .subsp.*turcicum*(	Gentianaceae	Centaury, feverwort	May–August	Branches with flowers	Branches with flowers mixed with honey	Appetizer, gallbladder diseases, diaphoretic, antipyretic, stomach diseases, hepatoprotective, make digestion easier

(Continued)

TABLE 19.1 (CONTINUED)
Plant Species Belonging to the Aegean Region of Türkiye and Method Used

Species	Family	English Name	Sporing/Flowering Time	Parts Used	Method Used	Purposes in Literature
Ceratonia siliqua	Fabaceae	Carob, St. John'sbread	September–November	Fruits	Raw, infusion tea	Hemostatic, diarrhea, laxative, anti-inflammatory, antioxidant, sedative
Cytinushypocitus subsp. *orientalis*	Cytinaceae	Cytinus	March–June	Parts above the soil	Juice is used after squashing of the parts	Tumor treatments, hemostatics, dysentery
Datura stromonium	Solanaceae	Thornapple, jimsonweed	May–November	Leaves	İnfusion tea	Asthma, analgesic, anti-inflammatory, antispasmodic,antidandruff, wolfthrower
Dioscoreacommunis	Dioscoreaceae	Black bryony	April–June	Roots	Squashed roots as a cream	Laxative, diutretic, emetic, rheumatism
Dracunculus vulgaris	Araceae	Common dracunculus, Dragon	May, June	Fruits	Squashed fruits as a cream	Cancer treatment, rheumatism, headache, hemorrhoids, eczema
Ecballium elaterium.	Cucurbitaceae	Squirting cucumber	April–October	Fruits, fruit juice, roots	Fruits are used by immersing in olive oil for about 1-2 months,. Diluted fruit juice, squashed roots	Diuretic, analgesic,woundhealer, hemorrhoids, rheumatism, eczema, laxative, sinusitis, abscesstreatment, cold and flu treatments
Echinophora tenuifolia subsp. *sibthorbiana*	Apiaceae	Turkish pickling herb	July–September	Parts above the soil	Infusion tea	Antispasmodic, digestive system diseases
Erica arborea	Ericaceae	Tree heath	March–July	Leaves, flowers	Infusion tea	Asthma, bloodpressure
Euphorbia apios	Euphorbiaceae	Spurge	March–June	Roots	Directly eaten or used (little amounts)	Dermatology, eczema, migraine, intestinal diseases, wart treatment
Foeniculum vulgare	Apiaceae	Fennel	May–September	Parts above the soil	Boiled parts with water	Analgesic, milk secretion enhancer, diuretic, antispasmodic, stimulant, expectorant, anti-inflammatory, stomach diseases

(Continued)

TABLE 19.1 (CONTINUED)
Plant Species Belonging to the Aegean Region of Türkiye and Method Used

Species	Family	English Name	Sporing/Flowering Time	Parts Used	Method Used	Purposes in Literature
Helichrysum plicatum subsp.*plicatum*	Asteraceae	Turkish helichrysum	June–August	Parts above the soil	Infusion tea	Laxative, cough suppressant, anti-inflammatory
Inula viscosa	Asteraceae	False yellowhead	June–November	Leaves	Infusion tea	Cancer treatment, analgesic, bloodstopper
Lavandula stoechas subsp. stoechas	Lamiaceae	Lavander	March–June	Flowers	Infusion tea	Asthma, expectorant, digestive organizer, antispasmodic,bronchitis, cholesterol, diarrhea
Malva sylvestris	Malvaceae	Mallow	May–October	Parts above the soil	Infusion tea	Antiseptic, analgesic, bloodpressure lowering, stomach diseases, anti-inflammatory
Marrubium vulgareL.	Lamiaceae	Horehound	April–August	Flowers	Infusion tea	Antiseptic, anti-inflammatory, analgesic, stomach diseases, bloodpressure lowering
Melissa officinalis subsp. *officinalis*	Lamiaceae	Common balm	June–July	Parts above the soil	Infusion tea	Sedative, antiseptic, asthma, cancer, diarrhea, bloodpressure lowering, cholesterol, migraine, stomach diseases, antitrombotic effects, acne treatment
Momordica charantia	Cucurbitaceae	Bitter gourd	July–August	Fruits	Fruits are used by storing in olive oil for about 2-4 weeks under sun, then the fruits are squashed with the oil	Tumors, skin diseases, eczema, scabies, rheumatism, malaria, diabetes mellitus, intestinal diseases
Muscari comosum	Asparagaceae	Tassel hyacinth	March–July	Bulbs	Bulbs can be boiled in water	Diuretic, expectorant, stomach diseases, wound treatment
Muscari neglectum	Asparagaceae	Grape hyacinth	March–May	Bulbs	Bulbs can be boiled in water and used as a cream	Rheumatism, analgesic

(Continued)

TABLE 19.1 (CONTINUED)
Plant Species Belonging to the Aegean Region of Türkiye and Method Used

Species	Family	English Name	Sporing/Flowering Time	Parts Used	Method Used	Purposes in Literature
Origanum onites	Lamiaceae	Pot marjoram	April–August	Parts above the soil	Infusion tea	Stomach diseases, cough suppressant, bloodpressere balancing, disinfectant, expectorant, digestive system diseases, diaphoretic, toothache, gallbladder diseases
Plantago lanceolata	Plantaginaceae	Narrow-leaved plantain	April–October	Leaves	Infusion tea	Hemostatic, sedative, expectorant, laxative, wound treatment, nervous diseases, asthma, cancer treatment, stomach diseases, abscess treatment
*Pistacia terebinthus*L. subsp. *terebinthus*	Anacardiaceae	Terebinthtree	March–May	Parts above the soil	Decoction	Asthma, bronchitis, antiseptic, stomach diseases, fungal foot treatment, cold and flu treatment
Rosmarinus officinalis	Lamiaceae	Rosemary	February–May	Flowers, leaves	Infusio ntea	Appetite stimulant Digestive system diseases, diarrhea, liver diseases, asthma, migraine, antiseptic, diuretic, bloodsugar lowering, diaphoretic, rheumatism
Ruscus aculeatus	Asparagaceae	Kneeholly	March–May	Fruits	Dried fruits after squashing can be used as pills	Diuretic, hemorrhoids, for kidneys, appetite stimulants, antipyretic
Salvia tomentosa	Lamiaceae	Balsamic sage	April–August	Leaves	Infusion tea	Cold and flu treatments, antiseptic
Scolymus hispanicus	Asteraceae	Common golden thistle	June–September	Roots	Decoction	Kidney stone removal, bloodsugar balancing
Silybum marianum subsp. *marianum*	Asteraceae	Milkthistle	April–May	Stem, leaves, seeds	Plant can be eaten after thorns removed, seeds can be eaten after grinding	Hemostatic, stimulant, diuretic, liverprotective, gallbladder diseases, diaphoretic, stomach diseases

(Continued)

TABLE 19.1 (CONTINUED)
Plant Species Belonging to the Aegean Region of Türkiye and Method Used

Species	Family	English Name	Sporing/Flowering Time	Parts Used	Method Used	Purposes in Literature
Sinapis alba subsp. *alba*	Brassicaceae	White mustard	February–April	Flowers	Infusion tea	Appetite stimulant, diuretic, emetic, expectorant, laxative, sedative, asthma, stomach diseases
Stachys cretica subsp. *smyrnaea*	Lamiaceae	Cretan hedge nettle	April–August	Branches with flowers	Infusion tea	Antiseptic, stomach and intestinal diseases, anti-inflammatory, anti-anxiety, antinephritis, antitumor, dermatology
Tordylium apulum	Apiaceae	Roman pimpernel, Mediterranean hartwort	April–May	Leaves	Chewing in the mouth	Stimulant, hair loss inhibitor, expectorant, appetite stimulant
Urtica dioica L. subsp. *dioica*	Urticaceae	Stinging nettle	June–September	Roots, parts above the soil	Boiled root water can be drunk, parts above the soil can be used for infusion tea	Asthma, diabetes, stomach diseases, expectorant, cough suppressant, hemorrhoids, diuretic, rheumatism, eczema, kidney and skin diseases, tuberculosis, bronchitis, hairloss
Viscum album subsp. *austriacum*	Santalaceae	Mistletoe	March–June	Leaves	Infusion tea	Hemorrhoids, antispamodic, eczema, cancer treatment, heart diseases, gastritis, ulcer, bloodsugar lowering, bloodpressure lowering

the species listed in Table 19.1 may cause toxic effects due to the alkaloids, phenolics, and many other biologically active components they contain. Therefore, they should generally be used slowly, with rapid use not being recommended for cardiac and metabolic reasons.

Within the scope of this chapter and the medicinal plants given, it can readily be seen that "Aegean Cuisine" is especially famous for its healing herb dishes and popular with its geography and related vegetation in the world.

REFERENCES

Büyükoğlan, F. 2010. Honaz Dağı ve Çevresinin Bitki Örtüsü. *Kastamonu Eğitim Dergisi*, 18(2), 631–652.

Ceylan, F., and S. Akar Şahingöz. 2019. Yenilebilir Otlar Tüketim Alışkanlığı: Düziçi Örneği. *Journal of Tourism and Gastronomy Studies*, 7(3), 2204–2225.

Davis, P. H. (Ed.). 1965–1985. *Flora of Turkey and the East Aegean Islands*. Vol. 1–9. Edinburgh: University Press.

Davis, P. H. (Ed.). 1988. *Flora of Turkey and the East Aegean Islands* (Supplement). Edinburgh: University Press.

Dönmez, Y. 2014. Türkiye Bitki Coğrafyası Çalışmaları. İstanbul Üniversitesi Edebiyat Fakültesi Coğrafya Bölümü, Coğrafya Dergisi. Sayı 29, 1–27. İstanbul, Turkey.

Günal, N. 2003. *Yukarı Gediz Havzasının Bitki Coğrafyası*. İstanbul, Turkey: Çantay Kitabevi.

Güner, A., N. Özhatay, T. Ekim, and K. H. C. Baser (Eds.). 2000. *Flora of Turkey and the East Aegean Islands* (Supplement). Vol. 11, pp. 271–274, 343–344. Edinburgh: University Press.

Güner, A. 2012. *Türkiye Bitkileri Listesi; Damarlı Bitkiler*. İstanbul, Turkey: ANG Vakfı Baskısı.

Güner, A. (Ed.). 2014. *Resimli Türkiye Florası, Cilt 2*. İstanbul, Turkey: Türkiye İş Bankası Kültür Yayınları.

Güner, A. (Ed.). 2018. *Resimli Türkiye Florası, Cilt 2*. İstanbul, Turkey: ANG Vakfı Nezahat Gökyiğit Botanik Bahçesi Yayınları.

Karaca, O., O. Yıldırım, and C. Çakıcı. 2015. Gastronomi Turizminde Otlar, Ot Yemekleri ve Sağlıkla İlişkisi Üzerine Bir Değerlendirme. *Journal of Tourism and Gastronomy Studies*, 27–42.

Kök, A., A. Kurnaz, H. Akyurt Kurnaz, and S. Karahan. 2020. Use of Aegean Herbs in Local Cuisines. *Journal of Tourism Intelligence and Smartness*, 3(2), 152–168.

Regel, C. V. 1963. Türkiye' nin Flora ve Vejetasyonuna Genel Bir Bakış. Ege Üniversitesi Fen Fakültesi Monografileri Serisi No: 1. İzmir, Turkey.

Satıl, F., and N. M. Kaya. (Eds.). 2021. Kaz Dağları' Nın Şifalı Bitkileri. Balıkesir Büyükşehir Belediyesi, Kent Arşivi Yayınları No: 25. Balıkesir, Turkey.

Solmaz, Y., and D. Altıner. 2018. Türk Mutfak Kültürü ve Beslenme Alışkanlıkları Üzerine Bir Değerlendirme. *Safran Kültür ve Turizm Araştırmaları Dergisi*, 1(3), 108–124.

Tuzlacı, E. 2011. *Türkiye'nin Yabani Besin Bitkileri ve Ot Yemekleri*. Ed: Rana Alpöz. İstanbul, Turkey: Alfa Basım Yayın.

20 Evaluation of Medicinal and Ecological Aspects of Pteridophytes in Turkiye

Mustafa Keskin, Volkan Altay, and Munir Ozturk

CONTENTS

20.1 INTRODUCTION

Approximately 10,000 taxa of ferns and 2,000 fern allies have been reported in the world. These occupy a wide variety of different ecological niches, representing just 5–7% of all vascular plants (Goswami et al., 2016; Ozturk et al., 2018). Although these plants have assumed nearly all forms of growth and adaptation, acting as essential biotic elements in ecosystems and occupying all types of terrestrial communities, they are most abundant in tropical and temperate areas (Mehltreter et al., 2010; Shah et al., 2022). The awareness of this taxonomy has recently increased globally (Ozturk et al., 2018; Shah et al., 2022) and we have therefore prepared the current chapter to summarize the taxonomical and ecological aspects, together with the medicinal uses, of pteridophytes in Turkiye.

20.2 CURRENT STATUS OF THE SYSTEMATICS OF PTERIDOPHYTES IN TURKIYE

In Turkiye, the first written record of this group was published in the "*Flora of Turkey and the East Aegean Islands*" Vol. 1 (Coode and Cullen, 1965; Henderson, 1965; Jermy, 1965), followed by Vol. 10 (Davis 1988) and Vol. 11 (Byfield, 2000; Güner, 2000; Sida and Tan, 2000; Tan and Suda, 2000) and Kaynak et al. (1996). The taxonomical list published by Parris and Fraser-Jenkins (1980) has remained the same for decades. The distribution maps for pteridophytes in Turkiye was given by Yıldırımlı (1996) with some newly added taxa. This list has been reviewed and revised by many different authors in the second volume of the work entitled "*Illustrated Flora of Turkey*," with a new, updated list (Aydınkal, 2018; Bona, 2018; Doğan-Güner, 2018; Ecevit-Genç, 2018; Erol, 2018;

DOI: 10.1201/9781003226925-20

TABLE 20.1
Endemic Pteridophytes From Different Families in Turkiye.

Isoetaceae	*Isoetes anatolica*
	Isoetes vanensis
Pteridaceae	*Cryptogramma bithynica*
Aspleniaceae	*Asplenium tadei*
	Asplenium x alternatifolium nothosubsp. *ponticum*
	Asplenium adiantum-nigrum nothosubsp. x*demirizii*
	Asplenium x reuteri
Dryopteridaceae	*Dryopteris x euxinensis*
	Polystichum asiae-minoris

Güner, 2018; Karabacak, 2018; Kaynak, 2018; Keskin, 2018; Özbek, 2018; Öztekin, 2018; Şenol, 2018; Uzunhisarcıklıoğlu, 2018; Vural, 2018; Yılmaz, 2018).

Fraser-Jenkins and Parris (2020) have since published a new list with a different, systematic understanding but this list does not go beyond reflecting their own systematic views. Moreover, this list contains some deficiencies such as providing no morphological or geographical differences, and incomplete herbarium records and locality information of the specimens, which make it rather a non-scientific record. Thus, pteridophyes in Turkiye have been reclassified according to the APG III (Angiosperm Phylogeny Group III) system, taking into account the systematic recommendations put forward by the International Standards, especially the Fern Phylogeny Group (PPG 1, 2016).

In all, 21 families with 38 genera and 115 taxa have been recorded in the re-revised checklist prepared for the flora of Turkiye (Appendix 20.1); nine of which are endemics, four being hybrids (Table 20.1).

The Pteridaceae family, which includes seven genera, stands out as having the largest number of genera. The *Asplenium* (27) and *Dryopteris* (17) genera (Table 20.2) are represented by the highest number of taxa.

20.3 HABITAT-BASED ECOLOGICAL GROUPING OF PTERIDOPHYTES IN TURKIYE

Pteridophytes generally occupy aquatic, semi-aquatic, rocky, epiphytic, terrestrial, and meadow habitats.

20.3.1 Aquatic Forms (Free-Floating)

Some of these taxa pass their entire life in water, where they are observed as swimming forms, usually with fringe-shaped root systems. However, in Turkiye, very few taxa have been reported from this group. These include the species belonging to *Salvinia*, *Azolla*, and *Marsilea* genera (Figure 20.1), where the genus *Marsilea* is known only from herbarium records, and its existence needs confirmation since its locality record information is not clear.

20.3.2 Semi-Aquatic Forms

While not living entirely in water, plants from this group do require extensive access to water during their life cycle, especially during spore formation, where they are always in water during this period, but appear especially in the rainy season. These plants can form spores and complete their life cycle when the rains stop and the water recedes. The root systems of these plants are usually small tubers or short, scaleless rhizomes. The best representatives are *Equisetum*, *Isoetes*, and *Ophioglossum*

TABLE 20.2
The Pteridophyte Families, Genera and Their Number of Taxa in Turkiye.

Family	Genera	Number of taxa
Aspleniaceae	*Asplenium*	27
Dryopteridaceae	*Dryopteris*	17
Equisetaceae	*Equisetum*	8
Dryopteridaceae	*Polystichum*	8
Isoetaceae	*Isoetes*	6
Polypodiaceae	*Polypodium*	4
Pteridaceae	*Pteris*	3
Pteridaceae	*Allosorus*	3
Pteridaceae	*Cystopteris*	3
Lycopodiaceae	*Diphasiastrum*	2
Selaginellaceae	*Selaginella*	2
Ophioglossaceae	*Ophioglossum*	2
Pteridaceae	*Cryptogramma*	2
Athyriaceae	*Athyrium*	2
Lycopodiaceae	*Lycopodium*	1
Lycopodiaceae	*Spinulum*	1
Lycopodiaceae	*Huperzia*	1
Lycopodiaceae	*Lycopodiella*	1
Ophioglossaceae	*Botrychium*	1
Osmundaceae	*Osmunda*	1
Hymenophyllaceae	*Hymenophyllum*	1
Marsileaceae	*Pilularia*	1
Marsileaceae	*Marsilea*	1
Salviniaceae	*Salvinia*	1
Salviniaceae	*Azolla*	1
Dennstaedtiaceae	*Pteridium*	1
Pteridaceae	*Anogramma*	1
Pteridaceae	*Cosentinia*	1
Pteridaceae	*Paragymnopteris*	1
Pteridaceae	*Adiantum*	1
Pteridaceae	*Gymnocarpium*	1
Thelypteridaceae	*Thelypteris*	1
Thelypteridaceae	*Oreopteris*	1
Thelypteridaceae	*Phegopteris*	1
Woodsiaceae	*Woodsia*	1
Onocleaceae	*Onoclea*	1
Blechnaceae	*Blechnum*	1
Nephrolepidaceae	*Nephrolepis*	1

(Figure 20.2). The genus *Isoetes* has two ecotypes: those that are partially submerged but actually grow on land, and those that grow entirely in or near shallow waters.

20.3.3 Rock Forms

These plants basically grow on rocks, playing a major ecological role in their biological weathering of rocks and turning them into soil over time. The rock forms possess creeping rhizomes, attached

FIGURE 20.1 Free-floating water forms in Turkiye [*Azolla filiculoides* (left); *Salvinia natans* (right)].

FIGURE 20.2 *Ophioglossum lusitanicum* (left), and *Isoetes durieui* (right).

firmly to the rocks, and these extend into the crevices or spread over moss-covered patches. These may also show short, erect rootstocks ascending up into rock crevices (Shah et al., 2022). The main representatives in Turkiye are from the genera *Cheilanthes*, *Adiantum*, *Lycopodium*, *Asplenium*, and *Polypodium* (Figure 20.3).

20.3.4 Epiphytic Forms

The best example in this group is the genus *Polypodium* (Figure 20.4). Plants of this genus can grow either in soil or on rock. Therefore, this group is morpho-structurally similar to the rock and terrestrial forms.

20.3.5 Terrestrial Forms

The most common habitats for pteridophytes are forest areas with longer wet seasons. For this reason, a great diversity of the pteridophytes exist in the Black Sea region of Turkiye. This region is very rich in terrestrial ferns, with elevated numbers of species and high population densities. The terrestrial pteridophytes are taller, with broader fronds compared with those found in other habitats. The taxa representing this group are quite similar to each other in terms of their external appearance. However, they can be easily distinguished from each other by the features of the spore sacs. The richness depending on light, together with the occurrence of distinctive forests. The pteridophytes in the forest have short rootstocks with erect or short creeping rhizomes.

In Turkiye, the terrestrial forms flourishing on the forest floor, including taxa like *Oreopteris limbosperma*, *Dryopteris pallida*, *Dryopteris pseudomas*, *Polystichum setiferum*, *Polystichum braunii*, and *Pteridium aquilinum* (Figure 20.5).

FIGURE 20.3 Some rock forms in Turkiye. [A: *Polypodium vulgare*; B: *Asplenium scolopendrium*; C: *Asplenium trichomanes*; D: *Asplenium ceterach*; E: *Lycopodium clavatum*; F: *Diphasiastrum tristachyum*].

FIGURE 20.4 Epiphytic *Polypodium vulgare* growing on a tree trunk.

20.3.6 Meadow Forms

The best example in this group is the genus *Isoetes*. The terrestrial species of this genus are small in appearance but need seasonal access to moisture. These plants flourish in meadow environments. In Turkiye, we find three terrestrial *Isoetes* species (Keskin and Musselman, 2014; Keskin, 2018) namely *I. durieui* Bory, *I. gymnocarpa* (Gennari) A.Braun, and *I. histrix* Bory (Figure 20.6). All these species are very similar to each other in morpho-structural features. Usually, the main diagnostic character is macrospores.

20.4 TRADITIONAL MEDICINAL USES

The pteridophytes have long been known and used for their medicinal and therapeutic purposes, being prescribed as a herbal extract for the cure of several diseases in ancient times. Theophrastus

FIGURE 20.5 Some terrestrial forms in Turkiye. [A: *Oreopteris limbosperma*; B: *Dryopteris pallida*; C: *Dryopteris pseudomas*; D: *Polystichum setiferum*; E: *Polystichum braunii*; F: *Pteridium aquilinum*].

FIGURE 20.6 Some meadow forms in Turkiye. [A: *Isoetes* meadow; B: *Isoetes durieui*; C: *Isoetes histrix*; D: *Isoetes gymnocarpa*].

(327–287 BC) and Dioscorides (50 AD) listed many pteridophytes as potential herbal formulations to cure more serious disorders (Goswami et al., 2016; Ozturk et al., 2018).

In the twentieth century, Caius (1935) is recognized as the first researcher to take an interest in pteridophytes and presented their medicinal uses in India. Since then, tremendous work has been done to determine the potential of pteridophytes as a chemical source. Today, the pteridophytes are well known as a rich source of glycosides, flavonoids, terpenoids, alkaloids, and several primary vis-à-vis secondary metabolites (Ozturk et al., 2018).

Additional information on medicinal pteridophytes in Turkiye has recently been published by Ozturk et al. (2018), where, based on their findings, traditionally, 19 pteridophyte taxa are commonly used in Turkiye for medicinal purposes (Figure 20.7). These are listed below in alphabetical order in detail.

1 **Species**: *Adiantum capillus-veneris* L. (Pteridaceae)
Part used: Aerial parts, leaves, and underground parts.
Preparation: Boiled, decoction, juice, infusion, pounded, and poultice.
Diseases treated: Anti-dandruff, appetite stimulants, bronchitis, chest tightness, cough, diarrhea, diuretic, expectorant, gastrointestinal diseases, kidney problems (especially kidney stones and sand), male aphrodisiac, menstrual regulation, prostate, shortness of breath, stomachache, swollen testicles, tonic, and urinary disorders.

2 **Species**: *Asplenium adiantum-nigrum* L. (Aspleniaceae)
Part used: Fronds, aerial parts, and whole plant.
Preparation: Boiled, decoction, and infusion.
Diseases treated: Kidney stone, stomachache, shortness of breath, hemorrhoids, diuretic, eczema, sinusitis, menstrual regulation, and tension regulator.

3 **Species**: *Asplenium scolopendrium* L. (Aspleniaceae)
Part used: Aerial parts, leaves, and whole plant.
Preparation: Decoction and infusion.
Diseases treated: Anemia, arteriosclerosis, burns, chest tightness, constipation, diabetes, diuretic, hemorrhoids, stomachache, and wounds.

4 **Species**: *Asplenium trichomanes* L. (Aspleniaceae)
Part used: Aerial parts, and whole plant.
Preparation: Boiled, and cooked.
Diseases treated: Eczema, hemorrhoids, and stomachache.

5 **Species**: *Ceterach officinarum* Willd. (Aspleniaceae)
Part used: Aerial parts, leaves, spores, and whole plant.
Preparation: Boiled, decoction, infusion, and poultice.
Diseases treated: Abdominal pain, bronchitis, cold, constipation, diuretic, gallbladder diseases, gallstones, gastrointestinal diseases, gynecological diseases, hemorrhoids, kidney diseases (especially kidney stone and sand), shortness of breath, stomachache, stomach diseases, ulcer, urinary antiseptic, urinary tract disorders, and wounds.

6 **Species**: *Dryopteris filix-max* (L.) Schott (Dryopteridaceae)
Part used: Aerial parts and rhizomes.
Preparation: Decoction
Diseases treated: Intestinal parasites, and stomachache.

7 **Species**: *Dryopteris raddeana* (Fomin) Fomin (Dryopteridaceae)
Part used: Leaves, and whole plant.
Preparation: Infusion
Diseases treated: Kidney stones.

FIGURE 20.7 Therapeutic uses of some pteridophytes in Turkiye. [1: *Adiantum capillus-veneris*, 2: *Asplenium adiantum-nigrum*, 3: *Asplenium scolopendrium*, 4: *Asplenium trichomanes*; 5: *Asplenium ceterach*, 6: *Dryopteris filix-mas*, 7: *Equisetum arvense*, 8: *Equisetum hyemale*, 9: *Equisetum ramosissimum*, 10: *Equisetum telmateia*, 11: *Lycopodium clavatum*, 12: *Osmunda regalis*, 13: *Polypodium vulgare*, 14: *Polystichum aculeatum*; 15: *Pteridium aquilinum*, 16: *Pteris cretica*].

8 **Species**: *Equisetum arvense* L. (Equisetaceae)
Part used: Aerial parts, leaves, shoot, stem, and whole plant.
Preparation: Boiled, decoction, and infusion.
Diseases treated: Antiperspirant, arteriosclerosis, blood purification, bronchitis, calcification, cancer, circulatory disorders, cough, cystitis, diabetes, diuretic, eczema, edema, gallbladder diseases, gastrointestinal diseases, gum inflammations, hair straightening, hemorrhoids, hypertension, internal bleeding, kidney diseases (especially kidney stone and sand), liver diseases, lung injury, menstrual regulation, nephralgia, nose bleeds, ovaritis, prostate diseases, rheumatism, shortness of breath, skin diseases, tachycardia, tonsillitis, tuberculosis, urinary tract diseases, weakness, and wounds.

9 **Species**: *Equisetum fluviatile* L. (Equisetaceae)
Part used: Aerial parts
Preparation: Boiled
Diseases treated: Kidney diseases and tension.

10 **Species**: *Equisetum hyemale* L. (Equisetaceae)
Part used: Aerial parts
Preparation: Boiled and decoction.
Diseases treated: Stomach disorders, as diuretic, and for kidney stones.

11 **Species**: *Equisetum palustre* L. (Equisetaceae)
Part used: Aerial parts and whole plant.
Preparation: Decoction and infusion.
Diseases treated: Hemorrhoids, kidney stones, and peptic ulcer.

12 **Species**: *Equisetum ramosissimum* Desf. (Equisetaceae)
Part used: Aerial parts, leaves, shoot, stem, and whole plants.
Preparation: Boiled, decoction, and infusion.
Diseases treated: Anticoagulant, anti-leptotrichia, arteriosclerosis, cancer, cardiovascular diseases, diabetes, diuretic, eczema, gum inflammations, hemostatic at inner bleeding, kidney stone and sand, mouth and throat inflammation, rheumatism, skin diseases, stomachache, stomach disorders, tonsillitis, tuberculosis, urinary tract diseases, and wounds.

13 **Species**: *Equisetum telmateia* Ehrh. (Equisetaceae)
Part used: Aerial parts, leaves, and whole plants.
Preparation: Boiled, decoction, and infusion.
Diseases treated: Acne, antifungal, arteriosclerosis, cardiac deficiency, chronic eczema, cystitis, diuretic, expectorant, infections in the mouth, internal bleeding, kidney diseases (especially kidney stone and sand), nephritis, pain, pain in former broken bones, prostate diseases, rheumatism, shortness of breath, skin and nails, stomachache, strengthen hair, urinary tract diseases, vasodilating, and wounds.

14 **Species**: *Lycopodium clavatum* L. (Lycopodiaceae)
Part used: Aerial parts, leaves, spores, and stem.
Preparation: Decoction and infusion.
Diseases treated: Carminative (in infants), for irritated body region, hepatitis, intertrigo, liver diseases, rheumatism, and skin diseases.

15 **Species**: *Osmunda regalis* L. (Osmundaceae)
Part used: Rhizome
Preparation: Infusion
Diseases treated: Constipation, diuretic, and roborant.

16 **Species**: *Polypodium vulgare* L. (Polypodiaceae)
Part used: Aerial parts and root.
Preparation: Decoction
Diseases treated: Abdominal pain, carminative, gallstones, headache, kidney stones, and tonsillitis.

17 **Species**: *Polystichum aculeatum* (L.) Roth (Dryopteridaceae)
Part used: Aerial parts
Preparation: Decoction
Diseases treated: Abdominal pain.

18. **Species**: *Pteridium aquilinum* (L.) Kuhn. (Dennstaedtiaceae)
Part used: Aerial parts, leaves, and root.
Preparation: Boiled, decoction, and infusion.
Diseases treated: Constipation, diuretic, eczema, intestinal parasites, kidney diseases, rheumatism, toothache, and wounds.

19. **Species**: *Pteris cretica* L. (Pteridaceae)
Part used: Aerial parts
Preparation: Boiled
Diseases treated: Hemorrhoids.

The majority of taxa among the pteridophytes are used for more than one disease, with only three taxa being used for treatment of a single disease (Appendix 20.1). In agreement with the information gathered here, on the basis of disease groups, we find that the most common use of pteridophytes is in the digestive, dermal, and respiratory systems to treat stomachache, wounds, eczema, and shortness of breath. Other treatments include kidney-related diseases (especially kidney stones and sands), as diuretic, to treat hemorrhoids and urinary tract diseases in the urinogenital system, arteriosclerosis in the cardiovascular system, rheumatism in the skeletal-muscular system, menstrual regulatory in the gynecological system, and tonsillitis in the ENT system.

The most common medicinal or therapeutic use of pteridophytes in Turkiye are as a diuretic, for kidney stones, to treat stomachache, and hemorrhoids (Appendix 20.1).

20.5 CONCLUSIONS

Pteridophytes appear to possess great potential for the development of allopathic medicines. These can thus be potentially used in the treatment of some dangerous and epidemic diseases. Medicines derived from pteridophytes can be produced in large quantities for common use via simple means so that their supply can be arranged to meet global scale requirements. An advantage of such herbal medicines is that they hardly show any side effects (Ozturk et al., 1996a, 1996b). Since pteridophytes can be prescribed as herbal formulations to cure many diseases, there is great scope for significant financial earnings. However, while several uses of pteridophytes have been recorded locally for the treatment of a number of health problems, we still need proper screening, testing, characterization, validation, and quality control of compounds, together with confirmation of their therapeutic values (Singh et al., 2010; Ozturk et al., 2018).

Singh et al. (2010), Goswami et al. (2016), Ozturk and Hakeem (2018, 2019a, 2019b), and Ozturk et al. (2018, 2020) have all strongly emphasized the fact that there is an immediate need for multidisciplinary collaboration among biochemists, botanists, molecular geneticists, organic chemists, and pharmaceutical faculty researchers to concentrate on plants in general and mosses-bryophytes-pteridophytes in particular to evaluate their tremendous healing potential. Moreover, in studies of biological activities of plants from this group, some species have shown cytotoxicity (with possible anticancer uses), or hepatoprotective, antihyperglycemic, leishmanicidal, trypanocidal, antinociceptive, anti-inflammatory, and immunomodulatory activities (Ozturk et al. 2018). The pteridophytes can serve us in the treatment of current, important diseases due to the action of their secondary metabolites (Cao et al., 2017; Ozturk et al., 2018).

REFERENCES

Aydınkal, R.M. 2018. *Blechnum* L. In: Güner, A., A. Kandemir, Y. Menemen, H. Yıldırım, S. Aslan, G. Ekşi, I. Güner and A.Ö. Çimen (Eds.), *Resimli Türkiye Florası, Vol. 2.* ANG Vakfı Nezahat Gökyiğit Botanik Bahçesi Yayınları, İstanbul, pp. 231–232.

Bona, M. 2018. Pteridaceae E.D.M. Kirchm. In: Güner, A., A. Kandemir, Y. Menemen, H. Yıldırım, S. Aslan, G. Ekşi, I. Güner and A.Ö. Çimen (Eds.), *Resimli Türkiye Florası, Vol. 2.* ANG Vakfı Nezahat Gökyiğit Botanik Bahçesi Yayınları, İstanbul, pp. 154–176.

Byfield, A.J. 2000. *Lycopodiella* J. Holub.;*Pilularia* L. In: Güner, A., N. Özhatay, T. Ekim and K.H.C. Başer (Eds.), *Flora of Turkey and the East Aegean Islands, Vol. 11.* University Press, Edinburgh, pp. 1–4.

Caius, J.F. 1935. The medicinal and poisonous ferns of India. *Journal of the Bombay Natural History Society* 38: 341–361.

Cao, H., T.T. Chai, X. Wang, M.F.B. Morais-Braga, J.H. Yang, F.C. Wong, R. Wang, H. Yao, J. Cao, L. Cornara, B. Burlando, Y. Wang, J. Xiao and H.D.M. Coutinho. 2017. Phytochemicals from fern species: Potential for medicine applications. *Phytochemistry Reviews: Proceedings of the Phytochemical Society of Europe* 16(3): 379–440.

Coode, M.J.E. and J. Cullen. 1965. *Lycopodium* L.;*Selaginella* Beauve. In: Davis, P.H. (Ed.), *Flora of Turkey and the East Aegean Islands, Vol. 1.* University Press, Edinburgh, pp. 34–36.

Davis, P.H. 1988. Pteridophyta. In: Davis, P.H. (Ed.), *Flora of Turkey and the East Aegean Islands, Vol. 10.* University Press, Edinburgh, pp. 1–11.

Doğan-Güner, E. 2018. Cystopteridaceae Schmakov.;*Osmunda* L. In: Güner, A., A. Kandemir, Y. Menemen, H. Yıldırım, S. Aslan, G. Ekşi, I. Güner and A.Ö. Çimen (Eds.), *Resimli Türkiye Florası, Vol. 2.* ANG Vakfı Nezahat Gökyiğit Botanik Bahçesi Yayınları, İstanbul, pp. 136–138 and 177–181.

Ecevit-Genç, G. 2018. *Salvinia* Ség. In: Güner, A., A. Kandemir, Y. Menemen, H. Yıldırım, S. Aslan, G. Ekşi, I. Güner and A.Ö. Çimen (Eds.), *Resimli Türkiye Florası, Vol. 2.* ANG Vakfı Nezahat Gökyiğit Botanik Bahçesi Yayınları, İstanbul, pp. 144–145.

Erol, O. 2018. *Marsilea* L. In: Güner, A., A. Kandemir, Y. Menemen, H. Yıldırım, S. Aslan, G. Ekşi, I. Güner and A.Ö. Çimen (Eds.), *Resimli Türkiye Florası, Vol. 2.* ANG Vakfı Nezahat Gökyiğit Botanik Bahçesi Yayınları, İstanbul, p. 142.

Fraser-Jenkins, J.R. and B.S. Parris. 2020. Pteridophytes of Turkey - A revised check-list and their relationships to Europe and the west Himalaya. *Indian Journal of Forestry* 43(3): 193–214.

Güner, A. 2000. *Asplenium* L.;*Azolla* Lam. In: Güner, A., N. Özhatay, T. Ekim and K.H.C. Başer (Eds.), *Flora of Turkey and the East Aegean Islands, Vol. 11.* University Press, Edinburgh, pp. 2–5.

Güner, A. 2018. *Hymenophyllum* Sm. In: Güner, A., A. Kandemir, Y. Menemen, H. Yıldırım, S. Aslan, G. Ekşi, I. Güner and A.Ö. Çimen (Eds.), *Resimli Türkiye Florası, Vol. 2.* ANG Vakfı Nezahat Gökyiğit Botanik Bahçesi Yayınları. İstanbul, pp. 138–140.

Goswami, H.K., K. Sen and R. Mukhopadhyay. 2016. Pteridophytes: Evolutionary boon as medicinal plants. *Plant Genetic Resources – Characterization and Utilization* 14(4): 328–355.

Henderson, D.M. 1965. Filicales. In: Davis, P.H. (Ed.), *Flora of Turkey and the East Aegean Islands, Vol. 1.* University Press, Edinburgh, pp. 38–63.

Jermy, A.C. 1965. *Isoetes* L. In: Davis, P.H. (Ed.), *Flora of Turkey and the East Aegean Islands, Vol. 1.* University Press, Edinburgh, pp. 37–38.

Karabacak, E. 2018. Aspleniaceae Newman. In: Güner, A., A. Kandemir, Y. Menemen, H. Yıldırım, S. Aslan, G. Ekşi, I. Güner and A.Ö. Çimen (Eds.), *Resimli Türkiye Florası, Vol. 2.* ANG Vakfı Nezahat Gökyiğit Botanik Bahçesi Yayınları, İstanbul, pp. 182–220.

Kaynak, G. 2018. *Pteridium* Gled. ex Scop.;*Matteuccia* Tod.; *Polypodium* L. In Güner, A., A. Kandemir, Y. Menemen, H. Yıldırım, S. Aslan, G. Ekşi, I. Güner and A.Ö. Çimen (Eds.), *Resimli Türkiye Florası, Vol. 2.* ANG Vakfı Nezahat Gökyiğit Botanik Bahçesi Yayınları, İstanbul, pp. 148–152, 230–231 and 275–284.

Kaynak, G., O. Benlioğlu and G. Tarımcılar. 1996. Türkiye eğrelti florasına katkılar. *OT Sistematik Botanik Dergisi* 3(1): 25–54.

Keskin, M. and L.J. Musselman. 2014. Türkiye'nin Çimeğreltisi [*Isoetes* L. (Isoetaceae)] türleri. *Bağbahçe Bilim Dergisi* 1(2): 1–13.

Keskin, M. 2018. Lycopodiaceae P. Beauve ex Mirb.;Selaginellaceae Willk.; Isoeteaceae Rchb.; Ophioglossaceae Martinov. In: Güner, A., A. Kandemir, Y. Menemen, H. Yıldırım, S. Aslan, G. Ekşi, I. Güner and A.Ö. Çimen (Eds.), *Resimli Türkiye Florası, Vol. 2.* ANG Vakfı Nezahat Gökyiğit Botanik Bahçesi Yayınları, İstanbul, pp. 82–110 and 128–135.

Mehltreter, K., L.R. Walker and J.M. Sharpe. 2010. *Fern Ecology.* Cambridge University Press, New York.

Özbek, M.U. 2018. Thelypteridaceae Pic. Serm. In: Güner, A., A. Kandemir, Y. Menemen, H. Yıldırım, S. Aslan, G. Ekşi, I. Güner and A.Ö. Çimen (Eds.), *Resimli Türkiye Florası, Vol. 2.* ANG Vakfı Nezahat Gökyiğit Botanik Bahçesi Yayınları, İstanbul, pp. 220–226.

Öztekin, M. 2018. Dryopteridaceae Herter. In: Güner, A., A. Kandemir, Y. Menemen, H. Yıldırım, S. Aslan, G. Ekşi, I. Güner and A.Ö. Çimen (Eds.), *Resimli Türkiye Florası, Vol. 2.* ANG Vakfı Nezahat Gökyiğit Botanik Bahçesi Yayınları, İstanbul, pp. 239–275.

Ozturk, M., O. Secmen and G. Gork. (Eds.). 1996a. *Plant Life in Southwest and Central Asia, Vol. I.* Ege University Press, Izmir, Turkiye, pp. 1–499.

Ozturk, M., O. Secmen and G. Gork. (Eds.). 1996b. *Plant Life in Southwest and Central Asia, Vol. II.* Ege University Press, Izmir, Turkey, p. 1093.

Ozturk, M., V. Altay, A. Latiff, T. Salman and I. Choudhry. 2018. A comparative analysis of the medicinal pteridophytes in Turkey, Pakistan, and Malaysia. In: Ozturk, M. and K.R. Hakeem (Eds.), *Plant and Human Health, Vol. 1.* Chapter 9. Springer International Publishing AG, New York, pp. 349–390.

Ozturk, M. and K.R. Hakeem. 2018. *Plant and Human Health - Vol. 1. Ethnobotay and Physiology.* Springer Science+Business Media, New York, pp. 1–805.

Ozturk, M. and K.R. Hakeem. 2019a. *Plant and Human Health - Vol. 2. Phytochemistry and Molecular Aspects.* Springer Science+Business Media, New York, pp. 1–697.

Ozturk, M. and K.R. Hakeem. 2019b. *Plant and Human Health - Vol. 3. Phytochemistryand Therapeutic Uses.* Springer Science+Business Media, New York, pp. 1–385.

Ozturk, M., D. Egamberdieva and P. Milica. 2020. *Biodiversity and Biomedicine - Our Future.* Academic Press-Elsevier Inc., London, pp. 1–561.

Parris, B.S. and C.R. Fraser-Jenkins. 1980. A provisional checklist of Turkish Pteridophyta. *Notes from the Royal Botanic Garden, Edinburgh* 38(2): 273–281.

PPG I. 2016. A community-derived classification for extant lycophytes and ferns. *Journal of Systematics and Evolution* 54(6): 563–603.

Shah, S.N., M. Ahmad and S.M. Khan. 2022. The ecology of Pakistani ferns and lycophytes. In: Ozturk, M., S.M. Khan, V. Altay, R. Efe, D. Egamberdieva and F. Khassanov (Eds.), *Biodiversity, Conservation and Sustainability in Asia: Volume II: Prospects and Challenges in South and Middle Asia.* Springer Nature Switzerland AG, Switzerland, pp. 85–101.

Sida, O. and K. Tan. 2000. *Polypodium* L. In: Güner, A., N. Özhatay, T. Ekim and K.H.C. Başer (Eds.), *Flora of Turkey and the East Aegean Islands, Vol. 11.* University Press, Edinburgh, p. 3.

Singh, A.P., V.K. Rawat, S.K. Behera and P.B. Khare. 2010. Perspectives of pteridophytes biodiversity: A source of economy elevation. In: *National Conference on Biodiversity, Development and Poverty Alleviation, 22nd May*, India, pp. 46–49.

Şenol, S.G. 2018. *Azolla* Lam.;*Pilularia* L. In: Güner, A., A. Kandemir, Y. Menemen, H. Yıldırım, S. Aslan, G. Ekşi, I. Güner and A.Ö. Çimen (Eds.), *Resimli Türkiye Florası, Vol. 2.* ANG Vakfı Nezahat Gökyiğit Botanik Bahçesi Yayınları, İstanbul, pp. 141–142 and 146–147.

Tan, K. and J. Suda. 2000. *Chelianthes* Swartz. In: Güner, A., N. Özhatay, T. Ekim and K.H.C. Başer (Eds.), *Flora of Turkey and the East Aegean Islands, Vol. 11.* University Press, Edinburgh, pp. 1–2.

Uzunhisarcıklı, M.E. 2018. Woodsiaceae Herter. In: Güner, A., A. Kandemir, Y. Menemen, H. Yıldırım, S. Aslan, G. Ekşi, I. Güner and A.Ö. Çimen (Eds.), *Resimli Türkiye Florası, Vol. 2.* ANG Vakfı Nezahat Gökyiğit Botanik Bahçesi Yayınları, İstanbul, pp. 226–229.

Vural, M. 2018. *Athyrium* Roth. In: Güner, A., A. Kandemir, Y. Menemen, H. Yıldırım, S. Aslan, G. Ekşi, I. Güner and A.Ö. Çimen (Eds.), *Resimli Türkiye Florası, Vol. 2.* ANG Vakfı Nezahat Gökyiğit Botanik Bahçesi Yayınları, İstanbul, pp. 234–238.

Yıldırımlı, Ş. 1996. Türkiye eğreltilerin yayılışı. *OT Sistematik Botanik Dergisi* 3(1): 93–104.

Yılmaz, Ö. 2018. *Equisetum* L. In: Güner, A., A. Kandemir, Y. Menemen, H. Yıldırım, S. Aslan, G. Ekşi, I. Güner and A.Ö. Çimen (Eds.), *Resimli Türkiye Florası, Vol. 2.* ANG Vakfı Nezahat Gökyiğit Botanik Bahçesi Yayınları, İstanbul, pp. 114–128.

APPENDIX 20.1 RE-REVISED CHECKLIST OF PTERIDOPHYTES OF TURKIYE.

Class	Order	Family	Genera	Species
Lycopodiopsida	Lycopodiales	Lycopodiaceae	Lycopodiella	*Lycopodiella inundata* (L.) Holub
			Diphasiastrum	*Diphasiastrum alpinum* (L.) Holub
				D. tristachyum (Pursh) Holub
			Lycopodium	*Lycopodium clavatum* L.
			Spinulum	*Spinulum annotinum* (L.) A.Haines
			Huperzia	*Huperzia selago* (L.) Bernh. ex Schrank & Mart.
	Isoëtales	Isoëtaceae	Isoetes	*Isoetes histrix* Bory
				I. gymnocarpa (Gennari) A.Braun
				I. durieui Bory
				I. olympica A.Braun
				I. anatolica Prada & Rolleri
				I. vanensis Keskin & G.Zare
	Selaginellales	Selaginellaceae	Selaginella	*Selaginella denticulata* (L.) Spring
				S. helvetica (L.) Spring
		Equisetaceae	Equisetum	*Equisetum hyemale* L.
				E. ramosissimum Desf.
				E. variegatum Schleich. ex F.Wever & D.Mohr
				E. fluviatile L.
				E. palustre L.
				E. sylvaticum L.
				E. arvense L.
				E. telmateia Ehrh.
				E. × *litorale* Kühlew ex Rupr.
		Ophioglossaceae	Ophioglossum	*Ophioglossum lusitanicum* L.
				O. vulgatum L.
			Botrychium	*Botrychium lunaria* (L.) Sw.
	Osmundales	Osmundaceae	Osmunda	*Osmunda regalis* L.
	Hymenophyllales	Hymenophyllaceae	Hymenophyllum	*Hymenophyllum tunbrigense* (L.) Sm.
	Salviniales	Salviniaceae	Salvinia	*Salvinia natans* (L.) All.
			Azolla	*Azolla filiculoides* Lam.
		Marsileaceae	Marsilea	*Marsilea quadrifolia* L.
			Pilularia	*Pilularia minuta* Durieu ex A.Braun
	Polypodiales	Pteridaceae	Cryptogramma	*Cryptogramma crispa* (L.) R.Br. ex Hook.
				C. bithynica S.Jess., L.Lehm. & Bujnoch
			Anogramma	*Anogramma leptophylla* (L.). Link
			Cosentinia	*Cosentinia vellea* (Aiton) Tod.
			Pteris	*Pteris cretica* L.
				P. vittata L.
				P. multifida Poit.
			Allosorus	*Allosorus acrosticus* (Balb.) Christenh.
				A. persicus (Bory) Christenh.

(Continued)

Class	Order	Family	Genera	Species
				A. tinaei (Tod.) Christenh.
			Adiantum	*Adiantum capillus-veneris* L.
			Paragymnopteris	*Paragymnopteris maranthae* (L.) K.H.Shing
		Dennstaedtiaceae	Pteridium	*Pteridium aquilinum* (L.) Kuhn. subsp. *aquilinum*
		Cystopteridaceae	Cystopteris	*Cystopteris fragilis* (L.) Bernh.
				C. alpina Desv.
				C. x vidae Fraser-Jenk. & Parris
			Gymnocarpium	*Gymnocarpium dryopteris* (L.) Newman
		Aspleniaceae	Asplenium	*Asplenium septentrionale* (L.) Hoffm.
				A. trichomanes L. subsp. *trichomanes*
				A. trichomanes L. subsp. *quadrivalens* D.E.Mey.
				A. viride Huds.
				A. tadei Fraser-Jenk. & Schneller
				A. bourgaei Milde
				A. adiantum-nigrum L. subsp. *adiantum-nigrum*
				A. adiantum-nigrum L. nothosubsp. x*demirizii* (Viane, Rasbach, Schneller & Reichst.) Fraser-Jenk. & Parris
				A. adiantum-nigrum L. nothosubsp. x*ticinense* (D.E.Mey) Fraser-Jenk. & Parris
				A. adiantum-nigrum L. nothosubsp. x *centovallense* (D.E.Mey.) Fraser-Jenk. & Parris
				A. cuneifolium Viv.
				A. woronowii Christ.
				A. onopteris L.
				A. obovatum Viv.
				A. obovatum Viv. subsp. *Protobillotii* Demiriz, Viane & Reichst.
				A. obovatum Viv. subsp. *Deltoideum* Demiriz, Viane & Reichst.
				A. aegaeum Lovis, Reichst. & Greuter
				A. lepidum C.Presl subsp. *haussknechtii* (Godet & Reuter) Brownsey
				A. ruta-muraria L.
				A. x alternatifolium Wulfen nothosubsp. *alternatifolium*

(*Continued*)

Class	Order	Family	Genera	Species
				A. x alternatifolium Wulfen nothosubsp. *ponticum* Fraser-Jenkins., Lovis & Boudrie
				A. x heufleri Reichatdt.
				A. x reuteri Milde
				A. ceterach L. subsp. *ceterach*
				A. ceterach L. nothosubsp. *mantoniae* (Váróczky & Vida) Jermy
				A. scolopendrium L.
				A. sagittatum (DC.) Bange
		Woodsiaceae	Woodsia	*Woodsia alpina* (Bolton) Gray
		Onocleaceae	Matteuccia	*Matteuccia struthiopteris* (L.) Tod.
		Blechnaceae	Blechnum	*Blechnum spicant* (L.) Roth
		Athyriaceae	Athyrium	*Athyrium filix-femina* (L.) Roth
				A. alpestre (Hoppe) Clairv.
		Thelypteridaceae	Oreopteris	*Oreopteris limbosperma* (All.) Holub
			Thelypteris	*Thelypteris palustris* Schott
			Phegopteris	*Phegopteris connectilis* (Michx.) Watt
		Dryopteridaceae	Dryopteris	*Dryopteris filix-mas* (L.) Schott
				D. affinis (Lowe) Fraser-Jenk.subsp. *affinis* var. *disjuncta* (Fomin) Fraser-Jenk.
				D. affinis (Lowe) Fraser-Jenk.subsp. *affinis* var. *punctata* Oberholzer & Tavel ex Fraser-Jenk.
				D. affinis (Lowe) Fraser-Jenk.subsp. *borreri* (Newman) Fraser-Jenk.
				D. affinis (Lowe) Fraser-Jenk.subsp. *coriacea* Fraser-Jenk.
				D. affinis (Lowe) Fraser-Jenk.subsp. *pontica* Fraser-Jenk.
				D. caucasica (A.Braun) Fraser-Jenk. & Corley
				D. raddeana (Fomin) Fomin
				D. dilatata (Hoffm.) A.Gray
				D. aemula (Aiton) Kuntze
				D. carthusiana (Vill.) H.P. Fuchs
				D. expansa (C.Presl) Fraser-Jenk. & Jermy
				D. pseudomas (Wol.) Holub & Pouzar
				D. remota (A.Braun) Hayek
				D. submontana(Fraser-Jenk. & Jermy) Fraser-Jenk.
				D. x deweveri (Jansen) Jansen & Wachter

(*Continued*)

Class	Order	Family	Genera	Species
				D. x euxinensis Fraser-Jenk. & Corley
				D. x initialis Fraser-Jenk. & Corley
			Polystichum	*Polystichum setiferum* (Forssk.) Moore ex Woyn.
				P. acuelatum (L.) Roth.
				P. lonchitis (L.) Roth.
				P. braunii (Spenn.) Fee
				P. woronowii Fomin
				P. asiae-minoris Tunçkol & Li Bing Zhang
				P. x bicknelii (Christ) Hahne
				P. x luerssenii (Dörfl.) Hahne
		Nephrolepidaceae	Nephrolepis	*Nephrolepis exaltata* (L.) Schott
		Polypodiaceae	Polypodium	*Polypodium vulgare* L.
				P. interjectum Shivas
				P. cambricum L.
				P. x mantoniae Rothm.

Index

T

U

V

W